T0207321

Lecture Notes in Computer Science 13638

More information about this series at https://link.springer.com/bookseries/558

Sihem Mesnager · Zhengchun Zhou (Eds.)

Arithmetic of Finite Fields

9th International Workshop, WAIFI 2022
Chengdu, China, August 29 – September 2, 2022
Revised Selected Papers

Editors
Sihem Mesnager
Department of Mathematics
University of Paris VIII
Paris, France

LAGA
University Sorbonne, CNRS
Villetaneuse, France

Zhengchun Zhou
School of Mathematics
Southwest Jiaotong University
Chengdu, China

ISSN 0302-9743 ISSN 1611-3349 (electronic)
Lecture Notes in Computer Science
ISBN 978-3-031-22943-5 ISBN 978-3-031-22944-2 (eBook)
https://doi.org/10.1007/978-3-031-22944-2

This Springer imprint is published by the registered company Springer Nature Switzerland AG
The registered company address is: Gewerbestrasse 11, 6330 Cham, Switzerland

Preface

These are the proceedings of the 9th edition of the International Workshop on the Arithmetic of Finite Fields (WAIFI 2022). The conference should have taken place in the beautiful city of Chengdu (China) from August 29 to September 2, 2022. However, because of the COVID-19 pandemic and the related situation in China, it finally took place online on the same dates.

We are very grateful to the Program Committee members and the external reviewers for their hard and professional work! The conference received 25 submissions, of which 19 contributed papers were finally selected for presentation after a single-blind peer review. Each paper was refereed by at least two reviewers. All final decisions were taken only after a clear position could be clarified through additional reviews and comments.

The Organizing Committee also invited Lilya Budaghyan, Cunsheng Ding, and Sylvain Duquesne to speak on topics of their choice, and we thank them for having accepted! Each speaker submitted an invited paper, and these were handled by Sihem Mesnager. Special compliments go out to the Chinese team, the local organizers of WAIFI 2022, who brought the workshop much success. We also would like to thank José Luis Imaña for his great help with publicity and guidance regarding the website. The submission and selection of papers were made using the Easychair software. We also thank Avik Adhikary for his precious help in this matter.

October 2022

Sihem Mesnager
Zhengchun Zhou

Organization

General Chairs

Sihem Mesnager	University of Paris VIII, France
Zhengchun Zhou	Southwest Jiaotong University, China

Program Committee Chairs

Sihem Mesnager	University of Paris VIII, France
Zhengchun Zhou	Southwest Jiaotong University, China

Publicity Chair

José Luis Imaña	Complutense University of Madrid, Spain

Organizing Committee

Dongchun Han	Southwest Jiaotong University, China
Haode Yan	Southwest Jiaotong University, China
Yang Yang	Southwest Jiaotong University, China

Steering Committee

Lilya Budaghyan	University of Bergen, Norway
Claude Carlet	Universities of Paris VIII, France, and University of Bergen, Norway
Anwar Hasan	University of Waterloo, Canada
José Luis Imaña	Complutense University of Madrid, Spain
Çetin Kaya Koç	University of California, Santa Barbara, USA
Sihem Mesnager	University of Paris VIII, France
Ferruh Özbudak	Middle East Technical University, Turkey
Svetla Petkova-Nikova	KU Leuven, Belgium
Francisco Rodríguez-Henríquez	CINVESTAV-IPN, Mexico
Erkay Sava	Sabanc University, Turkey

Program Committee

Claude Carlet	Universities of Paris VIII, France, and University of Bergen, Norway
Xiwang Cao	Nanjing University, China
Rongquan Feng	Peking University, China
Luca De Feo	University of Versailles Saint-Quentin-en-Yvelines, France
Tor Helleseth	University of Bergen, Norway
Çetin Kaya Koç	University of California Santa Barbara, USA
Kwang Ho Kim	State Academy of Sciences, South Korea
Gohar Kyureghyan	University of Rostock, Germany
Nian Li	Hubei University, China
Petr Lisonek	Simon Fraser University, Canada
Sihem Mesnager	University of Paris VIII, France
Ying Miao	University of Tsukuba, Japan
Svetla Nikova	KU Leuven, Belgium
Ferruh Özbudak	Middle East Technical University, Turkey
Daniel Panario	Carleton University, Canada
Alexander Pott	Otto von Guericke University Magdeburg, Germany
Joachim Rosenthal	University of Zurich, Switzerland
Erkay Savas	Sabanc University, Turkey
Leo Storme	Ghent University, Belgium
Arnaud Tisserand	CNRS, University of Rennes 1, France
Alev Topuzoğlu	Sabanc University, Turkey
Zhengchun Zhou	Southwest Jiaotong University, China

Additional Reviewers

Farzane Amirzade Dana
Kit Battarbee
Andrea Ferraguti
Ahmet Can Mert
Giacomo Micheli
Ali Sah Ozcan
Maria Sabitova
Simrin Tinani
Gaofei Wu
Giovanni Zini

Contents

Cryptography

Sequences

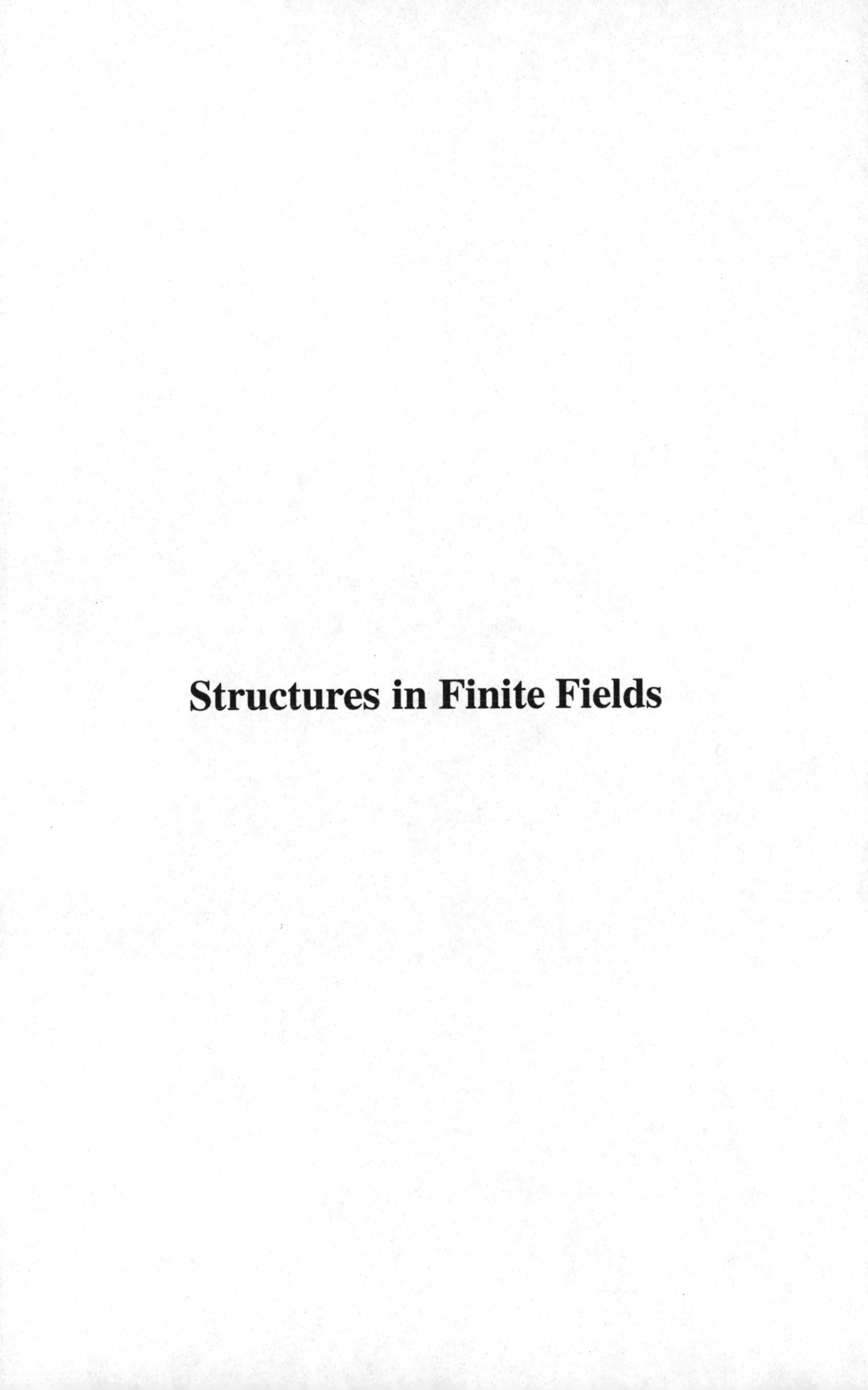

Structures in Finite Fields

On a Conjecture on Irreducible Polynomials over Finite Fields with Restricted Coefficients

Andrea Ferraguti[1,2](✉) and Giacomo Micheli[3,4]

[1] Scuola Normale Superiore di Pisa, Piazza dei Cavalieri 7, 56126 Pisa, Italy
[2] DICATAM, Università degli Studi di Brescia, via Branze 43, 25123 Brescia, Italy
andrea.ferraguti@unibs.it
[3] University of South Florida, 4202 E Fowler Ave, Tampa, FL 33620, USA
[4] Center for Cryptographic Research at USF, Tampa, USA
gmicheli@usf.edu

Abstract. Let q be a prime power, $\mathbb{F}_q$ be the finite field of order q and let n, d be positive integers. Munemasa and Nakamura conjectured at WAIFI 2016 that there exist $f \in \mathbb{F}_q[x]$ of degree n and $\alpha \in \mathbb{F}_{q^d}$ not lying in any proper subfield such that $f - \alpha$ is irreducible in $\mathbb{F}_{q^d}[x]$. In this paper, we prove that the conjecture holds true for every triple (q, n, d) such that d is larger than a constant that depends only on n. As a subproduct of our proofs we deduce that if $F \in \mathbb{F}_q[x]$ is a polynomial such that $F - t_0$ has a certain special factorization pattern for some $t_0 \in \mathbb{F}_q$, then the statistics of *all* the factorization patterns of $F - t_1$, where t_1 ranges in $\mathbb{F}_{q^d}$, are entirely determined up to an explicit error term independent of the size of the base field. At the end of the paper we provide some experimental results to show how sharp our statistics are.

Keywords: Finite fields · Irreducible polynomials · Densities · Factorization patterns

1 Introduction

Let q be a prime power and $\mathbb{F}_q$ be the finite field of size q. In this paper we prove an eventual version of the following conjecture by Munemasa and Nakamura [9, Conjecture 1], that has been formulated at WAIFI 2016.

Conjecture 1. Let q be a prime power and $n, d \in \mathbb{N}$. Let $\mathcal{G}$ be the set of elements $\alpha \in \mathbb{F}_{q^d}$ such that $\mathbb{F}_q(\alpha) = \mathbb{F}_{q^d}$. Then there exists a polynomial $f \in \mathbb{F}_q[x]$ of degree n and an element $\alpha \in \mathcal{G}$ such that $f - \alpha \in \mathbb{F}_{q^d}[x]$ is irreducible.

Remark 1. By Capelli's Lemma (see for example [3]), the above conjecture is equivalent to the following statement: for every prime power q and any pair of positive integers (m, n) there exist two polynomials $f, g \in \mathbb{F}_q[x]$ such that $\deg f = n$, $\deg g = m$ and $g \circ f$ is irreducible.

This work is supported by NSF grant 2127742.

S. Mesnager and Z. Zhou (Eds.): WAIFI 2022, LNCS 13638, pp. 3–13, 2023.
https://doi.org/10.1007/978-3-031-22944-2_1

Our main result is the following.

Theorem 1. *For every positive integer n, there exists an effective constant $C(n)$ such that Conjecture 1 holds true for any triple (q, n, d) with $d \geq C(n)$. Moreover, for any such triple there is an explicit polynomial $f \in \mathbb{F}_q[x]$ of degree n such that $f - \alpha$ is irreducible in $\mathbb{F}_{q^d}[x]$ for some α generating $\mathbb{F}_{q^d}$.*

Notice that this result gives also a direct method to verify the conjecture for a fixed pair (q, n), since it leaves out only finitely many d's to check.

Our arguments also allow us to deduce the following fact. If $F \in \mathbb{F}_q[x]$ and there exists $t_0 \in \mathbb{F}_q$ such that $F - t_0$ has a certain specific factorization pattern, then for every $d \geq 1$ the entire statistics of the factorization patterns of $F - t$ as t ranges over $\mathbb{F}_{q^d}$ are entirely determined (cf. Theorem 7).

The paper is organized as follows. In Sect. 2 we recall the basic number theoretical facts we need to prove our results and describe the strategy we employ to tackle this type of problems. In Sect. 3 we state and prove the main results of the paper. In Sect. 4 we provide experimental results to show how the asymptotics we predict agree with the actual statistics.

Notation

Let q be a prime power and $\mathbb{F}_q$ be the finite field of order q. For $n \in \mathbb{N}$, we denote by $\mathcal{S}_n$ the symmetric group on n symbols. We say that $\sigma \in \mathcal{S}_n$ has cycle decomposition $\Delta = (d_1, \ldots, d_n)$ if, when one writes σ in disjoint cycles, there are d_i cycles of length i for any $i \in \{1, \ldots, n\}$. Analogously, for a polynomial $H \in \mathbb{F}_q[x]$ we say that H splits according to Δ or H has factorization pattern Δ if H has d_i distinct irreducible factors of degree i for every $i \in \{1, \ldots, n\}$.

For a polynomial $F \in \mathbb{F}_q[x]$ we denote by M_F the splitting field of the polynomial $F - t$ over $\mathbb{F}_q(t)$ and by k_F its field of constants, where t is an indeterminate over $\mathbb{F}_q$. Moreover, we denote by G_F and N_F the Galois group of the extensions $M_F : \mathbb{F}_q(t)$ and $M_F : k_F\mathbb{F}_q(t)$, respectively. For a positive integer d, we denote by $G_{F,d}$ and $N_{F,d}$ the analogous groups when F is thought of as an element of $\mathbb{F}_{q^d}[x]$.

2 Chebotarev Density Theorem for Function Fields

In this section we recall certain techniques that are also used in [8].

Theorem 2. *Let $L : K$ be a finite separable extension of global function fields and let M be its Galois closure with Galois group G. Let P be a place of K and $\mathcal{Q}$ be the set of places of L lying over P. Let R be a place of M lying over P. There is a natural bijection between $\mathcal{Q}$ and the set of orbits of $H = \mathrm{Hom}_K(L, M)$ under the action of the decomposition group $D(R|P) = \{g \in G \,|\, g(R) = R\}$. In addition, let $Q \in \mathcal{Q}$ and let H_Q be the orbit corresponding to Q. Then $|H_Q| = e(Q|P)f(Q|P)$ where $e(Q|P)$ and $f(Q|P)$ are ramification index and relative degree of Q over P, respectively.*

Remark 2. Throughout the whole paper, when we talk about cycles we are considering the action of the Galois group on $\mathrm{Hom}_{\mathbb{F}_q(t)}(L, M)$. If M is the splitting field of a separable polynomial $f - t \in \mathbb{F}_q(t)[x]$, this is equivalent to looking at the action of the Galois group of $f - t$ on the set of its roots.

Let $M : K$ be a finite Galois extension of global function fields with Galois group G. Fix a place P of K of degree 1, and let R be a place of M lying above P. Let $h \in D(R|P)$ be a Frobenius for the finite field extension $\mathcal{O}_R/R : \mathcal{O}_P/P$, i.e. an element which acts as $\phi_q\colon u \mapsto u^q$ when restricted to $\mathcal{O}_R/R$.

The set $D_{\phi_q}(R, P)$ of all such elements coincides with the coset $I(R|P)h$ in $D(R|P)$, where $I(R|P)$ is the ramification group.

We define now the map

$$\alpha_P : G \longrightarrow [0,1]$$
$$g \mapsto \frac{\#\Gamma(g) \cap D_{\phi_q}(R|P)}{\#\Gamma(g) \cdot \#I(R|P)}$$

where $\Gamma(g)$ is the conjugacy class of g in G.

The following version of Chebotarev density theorem is due to Kosters [7].

Theorem 3 (Chebotarev Density Theorem). *Let $M : K$ be a finite Galois extension of function fields over a finite field k of cardinality q and let $\tilde{k}$ be the constant field of M. Let $G = \mathrm{Gal}(M : K)$ and $N = \mathrm{Gal}(M : \tilde{k}K)$. Let $\gamma \in G$ such that γ acts as $u \mapsto u^q$ when restricted to $\tilde{k}$. Let $g \in N\gamma$, Γ be the conjugacy class of g in G and let $\mathcal{P}^1$ be the set of places of degree 1 of K. Let $\mathfrak{g}$ be the genus of M. Then we have*

$$\left| \sum_{P \in \mathcal{P}^1} \alpha_P(g) - \frac{\#\Gamma}{\#N}(q+1) \right| \le 2\frac{\#\Gamma}{\#N}\mathfrak{g}q^{1/2}.$$

Notice that, in contrast with the number field theoretic version, there is a necessary condition for g to be a Frobenius, namely to lie in $N\gamma$; this reflects the fact that in order to be a Frobenius at some place it is necessary to be a Frobenius for the field of constants.

Corollary 1. *With the notation of Theorem 3, let $\mathcal{P}^1_{ur}$ be the set of places of degree 1 in K which are unramified in M. Set*

$$T(g) = \{P \in \mathcal{P}^1_{ur} | \text{a conjugate of } g \text{ is a Frobenius for } R|P \text{ , for some } R \text{ above } P\}.$$

Then we have

$$2\frac{\#\Gamma}{\#N}\mathfrak{g}q^{1/2} - \#\,\mathrm{Ram}^1(M:K) \le \#T(g) - \frac{\#\Gamma}{\#N}(q+1) \le 2\frac{\#\Gamma}{\#N}\mathfrak{g}q^{1/2} + \#\,\mathrm{Ram}^1(M:K),$$

where $\mathrm{Ram}^1(M : K)$ is the set of ramified places of degree 1 of K.

Proof. Simply write $\mathcal{P}^1 = \mathcal{P}^1_{\mathrm{ur}} \sqcup \mathcal{P}^1_{\mathrm{ram}}$, where $\mathcal{P}^1_{\mathrm{ram}}$ is the set of degree 1 places of K that ramify in M. Then use Theorem 3 together with the fact that if g is a Frobenius for $R|P$, then all of its conjugates are Frobeniuses for some $R'|P$. Moreover, if $P \in \mathcal{P}^1_{\mathrm{ur}}$ then $\alpha_P(g) \in \{0, 1/\#\Gamma\}$, while if $P \in \mathcal{P}^1_{\mathrm{ram}}$ we can only say that $0 \le \alpha_P(g) \le 1$.

The Method. Let us briefly explain how to combine Theorem 3 (or Corollary 1) with Theorem 2 in the framework we are interested in, namely the one where M is the splitting field of a polynomial of the form $f(x) - t$ with $f(x) \in \mathbb{F}_q[x]$.

Fix a certain squarefree factorization pattern Δ and suppose we are interested in the number N of $t_0 \in \mathbb{F}_{q^d}$ for which $f - t_0$ splits exactly as Δ, for q^d large enough with respect to $\deg(f)$. We then proceed as follows:

- Compute the arithmetic and geometric monodromy groups G_f and N_f and an element $\gamma \in G_f$ as in Theorem 3.
- Write G_f as a permutation group acting on the set of roots of $f - t$, and find all the elements in the coset $N_f\gamma$ that have the cycle decomposition Δ.
- Partition these elements into conjugacy classes $\Gamma_1, \ldots, \Gamma_s$.
- Consider the set T_i of u's in $\mathbb{F}_q$ such that Γ_i contains a Frobenius for u (identifying u with the corresponding place of degree 1 of $\mathbb{F}_q(t)$). Observe that $T_i \cap T_j = \emptyset$ and $\sum_i \#T_i = N$.
- Estimate the size of each of the T_i's with Theorem 3.
- Add up all the quantities obtained to get an estimate for N.

3 Proof of the Main Results

From now on, if $F \in \mathbb{F}_q[x]$ we will denote by G_F and N_F respectively the arithmetic and geometric Galois groups of $F - t$, and by k_F the field of constants of the splitting field of $F - t$. The following lemma can be stated more generally, but we will only need it in this form.

Lemma 1. *Let $F \in \mathbb{F}_q[x]$, $t_0 \in \mathbb{F}_q$, and $g_1, \ldots, g_\ell$ be the distinct irreducible factors of $F - t_0$. Suppose that $k_F = \mathbb{F}_{q^2}$. Then $2 \mid \deg(g_i)$ for some $i \in \{1, \ldots, \ell\}$.*

Proof. Suppose by contradiction that $\deg(g_i)$ is odd for every i. Let $\mathbb{F}_{q^m}$ be the smallest extension of $\mathbb{F}_q$ where all the g_i's split. Since they all have odd degree, then m is odd. Now observe that $G_{F,m} := \mathrm{Gal}(F(x) - t \mid \mathbb{F}_{q^m}(t)) \cong G_F$ and $N_{F,m} := \mathrm{Gal}(F(x) - t \mid \mathbb{F}_{q^{2m}}(t)) \cong N_F$, as we left the field of constant untouched by the fact that m is odd (see for example [8, Lemma 9]). Since the field of constants is non-trivial, $N_F \subsetneq G_F$. By hypothesis, $f - t_0$ splits into linear factors in $\mathbb{F}_{q^m}[x]$. It follows that the identity in $G_{F,m}$ is a Frobenius for t_0, but the only case in which the identity can be a Frobenius is when $G_{F,m} = N_{F,m}$, which is a contradiction.

Lemma 2. *Let $G \subseteq \mathcal{S}_n$ be a transitive subgroup containing a cycle of prime length $r > n/2$ and a transposition. Then $G = \mathcal{S}_n$.*

Proof. See [6].

Similary to [1,5], we deduce the structure of the Galois groups in terms of factorization patterns.

Theorem 4. *Let $F \in \mathbb{F}_q[x]$, $n = \deg(F) \geq 8$, and r be a prime in $\{\lfloor n/2 \rfloor + 1, \ldots, n-3\}$ (whose existence is guaranteed by Bertrand's Postulate). Suppose that there exists $t_0 \in \mathbb{F}_q$ such that*

$$F(x) - t_0 = \begin{cases} (x-a)^2 \cdot g(x) \cdot h(x) & \text{if } n \text{ is even} \\ (x^2+x+1) \cdot x^2 \cdot g(x) & \text{if } n \text{ is odd, } q = 2 \text{ and } r = n-4\,, \\ (x-a)^2 \cdot (x-b) \cdot g(x) \cdot h(x) & \text{otherwise} \end{cases}$$

where $a, b \in \mathbb{F}_q$ and $a \neq b$ in the third case, $g(x), h(x) \in \mathbb{F}_q[x]$ are irreducible and $\deg g(x) = r$. Then F is separable and $G_F = N_F = \mathcal{S}_n$.

Remark 3. Of course there are other nice criterions to establish whether a polynomial has full symmetric Galois group [2,11] but often they do not work in characteristic 2 or they are not explicit enough. We prefer then to use a uniform and simple approach which includes all degrees and characteristics, at least for $n \geq 8$.

Proof (Proof of Theorem 4). Separability follows immediately from the factorization pattern of $F(x) - t_0$.

First assume that $q > 2$ or $r < n-4$, so that we are not in the second case. The factorization pattern of $F - t_0$ shows that G_F contains an element δ which is a product of three disjoint cycles of lenght $2, r$, and $n-r-2$ when n is even, and a product of four disjoint cycles of length $2, 1, r$, and $n-r-3$ when n is odd by Theorem 2. Notice that our choice of r ensures that $(r, n-r-2) = (r, n-r-3) = 1$. Thus there are appropriate powers of δ that are a cycle of order r and a transposition, respectively. Then by Lemma 2 it follows that $G_F = \mathcal{S}_n$ and by Lemma 1 it follows that $G_F = N_F$, since $N_F \trianglelefteq G_F$ and all irreducible factors of $F(x) - t_0$ have odd degree.

When $q = 2$ and $r = n-4$ then the factorization of $F - t_0$ over $\mathbb{F}_{q^2}$ shows that N_F contains an element δ that is a product of a transposition and a cycle of prime degree r. It follows by Lemma 2 that $N_F = \mathcal{S}_n$, and hence $G_F = \mathcal{S}_n$.

Theorem 5. *Let k be a field of characteristic different from 2 and $g \in k[X]$. Suppose that the derivative g' of g has at least a simple root and for any pair of distinct roots α, β of g' in $\overline{k}$ we have that $g(\alpha) \neq g(\beta)$. In addition suppose that $char(k) \nmid \deg(g)$. Then the Galois group of $g - t$ over $k(t)$ is $\mathcal{S}_{\deg(g)}$.*

Proof. See [11, Theorem 3.6].

Theorem 6. *For any prime power q and any positive integer n, there exists an effective constant $C(n)$ and an explicit polynomial $f \in \mathbb{F}_q[x]$ of degree n such that for every positive integer $d \geq C(n)$ there exists $\alpha \in \mathbb{F}_{q^d}$ with $\mathbb{F}_q(\alpha) = \mathbb{F}_{q^d}$ such that $f - \alpha \in \mathbb{F}_{q^d}[x]$ is irreducible.*

Proof. Let us start by noticing that it is enough to prove the theorem for q prime. In fact suppose we have proven it for q prime, let $q = p^r$ for some p prime and let $n \geq 1$. Let $C(n)$ be the constant determined by the theorem for p. Then for every natural number k larger than $C(n)/r$ we have that $kr > C(n)$ and hence there is some generator $\alpha \in \mathbb{F}_{p^{kr}} = \mathbb{F}_{q^k}$ such that $f - \alpha$ is irreducible in $\mathbb{F}_{q^k}$, that is precisely the statement of the theorem.

The strategy is the following: first for every pair (q, n) we construct explicitly a separable polynomial $f \in \mathbb{F}_q[x]$ of degree n which has the property that $G_f = N_f = \mathcal{S}_n$. Then we will use Corollary 1 to produce a constant $C = C(n)$ such that whenever $d \geq C$, the set of places of degree 1 of $\mathbb{F}_{q^d}(t)$ that do not come from a proper subextension and whose Frobenius is a maximal cycle in $N_{f,d} = \mathrm{Gal}(f - t|\mathbb{F}_{q^d}(t))$ is non-empty. This allows to conclude thanks to Theorem 2.

Let us assume $n \geq 8$, we will deal with the cases $n < 8$ separately. Let us fix $f = F$ as in the statement of Theorem 4 for $t_0 = 0$, i.e.

$$F(x) = \begin{cases} (x-a)^2 \cdot g(x) \cdot h(x) & \text{if } n \text{ is even} \\ (x^2+x+1) \cdot x^2 \cdot g(x) & \text{if } n \text{ is odd, } q = 2 \text{ and } r = n-4 \text{ ,} \\ (x-a)^2 \cdot (x-b) \cdot g(x) \cdot h(x) & \text{otherwise} \end{cases}$$

where $\deg g(x) = r$ and both $g(x), h(x)$ are irreducible. Then $G_f = N_f = \mathcal{S}_n$.

When $n \in \{3, 5, 7\}$ let $f_n(x) = (x-a)^2 \cdot g_n(x)$, where $a \in \mathbb{F}_q$ and $g_n(x)$ is an irreducible polynomial of degree $n-2$. Then Lemmas 1 and 2 show immediately that $G_f = N_f = \mathcal{S}_n$, as the factorization pattern of f_n forces G_f to contain a p-cycle for a prime $p > n/2$ and a transposition.

When $n = 2$, let $f = x(x+1)$. It is immediate to see that there is some $t_0 \in \mathbb{F}_q$ such that $f + t_0$ is irreducible (see for example [4, p. 9]). It follows immediately that $G_f = N_f = \mathcal{S}_2$.

When $n \in \{4, 6\}$ and $(q, n) = (q, n-1) = 1$, let $f = x^n - \dfrac{n}{n-1}x^{n-1}$. Then f' has only two roots, namely 0 and 1 (1 being a simple root). Clearly $f(0) \neq f(1)$. Theorem 5 shows that $G_f = N_f = \mathcal{S}_n$. The same theorem can be used for $(q, n) \in \{(3, 4), (5, 6)\}$: when $q = 3$ and $n = 4$, use the polynomial $f = x^4 - x^2 + x$, when $q = 5$ and $n = 6$, use $f = x^6 - x^5 - 2x^3$.

This leaves out the pairs $(q, n) = (2, 4), (2, 6), (3, 6)$.

When $q = 3$ and $n = 6$ let $f = x^5(x+2)$. One checks that $f + 1 = (x+1)(x^2+2x+2)(x^3+2x^2+2x+2)$, showing that G_f contains a 5-cycle and a transposition. Hence $G_f = \mathcal{S}_6$.

Finally, let $q = 2$. When $n = 4$, let $f = x^3(x+1)$. Then $f+1$ is irreducible, proving that G_f contains a cycle of order 4 and one of order 3. The only subgroup of $\mathcal{S}_4$ containing two such elements is $\mathcal{S}_4$. When $n = 6$, let $f = x(x^5 + x^3 + 1)$. Then $f + 1 = (x+1)(x^2+x+1)(x^3+x+1)$. Thus G_f contains a cycle of order 5 and a transposition.

Thus for any pair (n, q) we have constructed $f \in \mathbb{F}_q[x]$ of degree n with $G_f = N_f = \mathcal{S}_n$. It follows immediately that for every $d \in \mathbb{N}$ we also have $G_{f,d} = N_{f,d} = \mathcal{S}_n$. Our purpose is now to show that any cycle of length n is a Frobenius for some place $t_0 \in \mathbb{F}_{q^d}$, when d is large enough compared with a constant that depends only on n. For any $d \in \mathbb{N}$, by Chebotarev Density Theorem, the number $I(n, d)$ of t_0's for which this happens satisfies the relation

$$\left| I(n,d) - \frac{\#\Gamma}{\#\mathcal{S}_n} q^d \right| \leq \frac{2}{n} \mathfrak{g} q^{\frac{d}{2}} + R,$$

where R is the number of ramified places of degree 1 and $\mathfrak{g}$ is the genus of the splitting field M_f of $f - t$. Now, since the place at infinity is ramified and the number of finite, ramified places can be bounded by the number of zeroes of f' over the algebraic closure of $\mathbb{F}_q$, this gives $R \leq n$.

Moreover, the conjugacy class of a cycle of maximal length has size $(n-1)!$, and therefore we get:

$$I(n,d) \geq \frac{1}{n} q^d - \frac{2}{n} \mathfrak{g} q^{\frac{d}{2}} - n.$$

Clearly, $M_f = \mathbb{F}_q(t, \alpha_1, \dots, \alpha_n)$ where the α_i's are all the roots of $f - t$ in an algebraic closure of $\mathbb{F}_q(t)$. It follows that the genus $\mathfrak{g}$ can be bounded by a constant $g(n)$ which is independent of d and q via (for example) a recursive application of Castelnuovo Inequality [10, Theorem 3.11.3], seeing M_f as the compositum $\mathbb{F}_q(t,\alpha_1)\mathbb{F}_q(t,\alpha_2)\cdots\mathbb{F}_q(t,\alpha_n)$. Therefore for any fixed n the function $\frac{1}{n}q^d - \frac{2}{n}g(n)q^{\frac{d}{2}} - n$ is asymptotic to $\frac{1}{n}q^d$ as $d \to \infty$.

In order to get the final claim and produce the t_0 not lying in any subextension and such that $f - t_0$ is irreducible in $\mathbb{F}_{q^d}$, it is enough to have $I(n,d)$ being strictly bigger than $\sum_{i|d} q^i$. The latter quantity is strictly less than $dq^{d/2}$, and therefore it is a sufficient condition that:

$$\frac{1}{n} q^d - \frac{2}{n} g(n) q^{\frac{d}{2}} - n \geq dq^{d/2}. \tag{1}$$

This holds whenever $q^d \geq \frac{n}{2}d + g(n) + \frac{n}{2}\sqrt{\frac{4}{n^2}g(n)^2 + d^2 - \frac{4g(n)d}{n} + 4}$, and again, for this to hold it is sufficient that:

$$d \geq \max\left\{0, \log_2\left(\frac{n}{2}d + g(n) + \frac{n}{2}\sqrt{\frac{4}{n^2}g(n)^2 + d^2 - \frac{4g(n)d}{n} + 4}\right)\right\}.$$

Remark 4. From the proof above, the constant $C(n)$ can be made explicit as all the inequalities have effective constants.

Example 1. Whenever the characteristic is larger than the degree of f the constant $C(n)$ is much better than in the worst case scenario (i.e. when one can use only Castelnuovo Inequality), so we explain how to produce it explicitly here using Hurwitz formula. Since the constant field extension of $M_f : \mathbb{F}_q(t)$ is trivial we can estimate the genus $\mathfrak{g}$ of $M'_f = \overline{\mathbb{F}}_q M_f$. Notice also that $\mathrm{Gal}(M'_f : \overline{\mathbb{F}}_q(t)) \cong \mathcal{S}_n$. Using [10, Corollary 3.4.14.] on $M'_f : \overline{\mathbb{F}}_q(t)$ we get

$$2\mathfrak{g} - 2 = -2(n!) + \sum_{P \in \mathcal{P}^1_{\overline{\mathbb{F}}_q}} \sum_{P' \in M'_f :\, P'|P} d(P').$$

Using the fact that in a Galois extension $e(P'|P)$ divides the order of the Galois group together with Hilbert's Different Exponent Theorem [10, Theorem 3.5.1], we get that

$$2\mathfrak{g} - 2 \leq -2(n!) + \sum_{P \in \mathrm{Ram}(\mathrm{M_f}:\overline{\mathbb{F}}_q(t))} \sum_{P' \in M'_f \,:\, P'|P} e(P'|P).$$

The fundamental equality [10, Theorem 3.1.11] now ensures that:

$$2\mathfrak{g} - 2 \leq -2(n!) + \sum_{P \in \mathrm{Ram}(\mathrm{M_f}:\overline{\mathbb{F}}_q(t))} n!,$$

which in turn leads to

$$\mathfrak{g} \leq n(n!)/2,$$

since $\#\mathrm{Ram}(\mathrm{M_f} : \overline{\mathbb{F}}_q(t)) \leq n$. Therefore we get that:

$$\frac{1}{n}q^d - \frac{2}{n}\mathfrak{g}q^{\frac{d}{2}} - n \geq \frac{1}{n}q^d - n!q^{\frac{d}{2}} - n.$$

Thus, in order to get a t_0 not lying in a subextension and such that $f - t_0$ is irreducible we again require that (in the notation of Theorem 6) $I(n, d) \geq dq^{d/2}$. This leads to

$$\frac{1}{n}q^d - n!q^{\frac{d}{2}} - n > dq^{d/2},$$

which holds whenever $d \geq 2\log(n!)$.

We now want to emphasize how, using our arguments, one can show that the existence of a *certain* factorization pattern for a polynomial $f \in \mathbb{F}_q[x]$ determines the statistics of *all* factorization patterns of $f - t_0$ where t_0 ranges over $\mathbb{F}_{q^d}$.

Theorem 7. *Let $F \in \mathbb{F}_q[x]$ have degree n and suppose that there exists t_0 such that*

$$F(x) - t_0 = \begin{cases} (x-a)^2 \cdot g(x) \cdot h(x) & \text{if } n \text{ is even} \\ (x^2+x+1) \cdot x^2 \cdot g(x) & \text{if } n \text{ is odd, } q=2 \text{ and } r = n-4 \\ (x-a)^2 \cdot (x-b) \cdot g(x) \cdot h(x) & \text{otherwise} \end{cases},$$

where $a, b \in \mathbb{F}_q$ and $a \neq b$ in the third case, $g(x), h(x) \in \mathbb{F}_q[x]$ are irreducible and $\deg g(x)$ is a prime larger than $n/2$. Then for any $i \in \mathbb{N}$ we have that the number of $t_1 \in \mathbb{F}_{q^i}$ such that $F - t_1$ has squarefree factorization pattern $\Delta = (d_1, d_2, \ldots, d_n)$ is

$$\frac{1}{\prod_{j=1}^{n} j^{d_j}(d_j!)} q^i + O(q^{i/2}), \tag{2}$$

where the implied constant can be made explicit and dependent only on n. Moreover, the number of t_1's such that $F - t_1$ is not squarefree is bounded by an absolute constant depending only on the degree of F.

Proof. The second part of the statement comes immediately from the fact that the t_0's in $\mathbb{F}_q$ such that $F - t_0$ has a factorization pattern that is not squarefree are zeroes of the discriminant $d(t) \in \mathbb{F}_q[t]$ of $F - t \in \mathbb{F}_q(t)[x]$ over $\mathbb{F}_q(t)$, whose degree is independent of q.

Let us prove the rest of the statement. Using Theorem 4 we immediately get that $N_F = G_F = \mathcal{S}_n$ for any base field $\mathbb{F}_{q^i}$ (i.e. $\mathrm{Gal}(\overline{\mathbb{F}}_q M_F : \overline{\mathbb{F}}_q)) = \mathcal{S}_n$. To show the rest of the statement, it is sufficient to apply Corollary 1 to any element $g \in G_F$ having the disjoint cycle decomposition prescribed by Δ

$$\underbrace{(-)(-)\cdots(-)}_{d_1}\underbrace{(--)(--)\cdots(--)}_{d_2})\cdots\underbrace{\underbrace{(--\cdots--)}_{n}\cdots\underbrace{(--\cdots--)}_{n}}_{d_n}$$

as its conjugacy class has $\dfrac{n!}{\prod_{j=1}^{n} j^{d_j}(d_j!)}$ elements.

4 Experimental Results

In this section we provide experimental results to show how close the statistics predicted by Theorems 6 and 7 are to the actual statistics of polynomial.

In Table 1a we fix the polynomial $x^2(x^5 + 4x^2 + 2)(x - 3)$ (which verifies the hypotheses of Theorem 7) and then we let t_0 range over $\mathbb{F}_{q^d}$ for $q = 11$ and $d \in \{1, 2, 3, 4, 5\}$ to see how many times $F - t_0 \in \mathbb{F}_{q^d}[x]$ is an irreducible polynomial. Using formula (2), in this case the predicted statistic is given by $11^d/4!$.

In Table 1b we fix the polynomial $F = x^2(x^5 + 4x + 11)(x - 3)$ verifying the hypothesis of Theorem 7 and then we let t_0 range over $\mathbb{F}_{q^d}$ for $q = 13$ and $d \in \{1, 2, 3, 4, 5\}$ to see the number of occurrences of the factorization pattern of the form $F - t_0 = u_{t_0}(x)v_{t_0}(x) \in \mathbb{F}_{q^d}[x]$ where $u_{t_0}(x)$ is an irreducible polynomial of degree 6 and $v_{t_0}(x)$ is an irreducible polynomial of degree 2. Using formula (2), in this case the predicted statistic is given by $13^d/12$.

Table 1. .

d	# of t_0's for which $f - t_0$ is irreducible	Expected # of t_0's for which $f - t_0$ is irreducible	Agreement (column 2/column 1)
1	2	1	0.5
2	14	17	1.21
3	143	190	1.32
4	1764	2091	1.18
5	19967	23007	1.15

(a) Statistics of the number of t_0's such that $x^2(x^5 + 4x^2 + 2)(x - 3) - t_0 \in \mathbb{F}_{11^d}[x]$ is irreducible.

d	# of t_0's with fact. pattern $u_{t_0}(x)v_{t_0}(x)$	Expected # of t_0's with fact. pattern $u_{t_0}(x)v_{t_0}(x)$	Agreement (column 2/column 1)
1	1	1	1
2	10	14	1.4
3	255	183	0.717
4	2300	2380	1.03
5	30366	30941	1.01

(b) Statistics of the number of t_0's for which the polynomial $x^2(x^5 + 4x + 11)(x - 3) - t_0 \in \mathbb{F}_{13^d}[x]$ has factorization pattern $u_{t_0}(x)v_{t_0}(x)$ with $u_{t_0}(x)$ irreducible of degree 6 and $v_{t_0}(x)$ irreducible of degree 2.

References

1. Bartoli, D., Zini, G., Zullo, F.: Investigating the exceptionality of scattered polynomials. Finite Fields Appl. **77**, Paper No. 101956, 27 (2022)
2. Birch, B.J., Swinnerton-Dyer, H.P.F.: Note on a problem of Chowla. Acta Arith. **5**(4), 417–423 (1959)
3. Fein, B., Schacher, M.: Properties of iterates and composites of polynomials. J. Lond. Math. Soc. **54**((2)3), 489–497 (1996)
4. Ferraguti, A., Micheli, G.: On the existence of infinite, non-trivial F-sets. J. Number Theory **168**, 1–12 (2016)
5. Ferraguti, A., Micheli, G.: Exceptional scatteredness in prime degree. J. Algebra **565**, 691–701 (2021)
6. Gallagher, P.X.: The large sieve and probabilistic Galois theory, pp. 91–101 (1973)
7. Kosters, M.: A short proof of the Chebotarev density theorem for function fields. Math. Commun. **22**(2), 227–233 (2017)
8. Micheli, G.: On the selection of polynomials for the DLP quasi-polynomial time algorithm for finite fields of small characteristic. SIAM J. Appl. Algebra Geom. **3**(2), 256–265 (2019)

9. Munemasa, A., Nakamura, H.: A note on the Brawley-Carlitz theorem on irreducibility of composed products of polynomials over finite fields. In: Duquesne, S., Petkova-Nikova, S. (eds.) WAIFI 2016. LNCS, vol. 10064, pp. 84–92. Springer, Cham (2016). https://doi.org/10.1007/978-3-319-55227-9_7
10. Stichtenoth, H.: Algebraic function fields and codes, vol. 254. Springer Science & Business Media, Heidelberg (2009). https://doi.org/10.1007/978-3-540-76878-4
11. Turnwald, G.: On Schur's conjecture. J. Aust. Math. Soc. Ser. A **58**(3), 312–357 (1995)

On Two Applications of Polynomials $x^k - cx - d$ over Finite Fields and More

Canberk İrimağzı[1,2] and Ferruh Özbudak[1,2](✉)

[1] Department of Mathematics, Middle East Technical University, Dumlupınar Bul., No:1, 06800 Ankara, Turkey
{canberk,ozbudak}@metu.edu.tr

[2] Institute of Applied Mathematics, Middle East Technical University, Dumlupınar Bul., No:1, 06800 Ankara, Turkey

Abstract. For integers $k \in [2, q-2]$ coprime to $q-1$, we first bound the number of zeroes of the family of polynomials $x^k - cx - d \in \mathbf{F}_q[x]$ where $q = 2^n$ such that $q-1$ is a prime or $q = 3^n$ such that $(q-1)/2$ is a prime. This gives us bounds on cross-correlation of a subfamily of Golomb Costas arrays.

Next, we show that the zero set of $x^k - cx - d$ over $\mathbf{F}_q$ is a planar almost difference set in $\mathbf{F}_q^*$ and hence for some set of pairs (c, d), they produce optical orthogonal codes with $\lambda = 1$.

More generally, we give an algorithm to produce optical orthogonal codes (OOCs) from $P(x) = x^{\ell_1} + c_{\ell_2} x^{\ell_2} + c_{\ell_2 - 1} x^{\ell_2 - 1} + \cdots + c_1 x \in \mathbf{F}_q[x]$ where interestingly $\ell_1 \gg \ell_2$. We focus on the case $\ell_2 \in \{2, 3\}$ and provide examples of $(q-1, w, \lambda)$-OOCs with $\lambda \in \{2, 3\}$.

Keywords: Golomb costas permutations · Planar cyclic almost difference sets · Almost difference families · Optical orthogonal codes · Radar · Sonar · Optical CDMA

1 Introduction

Costas arrays have applications in sonar and radar systems as they have optimal autocorrelation properties. Their study centres around two problems: searching for methods to create Costas arrays and studying cross-correlation of families of Costas arrays. The study of cross-correlation of Costas arrays boils down to finding a suitable family with good cross-correlation properties. One such family is considered by Gómez-Pérez and Winterhof in [8].

Optical orthogonal codes are primarily used in optical CDMA communication systems. It is important to construct such codes with good parameters and large size. There are constructions of optimal optical orthogonal codes with parameters (n, w, λ) in case $\lambda = 1$ in the literature ([2,11]), however either the number of such codes (even if optimal) is limited or these codes have low weight. In [4], Ding and Xing considers the next case where $\lambda = 2$.

Freedman and Levanon proved in [7] that any two distinct Costas arrays of the same size > 3 have cross-correlation of at least 2. In Subsect. 2.2, we

S. Mesnager and Z. Zhou (Eds.): WAIFI 2022, LNCS 13638, pp. 14–32, 2023.
https://doi.org/10.1007/978-3-031-22944-2_2

show that there are ℓ Golomb Costas arrays of size $q-2$ whose maximal cross-correlation achieves this lower bound where $q = 2^\ell$ and ℓ is a prime. More generally, let p be a prime, $n \geqslant 2$ a positive integer, and t denote the smallest prime divisor of n. We show that there is a collection of t distinct Golomb Costas arrays of size $q-2$ whose maximal cross-correlation is at most p where $q = p^n$.

The maximal cross-correlation $C(\mathcal{G}_q)$ of the set $\mathcal{G}_q = \{\pi_{g_1,g_2} \mid g_1 \in \mathbf{F}_q \text{ is a primitive element}\}$ (considered first by Gómez-Pérez and Winterhof in [8]) of Golomb Costas permutations where $g_2 \in \mathbf{F}_q$ is a fixed primitive element is expressed as follows:

$$\max_{\substack{2\leqslant k\leqslant q-2\\ \gcd(k,q-1)=1}} \max_{\substack{c,d\in\mathbf{F}_q\\ c\neq 0}} |\{x \in \mathbf{F}_q \setminus \{0,1\} \mid x^k - cx - d = 0\}|.$$

In [8], Gómez-Pérez and Winterhof showed that $C(\mathcal{G}_q)$ of the subfamily $\mathcal{G}_q$ of Golomb Costas permutations of size $q-2$ when $q-1 = 2^n - 1$ is a Mersenne prime is bounded above by $\lfloor (1-1/(q-1))(1+q^{1/2}) \rfloor$. We call this Case I. They also show that in case q is an odd prime power and $(q-1)/2$ is prime, $C(\mathcal{G}_q)$ is bounded above by $1 + \lfloor (1-2/(q-1))q^{1/2} \rfloor$. We call this Case II, and it consists of two subcases when

(a) q is a power of 3 and $(q-1)/2$ is prime (see Lemma 1 in [6]), and
(b) q is a safe prime, that is, both q and $(q-1)/2$ are prime.

In Part I, we focus on Case I and Case II(a). Using a combinatorial argument, we obtain two new bounds for each case: conditional bound (on computing some values with the help of a computer) and unconditional bound. We prove that our unconditional bounds (at worst) recover Gómez-Pérez and Winterhof's bounds while the numerics suggest a mild improvement. In either case, numerics show that the conditional bounds (whenever computed) significantly improve the bounds given by Gómez-Pérez and Winterhof.

The combinatorial nature of zero-sets of polynomials $x^k - cx - d$ led us to produce optical orthogonal codes with $\lambda = 1$. In [4], Ding and Xing construct optical orthogonal codes with parameters $(2^m - 1, w, 2)$ where $w \in \{5, 9, 11, 13\}$ using cyclotomy (also, for odd primes $w \geqslant 11$ for which 2 is a primitive element in the prime field $\mathbf{F}_w$). Although these codes are non-optimal, they are of large size and thus very promising for applications. Motivated by their work, we provide the following examples:

(i) $(2^{21} - 1, 32, 2)$ optical orthogonal code with size $2^{16} - 1$, and
(ii) $(3^{13} - 1, 81, 3)$ optical orthogonal code with size $\frac{3^9-1}{2}$,
(iii) $(2^{14} - 1, 16, 2)$ optical orthogonal code with size $2 \cdot (2^{10} - 1)$,
(iv) $(3^7 - 1, 5, 2)$ optical orthogonal code with size 14329,
(v) $(2110, 5, 2)$ optical orthogonal code with size 13600.

The first three codes are members of some infinite classes and they arise from linearized polynomials while the last two arise from non-linearized polynomials. We provide a general algorithm to construct such codes in Part III.

The details of our results and the organization of the paper is as follows. Our main results in Part I are Theorem 3 and Theorem 4 from which we derive Corollary 1 and Corollary 2. The strength of the conditional bounds we obtain is illustrated in Table 1 and Table 2. Moreover, in the appendix (Sect. A), the unconditional bounds are proved to at worst recover the corresponding bounds given by Gómez-Pérez and Winterhof.

The elementary proof of Theorem 3 and Theorem 4 led us to investigate further to uncover the mathematical reason behind the picture and thus in Part II, we study the combinatorial nature of the zero sets of the polynomials $x^k - cx - d$. Our main results of Part II are Theorem 6 and Theorem 7. We discover that the zero sets of the polynomials $x^k - cx^{p^i} - d$ are planar almost difference sets in $\mathbf{F}_q^*$. Some collections of these zero sets in fact form almost difference families yielding optical orthogonal codes. Corollary 3 turns out to be a result of Moreno et al. [11]. We end this section by discussing the importance of computing the multiplicity distribution of low-degree monomials (see Remark 17).

Our efforts culminate in Part III and we present an algorithm to construct optical orthogonal codes from polynomials and provide examples.

2 Part I: Cross-Correlation of Golomb Costas Arrays

2.1 Golomb Costas Arrays

Let $q \geqslant 4$ be a prime power and $\mathbf{F}_q$ denote the finite field of q elements and with characteristic p. For an integer $m \geqslant 1$, let $[m]$ denote the set $\{1, 2, \ldots, m\}$.

Definition 1 *(Definition 3, [6]). Fix two primitive elements g_1, g_2 of the field $\mathbf{F}_q$. Define a permutation $\pi_{g_1,g_2} : [q-2] \to [q-2]$ by*

$$\pi_{g_1,g_2}(i) = j \text{ if and only if } g_1^i + g_2^j = 1.$$

Such a permutation is called Golomb Costas permutation. Note that $\pi_{g,h} = \pi_{g_1,g_2}$ if $g = \sigma(g_1)$ and $h = \sigma(g_2)$ where $\sigma \in Gal(\mathbf{F}_q/\mathbf{F}_p)$, so the cardinality of the set of all Golomb Costas permutations is $\varphi(q-1)^2/n$. Here, φ denotes Euler's phi function.

Definition 2 *(Definition 4, [6]). Let $f, g : [n] \to [n]$ be two maps. The cross-correlation between f and g at $(u, v) \in \mathbf{Z}^2$ is*

$$C_{f,g}(u, v) := |\{(i + u, f(i) + v) \mid i \in [n]\} \cap \{(i, g(i)) \mid i \in [n]\}|.$$

Note that $C_{f,g}(u, v) = 0$ for pairs (u, v) such that $|u|, |v| \geqslant n$. The maximal cross-correlation $C(\mathcal{F})$ of a family $\mathcal{F}$ of maps (of cardinality at least 2) is $\max\limits_{\substack{f,g \in \mathcal{F} \\ f \neq g}} \max\limits_{(u,v) \in \mathbf{Z}^2} C_{f,g}(u, v)$.

2.2 A Small Family of Golomb Costas Permutations with Low Cross-Correlation

Let n denote the degree of the extension field $\mathbf{F}_q$ over the prime field $\mathbf{F}_p$. Let t denote the smallest prime divisor of n.

Proposition 1. *Let us fix two primitive elements g_1, g_2 of the field $\mathbf{F}_q$ and $\sigma \in Gal(\mathbf{F}_q/\mathbf{F}_p)$ denote the Frobenius automorphism. Then, the maximal cross-correlation of the subfamily*

$$\mathcal{G} = \{\pi_{g,h} \mid g = \sigma^r(g_1), h = g_2 \text{ where } 0 \leqslant r < t\}$$

(where t is the smallest prime divisor of n) of Golomb Costas permutations is at most p.

Proof. Let $\pi_1 := \pi_{\sigma^{r_1}(g_1),g_2}$ and $\pi_2 := \pi_{\sigma^{r_2}(g_1),g_2}$ be distinct permutations in $\mathcal{G}$. Here, $r_1 \neq r_2$ and without loss of generality we may assume $r_1 > r_2$. Then, $C_{\pi_1,\pi_2}(u,v)$ is the number of solutions of the equation

$$g_2^v(1 - g_1^{p^{r_1}x}) = (1 - g_1^{p^{r_2}(x+u)})$$

where $x, x+u \in [q-2]$. This number is bounded above by the number of $\mathbf{F}_q$-solutions of the polynomial

$$b(1-y)^{p^{r_1}} = (1-ay)^{p^{r_2}},$$

or equivalently

$$b^{p^{-r_2}} y^{p^{r_1-r_2}} - ay + 1 - b^{p^{-r_2}} = 0.$$

If we denote one of its zeroes by c, then all of its zeroes are of the form $c+dz$ where $c, d \in \overline{\mathbf{F}_q}$ are fixed and $z \in \mathbf{F}_{p^k}$ where $k = r_1 - r_2$. Suppose it has three zeroes $c, c+dz_1, c+dz_2$ in $\mathbf{F}_q$, then $z_2/z_1 \in \mathbf{F}_q \cap \mathbf{F}_{p^k}^* = \mathbf{F}_p^*$ (note that $0 < k = r_1 - r_2 < t$ and $\gcd(k,n) = 1$.). This forces that there are at most p zeroes in $\mathbf{F}_q$ and this completes the proof.

Remark 1. Let $0 < d_1 < d_2$ be two consecutive divisors of n. One can more generally prove that the maximal cross-correlation of the subfamily

$$\mathcal{G} = \{\pi_{g,h} \mid g = \sigma^r(g_1), h = g_2 \text{ where } 0 \leqslant r < d_2\}$$

of Golomb Costas permutations is at most p^{d_1}.

2.3 Cross-Correlation of a Subfamily of Golomb Costas Arrays

Notations 1. For a fixed primitive element $g_2 \in \mathbf{F}_q$, let $\mathcal{G}_q$ denote the set

$$\{\pi_{g_1,g_2} \mid g_1 \text{ is a primitive element of } \mathbf{F}_q\}$$

of Golomb Costas permutations. The maximal cross-correlation of this subfamily is studied in [8].

Let $\pi_1 = \pi_{g_1^r, g_2}$ and $\pi_2 = \pi_{g_1^s, g_2}$ be two distinct Golomb Costas permutations where $1 \leqslant r, s \leqslant q-2$ coprime to $q-1$ and $r \neq s$. Then, $C_{\pi_1,\pi_2}(u,v)$ is the number of nonzero solutions to the equation

$$g_2^v(1 - g_1^{rx}) = (1 - g_1^{s(x+u)})$$

where $x, x+u \in [q-2]$ so that $\max_{(u,v)\in\mathbf{Z}^2} C_{\pi_1,\pi_2}(u,v)$ is the number of $\mathbf{F}_q$-solutions other than 0 and 1 of the polynomial

$$b(1 - y^r) = 1 - ay^s$$

where $a, b \in \mathbf{F}_q^*$ are arbitrary. Composing this polynomial with the permutation polynomial $x^{1/r}$ where $1/r$ denotes the multiplicative inverse of r modulo $q-1$, we get the polynomial

$$ay^{s/r} - by + b - 1.$$

Hence, $C(\mathcal{G}_q)$ is equal to

$$\max_{\substack{2\leqslant k\leqslant q-2\\ \gcd(k,q-1)=1}} \max_{\substack{c,d\in\mathbf{F}_q\\ c\neq 0}} |\{x \in \mathbf{F}_q \setminus \{0,1\} \mid x^k - cx - d = 0\}|. \qquad (\bigstar)$$

2.4 Golomb Costas Arrays of Size $q-2$ Where $q-1$ Is a Mersenne Prime

Throughout Sect. 2.4, let q denote a power of 2 such that $q-1$ is a prime, i.e., $q-1 = 2^n - 1$ is a Mersenne prime. In [8], Gómez-Pérez and Winterhof showed that the maximal cross-correlation $C(\mathcal{G}_q)$ of the subfamily $\mathcal{G}_q$ of Golomb Costas permutations of size $q-2$ when $q-1$ is a Mersenne prime is bounded above by $\lfloor (1 - 1/(q-1))(1 + q^{1/2}) \rfloor$.

Lemma 1. *Suppose $n > 2$ is a positive integer such that $q-1 = 2^n - 1$ is a (Mersenne) prime. Then, we have*

$$C(\mathcal{G}_q) = \max_{2\leqslant k\leqslant q-2} \max_{\substack{c,d\in\mathbf{F}_q\\ c\neq 0}} |\{x \in \mathbf{F}_q \setminus \{0,1\} \mid x^k - cx - d = 0\}| = \max_{2\leqslant k\leqslant q-2} \max_{d\in\mathbf{F}_q^*} |\{x \in \mathbf{F}_q \mid x^k - x - d = 0\}|.$$

Proof. Note that the polynomials $x^k - cx - d$ and $(x/\alpha)^k - c/\alpha^{k-1}x/\alpha - d/\alpha^k$ have the same number of distinct zeroes where $\alpha \in \mathbf{F}_q^*$. Setting $\alpha = c^{\frac{1}{1-k}}$, we prove the statement. (Here, $\frac{1}{1-k}$ denotes the multiplicative inverse of $1-k$ modulo $q-1$.)

Remark 2. With the help of Lemma 1, we were able to compute that $C(\mathcal{G}_q) = 13$ for $n = 13$ using Magma [1] within two days.

Remark 3. For a Mersenne prime $q-1 = 2^n - 1$, let $\alpha \in \mathbf{F}_q^*$ be a fixed element other than 1. Since the multiplicative group $\mathbf{F}_q^*$ is generated by any element other than 1, both α and $\alpha + 1$ are primitive elements. As α is primitive, we have $\alpha^k = \alpha + 1$ for some $2 \leqslant k \leqslant q-2$, i.e., α is a zero of the polynomial $x^k - x - 1 \in \mathbf{F}_p[x]$. As α is primitive, it has n distinct Galois conjugates over $\mathbf{F}_p$ so that $C(\mathcal{G}_q) \geqslant n$.

Definition 3 *(Definition 1.1, [10]). Let $f \in \mathbf{F}_q[x]$ be a nonzero polynomial. Throughout this definition, $\mathbf{F}_q$ denotes any finite field. Let $\nu_i(f)$ denote the cardinality of the set*

$$\{(c,d) \in \mathbf{F}_q^2 \mid \text{ the polynomial } f(x) - cx - d \text{ has } i \text{ distinct zeroes in } \mathbf{F}_q\}.$$

The sequence $(\nu_i(f))_{i=0}^q$ is called the intersection distribution of f.

For $c \in \mathbf{F}_q$, let $\mathcal{M}_i(f,c)$ denote the set $\{d \in \mathbf{F}_q \mid f(x) - cx - d$ has i solutions in $\mathbf{F}_q\}$ and $M_i(f,c)$ be its cardinality. The sequence $(M_i(f,c))_{i=0}^q$ is called the multiplicity distribution of f at c. We will use the multiplicity distribution in Part II.

Notations 2. Fix a polynomial $f \in \mathbf{F}_q[x]$. For $c, d \in \mathbf{F}_q$, let $S_{c,d}(f)$ denote the set $\{x \in \mathbf{F}_q \mid f(x) - cx - d = 0\}$. Note that for an automorphism $\sigma \in \text{Gal}(\mathbf{F}_q/\mathbf{F}_p)$, we have $|S_{c,d}(f)| = |S_{\sigma(c),\sigma(d)}(f)|$. If f is clear from the context, we will simply write $S_{c,d}$ in place of $S_{c,d}(f)$.

Theorem 3. *Suppose $n > 2$ is a positive integer such that $q - 1 = 2^n - 1$ is a (Mersenne) prime. Let $f : \mathbf{F}_{2^n} \to \mathbf{F}_{2^n}$ be defined by $f(x) = x^k$ where for some $2 \leqslant k \leqslant q-2$. Then, $\nu_i(f) = 0$ for $i > \max\left\{\left\lfloor\sqrt{\frac{q-2}{n} + \frac{1}{4}} + \frac{1}{2}\right\rfloor, \mathbf{S}_{1,1}\right\}$ where $\mathbf{S}_{1,1} = \max_{2 \leqslant k \leqslant q-2} |S_{1,1}(x^k)|$.*

Proof. Consider the polynomial $f(x) - cx - d$. By Lemma 1, we may assume that $c = 1$ and $d \in \mathbf{F}_q^*$. Moreover, it suffices to show that the number of zeroes of $f(x) - x - d$ in $\mathbf{F}_q$ is bounded above by $\left\lfloor\sqrt{\frac{q-2}{n} + \frac{1}{4}} + \frac{1}{2}\right\rfloor$ whenever $d \neq 1$. In other words, we may assume that d is a primitive element.

Let $\beta, \beta + \alpha \in \mathbf{F}_{2^n}$ be two distinct zeroes of the polynomial $f(x) - x - d$ where $d \in \mathbf{F}_{2^n} \setminus \mathbf{F}_2$ (so, $\alpha \neq 0$). Then,

$$f(\beta) - \beta - d = 0$$
$$f(\beta + \alpha) - (\beta + \alpha) - d = 0$$

This implies that β and $\beta + \alpha$ are solutions to the equation $f(x+\alpha) + f(x) = \alpha$. Dividing both sides by α^k, we observe that

$$D_1 f(\beta/\alpha) = \alpha^{1-k}$$

where $D_1 f(x) = (x+1)^k + x^k$ is the derivative of f at 1. In other words, α^{1-k} is in the image of $D_1 f(x)$ and β/α and $\beta/\alpha + 1$ are in the corresponding preimage set. Now, let us denote by Δ the set $\{(x_1, x_2) \in S_{1,d} \times S_{1,d} \mid x_1 = x_2\}$ and define a map Φ from the set

$$(S_{1,d} \times S_{1,d}) \setminus \Delta = \{(x_1, x_2) \in \mathbf{F}_q^2 \mid f(x_1) - x_1 - d = 0, f(x_2) - x_2 - d = 0 \text{ and } x_1 \neq x_2\}$$

to the graph

$$\{(y,z) \in \mathbf{F}_q^2 \mid D_1 f(y) = z, \text{ and } y \neq 0, 1\}$$

of the derivative of f at 1 by $(x_1, x_2) \mapsto (x_1/(x_1 - x_2), (x_1 - x_2)^{1-k})$. Note that this map is injective as $\gcd(1-k, q-1) = 1$. Let

$$(1, b) \sim (1, d)$$

if $b = \sigma^r(d)$ for some $0 \leqslant r < n$. If $x \in S_{1,d}$, then $x^{p^r} \in S_{1,\sigma^r(d)}$ and since $d \notin \mathbf{F}_2$, we have $S_{1,d} \cap S_{1,\sigma^r(d)} = \varnothing$ for any $r \neq 0$. Note that Φ injectively extends to the domain $\bigsqcup_{(1,b)\sim(1,d)} ((S_{1,b} \times S_{1,b}) \setminus \Delta)$. Here, $|S_{1,b}| = |S_{1,d}|$ for any $(1,b) \sim (1,d)$ and there are n such pairs $(1,b)$ as d is primitive. The target has cardinality $q-2$ and the domain has $n\ell(\ell-1)$ elements where ℓ is the number of distinct zeroes of the polynomial $f(x) - x - d$. This implies that

$$n\ell(\ell - 1) \leqslant q - 2$$

so that

$$\ell \leqslant \sqrt{\frac{q-2}{n} + \frac{1}{4}} + \frac{1}{2}.$$

and this finishes the proof.

Remark 4. Note that since $\mathbf{S}_{1,1}$ is divisible by n, the proof in fact shows that $\mathbf{S}_{1,1} \leqslant \left\lfloor \sqrt{q - 2 + \frac{1}{4}} + \frac{1}{2} \right\rfloor_n$ where $\lfloor x \rfloor_n$ denotes the largest integer divisible by n which is less than or equal to x.

Corollary 1. *Suppose $n > 2$ is a positive integer such that $q - 1 = 2^n - 1$ is a (Mersenne) prime. Then, we have*

$$\mathbf{S}_{1,1} \leqslant C(\mathcal{G}_q) \leqslant \max\left\{ \left\lfloor \sqrt{\frac{q-2}{n} + \frac{1}{4}} + \frac{1}{2} \right\rfloor, \mathbf{S}_{1,1} \right\}.$$

Moreover,

$$\mathbf{S}_{1,1} \leqslant \left\lfloor \sqrt{q - 2 + \frac{1}{4}} + \frac{1}{2} \right\rfloor_n.$$

Proof. It follows from Lemma 1, Theorem 3 and Remark 4.

Remark 5. We denote $\max\left\{ \left\lfloor \sqrt{\frac{q-2}{n} + \frac{1}{4}} + \frac{1}{2} \right\rfloor, \mathbf{S}_{1,1} \right\}$ by Bound A, and $\left\lfloor \sqrt{q - 2 + \frac{1}{4}} + \frac{1}{2} \right\rfloor_n$ by Bound B.

Remark 6. In the appendix, we prove that Bound B is at worst recovers Gómez-Pérez and Winterhof's bound. Numerics suggest that although Bound B is only slightly better than that of Gómez-Pérez and Winterhof, Bound A gives a significant improvement.

Remark 7. Computation of $C(\mathcal{G}_q)$ for $n \geqslant 17$ is beyond our reach even with the help of Lemma 1. It would be interesting to tackle the first instance (if any) of n for which $\mathbf{S}_{1,1} \neq C(\mathcal{G}_q)$.

Table 1. Comparison of our bounds and that of Gómez-Pérez and Winterhof's in Case I.

n	$\mathbf{S}_{1,1}$	$C(\mathcal{G}_q)$	Bound A	Bound B	Gómez-Pérez and Winterhof's bound
3	3	3	3	3	3
5	5	5	5	5	6
7	7	7	7	7	12
13	13	13	25	91	91
17	51	*	88	357	363
19	57	*	166	722	725

2.5 Golomb Costas Arrays of Size $3^n - 2$ Where $(3^n - 1)/2$ Is a Prime

Throughout Sect. 2.5, let q be a power of 3 such that $(q-1)/2 = (3^n - 1)/2$ is a prime, i.e. q is a strict safe prime power as defined in [6]. Such n is necessarily an odd prime and first few values for it are $3, 7, 13, 71$. Recall that $C(\mathcal{G}_q)$ is equal to

$$\max_{\substack{2 \leqslant k \leqslant q-2 \\ \gcd(k,q-1)=1}} \max_{\substack{c,d \in \mathbf{F}_q \\ c \neq 0}} |\{x \in \mathbf{F}_q \setminus \{0,1\} \mid x^k - cx - d = 0\}|.$$

Lemma 2. *Suppose $n \geqslant 3$ is a positive integer such that $(q-1)/2 = (3^n - 1)/2$ is a prime. Then, we have*

$$C(\mathcal{G}_q) = \max_{\substack{2 \leqslant k \leqslant q-2 \\ gcd(k,q-1)=1}} \max_{\substack{c,d \in \mathbf{F}_q \\ c \neq 0}} |\{x \in \mathbf{F}_q \setminus \{0,1\} \mid x^k - cx - d = 0\}| = \max_{\substack{2 \leqslant k \leqslant q-2 \\ gcd(k,q-1)=1}} \max_{\substack{d \in \mathbf{F}_q^* \\ c \in \{1,-1\}}} |\{x \in \mathbf{F}_q \mid x^k - cx - d = 0\}|.$$

Proof. Note that the polynomials $x^k - cx - d$ and $(x/\alpha)^k - c/\alpha^{k-1}x/\alpha - d/\alpha^k$ have the same number of distinct zeroes where $\alpha \in \mathbf{F}_q^*$. Either c or $-c$ is a square in $\mathbf{F}_q$ and $\gcd(k-1, q-1) = 2$ so that

$$\alpha^{k-1} = c \text{ or}$$

$$\alpha^{k-1} = -c$$

has a zero in $\mathbf{F}_q$. Therefore, we may assume that $c \in \{1, -1\}$.
Let us now argue why 0 and 1 can be excluded: First note that polynomials $x^k - cx$ can be excluded from the list as we already know that $C(\mathcal{G}_q) \geqslant n \geqslant 3$. Moreover, 0 is not a zero of any $x^k - cx - d$ where $d \in \mathbf{F}_q^*$. Clearly, $1 \notin S_{1,d}$ where $d \in \mathbf{F}_q^*$ and even if $1 \in S_{-1,-1}$, we have $S_{-1,-1} = S_{-1,1}$ and $1 \notin S_{-1,1}$.

Remark 8. With the help of Lemma 2, we were able to compute that $C(\mathcal{G}_q) = 14$ for $n = 7$ using Magma [1] within minutes.

Theorem 4. *Suppose $n \geqslant 3$ is a positive integer such that $(q-1)/2 = (3^n - 1)/2$ is a prime. Let $f : \mathbf{F}_{2^n} \to \mathbf{F}_{2^n}$ be the (permutation) polynomial defined by*

$f(x) = x^k$ where for some $2 \leqslant k \leqslant q-2$ and $gcd(k, q-1) = 1$. Then, $\nu_i(f) = 0$ for

$$i > \max\left\{\sqrt{\frac{q-3}{n} + \frac{1}{4}} + \frac{1}{2}, \mathbf{S}_{1,1}, \mathbf{S}_{-1,1}\right\}$$

where $\mathbf{S}_{u,v} = \max_{\substack{2 \leqslant k \leqslant q-2 \\ gcd(k,q-1)=1}} |S_{u,v}(x^k)|$ *and* $u, v \in \{1, -1\}$.

Proof. Consider the polynomial $f(x) - cx - d$. Note that $\mathbf{S}_{1,1} = \mathbf{S}_{1,-1}$ and $\mathbf{S}_{-1,1} = \mathbf{S}_{-1,-1}$, so by Lemma 2 we may assume $c \in \{1, -1\}$ and $d \in \mathbf{F}_q \setminus \mathbf{F}_p$. Now, we imitate the proof of Theorem 3 by first considering the map Φ from the set $(S_{c,d} \times S_{c,d}) \setminus \Delta$ to the graph $\{(y, z) \in \mathbf{F}_q^2 \mid D_1 f(y) = z \text{ and } y \neq 0, 1\}$ defined by $(x_1, x_2) \mapsto (x_1/(x_1 - x_2), (x_1 - x_2)^{1-k})$ where $D_1 f(x) = f(x+1) - f(x)$. Note that now we have $\gcd(1-k, q-1) = 2$, so the injectivity of Φ requires a new argument: Suppose $\Phi(\beta_1, \beta_2) = \Phi(\beta_3, \beta_4)$. This implies that $(\beta_2 - \beta_1)^{1-k} = (\beta_4 - \beta_3)^{1-k}$, i.e.,

$$\beta_4 - \beta_3 = \xi(\beta_2 - \beta_1)$$

for some $\xi \in \mathbf{F}_q$ such that $\xi^{1-k} = 1$. Since $\gcd(k-1, q-1) = 2$, we must have $\xi = \pm 1$.

Note that $\Phi(\beta_1, \beta_2) = \Phi(\beta_3, \beta_4)$ also forces that $\beta_3 = \xi\beta_1$. However, we have

$$\beta_3^k - c\beta_3 - d = (\xi\beta_1)^k - c\xi\beta_1 - d = \xi(\beta_1^k - c\beta_1) - d = 0$$

as $\xi^{1-k} = 1$. This implies that $d = \xi d$, so we must have $\xi = 1$. This implies that $(\beta_1, \beta_2) = (\beta_3, \beta_4)$ proving the injectivity.

Next, we consider the equivalence relation defined in Theorem 3:

$$(1, b) \sim (1, d)$$

if $b = \sigma^r(d)$ for some $0 \leqslant r < n$ where σ denotes the Frobenius automorphism. Note that we have $|S_{1,b}| = |S_{1,d}|$ for any $(1, b) \sim (1, d)$ and moreover these $S_{1,b}$'s are all pairwise disjoint. Note that Φ injectively extends to the domain $\bigsqcup_{(1,b)\sim(1,d)} ((S_{1,b} \times S_{1,b}) \setminus \Delta)$ because $-d$ cannot be a Galois conjugate of d as the extension degree of $\mathbf{F}_q$ over $\mathbf{F}_p$ is necessarily an odd prime. There are n such pairs $(1, b)$ since $d \in \mathbf{F}_q \setminus \mathbf{F}_p$ and n is prime. The target has cardinality $q-2$ and the domain has $n\ell(\ell-1)$ elements where ℓ is the number of distinct zeroes of the polynomial $f(x) - cx - d$. This implies that $n\ell(\ell-1) \leqslant \lfloor q-2 \rfloor_n = q-3$ so that

$$\ell \leqslant \sqrt{\frac{q-3}{n} + \frac{1}{4}} + \frac{1}{2}$$

and this finishes the proof.

Remark 9. Note that since $\mathbf{S}_{-1,1} = 1 \,(\mathrm{mod}\, n)$ and $\mathbf{S}_{1,1}$ is divisible by n, the proof shows that $\mathbf{S}_{1,1} \leqslant \left\lfloor \sqrt{q-3+\frac{1}{4}} + \frac{1}{2} \right\rfloor_n$ and $\mathbf{S}_{-1,1} \leqslant \left\lfloor \sqrt{q-3+\frac{1}{4}} + \frac{1}{2} \right\rfloor_{n,1}$. Here, $\lfloor x \rfloor_{n,1}$ denotes the maximum of the two integers that are not exceeding x and equal to 0 or 1 modulo n.

Corollary 2. *Suppose $n \geqslant 3$ is a positive integer such that $(q-1)/2 = (3^n-1)/2$ is prime. Then, we have*

$$\max\{\mathbf{S}_{1,1}, \mathbf{S}_{-1,1}\} \leqslant C(\mathcal{G}_q) \leqslant \max\left\{\left\lfloor\sqrt{\frac{q-3}{n}+\frac{1}{4}}+\frac{1}{2}\right\rfloor, \mathbf{S}_{1,1}, \mathbf{S}_{-1,1}\right\}.$$

Moreover,

$$\max\{\mathbf{S}_{1,1}, \mathbf{S}_{-1,1}\} \leqslant \left\lfloor\sqrt{q-3+\frac{1}{4}}+\frac{1}{2}\right\rfloor_{n,1}.$$

Proof. Immediate.

Remark 10. We call the expression $\max\left\{\left\lfloor\sqrt{\frac{q-3}{n}+\frac{1}{4}}+\frac{1}{2}\right\rfloor, \mathbf{S}_{1,1}, \mathbf{S}_{-1,1}\right\}$ Bound A, and $\left\lfloor\sqrt{q-3+\frac{1}{4}}+\frac{1}{2}\right\rfloor_{n,1}$ is called Bound B.

Remark 11. We prove in the appendix that Bound B is at worst recovers Gómez-Pérez and Winterhof's bound.

Table 2. Comparison of our bounds and that of Gómez-Pérez and Winterhof's in Case II(a).

n	$\mathbf{S}_{1,1}$	$\mathbf{S}_{-1,1}$	$C(\mathcal{G}_q)$	Bound A	Bound B	Gómez-Pérez and Winterhof's bound
3	3	4	4	4	4	5
7	14	8	14	18	43	47
13	*	*	*	*	1262	1263

Remark 12. Computation of $C(\mathcal{G}_q)$ for $n \geqslant 13$ is beyond our reach even with the help of Lemma 2. It would be interesting to tackle the first instance (if any) of n for which $\max\{\mathbf{S}_{1,1}, \mathbf{S}_{-1,1}\} \neq C(\mathcal{G}_q)$.

Remark 13. Using the idea in Theorem 3 and Theorem 4, one can obtain an analogue of Bound B for the subfamily $\mathcal{G}_p$ of Golomb Costas permutations and the family $\mathcal{W}_p$ of Welch Costas permutations where p is a safe prime. However, there is no analogue of Bound A in these cases due to the lack of nontrivial automorphisms.

3 Part II: Almost Difference Families Arising from $x^k - cx - d$

3.1 Planar Almost Difference Sets Arising from the Polynomials $x^k - cx^{p^i} - d$

Definition 4 *([3]). Let $(A, +)$ be an abelian group of order n. A subset of $D \subset A$ of cardinality w is an (n, w, λ, t) almost difference set in A if, for t times, the* *difference function diff*: $A\setminus\{0\} \to \mathbf{Z}^{\geqslant 0}$ *takes the value* λ *and, for* $n-1-t$ *times, it takes the value* $\lambda+1$ *where*

$$\mathit{diff}(\alpha) = |(D+\alpha) \cap D|.$$

Notations 5. Let $\mathbf{F}_q$ be a finite field where $q \geqslant 4$ and for fixed $c, d \in \mathbf{F}_q^*$ and $0 \leqslant i < n$, consider the set $S_{i,c,d}(k) = \{x \in \mathbf{F}_q \,|\, x^k - cx^{p^i} - d = 0\}$. Let $2 \leqslant k \leqslant q-2$ be an integer such that $\ell := |S_{i,c,d}(k)| \geqslant 2$. For brevity, we write $S_{i,c,d}$ in place of $S_{i,c,d}(k)$ if k is clear from the context. Moreover, if $i = 0$, we write $S_{c,d}$ in place of $S_{0,c,d}$ (see Notation 2).

Lemma 3. *The map* $G : (S_{i,c,d} \times S_{i,c,d}) \setminus \Delta \to \mathbf{F}_q$ *defined by* $G(\beta_1, \beta_2) = \dfrac{\beta_1}{\beta_2}$ *is injective.*

Proof. Let us first consider the map $\Phi : (S_{i,c,d} \times S_{i,c,d}) \setminus \Delta \to \mathbf{F}_q^2$ defined by

$$\Phi(x_1, x_2) = \left(\frac{x_1}{x_1 - x_2}, c(x_1 - x_2)^{p^i - k}\right).$$

This map is injective: For if $\Phi(\beta_1, \beta_2) = \Phi(\beta_3, \beta_4)$, then

$$c(\beta_1 - \beta_2)^{p^i-k} = c(\beta_3 - \beta_4)^{p^i-k}$$

implies that $\beta_1 - \beta_2 = \xi(\beta_3 - \beta_4)$ for some $\xi \in \mathbf{F}_q$ such that $\xi^{p^i-k} = 1$. This implies that $\beta_1 = \xi\beta_3$, but then

$$\beta_1^k - c\beta_1^{p^i} - d = (\xi\beta_3)^k - c\xi^{p^i}\beta_3^{p^i} - d = \xi^{p^i}(\beta_3 - c\beta_3^{p^i}) - d = 0$$

as $\xi^{p^i-k} = 1$. This implies that $d = \xi^{p^i} d$, so $\xi^{p^i} = 1$ and we conclude that $\xi = 1$. Hence, $(\beta_1, \beta_2) = (\beta_3, \beta_4)$. Moreover, the image of Φ is contained(!) in the graph $\{(x, y) \in \mathbf{F}_q^2 \,|\, y = (x+1)^k - x^k\}$ (see the proof of Theorem 3 for how we construct Φ) so that the map $F := \pi_1 \circ \Phi$ is also injective where $\pi_1 : \mathbf{F}_q^2 \to \mathbf{F}_q$ is the first projection. Note that $1 \notin \mathrm{Im}(F)$, so we may consider the composition $r \circ F$ where $r : \mathbf{F}_q \setminus \{1\} \to \mathbf{F}_q$ is the rational map $r(x) = \dfrac{x}{x-1}$. Observe that the map $r \circ F$ is the map G given in the statement above. The injectivity of r implies that G is injective, and we are done.

Theorem 6. $S_{i,c,d}(k)$ *is a* $(q-1, \ell, 0, q-2-\ell(\ell-1))$ *almost difference set in the group* $\mathbf{F}_q^*$.

Proof. For simplicity, we denote $S_{i,c,d}$ by D. Let $a \in \mathrm{Im}(G)$ (so, $a \neq 1$), then there exist distinct elements $\beta_1, \beta_2 \in D$ such that $a = \beta_1/\beta_2$. I.e. $\beta_1 \in aD \cap D$ so that $\mathrm{diff}(a) \geqslant 1$. Let us now show that in fact $\mathrm{diff}(a) = 1$ in this case. Let $\beta_1, \beta_1' \in aD \cap D$, then there exist $\beta_2, \beta_2' \in D$ such that

$$\beta_1 = a\beta_2$$
$$\beta_1' = a\beta_2',$$

i.e., $\frac{\beta_1}{\beta_2} = \frac{\beta_1'}{\beta_2'}$. This implies that $G(\beta_1, \beta_2) = G(\beta_1', \beta_2')$, so by the injectivity of G, we conclude that $\beta_1 = \beta_2$. Thus, $\mathrm{diff}(a) = 1$.

Now, suppose $a \notin \mathrm{Im}(G)$ and $a \neq 1$. We claim that $aD \cap D = \emptyset$, for if $\beta_1 \in aD \cap D$, then there would exist $\beta_2 \in D$ (with $\beta_2 \neq \beta_1$ as $a \neq 1$) such that $a\beta_2 = \beta_1$. This contradicts with our assumption that $a \notin \mathrm{Im}(G)$. Hence, D is a $(q-1, \ell, 0, q-2-\ell(\ell-1))$ almost difference set of the group $\mathbf{F}_q^*$.

3.2 Almost Difference Families Arising from the Polynomials $x^k - cx - d$

Definition 5 *([5]). Let $\mathcal{F} = \{D_1, D_2, \ldots, D_m\}$ be a family of w-subsets of a finite abelian group G of cardinality n. For $1 \leqslant j \leqslant m$, let ΔD_j denote the multiset*

$$\{a - b \mid a, b \in D_j,\ a \neq b\}.$$

Let $\Delta\mathcal{F}$ denote the formal sum of ΔD_j's. $\mathcal{F}$ is called an (n, w, λ, t) almost difference family of size m if some t nonzero elements of G occur in the multiset $\Delta\mathcal{F}$ with multiplicity λ, and the remaining $n-1-t$ nonzero elements of G occur in $\Delta\mathcal{F}$ with multiplicity $\lambda + 1$.

The setup of the next theorem is as follows:

Let $\mathbf{F}_q$ be a finite field. Let $c \in \mathbf{F}_q^*$, $d \in \mathbf{F}_q$ and $2 \leqslant k \leqslant q-2$ be an integer. Let $r := \gcd(k-1, q-1)$ and we denote the subgroup of $\mathbf{F}_q^*$ consisting elements of order dividing r by H_r. Note that there is a (faithful) group action

$$\begin{aligned} H_r \times \mathcal{M}_s(x^k, c) \setminus \{0\} &\to \mathcal{M}_s(x^k, c) \setminus \{0\} \\ (h, x) &\mapsto hx \end{aligned}$$

where $s := |S_{c,d}|$ by multiplication (see Definition 3).

Theorem 7. *Let R be a set of representatives of the orbit space $(\mathcal{M}_s(x^k, c) \setminus \{0\})/H_r$. Then, $\{S_{c,d}(k) \mid d \in R\}$ is an $(q-1, s, 0)$ almost difference family in $\mathbf{F}_q^*$ of size $\left\lfloor \frac{M_s(x^k, c)}{r} \right\rfloor$.*

The proof of this theorem will be given after Remark 15.

Remark 14. Note that if $x^k - cx$ has s distinct zeroes in $\mathbf{F}_q$, then $\left\lfloor \frac{M_s(x^k, c)}{r} \right\rfloor = \frac{M_s(x^k, c) - 1}{r}$, and otherwise $\left\lfloor \frac{M_s(x^k, c)}{r} \right\rfloor = \frac{M_s(x^k, c)}{r}$.

Note that $M_q(x^q, 1) = q^{n-1}$ ([10, Proposition B.1]) in $\mathbf{F}_{q^n}$ and this gives us the next immediate corollary known as the generalized Bose-Chowla construction of OOCs.

Corollary 3 *(Generalized Bose-Chowla construction, [11]). There is an almost difference family in* $\mathbf{F}_{q^n}^*$ *with parameters* $(q^n-1,q,0)$ *of size* $\frac{q^{n-1}-1}{q-1}$.

Proof. It immediately follows from Theorem 7 and the remark above.

Remark 15. Note that after choosing an isomorphism $\mathbf{F}_{q^n}^* \simeq \mathbf{Z}_{q^n-1}$, associated to the family in the corollary above, there is an optimal optical orthogonal code (OOC for short) with parameters $(q^n-1,q,1)$ for every prime power q and an integer $n \geqslant 2$ (see Definition 6 below). One may more generally consider the zero set of the polynomials $x^q - cx$ where c is a nonzero $(q-1)^{\text{th}}$ power, i.e. $c = \alpha^{q-1}$ for some $\alpha \in \mathbf{F}_{q^n}^*$. However, in the case, the two codewords (of weight q) associated to the zero sets of $x^q - x - d$ (where $d \in R$ is fixed) and $x^q - cx - \alpha^q d$ are cyclic shifts of each other. Hence, two relevant optical orthogonal codes would be related to each other by an already-known operation of replacing any codeword by a cyclic shift of itself (see Remark 18 below).

Proof (Proof of Theorem 7). Let us first consider the map

$$\Phi : \bigsqcup_{d\in R}((S_{c,d}\times S_{c,d})\setminus \Delta) \to \mathbf{F}_q^2$$

defined by

$$\Phi(x_1,x_2) = \left(\frac{x_1}{x_1-x_2}, c(x_1-x_2)^{1-k}\right).$$

This map is injective: For if $\Phi(\beta_1,\beta_2) = \Phi(\beta_3,\beta_4)$ where $\beta_1,\beta_2 \in S_{c,d_1}$ and $\beta_3,\beta_4 \in S_{c,d_2}$, then we have $\beta_3-\beta_4 = \xi(\beta_1-\beta_2)$ for some $\xi \in \mathbf{F}_q$ such that $\xi^{1-k}=1$. This implies that $\beta_3 = \xi\beta_1$ but note that

$$\beta_3^k - c\beta_3 - d_2 = \xi(\beta_1^k - c\beta_1) - d_2 = 0.$$

Therefore, $\xi d_1 = d_2$ and this forces that $\xi = 1$ since $d_1,d_2 \in R$, proving the injectivity of Φ.

The rest is identical to the arguments given in Lemma 3 and Theorem 6.

Remark 16. For r as above, let $C_0^{(r,q)}$ is the set of r−th powers in $\mathbf{F}_q^*$ and $C_i^{(r,q)} = \{\alpha^i x \,|, C_0^{(r,q)}\}$ where $\alpha \in \mathbf{F}_q$ is a fixed primitive element and $0 \leqslant i \leqslant r-1$. Note that the almost difference family in Theorem 7 can in fact be extended to the union $\bigcup_{i=1}^{r-1}\{S_{c_i,d}(k)\,|\, d\in R_i\}$ where $c_i \in C_i^{(r,q)}$ are fixed elements and R_i's are fixed sets of representatives of the orbit spaces $(\mathcal{M}_s(x^k,c_i)\setminus\{0\})/H_r$. In this way, the size of the extended family increases to $\left\lfloor \frac{\sum_{i=0}^{r-1} M_s(x^k,c_i)}{r} \right\rfloor$.

Remark 17. In [9], Kyureghyan-Li-Pott computed the multiplicity distribution of x^3 over arbitrary finite fields. Using their result and Remark 16 above one can isolate the prime powers q such that there exist $(q-1,3,1)$ optimal optical orthogonal codes arising from the zero sets of polynomials $x^3 - cx - d$. Note however that the existence problem of optimal OOCs of weight 3 is already settled (see Theorem 5, [2]).

If the multiplicity distribution of x^4 is studied, this might lead to a partial progress towards the open problem of existence of optimal OOCs of weight 4 and it might even be possible to formulate a conjecture about an exhaustive list of lengths for which such OOCs exist. We expect that such a task boils down to cyclotomy but we do not tackle this problem here.

4 Part III: An Algorithm to Produce OOCs

Definition 6. *An optical orthogonal code (OOC for short) $\mathcal{C}$ with parameters (n, w, λ) is a collection of sequences consisting 0s and 1s of length n and weight w such that*

(i) $\sum_{i=0}^{n-1} \mathbf{c}_i \mathbf{c}_{i+j} \leqslant \lambda$ *for any* $\mathbf{c} \in \mathcal{C}$ *and* $j \not\equiv 0 \,(\mathrm{mod}\, n)$, *and*

(ii) $\sum_{i=0}^{n-1} \mathbf{c}_i \mathbf{d}_{i+j} \leqslant \lambda$ *for any distinct* $\mathbf{c}, \mathbf{d} \in \mathcal{C}$.

Remark 18. For a codeword $\mathbf{c} = (c_1, c_2, \ldots, c_n)$, let $T(\mathbf{c})$ denote the cyclic shift $(c_n, c_1, c_2, \ldots, c_{n-1})$. Then, for an optical orthogonal codes $\mathcal{C}$, the collection $(\mathcal{C} \setminus \{\mathbf{c}\}) \cup \{T^m(\mathbf{c})\}$ is an OOC as well for any $0 \leqslant m \leqslant n$. That is, we can replace any codeword with some cyclic shift of itself.

This section is motivated by the work [4] of Ding-Xing (though we do not use cyclotomy). Here is our algorithm:

Step 1: Fix a polynomial of the form $P(x) = x^{\ell_1} + c_{\ell_2} x^{\ell_2} + c_{\ell_2 - 1} x^{\ell_2 - 1} + \cdots + c_1 x \in \mathbf{F}_q[x]$ where c_1 and c_2 are nonzero (or more generally, where c_i and c_j are nonzero for some $1 \leqslant i < j \leqslant \ell_2$ so that $\gcd(\ell_1 - i, \ell_1 - j) = 1$).

Step 2: Fix an extension $\mathbf{F}_{q^n}^*$ of $\mathbf{F}_q$ and element $d \in \mathbf{F}_{q^n}^*$ such that $P(x) - d$ has ℓ zeroes $\mathbf{F}_{q^n}^*$ where $\ell > \ell_2$. (Note that one can always let d be any nonzero element of $\mathbf{F}_q$ and set $\mathbf{F}_{q^n}$ as the splitting field of $P(x) - d$.)

Step 3: Find all elements $d \in \mathbf{F}_{q^n}$ such that $P(x) - d$ has ℓ non-zero zeroes in $\mathbf{F}_{q^n}$.

Output: Associated to the family

$$\{\{x \in \mathbf{F}_{q^n}^* \mid P(x) - d = 0\} \mid d \text{ is nonzero and } P(x) - d \text{ has } \ell \text{ zeroes in } \mathbf{F}_{q^n}^*\}$$

of sets (of cardinality ℓ), we have optical orthogonal codes with parameters $(q^n - 1, \ell, \ell_2)$ whose supports are obtained by taking discrete logarithm with respect to some primitive element of $\mathbf{F}_{q^n}$.

This algorithm will be extended in a way that the size of the code is as large as possible (yet likely still non-optimal) but we first provide some examples by considering linearized polynomials:

Proposition 2. *For any integer* $f \geqslant 1$, *there exists an* $(2^{21f} - 1, 32, 2)$ *optical orthogonal code with size* $2^{21f-5} - 1$.

Proof. It has been checked with Magma that the additive polynomial $L(x) = x^{32} - x^2 - x$ splits over $\mathbf{F}_{2^{21}}$. For nonzero elements $d_1, d_2 \in \mathbf{F}_{2^{21f}}$, denote by $D_i \subset \mathbf{F}^*_{2^{21f}}$ the zero set of $x^{32} - x^2 - x - d_i$ in $\mathbf{F}_{2^{21f}}$ for $i = 1, 2$. Note that for nonzero $\alpha \in \mathbf{F}_{2^{21f}}$,

$$\begin{aligned} |\alpha^{-1}D_1 \cap D_2| &= \deg\gcd(\alpha^{32}x^{32} - \alpha^2x^2 - \alpha x - d_1, x^{32} - x^2 - x - d_2) \\ &= \deg\gcd((\alpha^{32} - \alpha^2)x^2 + (\alpha^{32} - \alpha)x + \alpha^{32}d_2 - d_1, x^{32} - x^2 - x - d_2) \\ &\leqslant 2 \end{aligned}$$

since $(\alpha^{32} - \alpha^2)x^2 + (\alpha^{32} - \alpha)x + \alpha^{32}d_2 - d_1$ is a nonzero polynomial (of degree at most 2) where $d_1 = d_2$ and $\alpha \neq 1$, or $d_1 \neq d_2$.

Then, associated to the set of 32-subsets

$$\{\{x \in \mathbf{F}^*_{2^{21f}} \mid x^{32} - x^2 - x - d = 0\} \mid d \text{ is nonzero and in the image of } L : \mathbf{F}_{2^{21f}} \to \mathbf{F}_{2^{21f}}\},$$

we have an OOC with the desired properties.

Proposition 3. *For any integer* $f \geqslant 1$, *there exists an* $(3^{13f} - 1, 81, 3)$ *optical orthogonal code with size* $\frac{3^{13f-4}-1}{2}$.

Proof. It has been checked with Magma that the $\mathbf{F}_3$-linear polynomial $L(x) = x^{81} + x^3 + x$ splits over $\mathbf{F}_{3^{13}}$. As for a nonzero element $\alpha \in \mathbf{F}^*_{3^{13f}}$, that $\alpha^{81} = \alpha^3$ and $\alpha^{81} = \alpha$ implies $\alpha^2 = 1$, therefore associated to the set of 81-subsets

$$\{\{x \in \mathbf{F}^*_{3^{13f}} \mid x^{81} + x^3 + x - d = 0\} \mid d \text{ is a non-square element in the image of } L : \mathbf{F}_{3^{13f}} \to \mathbf{F}_{3^{13f}}\},$$

we have an OOC with the desired properties.

We will now give an example that will motivate the next subsection:

Proposition 4. *For an integer* $f \geqslant 1$, *there exists an* $(2^{14f} - 1, w(f), 2)$ *optical orthogonal code with size*

$$\begin{cases} 2 \cdot \left(\frac{2^{14f}}{w(f)} - 1\right) & \text{if } f \text{ is not divisible by 5,} \\ \frac{2^{14f}}{w(f)} - 1 & \text{if } f \text{ is divisible by 5} \end{cases}$$

where

$$w(f) = \begin{cases} 16 & \text{if } f \text{ is odd} \\ 32 & \text{if } f \text{ is even but not divisible by 4,} \\ 64 & \text{if } f \text{ is divisible by 4.} \end{cases}$$

Proof. Let $\theta \in \mathbf{F}_{2^{14}}$ be a primitive third root of unity. It has been checked with Magma that the additive polynomial $L_\theta(x) = x^{64} - x^2 - \theta x$ factorizes over $\mathbf{F}_{2^{14}}$ as a product of 16 linear factors, 8 quadratic factors and 8 quartic factors (so does $L_{\theta^2}(x) = x^{64} - x^2 - \theta^2 x$).

If f is not divisible by 5, then 62 is coprime to $2^{14f} - 1$ and associated to the union of the set of $w(f)$-subsets

$$\{\{x \in \mathbf{F}^*_{2^{14f}} \mid x^{64} - x^2 - \theta x - d = 0\} \mid d \text{ is nonzero and in the image of } L_\theta : \mathbf{F}_{2^{14f}} \to \mathbf{F}_{2^{14f}}\}$$

with the set of $w(f)$-subsets

$$\{\{x \in \mathbf{F}^*_{2^{14f}} \mid x^{64} - x^2 - \theta^2 x - d = 0\} \mid d \text{ is nonzero and in the image of } L_{\theta^2} : \mathbf{F}_{2^{14f}} \to \mathbf{F}_{2^{14f}}\}$$

we have an OOC with the desired properties. In case f is divisible by 5, only one of these sets yields an OOC (and has the desired properties).

4.1 Algorithm Continued: How to Extend the Size of the OOC

Note that the algorithm in the previous section does not guarantee that the OOC produced will be optimal, so it is important to extend the algorithm in a way that we get as many codes as possible as an output. Here are the remaining steps of the algorithm:

Step 4: Set $S = \{P(x)\}$ and W be the complement of S in the set of polynomials in $\mathbf{F}_q$ which are of the form specified in Step 1.
Step 5: Let $Q(x) = x^{\ell_1} + c_{\ell_2}x^{\ell_2} + c_{\ell_2-1}x^{\ell_2-1} + \cdots + c_1 x$ be in W. If
(i) $Q(x) - e\mathbf{F}_{q^n}$ has ℓ zeroes in $\mathbf{F}^*_{q^n}$ for any $e \in \mathbf{F}^*_{q^n}$, and
(ii) $Q(x) \neq \alpha^{-\ell_1} P(\alpha x)$ for any $\alpha \in \mathbf{F}^*_{q^n}$ and any $P(x) \in S$,
then add $Q(x)$ to the list S.
Step 6: Remove $Q(x)$ from the set W and return Step 5 until S stabilizes.
Output: Associated to the family

$$\{\{x \in \mathbf{F}^*_{q^n} \mid P(x) - d = 0\} \mid P(x) \in S, d \text{ is nonzero and } P(x) - d \text{ has } \ell \text{ zeroes in } \mathbf{F}^*_{q^n}\}$$

of sets (of cardinality ℓ), we have optical orthogonal codes with parameters $(q^n - 1, \ell, \ell_2)$ whose supports are obtained by taking discrete logarithm with respect to some primitive element of $\mathbf{F}_{q^n}$.

The proofs of the propositions in previous subsection illustrate the idea behind our algorithm. Nevertheless, we provide an argument to show that the algorithm works:

(Proof of the validity of the algorithm:) Note that by Step 2 and Step 5, the family

$$\{\{x \in \mathbf{F}^*_{q^n} \mid P(x) - d = 0\} \mid P(x) \in S, d \text{ is nonzero and } P(x) - d \text{ has } \ell \text{ zeroes in } \mathbf{F}^*_{q^n}\}$$

contains only subsets of cardinality ℓ. Let $D_1 = \{x \in \mathbf{F}_{q^n}^* \mid P(x) - d_1 = 0\}$ and $D_2 = \{x \in \mathbf{F}_{q^n}^* \mid Q(x) - d_2 = 0\}$ and $\alpha \in \mathbf{F}_{q^n}^*$, then

$$\begin{aligned} |\alpha^{-1}D_1 \cap D_2| &= \deg \gcd(P(\alpha x) - d_1, Q(x) - d_2) \\ &= \deg \gcd(Q(x) - \alpha^{-\ell_1} P(\alpha x) + d_2 - \alpha^{-\ell_1} d_1, Q(x) - d_2) \end{aligned}$$

If $P(x) = Q(x)$ and $d_1 \neq d_2$, the polynomial $Q(x) - \alpha^{-\ell_1} P(\alpha x) + d_2 - \alpha^{-\ell_1} d_1$ is a nonzero polynomial of degree at most ℓ_2 by Step 1. If $P(x) \neq Q(x)$, then $Q(x) - \alpha^{-\ell_1} P(\alpha x) + d_2 - \alpha^{-\ell_1} d_1$ is a nonzero polynomial of degree at most ℓ_2 by Step 5.

We end by providing examples of OOCs arising from non-additive polynomials.

Example 1. Associated to the family

$$\{\{x \in \mathbf{F}_{3^7}^* \mid x^{59} - x^2 - cx - d = 0\} \mid d \text{ is nonzero and } x^{59} - x^2 - cx - d \text{ has at least 5 zeroes in } \mathbf{F}_{3^7}^*\},$$

we have a variable-weight OOC with parameters $(2186, \{5, 6, 7, 8, 9\}, 2)$ of size 17143 (14329 of them have weight 5). Note that the assumption in Step 5 of our algorithm is satisfied as $\gcd(57, 2186) = 1$.

Example 2. Associated to the family

$$\{\{x \in \mathbf{F}_{2111}^* \mid x^{59} - x^2 - cx - d = 0\} \mid d \text{ is nonzero and } x^{59} - x^2 - cx - d \text{ has at least 5 zeroes in } \mathbf{F}_{2111}^*\},$$

we have a variable-weight OOC with parameters $(2110, \{5, 6, 7, 8, 9\}, 2)$ of size 16263 (13600 of them have weight 5).

Acknowledgement. We would like to thank the anonymous referees for their valuable suggestions and comments.

A Appendix

Proposition 5. *Bound B recovers Gómez-Pérez and Winterhof's bound when $q = 2^n$ and $q - 1$ is a prime for $n \geqslant 3$.*

Proof. For $n = 3$, we have an equality (see Table 1), so may assume $n \geqslant 5$. Then, we have

$$\begin{aligned} \left(1 - \frac{1}{q-1}\right)(1 + \sqrt{q}) - \left(\frac{1}{2} + \sqrt{q - \frac{7}{4}}\right) &= \frac{1}{2} - \frac{1}{q-1} + \sqrt{q} - \frac{\sqrt{q}}{q-1} - \sqrt{q - \frac{7}{4}} \\ &\geqslant \frac{1}{2} - \frac{1}{q-1} - \frac{\sqrt{q}}{q-1} \end{aligned}$$

Note also that

$$\left(\frac{1}{2}-\frac{1}{q-1}\right)^2-\left(\frac{\sqrt{q}}{q-1}\right)^2 \geqslant \frac{1}{4}-\frac{1}{q-1}+\frac{1}{(q-1)^2}-\frac{q}{(q-1)^2}$$
$$\geqslant \frac{q^2-10q+9}{4(q-1)^2}.$$

As $q \geqslant 32$, we have $q^2-10q+9 \geqslant 0$. Hence, $\left\lfloor \frac{1}{2}+\sqrt{q-\frac{7}{4}}\right\rfloor_n \leqslant \left\lfloor \left(1-\frac{1}{q-1}\right)(1+\sqrt{q})\right\rfloor$.

Proposition 6. *Bound B recovers Gómez-Pérez and Winterhof's bound when $q=3^n$ and $(q-1)/2$ is a prime.*

Proof. Assume the hypothesis on q. Then, we have

$$\left(1+\left(1-\frac{2}{q-1}\right)\sqrt{q}\right)^2-\left(\sqrt{q-3+\frac{1}{4}}+\frac{1}{2}\right)^2$$
$$=\left(1+2\sqrt{q}\frac{q-3}{q-1}+\left(1-\frac{4}{q-1}+\frac{4}{(q-1)^2}\right)q\right)-\left(q-3+\frac{1}{4}+\sqrt{q-3+\frac{1}{4}}+\frac{1}{4}\right)$$
$$=\left(1+q-\frac{4q}{q-1}+\frac{4q}{(q-1)^2}+2\sqrt{q}\frac{q-3}{q-1}\right)-\left(-\frac{5}{2}+q+\sqrt{q-\frac{11}{4}}\right)$$
$$=-\frac{1}{2}+\frac{4}{q-1}+\frac{4q}{(q-1)^2}+\sqrt{q}\left(\frac{2(q-3)}{q-1}-\sqrt{1-\frac{11}{4q}}\right)$$
$$\geqslant -\frac{1}{2}+\sqrt{q}\left(1-\frac{4}{q-1}\right)$$
$$\geqslant -\frac{1}{2}+\frac{\sqrt{q}}{2}$$

Hence, $\left\lfloor \sqrt{q-3+\frac{1}{4}}+\frac{1}{2}\right\rfloor_{n,1} \leqslant \left\lfloor 1+\left(1-\frac{2}{q-1}\right)\sqrt{q}\right\rfloor = 1+\left\lfloor \left(1-\frac{2}{q-1}\right)\sqrt{q}\right\rfloor$.

References

1. Bosma, W., Cannon, J., Playoust, C.: The Magma algebra system. I. The user language. J. Symbolic Comput. **24**(3-4), 235–265 (1997). https://doi.org/10.1006/jsco.1996.0125, http://dx.doi.org/10.1006/jsco.1996.0125. computational algebra and number theory (London, 1993)
2. Chung, F., Salehi, J., Wei, V.: Optical orthogonal codes: design, analysis and applications. IEEE Trans. Inf. Theory **35**(3), 595–604 (1989). https://doi.org/10.1109/18.30982
3. Ding, C., Feng, T.: Codebooks from almost difference sets. Des. Codes Cryptogr. **46**(1), 113–126 (2008)
4. Ding, C., Xing, C.: Several classes of (2m–1, w, 2) optical orthogonal codes. Discret. Appl. Math. **128**, 103–120 (2003)

5. Ding, C., Yin, J.: Constructions of almost difference families. Discret. Math. **308**(21), 4941–4954 (2008). https://doi.org/10.1016/j.disc.2007.09.017, https://www.sciencedirect.com/science/article/pii/S0012365X07007418. chongqing 2004
6. Drakakis, K., Gow, R., Rickard, S., Sheekey, J., Taylor, K.: On the maximal cross-correlation of algebraically constructed costas arrays. IEEE Trans. Inf. Theory **57**, 4612–4621 (2011). https://doi.org/10.1109/TIT.2011.2145890
7. Freedman, A., Levanon, N.: Any two n × n costas signals must have at least one common ambiguity sidelobe if n gt; 3-a proof. Proc. IEEE **73**(10), 1530–1531 (1985). https://doi.org/10.1109/PROC.1985.13329
8. Gómez-Pérez, D., Winterhof, A.: A note on the cross-correlation of costas permutations. IEEE Trans. Inf. Theory **66**(12), 7724–7727 (2020). https://doi.org/10.1109/TIT.2020.3009880
9. Kyureghyan, G., Li, S., Pott, A.: On the intersection distribution of degree three polynomials and related topics (2020). https://doi.org/10.48550/ARXIV.2003.10040, https://arxiv.org/abs/2003.10040
10. Li, S., Pott, A.: Intersection distribution, non-hitting index and Kakeya sets in affine planes. Finite Fields Their Appl. **66**, 101691 (2020). https://doi.org/10.1016/j.ffa.2020.101691, https://doi.org/10.1016/j.ffa.2020.101691
11. Moreno, O., Omrani, R., Kumar, P.V., Lu, H.F.: A generalized bose-chowla family of optical orthogonal codes and distinct difference sets. IEEE Trans. Inf. Theory **53**(5), 1907–1910 (2007). https://doi.org/10.1109/TIT.2007.894658

Effcient Finite Field Arithmetic

Polynomial Constructions of Chudnovsky-type Algorithms for Multiplication in Finite Fields with Linear Bilinear Complexity

Stéphane Ballet, Alexis Bonnecaze, and Bastien Pacifico(✉)

Institut de Mathématiques de Marseille, 169 Avenue de Luminy, 13009 Marseille, France
{stephane.ballet,alexis.bonnecaze,bastien.pacifico}@univ-amu.fr

Abstract. Chudnovsky-type algorithms for the multiplication in finite extensions of finite fields are well-known for having a good bilinear complexity, both asymptotically and at finite distance. More precisely, for every degree n of the extension, the existence of a family of algorithms with linear bilinear complexity in n has been proved using the original method applied to an explicit recursive tower of function fields. However, there is currently no method to build these algorithms in polynomial time. Nevertheless, one can construct in polynomial time a Chudnovsky-type algorithm over the projective line for the multiplication in any extension degree, with a quasi-linear bilinear complexity. In this paper, we prove that we can obtain algorithms both constructible in polynomial time and having a linear bilinear complexity by mixing up these two strategies.

1 Introduction

1.1 Multiplication in Finite Extension of Finite Fields

Let $\mathbb{F}_q$ be the finite field with q elements, q being a prime power. The search for algorithms computing multiplications in a finite extension $\mathbb{F}_{q^n}$ efficiently is an important research area. This problem can be approached according to different models of complexity. The algebraic complexity relies on counting the operations in the base field used by an algorithm. Many works use this model, and focus on obtaining the best asymptotic complexities, for example [14,19,22], and more recently [18], that proves that multiplication in $\mathbb{F}_{q^n}$ can be done with $\mathcal{O}(n \log n)$ operations in $\mathbb{F}_q$. Furthermore, one can consider that there are different types of operations in the base field. More precisely, let $\mathcal{B} = \{e_1, ..., e_n\}$ be a basis of $\mathbb{F}_{q^n}$ over $\mathbb{F}_q$. Then, for $x = \sum_{i=1}^{n} x_i e_i$ and $y = \sum_{j=1}^{n} y_j e_j$, the product of x and y is given canonically by

$$z = xy = \sum_{h=1}^{n} z_h e_h = \sum_{h=1}^{n} \left(\sum_{i,j=1}^{n} t_{ijh} x_i y_j \right) e_h, \quad (1)$$

S. Mesnager and Z. Zhou (Eds.): WAIFI 2022, LNCS 13638, pp. 35–52, 2023.
https://doi.org/10.1007/978-3-031-22944-2_3

where $e_ie_j = \sum_{h=1}^{n} t_{ijh}e_h$, $t_{ijh} \in \mathbb{F}_q$ being some constants. This formula involves additions and multiplications. One can moreover distinguish two types of multiplications. There are multiplications by a constant in $\mathbb{F}_q$ (by t_{ijh} in (1)), called scalar multiplications, and the bilinear multiplications that are depending of the two elements being multiplied (the x_iy_j). Each kind of operations has a different computational cost. Additions are known to be less expensive than multiplications. Among the multiplications, the scalar ones are their-selves known to be less expensive than the bilinear ones [23]. The last observation lead to the introduction of the bilinear complexity theory [10], that is the study of the complexity of the multiplication considering only bilinear multiplication in the base field.

Definition 1.1. *Let $\mathcal{U}$ be an algorithm for the multiplication in $\mathbb{F}_{q^n}$ over $\mathbb{F}_q$. Its number of bilinear multiplications is called its bilinear complexity, written $\mu(\mathcal{U})$. The bilinear complexity of the multiplication in $\mathbb{F}_{q^n}$ over $\mathbb{F}_q$, denoted by $\mu_q(n)$, is the quantity:*

$$\mu_q(n) = \min_{\mathcal{U}} \mu(\mathcal{U}),$$

where $\mathcal{U}$ is running over all multiplication algorithms in $\mathbb{F}_{q^n}$ over $\mathbb{F}_q$.

1.2 Known-Results

It is known that the method of D. V. and G. V. Chudnovsky [12] currently provides the best results on bilinear complexity. The Chudnovsky-Chudnovsky Multiplication Algorithm (CCMA) is an evaluation/interpolation algorithm using rational points of algebraic curves, i.e. rational places of a function field. As the degree of the extension is increasing, the algorithm requires more and more rational places for the evaluation. From the Hasse–Weil bound, it is known that the number of rational places of a function fields of genus g is bounded. Consequently, these algorithms use function fields of increasing genus according to the degree of the extension. Using an explicit recursive tower of function fields defined by Garcia and Stichtenoth [15], Ballet has proven the existence of CCMA having a linear bilinear complexity ([2], see [8]). The method had since been generalized. Ballet and Rolland [6] made it possible to use of places of degree 1 and 2. Arnaud [1] introduced the use of derivative evaluations. Cenk and Özbudak [11] extended these results. Finally, the last generalization is due to Randriambololona [21], allowing the construction of asymmetric algorithms. These works, as well as others on building the objects on which the algorithm is based (e.g. [5,20]), have allowed to improve the bounds for the bilinear complexity [8, Section 8.2].

However, the effectiveness of the asymptotic construction of these algorithms is unclear. According to Shparlinski, Tsfasman, and Vlăduţ [23, Remark 4.5], they can be constructed in polynomial time as long as a place of degree n of the function field, required to realize the extension as a residue class field, is given. For now, there is no method for constructing such a place other than looking through an exponentially large set, and that makes algorithms with a linear bilinear complexity not constructible in polynomial time. But the generalizations of CCMA lead to the introduction of another strategy for constructing an algorithm for the multiplication in any extension degree. Instead of using function fields of increasing genus according to the extension degree, one can fix the genus of the function field and use places of increasing degrees to evaluate. This strategy was first introduced in [4], fixing the genus to be equal to 1 (i.e. using elliptic curves). This way, algorithms for the multiplication in any extension degree can be constructed in polynomial time, and the bilinear complexity of these algorithms is asymptotically quasi-linear. More recently, the authors introduced in [7] a generic construction over the projective line (i.e. using the function field of genus 0), that also gives algorithms constructible in polynomial time, and with a uniform quasi-linear bound for their bilinear complexity.

1.3 New Results and Organization of the Paper

Thanks to the notion of tester, Bshouty obtained in [9] for the first time a family of multiplication algorithms that are constructible in polynomial time and have a linear bilinear complexity in $\mathcal{O}(n)$. It is a reasonable question whether the method of D.V. and G.V. Chudnovsky, and its generalizations, can also provide algorithms both constructible in polynomial time and having a linear bilinear complexity. The aim of this paper is to give a positive answer to this question. The key idea is to mix the two strategies discussed above. First, we construct a Chudnovsky-type algorithm over the projective line, using places of arbitrary degrees. Then, we use a CCMA over a tower of function fields to multiply the evaluation at the non-rational places. This method allows us to give an uniform bound for the bilinear complexity, that is linear according to the extension degree. This is the first time such an uniform bound is given for algorithms constructible deterministically and in polynomial time.

The paper is organized as follows. In Sect. 2, we recall the basics of function field theory and give the necessary information on Chudnovsky-type algorithms. In Sect. 3, we introduce a new construction called Hybrid Chudnovsky-type Algorithms. In Sect. 4, we prove the existence of such algorithms having a linear bilinear complexity while they are constructible deterministically and in polynomial time for any base field and any extension degree.

2 Chudnovsky-type Multiplication Algorithms

In this section, we first recall the basics of function field theory. Then, we give a version of the Chudnovsky method sufficient for our discussion. In a third time, we review the existing strategies for asymptotic construction of these algorithms.

2.1 Background and Notations

Let $F/\mathbb{F}_q$ be a function field of genus $g = g(F)$ over $\mathbb{F}_q$. For $\mathcal{O}$ a valuation ring, the place P is defined to be $P = \mathcal{O} \setminus \mathcal{O}^{\times}$. We denote by F_P the residue class field at the place P, that is isomorphic to $\mathbb{F}_{q^d}$, d being the degree of the place. A rational place is a place of degree 1. We also denote by $N_d(F/\mathbb{F}_q)$ the number of places of degree d of F over $\mathbb{F}_q$. A divisor $\mathcal{D}$ is a formal sum $\mathcal{D} = \sum_i n_i P_i$, where P_i are places and n_i are relative integers. The support *supp* $\mathcal{D}$ of $\mathcal{D}$ is the set of the places P_j for which $n_j \neq 0$, and $\mathcal{D}$ is effective if all the n_i are positive. The degree of $\mathcal{D}$ is defined by $\deg \mathcal{D} = \sum_i n_i$. The Riemann-Roch space associated to the divisor $\mathcal{D}$ is denoted by $\mathcal{L}(\mathcal{D})$. A divisor $\mathcal{D}$ is said to be non-special if $\dim \mathcal{L}(\mathcal{D}) = \deg(\mathcal{D}) + 1 - g$. Details about algebraic function fields can be found in [24]. In order to obtain an explicit formula, let us recall the following notation.

Definition 2.1. *Let q be a prime power and $d_1, \ldots, d_N$ be positive integers. The generalized Hadamard product in $\mathbb{F}_{q^{d_1}} \times \cdots \times \mathbb{F}_{q^{d_N}}$, denoted by $\odot$, is given for all $(a_1, \ldots, a_N), (b_1, \ldots, b_N) \in \mathbb{F}_{q^{d_1}} \times \cdots \times \mathbb{F}_{q^{d_N}}$ by*

$$(a_1, \ldots, a_N) \odot (b_1, \ldots, b_N) = (a_1 b_1, \ldots, a_N b_N).$$

2.2 Chudnovsky-type Multiplication Algorithm

Now, we can give a version of the method of D.V. and G.V. Chudnovsky, sufficient for our purpose, using only the generalization to the evaluation at places of arbitrary degrees. This follows directly from [8, Theorem 5.3].

Theorem 2.2. *Let q be a prime power and n be a positive integer. Let $F/\mathbb{F}_q$ be an algebraic function field of genus g, Q be a degree n place of $F/\mathbb{F}_q$, $\mathcal{D}$ be a divisor of $F/\mathbb{F}_q$, and $\mathcal{P} = \{P_1, \ldots, P_N\}$ be a set of places of arbitrary degrees of $F/\mathbb{F}_q$. We suppose that supp $\mathcal{D} \cap \{Q, P_1, \ldots, P_N\} = \emptyset$ and that*

(i) the evaluation map

$$\begin{aligned} Ev_Q : \mathcal{L}(\mathcal{D}) &\rightarrow F_Q \\ f &\mapsto f(Q) \end{aligned}$$

is surjective,

(ii) the evaluation map

$$\begin{aligned} Ev_{\mathcal{P}} : \mathcal{L}(2\mathcal{D}) &\rightarrow \mathbb{F}_{q^{\deg P_1}} \times \cdots \times \mathbb{F}_{q^{\deg P_N}} \\ f &\mapsto (f(P_1), \ldots, f(P_N)) \end{aligned}$$

is injective.

Then,

(1) we have a multiplication algorithm $\mathcal{U}_{q,n}^{F,\mathcal{P}}(\mathcal{D}, Q)$ such that for any two elements x, y in $\mathbb{F}_{q^n}$:

$$xy = E_Q \circ {Ev_{\mathcal{P}}|_{ImEv_{\mathcal{P}}}}^{-1} \left(E_{\mathcal{P}} \circ Ev_Q^{-1}(x) \odot E_{\mathcal{P}} \circ Ev_Q^{-1}(y) \right), \qquad (2)$$

where E_Q denotes the canonical projection from the valuation ring $\mathcal{O}_Q$ of the place Q in its residue class field F_Q, $E_{\mathcal{P}}$ the extension of $Ev_{\mathcal{P}}$ on the valuation ring $\mathcal{O}_Q$ of the place Q, ${Ev_{\mathcal{P}}|_{ImEv_{\mathcal{P}}}}^{-1}$ the restriction of the inverse map of $Ev_{\mathcal{P}}$ on its image, $\odot$ the generalized Hadamard product and $\circ$ the standard composition map;

(2) *the algorithm* $\mathcal{U}_{q,n}^{F,\mathcal{P}}(\mathcal{D},Q)$ *defined by* (2) *has bilinear complexity*

$$\mu(\mathcal{U}_{q,n}^{F,\mathcal{P}}(\mathcal{D},Q)) = \sum_{i=1}^{N} \mu(\mathcal{U}_{q,\deg P_i}(P_i)),$$

where $\mathcal{U}_{q,\deg P_i}(P_i)$ *is the algorithm used to multiply the evaluations at* P_i.

Existence of the objects satisfying the above conditions is ensured by the following numerical criteria:

(a) *a sufficient condition for the existence of a place* Q *in* $F/\mathbb{F}_q$ *of degree* n *is that* $2g+1 \leq q^{(n-1)/2}(q^{1/2}-1)$, *where* g *is the genus of* F,
(b) *a sufficient condition for (i) is that the divisor* $D - Q$ *is non-special,*
(c) *a necessary and sufficient condition for (ii) is that the divisor* $2\mathcal{D} - \mathcal{G}$ *is zero-dimensional:*

$$\dim \mathcal{L}(2\mathcal{D} - \mathcal{G}) = 0$$

where $\mathcal{G} = P_1 + \cdots + P_N$.

2.3 Existing Asymptotic Constructions

The topic of this paper is to show how algorithms for multiplication in an arbitrary extension degree can be constructed. In this purpose, we will give several families of algorithms.

Definition 2.3. *Let* q *be a prime power.*

(i) *A family of Chudnovsky-type algorithms* $\mathcal{U}_q = (\mathcal{U}_{q,2}, \mathcal{U}_{q,3}, \ldots, \mathcal{U}_{q,n}, \ldots)$ *is a collection of Chudnovsky-type algorithms such that for all integer* $n \geq 2$, *the algorithm* $\mathcal{U}_{q,n} = \mathcal{U}_{q,n}^{F,\mathcal{P}}(\mathcal{D},Q)$ *is an algorithm for the multiplication in* $\mathbb{F}_{q^n}$ *over* $\mathbb{F}_q$.

Let $\mathcal{U}_q = (\mathcal{U}_{q,2}, \mathcal{U}_{q,3}, \ldots, \mathcal{U}_{q,n}, \ldots)$ *be a family of Chudnovsky-type algorithms.*

(ii) *The family* $\mathcal{U}_q$ *admits an asymptotically bounded bilinear complexity if there exists* $f : \mathbb{N} \rightarrow \mathbb{R}$ *such that* $\limsup_{n\rightarrow+\infty} \frac{\mu(\mathcal{U}_{q,n})}{n} \leq f(n)$. *In this case, the bilinear complexity of* $\mathcal{U}_q$ *is said to be asymptotically bounded by* f, *which is denoted by* $\mu(\mathcal{U}_q) \in \mathcal{O}(f(n))$.
(iii) *Moreover, the family* $\mathcal{U}_q$ *admits an uniformly bounded bilinear complexity if there exists* $f : \mathbb{N} \rightarrow \mathbb{R}$ *such that for all integer* $n \geq 2$ *the bilinear complexity of* $\mathcal{U}_{q,n}$ *verifies* $\mu(\mathcal{U}_{q,n}) \leq f(n)n$. *Then, the bilinear complexity of the family* $\mathcal{U}_q$ *is said to be uniformly bounded by* f, *which is denoted it by* $\mu(\mathcal{U}_q) \leq f(n)$.

Note that for given q and n, when F, $\mathcal{D}$ and $\mathcal{Q}$ are fixed, the same construction can be applied to any place Q of degree n. Consequently, we refer to $\mathcal{U}_{q,n} = \mathcal{U}_{q,n}(P)$ depending on whether or not the choice of Q matters. Using these notations, the result that the bilinear complexity is linear according to the degree of the extension can be rephrased as the existence for any q of a family of algorithm $\mathcal{U}_q$ such that their bilinear complexities verify $\mu(\mathcal{U}_q) \leq C$, where C is a constant. Moreover, we are interested in the construction cost of these algorithms. In this sense, we introduce the following notation.

Definition 2.4. *Let $\mathcal{U}_{q,n}$ be a Chudnovsky-type algorithm for the multiplication in $\mathbb{F}_{q^n}$ over $\mathbb{F}_q$. Its complexity of construction, denoted by $\psi(\mathcal{U}_{q,n})$, is given by the number of operations in $\mathbb{F}_q$ required to build it deterministically.*

Now, let us recall the two main strategies of asymptotic constructions of families of Chudnovsky-type algorithms.

Increasing Genus Strategy. Originally, CCMA is an evaluation/interpolation algorithm using only evaluation at rational places of function fields. Consequently, the construction of the algorithm requires an increasing number of rational places, in accordance with the degree of the extension. From the Hasse–Weil bound, the number of rational places is bounded relatively to the genus of the function fields. Thus, the historical approach is to use an infinite family $\mathcal{F}$ of function fields of increasing genus in order to obtain an increasing number of rational places.

Example 2.5. *For any $q > 3$, Ballet proved in [2] the existence of a family of CCMA $\mathcal{U}^{\mathcal{F}}_{q^2} = (\mathcal{U}^{\mathcal{F}}_{q^2,n})_{n\geq 2}$ such that for all $n \geq 2$, the algorithm $\mathcal{U}^{\mathcal{F}}_{q^2,n} = \mathcal{U}^{F_i,\mathcal{P}}_{q^2,n}(\mathcal{D}, Q)$ is an algorithm from Theorem 2.2 such that*

- *F_k is the step with the smallest possible genus in $\mathcal{F} = (F_1, \ldots, F_k, \ldots)$, the recursively defined tower of function fields over $\mathbb{F}_{q^2}$ defined by Garcia and Stichtenoth in [15], such that*
  1. *$F_k/\mathbb{F}_q$ contains a place of degree n,*
  2. *$N_1(F_k/\mathbb{F}_q) > 2n + 2g(F_k) - 1$*
- *$\mathcal{P}$ is a set of $2n+g(F_k)-1$ rational places, $\mathcal{D}$ is a divisor of degree $n+g(F_k)-1$ such that $D - Q$ is non-special, and Q is a place of degree n.*

The bilinear complexity of this family verifies $\mu(\mathcal{U}^{\mathcal{F}}_{q^2}) \leq 2\left(1 + \frac{q}{q-3}\right)$. The existence of this family proves that the bilinear complexity is linear for the extensions of $\mathbb{F}_{q^2}$, where $q > 3$.

Remark 2.6. *This is an existence result, and there is no method to construct the algorithms of the family $\mathcal{U}^{\mathcal{F}}_{q^2}$ efficiently (for the moment). More precisely, there is no method to construct the place Q of degree n better than use an exhaustive search in an exponentially large set [23, Remark 4.5]. There is also method to construct such a divisor $\mathcal{D}$.*

Recursive Strategy Using Places of Increasing Degrees. The generalization of the method to the use of evaluations at places of arbitrary degrees made possible a new asymptotic construction strategy. It is in fact no longer necessary to use more and more rational places, since places of increasing degree can be used instead. But the evaluation at a place of degree $d > 1$ lies in an extension of degree d of $\mathbb{F}_q$. Then, one can recursively construct a Chudnovsky-type algorithm to multiply these evaluations. Consequently, it leads to a recursive construction of Chudnovsky-type algorithms. This strategy was introduced by Ballet, Bonnecaze and Tukumuli in [4], using functions fields of genus 1 only, i.e. elliptic curves. They proved what can be rephrased as follows

Proposition 2.7. *For any prime power q, there exists a family of Chudnovsky-type algorithms $\mathcal{U}_q^1 = (\mathcal{U}_{q,2}^1, \ldots, \mathcal{U}_{q,n}^1, \ldots)$, where for any positive integer $n \geq 2$, the algorithm $\mathcal{U}_{q,n}^1 = \mathcal{U}_{q,n}^{F,\mathcal{P}}(\mathcal{D}, Q)$ is an algorithm from Theorem 2.2 recursively constructed over a function field F of genus at most 1, such that*

(i) its bilinear complexity is asymptotically bounded, and verifies

$$\mu(\mathcal{U}_q^1) \in \mathcal{O}\left((2q)^{\log^*(n)}\right),$$

(ii) there exists $\alpha \in \mathbb{N}$ such that for all prime power q and all positive integer n, $\psi(\mathcal{U}_{q,n}^1) \in \mathcal{O}(n^\alpha)$.

In this result, the iterated logarithm $\log^*$ is defined as follows.

Definition 2.8. *For all integer n, the iterated logarithm of n, denoted by $\log^*(n)$, is defined by the following recursive function:*

$$\log^*(n) = \begin{cases} 0 & \text{if } n \leq 1 \\ 1 + \log^*(\log(n)) & \text{elsewhere.} \end{cases}$$

This value is the number of times the logarithm is iteratively applied to n until we obtain a result lower than or equal to 1.

This bilinear complexity is not linear, but is said to be quasi-linear since the iterated logarithm is a very slow-growing function, as it can be seen in [4, Table 2]. The bilinear complexity of these algorithms is not as good as that of algorithms provided by the increasing genus strategy, but unlike the latter, the algorithms of $\mathcal{U}_q^1$ are constructible in polynomial time. More recently, the authors applied this recursive strategy of construction to the rational function field $\mathbb{F}_q(x)$ in [7]. These algorithms will be at the heart of our new construction.

Definition 2.9. *Let q be a prime power and n be a positive integer. A recursive Chudnovsky-type algorithm $\mathcal{U}_{q,n}^{\mathcal{P}_n}(Q)$ over the projective line is an algorithm $\mathcal{U}_{q,n}^{F,\mathcal{P}}(\mathcal{D}, Q)$ satisfying the assumptions of Theorem 2.2 such that:*

- *$F/\mathbb{F}_q$ is the rational function field $\mathbb{F}_q(x)$, Q is a place of degree n of $\mathbb{F}_q(x)$,*
- *$\mathcal{D} = (n-1)P_\infty$, where P_∞ is the place at infinity of $\mathbb{F}_q(x)$,*
- *$\mathcal{P}_n$ is a set of places of degrees lower than n such that*

$$\sum_{P \subset \mathcal{P}_n} \deg P = 2n - 1,$$

- *the multiplication in $F_P \simeq \mathbb{F}_{q^d}$, where $d = \deg P$, is computed by $\mathcal{U}_{q,d}^{\mathcal{P}_d}(P)$, where $P \in \mathcal{P}_n$.*

The results from [7] can be expressed as follows.

Proposition 2.10. *For all prime power q, there exists a family of recursive Chudnovsky-type algorithms over the projective line* $\mathcal{U}_q^0 = (\mathcal{U}_{q,2}^0, \ldots, \mathcal{U}_{q,n}^0, \ldots)$, *where for any positive integer* $n \geq 2$, *the algorithm* $\mathcal{U}_{q,n}^0 = \mathcal{U}_{q,n}^{\mathcal{P}_n}(Q)$ *is an algorithm from Definition 2.9, such that*

(i) its bilinear complexity verifies $\mu(\mathcal{U}_q^0) \leq C\left(\frac{4q^2}{q-1}\right)^{\log^*_{\sqrt{q}}(2n)}$, *where* $C = 3$ *if* $q = 2$ *and* $C = 1$ *elsewhere, and the iterated logarithm in the basis* $\sqrt{q}$ *is defined as follows:*

$$\log^*_{\sqrt{q}}(n) = \begin{cases} 0 & \text{if } n \leq 1 \text{ and } q > 2, \\ 0 & \text{if } n \leq 5 \text{ and } q = 2, \\ 1 + \log^*_{\sqrt{q}}(\log_{\sqrt{q}}(n)) & \text{elsewhere.} \end{cases}$$

(ii) The algorithms of $\mathcal{U}_q^0$ *are constructible deterministically and in time* $\psi(\mathcal{U}_{q,n}^0) \in \mathcal{O}(n^4)$.

Compared with the recursive construction over function fields of genus one, the use of the rational function field allows the authors to prove a uniform bound for the bilinear complexity of the algorithms, and to give an estimation for the cost of construction. Moreover, this construction is generic unlike the case of elliptic curves, since the form of the divisor $\mathcal{D}$ and the bases of the Riemann-Roch spaces are fixed.

Remark 2.11. *The use of evaluation with multiplicity, using generalized evaluation maps, can also be considered in this construction.*

3 New Strategy of Asymptotic Constructions

In this section, we introduce a new strategy of construction of Chudnovsky-type algorithms, in order to obtain a family of algorithms that are constructible in polynomial time and have a linear bilinear complexity. This strategy is an hybrid construction, involving both strategies introduced in the Sect. 2.3. The clearest way to expose this construction is to introduce a tree representation of Chudnovsky type algorithms.

3.1 Tree Construction of Chudnovsky-type Algorithms

The algorithms provided by the recursive construction of Sect. 2.3 can be seen as in the following example.

Example 3.1. *Let us consider the multiplication in* $\mathbb{F}_{3^6}$ *over* $\mathbb{F}_3$. *The recursive Chudnovsky-type algorithm over the projective line* $\mathcal{U}_{3,6}^{\mathcal{P}_6}(Q_6)$ *evaluates at the 4 rational places of* $\mathbb{F}_3(x)$, *denoted by* P_0, P_1, P_2 *and* P_∞, *2 places of degree 2, denoted by* P_1^2 *and* P_2^2, *and a place of degree 3 denoted by* P^3. *The product of the evaluations at the places of degree 2 are computed using the recursively defined algorithm,* *with $\mathcal{P}_2 = \{P_0, P_1, P_\infty\}$ (that corresponds to the Karatsuba algorithm). The evaluations at P^3 are multiplied using $\mathcal{U}_{3,3}^{\mathcal{P}_3}(P^3)$, where $\mathcal{P}_3 = \{P_0, P_1, P_\infty, P_1^2\}$, that also involves the algorithm for the multiplication in the quadratic extension. The tree in Table 1 shows the sub-algorithms used by $\mathcal{U}_{3,6}^{\mathcal{P}_6}(Q_6)$, where $P_i = \mathcal{U}_{q,1}(P_i)$ is the algorithm of multiplication in $\mathbb{F}_q$, consisting of a bilinear multiplication in $\mathbb{F}_q$. In the langage of trees, the root is the main algorithm, used to multiply in $\mathbb{F}_3^6$. This algorithm reduce the problem to the multiplication at some places of degrees one, two or three, at the level 1 of the tree. The leaves of the tree are multiplications at rational places and thus the bilinear complexity of the algorithm is given by the breadth of the tree, i.e. its number of leaves.*

Table 1. Tree representation of $\mathcal{U}_{3,6}^{\mathcal{P}_6}(Q_6)$

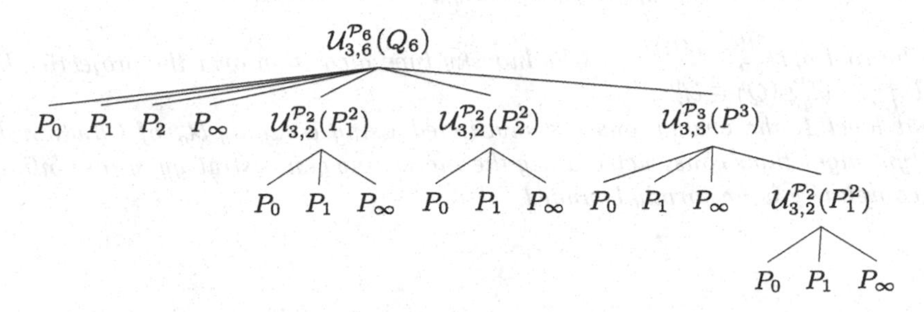

Using this formalism, one can define an algorithm using the structure of a tree.

Definition 3.2. *A tree-structure Chudnovsky-type algorithm is an algorithm of type Chudnovsky given by its tree representation.*

Consequently, one can express already known constructions as follows.

- **CCMA:** the original algorithm is fully defined by is root. It uses evaluation only at rational places and can be represented as a tree of depth one.
- **Recursive construction over the projective line**: these algorithms are given by recursively constructed trees such as in Table 1.

3.2 Hybrid Strategy

Now, let us introduce a new strategy of construction, that we call hybrid strategy. Our aim is to construct an algorithm for the multiplication in $\mathbb{F}_{q^n}$ with a linear bilinear complexity, and that is constructible efficiently, i.e. deterministically and in polynomial time with respect to n. The algorithms built over a tower of function fields as in Example 2.5 already have a linear bilinear complexity, but have an exponential cost to construct the place of degree n [23, Remark 5]. On the other side, the algorithms over the projective line are constructible in polynomial time, but lose the linearity of the bilinear complexity. The idea is to mix these strategies to obtain algorithms with at the same time a linear bilinear complexity and a construction that is deterministic and in polynomial time (Table 2).

To construct such an algorithm of multiplication in $\mathbb{F}_{q^n}$, we start with a Chudnovsky type algorithm over the projective line. In this case, the place of degree n is constructible efficiently, and we use the evaluation at places of degrees until some integer d. The multiplication of the evaluations at these places is computed using a family of algorithms of type Chudnovsky built over a tower of function fields whose bilinear complexity is uniformly bounded by a constant. We define such an algorithm as a Hybrid Chudnovsky-type Algorithm (HCA).

Definition 3.3. *Let q be a prime power and n be a positive integer. A Hybrid Chudnovsky-type Algorithm for the multiplication in $\mathbb{F}_{q^n}$ over $\mathbb{F}_q$ is a tree-structured Chudnovsky-type algorithm $\mathcal{U}_{q,n}^{\mathcal{H}(\mathcal{U}_{q,n}^0,\mathcal{U}_q)}$ such that*

- *the root of $\mathcal{U}_{q,n}^{\mathcal{H}(\mathcal{U}_{q,n}^0,\mathcal{U}_q)}$ is a Chudnovsky-type algorithm over the projective line $\mathcal{U}_{q,n}^0 = \mathcal{U}_{q,n}^{\mathcal{P}_n}(Q) \in \mathcal{U}_q^0$,*
- *at level 1, the evaluations are multiplied using a family $\mathcal{U}_q$ of Chudnovsky-type algorithms constructed using the increasing genus strategy, whose bilinear complexity is uniformly bounded.*

Table 2. Tree representation of a $\mathcal{U}_{q,n}^{\mathcal{H}(\mathcal{U}_{q,n}^0,\mathcal{U}_q)}$

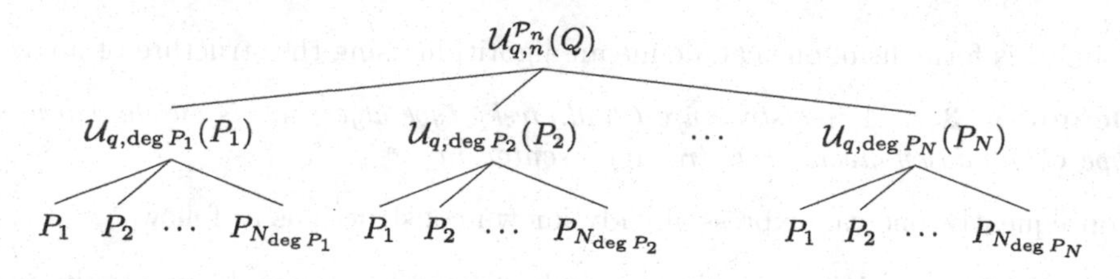

Proposition 3.4. *Let q be a prime power and n be a positive integer. Let $\mathcal{U}_q^{\mathcal{H}(\mathcal{U}_{q,n}^0,\mathcal{U}_q)} = (\mathcal{U}_{q,2}^{\mathcal{H}(\mathcal{U}_{q,2}^0,\mathcal{U}_q)}, \ldots, \mathcal{U}_{q,n}^{\mathcal{H}(\mathcal{U}_{q,n}^0,\mathcal{U}_q)}, \ldots)$ be a family of HCA as in Definition 3.3. Then,*

(i) its bilinear complexity verifies $\mu(\mathcal{U}_q^{\mathcal{H}(\mathcal{U}_{q,n}^0,\mathcal{U}_q)}) \leq 2\mu(\mathcal{U}_q)$,
(ii) for all n, if each algorithm $\mathcal{U}_{q,n} \in \mathcal{U}_q$ is constructible deterministically and in time $\psi(\mathcal{U}_{q,n})$, then for any positive integer n, the algorithm $\mathcal{U}_{q,n}^{\mathcal{H}(\mathcal{U}_{q,n}^0,\mathcal{U}_q)}$ is constructible deterministically and in time $\mathcal{O}\left(n^4 + \sum_{P\in\mathcal{P}} \psi(\mathcal{U}_{q,\deg P}(P))\right)$.

Proof. (i) By Theorem 2.2, the bilinear complexity of the algorithm $\mathcal{U}_q^{\mathcal{H}(\mathcal{U}_{q,n}^0,\mathcal{U}_q)}$ is given by $\sum_{P\in\mathcal{P}} \mu(\mathcal{U}_{q,\deg P})$. Let n_k denote the number of places of degree k in $\mathcal{P}$. By Definition 2.9, we have that $\sum_k kn_k = 2n - 1$ and $d = \max_{P \in \mathcal{P}}(\deg P)$. Then, for all positive integer n,

$$\mu\left(\mathcal{U}_q^{\mathcal{H}(\mathcal{U}_{q,n}^0, \mathcal{U}_q)}\right) = \sum_{k=1}^{d} n_k \mu(\mathcal{U}_{q,k}) \leq \sum_{k=1}^{d} kn_k \mu(\mathcal{U}_q) = (2n-1)\mu(\mathcal{U}_q).$$

It follows that $\mu\left(\mathcal{U}^{\mathcal{H}(\mathcal{U}_{q,n}^0, \mathcal{U}_q)}\right) \leq 2\mu(\mathcal{U}_q)$.

(ii) First, we need to construct the main part of a Chudnovsky-type over the projective line. For all positive integer n, there exists a Chudnovsky-type algorithm $\mathcal{U}_{q,n}^{\mathcal{P}}(Q)$ over $\mathbb{F}_q(x)$ that is constructible in time $\mathcal{O}(n^4)$ deterministically [7, Theorem 5.3]. This gives the cost of the construction of the root of $\mathcal{U}^{\mathcal{H}(\mathcal{U}_{q,n}^0, \mathcal{U}_q)}$. It remains to consider the construction of the sub-algorithms $\mathcal{U}_{q,\deg P}(P)$, for all $P \in \mathcal{P}$. By hypothesis, these algorithms are constructible deterministically and in time $\psi(\mathcal{U}_{q,\deg P}(P))$. Consequently,

$$\psi\left(\mathcal{U}^{\mathcal{H}(\mathcal{U}_{q,n}^0, \mathcal{U}_q)}\right) \in \mathcal{O}\left(n^4 + \sum_{P \in \mathcal{P}} \psi\left(\mathcal{U}_{q,\deg P}(P)\right)\right).$$

4 Explicit Construction of Algorithms with Linear Bilinear Complexity and Constructible in Polynomial Time

This section is devoted to prove for all prime power q the existence of a family $\mathcal{U}_q^{\mathcal{H}}$ of algorithms having a linear bilinear complexity, and constructible deterministically and in polynomial time. More precisely, we prove the following result.

Theorem 4.1. *For all prime power q, there exist a family $\mathcal{U}_q^{\mathcal{H}} = \mathcal{U}_q^{\mathcal{H}(\mathcal{U}_{q,n}^0, \mathcal{U}_q)}$ of Hybrid Chudnovsky-type algorithm such that*

- *the bilinear complexity of the family verifies $\mu(\mathcal{U}_q^{\mathcal{H}}) \leq C_q$,*
- *the algorithms of $\mathcal{U}_q^{\mathcal{H}}$ are constructible deterministically, and for all positive integer n, they are constructible in time $\psi(\mathcal{U}_{q,n}^{\mathcal{H}}) \in \mathcal{O}(n^4)$.*

To complete this proof, we first recall a property of the tower of algebraic function fields define by Garcia and Stichtenoth in [15]. In a second time, we give a specific family $\mathcal{U}_{q^2}^{\mathcal{F}_{\mathrm{sp}}}$ over this tower of function fields. Finally, the last section proves Theorem 4.1.

4.1 A Garcia-Stichtenoth Tower of Function Fields

Recall that a recursively defined tower of function fields is defined in [24, Definition 7.2.2]. In what follows, we consider the recursive tower $\mathcal{F} = (F_1, \ldots, F_k, \ldots)$ defined over $\mathbb{F}_{q^2}$ introduced by Garcia and Stichtenoth in [15] and defined by the equation $Y^q + Y = \frac{X^q}{X^{q-1}+1}$. This tower is the one used to prove the linearity of the bilinear complexity, as explained in Example 2.5 and Remark 2.6. Recall this result from [15].

Theorem 4.2. *The genus g_k of $F_k/\mathbb{F}_{q^2}$ is given by*

$$g_k = \begin{cases} q^k + q^{k-1} - q^{(k+1)/2} - 2q^{(k-1)/2} + 1, & \text{if } k \equiv 1 \mod 2 \\ q^k + q^{k-1} - \frac{1}{2}q^{(k/2)+1} - \frac{3}{2}q^{(k/2)} - q^{(k/2)-1} + 1, & \text{if } k \equiv 0 \mod 2 \end{cases}$$

Moreover, if N_k denotes the number of rational places of $F_k/\mathbb{F}_{q^2}$, then for all $k \geq 3$

$$N_1(F_k) \geq (q^2 - 1)q^{k-1} + 2q.$$

4.2 Specific Construction on the Garcia-Stichtenoth Tower

In our purpose, we need to introduce a specific construction of algorithms for the multiplication in any finite extension of a finite field. This construction relies on the explicit recursive tower of function fields defined by Garcia and Stichtenoth [15], in a manner similar to that of Ballet [2]. However, we are interested in having some informations on the complexity of the construction of these algorithms. In this purpose, we set some parameters generically, as in [23, Proposition 4.1]. Let us introduce this family of Chudnovsky-type algorithms.

Definition 4.3. *Let $q > 5$ be a prime power. Let $\mathcal{U}_{q^2}^{\mathcal{F}_{sp}} = (\mathcal{U}_{q^2,n}^{\mathcal{F}_{sp}}, \ldots, \mathcal{U}_{q^2,n}^{\mathcal{F}_{sp}}, \ldots)$ be a family of CCMA where for all $n \geq 2$, the algorithm $\mathcal{U}_{q^2,n}^{\mathcal{F}_{sp}} = \mathcal{U}_{q^2,n}^{F,\mathcal{P}}(\mathcal{D}, Q)$ is an algorithm as in Theorem 2.2 with the following parameters.*

- *$F = F_k$ is the step with the smallest possible genus of the Garcia-Stichtenoth tower $\mathcal{F}$ of Sect. 4.1 such that*
  1. *$2g_k + 1 \leq q^{n-1}(q-1)$, where $g_k = g(F_k)$,*
  2. *$N_1(F_k) \geq 2n + 4g_k$.*
- *$\mathcal{D} = (n + 2g - 1)P$, where P is a rational place of F_k.*
- *$\mathcal{P}$ is a set of $2n + 4g_k - 1$ rational places of F_k, distinct from P.*
- *Q is a degree n place of F_k.*

Proposition 4.4. *Let $q > 5$ be a prime power. The algorithm $\mathcal{U}_{q^2,n}^{\mathcal{F}_{sp}}$ exists for any positive integer n, and its bilinear complexity is given by*

$$\mu(\mathcal{U}_{q^2,n}^{\mathcal{F}_{sp}}) = 2n + 3g_k - 1.$$

Proof. The existence of such an algorithm is given by the existence of a step in the Garcia Stichtenoth recursive tower of function fields verifying the conditions 1. and 2. in Definition 4.3. Indeed, suppose that such a function fields exists. Then, the condition 1. is the condition (a) of Theorem 2.2. Moreover, the Divisor $\mathcal{D} - Q$ is of degree $2g_k - 1$, and thus trivially non special and (b) is verified. Finally, the divisor $2\mathcal{D} - \sum_{P \in \mathcal{P}} P$ is of degree -1 and hence zero dimensional, and this gives condition (c). Theorem 2.2 is hence verified and we can construct an algorithm $\mathcal{U}_{q^2,n}^{\mathcal{F}_{\mathrm{sp}}} = \mathcal{U}_{q^2,n}^{F_k,\mathcal{P}}(\mathcal{D}, Q)$. Since $\mathcal{L}(2\mathcal{D})$ is of dimension $2n + 3g_k - 1$, it is enough to use exactly this number of places of $\mathcal{P}$. Consequently, the bilinear complexity of the algorithm is given by $\mu(\mathcal{U}_{q^2,n}^{\mathcal{F}_{\mathrm{sp}}}) = 2n + 3g_k - 1$. It remains to prove that such a function field exists. In [2], Ballet proved its existence where the condition 2. was given by $N_1(F_k) \geq 2n+2g_k-2$, where q was supposed to be greater than or equal to 3. It has been possible because the Garcia-Stichtenoth tower reaches the Drinfeld-Vlăduţ Bound. Indeed, the parameters chosen by Ballet are possible since $\lim_{k \mapsto +\infty} \frac{N_1(F_k)}{g_k} = q-1 \geq (2n+2g_k-2)/g_k$, then $q-1 \geq 2$ and hence $q \geq 3$. Here, we need $\lim_{k \mapsto +\infty} \frac{N_1(F_k)}{g_k} = q-1 \geq (2n+4g_k)/g_k$, then $q-1 \geq 4$ and hence $q \geq 5$. What remains of the existence proof is similar to that of [2].

Proposition 4.5. *The bilinear complexity of* $\mathcal{U}_{q^2,n}^{\mathcal{F}_{sp}}$ *verifies*

$$\mu\left(\mathcal{U}_{q^2}^{\mathcal{F}_{sp}}\right) \leq 2\left(1+\frac{3q}{q-5}\right).$$

Proof. Recall that $N_1(F_k) \geq (q^2-1)q^{k-1} + 2q = M_k$. In the following, let

$$\Delta_{q,k} = M_k - 4g_k = (q^2-4q-5)q^{k-1} + f(k)$$

where by Theorem 4.2,

$$f(k) = \begin{cases} 4q^{(k+1)/2} + 8q^{(k-1)/2} + 2q - 4 & \text{if } k \equiv 1 \mod 2 \\ 2q^{(k/2)+1} + 6q^{k/2} + 4q^{(k/2)-1} + 2q - 4 & \text{if } k \equiv 0 \mod 2 \end{cases}$$

Note that since $q > 2$, the value of $f(k)$ is strictly positive for any q and k. For any integer n, let k be the smallest such that $(q^2-1)q^{k-1} \geq 2n+4g_k$, or equivalently $2n \leq (q^2-1)q^{k-1} - 4g_k$. Hence, we also have that $2n \geq (q^2-1)q^{k-2} - 4g_{k-1}$. i.e. $2n \geq \Delta_{q,k-1} \geq q^{k-2}(q^2-4q-5)$. Thus $k-1 \leq \log_q(2n) - \log_q(q^2-4q-5)+1$. Consequently, we have $\mu(\mathcal{U}_{q^2,n}^{\mathcal{F}_{sp}}) \leq 2n + 3g_k \leq 2n + 3(q^k + q^{k-1})$, and the latter bound on $k-1$ implies that

$$\mu\left(\mathcal{U}_{q^2,n}^{\mathcal{F}_{sp}}\right) \leq 2n + 3(q+1)q^{\log_q(2n)-\log_q(q^2-4q-5)+1} = 2n\left(1+\frac{3q}{q-5}\right).$$

These algorithms are constructible efficiently if a degree n place is given.

Proposition 4.6. *The algorithm* $\mathcal{U}_{q^2,n}^{\mathcal{F}_{sp}} = \mathcal{U}_{q^2,n}^{F_k,\mathcal{P}}(\mathcal{D}, Q)$ *is constructible deterministically and in polynomial time if the place* Q *of degree* n *of* F_k *is given.*

Proof. By the results of Elkies [13], the recursive tower of Garcia and Stichtenoth of Sect. 4.1 is a family of Drinfeld modular curves, and according to Tsfasman and Vlăduţ, we can work polynomially with points and linear systems on these curves [25]. Then we can construct F_k and its rational places in polynomial time. Moreover, the divisor $\mathcal{D}$ is set to be $\mathcal{D} = (n+2g_k-1)P$, for P a rational place of F_k, and $\mathcal{D}$ is also constructible in polynomial time. Consequently, if a place Q of degree n is given, all of the objects involved in the construction of $\mathcal{U}_{q^2,n}^{\mathcal{F}_{sp}}$ are constructible in polynomial time. It remains to compute the bases of $\mathcal{L}(\mathcal{D})$ and $\mathcal{L}(Q)$, and the evaluation maps, but this can be done polynomially according to [23, Proposition 4.1]. In particular, the $2n+3g_k-1$ places of $\mathcal{P}$ used to perform the inversion can be found by using Gaussian elimination in time $\mathcal{O}((2n+4g_k-1)^3)$.

Complexity of the Construction of the Degree n Place. In the language of function fields, the construction of a point of order n corresponds to the construction of a place of degree n, that is given by the orbit of a point of degree n under the action of the Frobenius. In fact, every algebraic function field $F/\mathbb{F}_q$ can be seen as the field of the functions defined on a curve $\mathcal{C}$. There is a 1-1 correspondence between rational places of $F/\mathbb{F}_q$ and the $\mathbb{F}_q$-rational points of $\mathcal{C}$ ([17], Theorem 3.2). Moreover, let $G = Gal(\overline{\mathbb{F}_q} \mid \mathbb{F}_q)$, the G-orbits of points of $\mathcal{C}(\overline{\mathbb{F}_q})$ are called closed points of $\mathcal{C}/\mathbb{F}_q$. The degree of a closed point is equal to the cardinality of the orbit. Then, Corollary 3.6.5 of [24] implies that there is a one-one correspondence between the places of degree d of a function field $F/\mathbb{F}_q$ and the closed points of degree d on the associated curve $\mathcal{C}/\mathbb{F}_q$. Following [23], we say that a place of degree n is constructed if the coordinates of one of the points of the corresponding orbit are given in some projective embedding. In the case of recursive towers of function fields, we consider the following.

Definition 4.7. *Let $\mathcal{F} = (F_0, F_1, \ldots, F_\ell, \ldots)$ be a recursive tower of function fields defined by the equation $f(Y) = h(X)$ over $\mathbb{F}_q$. A degree n place P of $F_\ell/\mathbb{F}_q$ is constructed if the coordinates of $x = (x_0, x_1, \ldots, x_\ell)$, where $x_i \in \mathbb{F}_{q^n}$ and for all $1 \leq i \leq \ell$ we have $f(x_i) = h(x_{i-1})$, and there exists one x_j, for $0 \leq j \leq \ell$, such that for all integers d dividing n, $x_j \notin \mathbb{F}_{q^d}$.*

This definition makes sense since once such coordinates are given, one can construct the orbit under the action of the Frobenius polynomially, using [16], Algorithm 14.26. Consequently, the problem of constructing a degree n place of a function field in polynomial time is the same as that of constructing the coordinates of a point of order n on the associated curve. One can estimate the complexity of the construction of such a place using brute force.

Lemma 4.8. *Let F_ℓ be the ℓ-th step of the recursively defined tower of function fields of Garcia-Stichtenoth of Section 4.1, given by the equation $f(Y) = h(X)$ and defined over $\mathbb{F}_{q^2}$. Assume that F_ℓ contains a place of degree n. Then, one can construct such a place with $\mathcal{O}((q^2)^{2n} M(n) \log n)$ operations in $\mathbb{F}_{q^2}$, with $M(n) \in \mathcal{O}(n^\omega)$, where $\omega = 2,373\ldots$ is the best exponent for the multiplication of two matrices of size $n \times n$, and using a table of precomputations of size $(q^2)^n \times q$.*

Proof. According to Definition 4.7, we are looking for an element $(x_0, x_1, \ldots, x_l) \in \mathbb{F}^{\ell+1}_{(q^2)^n}$ such that for all $0 \leq i \leq \ell - 1$, we have $f(x_{i+1}) = h(x_i)$ and whose orbit under the Frobenius action is of cardinal n. We precompute the table of couples that gives for all $x \in \mathbb{F}_{(q^2)^n}$ the list of elements $y \in \mathbb{F}_{(q^2)^n}$ such that $f(y) = h(x)$. There are $(q^2)^{2n}$ possible pairs of solutions. Each pair is computed in a constant number of operations in $\mathbb{F}_{(q^2)^n}$, since q is fixed. By [16], a division in $\mathbb{F}_{(q^2)^n}$ can be done with $\mathcal{O}(M(n) \log(n))$ operations in $\mathbb{F}_{q^2}$, and that is more than a multiplication, or an addition. Thus, the table can be computed in $\mathcal{O}\left((q^2)^{2n} M(n) \log n\right)$ operations in $\mathbb{F}_{q^2}$. This table allows us to obtain a tuple $(x_0, \ldots, x_\ell)$ verifying the recursive equation without any arithmetic operation in $\mathbb{F}_{q^2}$. Since f is of degree q, there at most q solutions y of the equation $f(y) = h(x)$ for a given $x \in \mathbb{F}_{(q^2)^n}$, and the table of precomputations is of size $(q^2)^n \times q$. Then, for each x_i there are at most q elements x_{i+1} solutions of the equation and overall, there are at most $(q^2)^n q^\ell$ tuples. For a given tuple, we apply the iterated Frobenius to compute its orbit, that costs at most $\mathcal{O}(M(n)^2 \log(n)^2)$ for all the $\ell+1$ coordinates, using Algorithm 14.26 in [16]. Therefore, the cost of the construction of a place of degree n in F_ℓ is bounded by $\mathcal{O}((q^2)^n q^\ell(\ell+1)M(n)^2\log(n)^2)$ when the table is given. Overall, the cost of the construction of a place is thus $\mathcal{O}\left((q^2)^{2n}M(n)\log n + (q^2)^n q^\ell(\ell+1)M(n)^2\log(n)^2\right) = \mathcal{O}\left((q^2)^{2n}M(n)\log n\right)$.

Remark 1. One can also estimate this complexity without using a precomputation table. The algorithm would in this case be as follows. For each $x_0 \in \mathbb{F}_{(q^2)^n}$, we find all the solutions of the recursive equation in time $\mathcal{O}((q^2)^n M(n)\log(n))$. At this step we found the (x_0, x_1) verifying the recursive equation. Then for the (at most) q solution x_1 found for a given x_0, we apply again this process to find the (x_0, x_1, x_2). The running time at this step is in $\mathcal{O}((q^2)^n((q^2)^n M(n)\log(n) + q \times (q^2)^n M(n)\log(n))$. This process has to be done ℓ times. Consequently, we obtain an algorithm in time

$$\mathcal{O}\Big((q^2)^n\Big((q^2)^n M(n)\log(n) + q\times\Big((q^2)^n M(n)\log(n) + q\times(\cdots)\cdots\Big)\Big)\Big).$$

Since the process is done ℓ times, this can be rewritten

$$\mathcal{O}\left((q^2)^n(\sum_{i=0}^{\ell} q^i)((q^2)^n M(n)\log(n))\right).$$

Consequently, the algorithm can be executed in time

$$\mathcal{O}\left((q^2)^n q^\ell (q^2)^n M(n)\log(n)\right) = \mathcal{O}((q^{4n+\ell}M(n)\log(n)).$$

Lemma 4.8 can easily be generalized to any recursive tower of function fields, with respect to the degree of f and the base field on which the function fields are defined.

4.3 Constructions of Hybrid Chudnovsky-type Algorithms

Now, let us introduce the family of Hybrid Chudnovsky-type Algorithms that allows us to prove Theorem 4.1.

Definition 4.9. *Let $q > 5$ be a prime power. Let $\mathcal{U}_{q^2}^{\mathcal{H}} = (\mathcal{U}_{q^2,2}^{\mathcal{H}}, \ldots, \mathcal{U}_{q^2,n}^{\mathcal{H}}, \ldots)$ be the family of Hybrid Chudnovsky-type Algorithms for the multiplication in the extensions of $\mathbb{F}_{q^2}$ such that for any $n \geq 2$, the algorithm $\mathcal{U}_{q^2,n}^{\mathcal{H}} = \mathcal{U}_{q^2}^{\mathcal{H}(\mathcal{U}_{q^2,n}^0, \mathcal{U}_{q^2}^{\mathcal{F}_{sp}})}$ is an HCA where :*

- *The root is a Chudnovsky-type algorithm $\mathcal{U}_{q^2,n}^0$ over the projective line from the family $\mathcal{U}_{q^2}^0$,*
- *The algorithms at level one are those from the family $\mathcal{U}_{q^2}^{\mathcal{F}_{sp}}$, given by the specific construction over the Garcia-Stichtenoth tower of function fields given in Sect. 4.2.*

Theorem 4.10. *Let $q \geq 5$ be a prime power and n be a positive integer. Then, the hybrid Chudnovsky-type algorithms $\mathcal{U}^{\mathcal{H}}_{q^2,n}$ for the multiplication in the degree n extension of $\mathbb{F}_{q^2}$ provides a family $\mathcal{U}^{\mathcal{H}}_{q^2}$ such that*

(i) its bilinear complexity is uniformly bounded by $\mu(\mathcal{U}^{\mathcal{H}}_{q^2}) \leq 4\left(1+\frac{3q}{q-5}\right)$,
(ii) for all $n \in \mathbb{N}$, the algorithms $\mathcal{U}^{\mathcal{H}}_{q^2,n}$ are constructible deterministically, and in time $\psi\left(\mathcal{U}^{\mathcal{H}}_{q^2,n}\right) \in \mathcal{O}(n^4)$.

Proof. We apply Proposition 3.4 with the algorithms given in Definition 4.9. By Proposition 4.5, we know that $\mu(\mathcal{U}^{\mathcal{F}_{\text{sp}}}_{q^2}) \leq 2\left(1+\frac{3q}{q-5}\right)$. It follows that $\mu(\mathcal{U}^{\mathcal{H}}_{q^2}) \leq 4\left(1+\frac{3q}{q-5}\right)$, and (i) is proven. For all positive integer n, recall that there exists a Chudnovsky-type algorithm $\mathcal{U}^{0}_{q^2,n} = \mathcal{U}^{\mathcal{P}_n}_{q^2,n}(Q)$ over $\mathbb{F}_{q^2}(x)$, where Q is a degree n place of $\mathbb{F}_{q^2}(x)$ and $\mathcal{P}_n$ is a set of places of arbitrary degrees such that $\sum_{P\in\mathcal{P}_n} \deg P = 2n-1$. Moreover, it is known that $(q^2)^d+1 = \sum_{k|d} kB_k(\mathbb{F}_{q^2}(x))$, where $B_k(\mathbb{F}_{q^2}(x))$ is the number of places of degree k of $\mathbb{F}_{q^2}(x)$. This implies that there exists a convenient set $\mathcal{P}_n$ such that the degree of each place in $\mathcal{P}_n$ is at most $d_m \in \mathcal{O}(\log_{q^2}(2n))$. Consequently, we have to construct the algorithms $\mathcal{U}^{\mathcal{F}_{\text{sp}}}_{q^2,d}$, where d is running through all the degrees of the places in $\mathcal{P}_n$. By Proposition 4.6, the algorithm $\mathcal{U}^{\mathcal{F}_{\text{sp}}}_{q^2,d}$, defined over F_k, is constructible in polynomial time in d as long as a place of degree d of F_k is given, i.e. time $\mathcal{O}(d^\alpha)$, for some positive integer α. Moreover, the place of degree d can be constructed in time $\mathcal{O}(\mathcal{O}((q^2)^{2d}M(d)\log d))$ according to Proposition 4.8. By Proposition 4.5, we have that $k-1 \leq \log_q(2d) - \log_q(q^2-4q-5)+1$, and since $q > 5$ is fixed, $k \leq \log_q(2d)$. It follows that the place of degree d and the algorithm $\mathcal{U}^{\mathcal{F}_{\text{sp}}}_{q^2,d}$ can both be constructed in time $\psi(\mathcal{U}^{\mathcal{F}_{\text{sp}}}_{q^2,d}) \in \mathcal{O}((q^2)^{2d}M(d)\log d + d^\alpha) = \mathcal{O}((q^2)^{2d}M(d)\log d)$. This process has to be done for all the degrees of the places in $\mathcal{P}_n$. Recall that $d \leq d_m \in \mathcal{O}(\log_{q^2}(2n))$. Therefore, it is required to construct $\mathcal{O}(d_m)$ algorithms $\mathcal{U}^{\mathcal{F}_{\text{sp}}}_{q^2,d}$, and the construction of each of them is bounded by $\mathcal{O}((q^2)^{2\log_{q^2}(2n)}M(\log_{q^2}(2n))\log\log_{q^2}(2n)) = \tilde{\mathcal{O}}(2n^2)$. Consequently, the construction of all these algorithms can be done in time $\tilde{\mathcal{O}}(2n^2d_m)$, i.e. $\sum_{P\in\mathcal{P}_n}\psi(\mathcal{U}^{\mathcal{F}_{\text{sp}}}_{q^2,n}) \in \tilde{\mathcal{O}}(2n^2)$. Finally by Proposition 3.4, the construction of the algorithm $\mathcal{U}^{\mathcal{H}}_{q^2,n}$ can be done in time $\mathcal{O}(n^4)$, and (ii) is proven.

Theorem 4.11. *Let q be a prime power and n be a positive integer. Then, there exists a family of Hybrid Chudnovsky-type algorithm $\mathcal{U}^{\mathcal{H}}_{q}$ such that*

(i) its bilinear complexity is uniformly bounded by $\mu(\mathcal{U}_q^{\mathcal{H}}) \leq C_q$*, where*

$$C_q = \begin{cases} 4\left(1 + \frac{3\sqrt{q}}{\sqrt{q}-5}\right) & \text{if } q > 25 \text{ is a square,} \\ 12\left(1 + \frac{3q}{q-5}\right) & \text{if } q > 5, \\ 152 & \text{if } q = 5, \\ 172 & \text{if } q = 4, \\ 279 & \text{if } q = 3, \\ 648 & \text{if } q = 2. \end{cases}$$

(ii) for any integer $n \geq 2$*,*

$$\psi\left(\mathcal{U}_{q,n}^{\mathcal{H}}\right) \in \mathcal{O}(n^4).$$

Proof. If $q > 25$ is a square, the result is directly given by Theorem 4.10. Elsewhere, we use the family of algorithms defined over $\mathbb{F}_{q^2}$ (or $\mathbb{F}_{q^4}$ or $\mathbb{F}_{q^6}$), and use the algorithms with the best known bilinear complexities to perform the multiplications in $\mathbb{F}_{q^2}$ (or $\mathbb{F}_{q^4}$ or $\mathbb{F}_{q^6}$) over $\mathbb{F}_q$. The construction time of these algorithms is negligible, and the cost of the construction remains asymptotically identical. If $q > 5$, then we use the family $\mathcal{U}_{q^2}^{\mathcal{H}}$, that gives algorithms over $\mathbb{F}_{q^2}$. Then, we compute each multiplication in $\mathbb{F}_{q^2}$ with $\mu_q(2) = 3$ bilinear multiplications. Thus, $\mu(\mathcal{U}_q^{\mathcal{H}}) = 3\mu(\mathcal{U}_{q^2}^{\mathcal{H}}) \leq 12\left(1 + \frac{3q}{q-5}\right)$. For $q = 3, 4, 5$, we use $\mathcal{U}_{q^4}^{\mathcal{H}}$. Since $\mu_3(4) = 9$ and $\mu_4(4) = \mu_5(4) = 8$, it provides the existence of a family of algorithms such that $\mu(\mathcal{U}_5^{\mathcal{H}}) = 8\mu(\mathcal{U}_{25^2}^{\mathcal{H}}) \leq 152$, $\mu(\mathcal{U}_4^{\mathcal{H}}) = 8\mu(\mathcal{U}_{16^2}^{\mathcal{H}}) \leq 172$ and $\mu(\mathcal{U}_3^{\mathcal{H}}) = 9\mu(\mathcal{U}_{9^2}^{\mathcal{H}}) \leq 279$. Finally, for $q = 2$ we use $\mathcal{U}_{2^6}^{\mathcal{H}} = \mathcal{U}_8^{\mathcal{H}}$ and $\mu_2(3) = 6$. According to what precedes $\mu(\mathcal{U}_8^{\mathcal{H}}) \leq 12\left(1 + \frac{24}{3}\right)$, thus $\mu(\mathcal{U}_2^{\mathcal{H}}) = 6\mu(\mathcal{U}_8^{\mathcal{H}}) \leq 648$.

The bounds given for the bilinear complexity can be refined, for instance using densified tower of function fields such as in [3]. Nevertheless, this latest result proves Theorem 4.1, i.e. we showed an explicit construction of algorithm constructible deterministically and in polynomial time, while giving a uniform linear bound for their bilinear complexity.

References

1. Arnaud, N.: Évaluations dérivées, multiplication dans les corps finis et codes correcteurs. Ph.D. thesis, Université de la Méditerranée, Institut de Mathématiques de Luminy (2006)
2. Ballet, S.: Curves with many points and multiplication complexity in any extension of $\mathbb{F}_q$. Finite Fields Appl. **5**, 364–377 (1999)
3. Ballet, S.: Low increasing tower of algebraic function fields and bilinear complexity of multiplication in any extension of $\mathbb{F}_q$. Finite Fields Appl. **9**, 472–478 (2003)
4. Ballet, S., Bonnecaze, A., Tukumuli, M.: On the construction of elliptic Chudnovsky-type algorithms for multiplication in large extensions of finite fields. J. Algebra Appl. **15**(1), 26 (2016)
5. Ballet, S., Pieltant, J.: On the tensor rank of multiplication in any extension of $\mathbb{F}_2$. J. Complex. **27**, 230–245 (2011)

6. Ballet, S., Rolland, R.: On the bilinear complexity of the multiplication in finite fields. In: Proceedings of the Conference Arithmetic, Geometry and Coding Theory (AGCT 2003), vol. 11, pp. 179–188. Société Mathématique de France, sér. Séminaires et Congrès (2005)
7. Ballet, S., Bonnecaze, A., Pacifico, B.: Multiplication in finite fields with Chudnovsky-type algorithms on the projective line. arXiv (2020)
8. Ballet, S., Chaumine, J., Pieltant, J., Rambaud, M., Randriambololona, H., Rolland, R.: On the tensor rank of multiplication in finite extensions of finite fields and related issues in algebraic geometry. Uspekhi Mathematichskikh Nauk **76**(1(457)), 31–94 (2021)
9. Bshouty, N.: Tester and their applications. Electron. Colloq. Comput. Complex. (ECCC) **19**(11) (2012)
10. Bürgisser, P., Clausen, M., Shokrollahi, A.: Algebraic Complexity Theory. Springer, Berlin, Heidelberg (1997). https://doi.org/10.1007/978-3-662-03338-8
11. Cenk, M., Özbudak, F.: On multiplication in finite fields. J. Complex. 172–186 (2010)
12. Chudnovsky, D., Chudnovsky, G.: Algebraic complexities and algebraic curves over finite fields. J. Complex. **4**, 285–316 (1988)
13. Elkies, N.: Explicit towers of Drinfeld modular curves. In: European Congress of Mathematics. Progress in Mathematics, vol. 202, pp. 189–198. Birkhäuser (2001), Proceedings of the 3rd European Congress of Mathematics, Barcelona, 10–14 July 2000
14. Fürer, M.: Faster integer multiplication. In: Proceedings of the 39th annual ACM Symposium on Theory of Computing (STOC) (2007)
15. Garcia, A., Stichtenoth, H.: A tower of Artin-Schreier extensions of function fields attaining the Drinfeld-Vlăduţ bound. Invent. Math. **121**, 211–222 (1995)
16. von zur Gathen, J., Gerhard, J.: Modern Computer Algebra. Cambridge University Press, Cambridge (2003). https://books.google.fr/books?id=NuEHj0wPwgIC
17. Hartshorne, R.: Algebraic Geometry. GTM, vol. 52. Springer, New York (1977). https://doi.org/10.1007/978-1-4757-3849-0
18. Harvey, D., van der Hoeven, J.: Polynomial multiplication over finite fields in time O(nlogn) (2019)
19. Karatsuba, A.: Multiplication of multidigit number on automata. Soviet Physics Doklady **7**, 595–596 (1963)
20. Rambaud, M.: Courbes de Shimura et algorithmes bilinéaires de multiplication dans les corps finis. Ph.D. thesis, Telecom ParisTech (2017). written in English
21. Randriambololona, H.: Bilinear complexity of algebras and the Chudnovsky-Chudnovsky interpolation method. J. Complex. **28**(4), 489–517 (2012)
22. Schönhage, A., Strassen, V.: Schnelle multiplikation großer zahlen [fast multiplication of large numbers]. Computing **7**(3–4), 281–292 (1971). https://doi.org/10.1007/BF02242355
23. Shparlinski, I.E., Tsfasman, M.A., Vladut, S.G.: Curves with many points and multiplication in finite fileds. In: Stichtenoth, H., Tsfasman, M.A. (eds.) Coding Theory and Algebraic Geometry. LNM, vol. 1518, pp. 145–169. Springer, Heidelberg (1992). https://doi.org/10.1007/BFb0087999
24. Stichtenoth, H.: Algebraic Function Fields and Codes, second edn. No. 254 in Graduate Texts in Mathematics, Springer-Verlag, Berlin, Heidelberg (2008). https://doi.org/10.1007/978-3-540-76878-4
25. Tsfasman, M., Vlăduţ, S.: Algebraic-Geometric Codes. Kluwer Academic Publishers, Dordrecht/Boston/London (1991)

Reduction-Free Multiplication for Finite Fields and Polynomial Rings

Samira Carolina Oliva Madrigal[1(✉)], Gökay Saldamlı[1], Chen Li[2], Yue Geng[2], Jing Tian[2], Zhongfeng Wang[2], and Çetin Kaya Koç[3,4,5]

[1] San José State University, San José, USA
scolivamadrigal@gmail.com, gokay.saldamli@sjsu.edu
[2] Nanjing University, Nanjing, China
{MG21230068,181180030}@smail.nju.edu.cn, {tianjing,zfwang}@nju.edu.cn
[3] Iğdır University, Iğdır, Turkey
[4] Nanjing University of Aeronautics and Astronautics, Nanjing, China
[5] University of California, Santa Barbara, Santa Barbara, USA
cetinkoc@ucsb.edu

Abstract. The complexity of the multiplication operation over polynomial rings and finite fields drastically changes with the selection of the defining polynomial of the respective mathematical structure. Trinomials and pentanomials are the most natural choices for the best arithmetic. In this paper, we first present a study in which a special type of trinomial does not require any reduction steps. We then introduce two new algorithms, FIKO and RF-FIKO, fully interleaved bit-parallel Karatsuba-Ofman multipliers where the latter is only concerned with the three Karatsuba-Ofman terms and is free from the bipartite reduction circuits. All algorithms are implemented in FPGA and ASIC, and detailed implementation results are presented, showing significant improvements to existing methods.

Keywords: Cryptography · Finite field arithmetic · Polynomials rings · Karatsuba-Ofman multiplication · Polynomial bi-partite multiplication · Montgomery multiplication · Interleaved multiplication · Mersenne polynomials · Pseudoprimes · Equally-spaced polynomials · Reduction-free trinomials · Reduction-free multiplication

1 Introduction

The study of irreducible polynomials is particularly important in cryptography. For modern schemes that employ modular arithmetic such as, RSA and ECDSA, the multiplication operation is the most expensive. In particular, Post-quantum Cryptography (PQC) lattice-based schemes and Fully Homomorphic Encryption (FHE) systems would benefit from efficient finite field arithmetic. For practicality, security, and efficiency, we are concerned with binary fields in the polynomial basis (PB).

S. Mesnager and Z. Zhou (Eds.): WAIFI 2022, LNCS 13638, pp. 53–78, 2023.
https://doi.org/10.1007/978-3-031-22944-2_4

The most efficient bit-parallel multipliers known all make use of special trinomials to obtain time and space complexity speedups. Moreover, they make efforts to simplify modular reduction as a means to obtain faster implementations. In this paper we present a solution to the problem $A(t) \times B(t) \pmod{F(t)}$ that does not require reduction modulo $F(t)$ but only three multiplications followed by some shifts and additions in the last clock cycle, when $F(t)$ has a special form - A fully interleaved reduction-free modular multiplier.

2 Preliminaries

This section provides an overview of the notation used throughout in the paper, concerning modular reduction with finite rings and fields where the elements are represented in the polynomial basis.

Let $\mathbb{Z}_p[t]/(F(t))$ denote the set of polynomials in variable t with coefficients over p, defined by a monic $F(t)$ of degree n which forms a finite field, for p prime and $F(t)$ irreducible modulo p; and a ring, $\mathcal{R}_{q=p^n}$, otherwise. We express $F(t)$ as $\sum_{i=0}^{n} = f_n t^n + f_{n-1} t^{n-1} + f_{n-2} t^{n-2} + \cdots + f_2 t^2 + f_1 t + f_0 t^0$ where the f_i are the coefficients. The same notation is used for all other polynomials, strings, and data registers in hardware to refer to the ith coefficient or data bit. Let f_i^{-1} denote the multiplicative inverse of $f_i \pmod{p}$. An element $A(t)$ in any such structure is expressed as $A(t) = \sum_{i=0}^{n} = a_{n-1} t^{(n-1)} + a_{n-2} t^{(n-2)} + \cdots + a_1 t + a_0 t^0$. Then, the product of two elements is $A(t) \times B(t)$, is expressed as $C(t) = A(t) \times B(t) \pmod{p, F(t)}$. Arithmetic concerning coefficients of elements $A(t)$ and $B(t)$ conforms to $\pmod{p}$. Polynomial arithmetic conforms to $\pmod{F(t)}$.

Let $C_{max} = 2(n-1)$ denote the maximum degree of a product of two elements in $\mathcal{R}_q$. Let F_d be the degree of defining polynomial $F(t)$ and C_d the degree of an arbitrary product of two elements, $C(t) = A(t) \times B(t) \in \mathcal{R}_q$. Then, $\rho = C_d - (F_d - 1)$ denotes the number of reductions required to reduce a product of degree C_d to become representable in the field or ring.

For a hardware register or string holding $A(t)$ of depth $k = 2 \times r$, let $A(t)_{[j:i]}$ denote the bit range of data from j to i inclusive, from the jth most significant bit to the ith least significant bit. For a register A of width $k = 2 \times r$, let A_{UR} denote the upper register $A_{[k-1:r]}$ and A_{LR} denote the lower register $A_{[r-1:0]}$. Similarly, $||$ denotes concatenation of strings. For example, $C = A||B$, would refer to the equivalent SystemVerilog assignment $C = \{A, B\}$ for C of width k and A and B both of width r.

Let $GF(p^n)$ and $\mathbb{F}_q$ with $q = p^n$, be a finite field for p prime and defining polynomial $F(t)$ of degree n irreducible in modulo p. Our work is primarily concerned with Galois fields of the form $GF(2^n)$ and is easily adaptable to rings where the coefficients vary over $\mathbb{Z}_p$. We are interested in prime odd binary curves of degree n that define different binary fields $GF(2^n)$ for varying $F(t)$. We use the notation $GF(2^n)$, $(n+1)$-bit curve, and $k = (n+1)$-bit field interchangeably, where $(n+1)$-bit curve refers to $F(t)$ and k-bit field to $GF(2^n)$ which is defined by k-bit $F(t)$.

3 Reduction with Polynomial Rings

Efficient arithmetic over polynomial rings and fields $\mathbb{Z}_p[t]/F(t)$ has originated from usual modular arithmetic routines. If polynomial coefficients and words of multi-precision numbers are considered as the atomic units of their representations, polynomials enjoy carry-free arithmetic over these units. Multiplication involving creative representations make the arithmetic over polynomial rings quite interesting.

Modular multiplication involves multiplication and reduction steps; these are often implemented as separate algorithms or interleaved into one algorithm. Both type of implementations are characterized by the reduction technique employed. Modular reduction can applied from left-to-right, right-to-left, or both directions; corresponding examples to these would be Blakely, Montgomery, and Bipartite, respectively [3,16,25].

In this section, we go over the reduction techniques when given an already computed product $C(t) = A(t) \times B(t)$ with $A(t), B(t) \in \mathbb{Z}_p[t]/F(t)$ for some p and $F(t)$ a monic of degree n. We consider the product $C(t)$ of maximum degree $m = C_{max}$ and reduction (mod $q, F(t)$).

3.1 Left-to-Right Reduction

Any algorithm that reduces a product $C(t)$ modulo $F(t)$ from the left falls into the left-to-right category, including the standard division algorithm.

Algorithm 1. Blakely Polynomial Reduction

Require: $C(t)$ of degree m
Ensure: $R(t) \equiv C(t)(\text{mod} p, F(t))$
1: $R(t) = C(t)$
2: $j = m - n$
3: **for** $i = m$ **downto** n **do**
4: $\quad q_j = r_i \bmod p$
5: $\quad R(t) = R(t) - q_j F(t) t^{i-n}$
6: $\quad j = j - 1$
7: **end for**
8: **return** $R(t)$

Algorithm 1 presents the Blakely reduction method from [3] which can be adapted to $GF(2^n)$ as in [20]. This is a bit-serial algorithm in which we loop $k = m - (n - 1)$ times reducing the degree of $C(t)$ from the most significant position until $i - n$ and we obtain a residue of degree $\leq n - 1$. In line 4 we compute the jth digit of the full quotient $Q(t)$ as the jth digit of $R(t)$ modulo p, starting from the most significant position. In line 5 we subtract an aligned multiple of the modulus, to continue reduction from left-to-right in each loop. The resulting residue satisfies the closed form of the division theorem, $R(t) = C(t) - F(t)Q(t)$.

3.2 Right-to-Left Reduction

Montgomery multiplication [25] demonstrates the unique example of right-to-left reduction which is shown in Algorithm 2.

Algorithm 2. Polynomial Montgomery Reduction

Require: $C(t)$ of degree m
Ensure: $S(t) \equiv C(t)(\text{mod} p, F(t))$
1: $S(t) = C(t)$
2: **for** $i = 0$ **to** $k - 1$ **do**
3: $\quad q_i' = f_0^{-1} s_0 \bmod p$
4: $\quad S(t) = (S(t) - q_i' F(t))/t$
5: **end for**
6: **return** $S(t)$

The residue is computed in a bit-serial fashion, where we loop $k = m-(n-1)$ times, reducing the product from the least significant position, one degree per loop. In line 3 we compute the ith quotient of the Montgomery quotient, q_i', as the multiplicative inverse of the least significant coefficient of $F(t)$ multiplied with the least significant digit of $S(t)$, modulo p. In line 4, subtracting $q_i'F(t)$ from $S(t)$ sets the constant coefficient, s_0, to zero and hence the division by t corresponds to a trivial division or a right-shift. This is the reason why Montgomery reduction is preferred in most repetitive multiply-reduce designs. The result is a residue $S(t) = C(t)t^{-k} \pmod{F(t)}$ with degree $\leq n - 1$.

3.3 Bipartite Reduction

The bipartite modular multiplication (BMM) method introduced by Kaihara and Takagi in [16], presents a method of modular reduction in which a left-to-right and a right-to-left technique can be applied in parallel to reduce a product from both ends simultaneously. This method is presented in Algorithm 3; for completeness we simply combine Algorithms 1 and 2.

Algorithm 3 executes in a sequential fashion but it loops $\rho = \lfloor n/2 \rfloor = \frac{m-(n-1)}{2}$ half the number of times as Blakely and Montgomery which require $\rho = m - (n - 1)$ reductions. In lines 4–5, we compute the standard and Montgomery quotients, respectively. In line 6, we apply Blakely reduction to $S(t)$ and in line 7 we apply Montgomery reduction. Lines 7–8 can be implemented as separate threads in software or functional units in hardware executing in parallel. When the coefficients are over $\mathbb{Z}_p$ we must account for the Montgomery domain and set the parameter R to be less than the modulus.

4 Interleaved Modular Reduction

This sections builds on the previous section by interleaving multiplication of a product and reduction using a simpler structure, $GF(2^n)$. We present the

Algorithm 3. Bipartite Polynomial Reduction

Require: $C(t)$ of degree m
Ensure: $S(t) \equiv C(t)(\text{mod} p, F(t))$
1: $S(t) = C(t)$
2: $k = \lfloor n/2 \rfloor$
3: **for** $i = 0$ **to** $\lfloor n/2 \rfloor - 1$ **do**
4: $\quad q_k = s_k \bmod p$
5: $\quad q'_i = f_0^{-1} s_0 \bmod p$
6: $\quad S(t) = S(t) - q_k F(t) t^{k-n}$
7: $\quad S(t) = (S(t) - q'_i F(t))/t$
8: $\quad k = k - 1$
9: **end for**
10: **return** $S(t)$

corresponding interleaved modular multiplication algorithms for Blakely, Montgomery, and BMM.

4.1 Blakely

Algorithm 4. Interleaved Blakely

Require: k-bit F(t), $(k-1)$-bit A(t) and B(t)
Ensure: $R(t) \equiv A(t) \times B(t)(\text{mod} F(t))$
1: $R(t) = 0$
2: **for** $i = k - 2$ **downto** 0 **do**
3: $\quad R(t) = R(t) \ll 1$
4: $\quad$**if** a_i **then**
5: $\quad\quad R(t) = R(t) \oplus B(t)$
6: $\quad$**end if**
7: $\quad$**if** r_{k-1} **then**
8: $\quad\quad R(t) = R(t) \oplus F(t)$
9: $\quad$**end if**
10: **end for**
11: **return** $R(t)$

Algorithm 4 shows the adapted version of the original interleaved algorithm in the binary basis [3, 20]. Multiplication and reduction are interleaved using the standard shift and add technique. In this case we observe the bits of $A(t)$ starting from the most significant bit; if the bit is set, we multiply or add $B(t)$ to $R(t)$. Similarly, if the most significant bit of $R(t)$ is set, we reduce $R(t)$ with $F(t)$. After a multiplication and a reduction, we shift out the degree that has been knocked down.

4.2 Montgomery

Algorithm 5 shows the interleaved binary version adapted from [2]. If we are only concerned with multiplication, we observe the bit of the multiplier from either end. However, when we are performing interleaved multiplication, these bits must be observed according to the technique; for Blakely and Montgomery it is according to the direction we are reducing from. In this case, we test if the least significant bit of the residue is set and reduce it with $F(t)$. Lastly, we shift out the knocked-down degree.

Algorithm 5. Interleaved Mongtomery

Require: k-bit F(t), $n = (k-1)$-bit A(t) and B(t)
Ensure: $R(t) \equiv A(t) \times B(t) \times 2^{-n} (\text{mod} F(t))$
1: $R(t) = 0$
2: **for** $i = 0$ **to** $k - 2$ **do**
3:   **if** b_i **then**
4:     $R(t) = R(t) \oplus A(t)$
5:   **end if**
6:   **if** r_0 **then**
7:     $R(t) = R(t) \oplus F(t)$
8:   **end if**
9:   $R(t) = R(t) \gg 1$
10: **end for**
11: **return** $R(t)$

4.3 Bipartite Modular Multiplication

Algorithm 8 adapts BMM from [16] to $GF(2^n)$. In hardware, the intermediate Blakely $S(t)$ and Montgomery $T(t)$ residues are computed in parallel. The algorithm uses Algorithm 6 and Algorithm 7 to compute interleaved modular multiplication with $A(t)$ and the upper and lower words of $B(t)$. For k-bit curves, we can compute the bipartite residue in $k/2$ CC without dependencies. The Blakely residue requires $r - 1$ reductions in Algorithm 6 while the Montgomery residue requires r reductions in Algorithm 7 since BH is one bit less. BMM ensures a residue $R(t) \equiv (A(t) \times BH(t) \pmod{F(t)} + (A(t) \times BL(t) \times 2^{-\lceil \frac{k-1}{2} \rceil} \pmod{F(t)})$ $(\text{mod } F)(t)$. This residue is also implicitly expressed as $A(t) \times B(t) \times 2^{-r}$ $(\text{mod } F(t))$ indicating the $r = \lceil \frac{k-1}{2} \rceil$ Montgomery degrees knocked down.

Algorithm 6. ibBlakely

Require: $A(t)_{[k-2:0]}$, $BH(t)_{[r-2:0]}$, $F(t)_{[k-1:0]}$
Ensure: $S(t) \equiv A(t) \times BH(t)(\text{mod}F(t))$
1: $S(t) = 0$
2: **for** $i = r - 2$ **downto** 0 **do**
3: $\quad S(t) = S(t) \ll 1$
4: $\quad$ **if** bh_i **then**
5: $\quad\quad S(t) = S(t) \oplus A(t)$
6: $\quad$ **end if**
7: $\quad$ **if** s_{k-1} **then**
8: $\quad\quad S(t) = S(t) \oplus F(t)$
9: $\quad$ **end if**
10: **end for**
11: **return** $S(t)$

Algorithm 7. ibMontgomery

Require: $A(t)_{[k-2:0]}$, $BL(t)_{[r-1:0]}$, $F(t)_{[k-1:0]}$
Ensure: $T(t) \equiv A(t) \times BL(t) \times 2^{-r}(\text{mod}F(t))$
1: $T(t) = 0$
2: **for** $i = 0$ **to** $r - 1$ **do**
3: $\quad$ **if** bl_i **then**
4: $\quad\quad T(t) = T(t) \oplus A(t)$
5: $\quad$ **end if**
6: $\quad$ **if** t_0 **then**
7: $\quad\quad T(t) = T(t) \oplus F(t)$
8: $\quad$ **end if**
9: $\quad T(t) = T(t) \gg 1$
10: **end for**
11: **return** $T(t)$

Algorithm 8. BMM

Require: k-bit $F(t)$, $(k-1)$-bit $A(t)$, $B(t)$, $r = \lceil(\frac{k-1}{2})\rceil$
Ensure: $R(t) \equiv A(t) \times B(t) \times 2^{-r}$ $(\text{mod } F)(t)$
1: $S(t) = ibBlakely(A(t), B(t)_{[k-2:r]}, F(t))$
2: $T(t) = ibMontgomery(A(t), B(t)_{[r-1:0]}, F(t))$
3: $R(t) = S(t) \oplus T(t) \oplus F(t)$
4: **return** $R(t)$

5 Partially Interleaved Karatsuba-Ofman

Modular multipliers fall under one of two types: multiply and reduce or interleaved multiply and reduce. Varying implementations of Montgomery, BMM, and Mastrovito are among the most efficient interleaved modular multipliers introduced [14,24]. In general, interleaving fast multi-digit multipliers, such as Schönhage-Strassen or Fürer is difficult and application-specific [12,34]. For example, Fürer is intended for huge numbers in the order of 10^{82} and becomes efficient when the operands are in that order. In this section, we present the Karatsuba-Ofman Algorithm (KOA) and the work from [33], a Partially Interleaved Karatsuba-Ofman (PIKO) modular multiplier.

5.1 Karatsuba-Ofman Multiplication

KOA is a recursive divide and conquer algorithm based on the observation by Babbage that two n-digit numbers can be expressed as binomials and multiplied out likewise requiring four multiplications [1,19]. In 1962, Karatsuba and Ofman observed that the middle term could actually be computed in one multiplication with some additions and subtractions from already computed terms requiring only three multiplications total. KOA is intended for very large numbers in the order of several thousand digits ranging from 10^3 to 10^4 digits with complexity $O(n^{log_2(3)})$ in the input size.

The algorithm is quite generic that it can easily be adapted to polynomial multiplication for binary fields where operands range from a few thousand bits to a few hundred thousand bits. In practice, KOA is used in conjunction with reduction routines such as Blakely or Montgomery [3,25]. Recursion can be set to any desired level. However, further recursion and improved KOA must account for platform constraints and is completely application-specific.

Now, consider two arbitrary polynomials $A(t) = a_p t^m + \cdots + a_2 t^2 + a_1 t + a_0$ and $B(t) = b_q t^n + \cdots + b_2 t^2 + b_1 t + b + 0$ of degrees m and n respectively, and without loss of generality, let $m \geq n$ and $r = \lfloor m/2 \rfloor$. For simplicity, let's say they are both expressed as the closest power of two and are split into half-size equal words. Let:

$$\begin{aligned} A(t) &= A_1(t) \cdot t^r + A_0(t), \\ B(t) &= B_1(t) \cdot t^r + B_0(t) \end{aligned}$$

where, $A_1(t)$ and $A_0(t)$ represent the upper and lower words of the polynomial $A(t)$, each of degree r. Standard multiplication of $A(t)$ and $B(t)$ can expressed as follows:

$$\begin{aligned} C(t) &= A(t) \cdot B(t) \\ &= (A_1(t)t^r + A_0(t))(B_1(t)t^r + B_0(t)) \\ &= (A_1(t)B_1(t))t^{2r} + (A_1(t)B_0(t) + A_0(t)B_1(t))t^r \\ &\quad + A_0(t)B_0(t) \\ &= C_2(t)t^{2r} + C_1(t)t^r + C_0(t). \end{aligned}$$

Notice that, the above calculation requires four different polynomial multiplications with operands of degree r. This has quadratic complexity in operand size. KOA however, achieves the same computation with only three multiplications as follows:

$$C_0(t) = A_0(t) \cdot B_0(t)$$
$$C_2(t) = A_1(t) \cdot B_1(t)$$
$$C_1(t) = (A_0(t) + A_1(t))(B_0(t) + B_1(t)) - C_0(t) - C_2(t)$$
$$C(t) = C_2(t)t^{2r} + C_1(t)t^r + C_0(t).$$

5.2 Interleaving Karatsuba-Ofman and Bipartite Reduction

In [33] Saldamlı et al. present PIKO, consisting of Karatsuba-Ofman multiplication and bipartite reduction circuits. The algorithm considers only the first layer of recursion requiring half-size words. In this section, we give a general overview of the algorithm as follows.

Let $F(t) = x^n + x^{\lfloor \frac{n}{2} \rfloor} + 1$ the defining polynomial of degree n for some odd prime curve with coefficients defined over $GF(2)$. Now, consider two n-bit elements $A(t)$ and $B(t)$ in the field, both of maximum degree $n-1$. First, we prefix the elements to consists of $k = n+1$ bits. This allows all operands $F(t)$, $A(t)$, and $B(t)$ to be split into equal half-size words of size $r = \frac{k}{2}$ bits. Note that the maximum degree for the lower word of an element is $d = r - 1$ but for the upper word it is $r - 2$. For uniformity, because we are working with r-bit registers, the operands are decomposed as follows.

$$F(t) = F_1(t) \cdot t^d + F_0(t),$$
$$A(t) = A_1(t) \cdot t^d + A_0(t),$$
$$B(t) = B_1(t) \cdot t^d + B_0(t),$$
$$Q(t) = Q_1(t) \cdot t^d + Q_0(t),$$
$$Q'(t) = Q'_1(t) \cdot t^d + Q'_0(t).$$

where $Q(t)$ and $Q'(t)$ represent the Blakely and Montgomery quotients.

Figure 1 illustrates the PIKO algorithm. The algorithm is the same as BMM except that multiplication is done using KOA. In the upper part, we compute the Karatsuba-Ofman terms $C_0(t)$, $C_1(t)$, and $C_2(t)$; the middle term is partially computed. In the middle part, we compute bipartite terms consisting of bipartite quotients and bipartite products. The quotients $Q'_0(t)$ and $Q_1(t)$ are computed using fully interleaved Montgomery and Blakely algorithms applied to the products $C_0(t)$ and $C_2(t)$ which are reduced with $F_0(t)$ and $F_1(t)$, respectively.

The reduction terms are then $(Q'_0(t)F_0(t))$, $(Q'_0(t) + Q_1(t))(F_0(t) + F_1(t))$, and $(Q_1(t)F_1(t))$. These terms are applied to $C_0(t)$, $C_1(t)$, and $C_2(t)$ to produce reduced products $C'_0(t)$, $C'_1(t)$, and $C'_2(t)$. Lastly, the final sum can be computed in different ways. In Fig. 1, the final sum is simply $S(t) = C'_{2LR}(t)||C'_{0UR}(t) + C'_1(t) + C'_{2LR}(t)||C'_{0UR}(t)$. This gives us a residue $A(t) \times B(t) \times t^{-r} \pmod{F(t)}$. The term t^{-r} is equivalently expressed as $(t^r)^{-1} = 2^{-r} = (2^r)^{-1}$ since in the binary base, t^{-r} is a shift right by r indicating the Montgomery degrees knocked down.

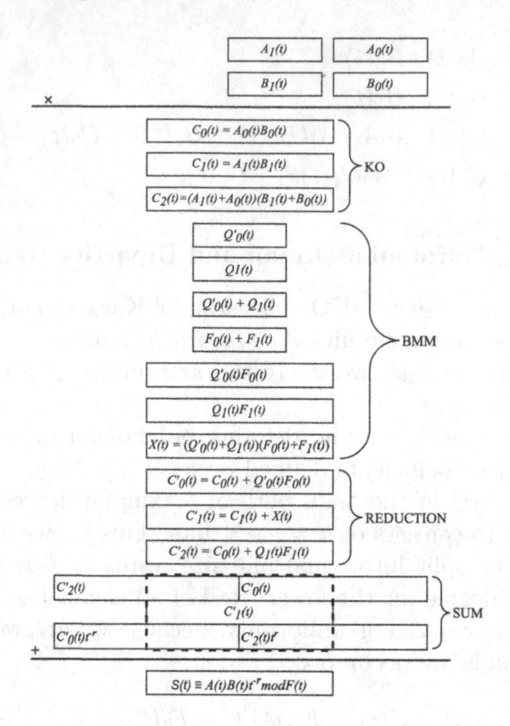

Fig. 1. Partially interleaved Karatsuba-Ofman algorithm.

If we compute all terms in PIKO in a bit-parallel fashion, bit-by-bit per clock cycle, it becomes fully interleaved. Then, a residue of k-bit operands can be computed in $\frac{k}{2}$ clock-cycles. We call this version the Fully Interleaved Karatsuba-Ofman (FIKO) algorithm. This follows by noting that when we compute a product $A \times B$, the bits of a multiplier B, can be observed from either the least significant or most significant position. In this manner, we can compute the Blakely and Montgomery quotients from opposite ends. Close attention to the half-size products, reveals no dependencies in this approach. For example, the product of two upper words $A_1(t) \times B_1(t)$ will always have $C_{max} = 2(r-2)$ because they are prefixed. Consequently, the most significant $(k-1) - C_{max}$ bits of $C(t)_2$ will always be zero. The upper bits of the product $C_2(t)$ become fixed as the amount by which we shift $A_1(t)$ by decreases.

The remaining multiplications such as $Q_1(t)F_1(t)$, can all be accomplished using shifts and adds, according to $F(t)$. The three types of $F(t)$ used are discussed in Sect. 7. Figure 2 shows the core of the FIKO bit-parallel algorithm in SystemVerilog. $C0$, $C1$, $C2$ correspond to KOA terms; $R1$ and $R0$ are the Blakely and Montgomery residues; and $Q1$ and $Q0$ are the Blakely and Montgomery quotients.

```
always@(posedge clk)
begin
    if(load) begin: loading
        C0 = 0; //koa
        C1 = 0;
        C2 = 0;
        R1 = 0;
        Q1 = 0;
        R0 = 0;
        Q0 = 0;
        i = r - 1;
        j = 0;
    end: loading
    else if(mul) begin: multiply
        C0 = C0 << 1;
        C0 = B0[i] ? C0 ^ A0: C0;
        C1 = C1 << 1;
        C1 = B_sum[i]? C1 ^ A_sum:C1;
        C2 = C2 << 1;
        C2 = B1[i] ? C2 ^ A1:C2;

        R1 = R1 << 1;//blakely terms
        R1 = A1[i]? (R1 ^ B1): R1; //from ith most singificant bit
        Q1[i] = R1[r-1];//blakely quotient
        R1 = R1[r-1]? (R1 ^ F1): R1;//blakely residue

        R0 = B0[j]? (R0 ^ A0): R0;//montgomery form jth least significant bit
        Q0[j] = R0[0];//montgomery quotient
        R0 = R0[0]? (R0 ^ F0): R0;
        R0 = R0 >> 1;//montgomery residue
        i = i - 1;
        j = j + 1;
    end: multiply
end
```

Fig. 2. Fully interleaved Karatsuba-Ofman algorithm core.

6 Reduction-Free FIKO

In this section we develop the FIKO algorithm into a reduction-free version. First, we present the special form of the defining polynomial which allows us to accomplish this.

6.1 Equally Spaced Polynomials

In [14], Koç provides a concise definition of Equally Spaced Polynomials (ESPs) and Equally Spaced Trinomials (ESTs). An ESP with degree $n = \delta k$ has form $x^{\delta k} + x^{\delta(k-1)} + \cdots + x^{\delta} + 1$ and is necessarily of even degree with all non-zero terms equally spaced by $\delta - 1$ zero terms. For example, $x^{\delta k} + x^{\delta(k-1)} + x^{\delta(k-2)} = x^6 + x^3 + x^0 = 1001001_2$ for $\delta = 3$ and $k = 2$, the terms are equally spaced by $\delta - 1 = 2$ zero terms. Similarly, $x^4 + 1$ is an ESP. A special case is the All-One-Polynomial (AOP), in which case $\delta = 1$. For example, $x^{(k-1)} + x^{(k-2)} + x^{(k-3)} + x^{(k-4)} + x^{(k-5)} + x^{(k-6)} = 1111111_2$ with $k = 6$. An EST is a trinomial with all non-zero terms equally spaced and necessarily of even degree; for example $x^4 + x^2 + 1 = 10101_2$.

6.2 Reduction-Free Trinomials

Ideally, we would like to work with an intermediary binomial or an EST where we can reduce a product from both ends in parallel without dependencies. However, for our work, ESTs cannot be used as they have an even degree, cannot be equally split, and have security considerations. Now, when the defining polynomial is an ESP-like trinomial of form $x^n + x^r + 1$ with n odd and $r = \lceil n/2 \rceil$, we can enjoy reduction-free multiplication. We define this special polynomial as a Reduction-Free Trinomial (RFT). Such polynomials can be split into equal half-size r-bit registers and are of necessarily odd prime degree. This trinomial is easily characterized as one whose lower register is a 1 and whose upper register is an ESP of desired binomial form. For symmetry, we zero-extend the input operands and work with r-bit registers.

When we work with half-size operands and an RFT, careful observation of the computations and results in FIKO reveal that the bipartite reduction circuits can be removed. The following example shows this.

Example 1. For simplicity, consider $GF(2^9)$, $F(t)$ an RFT that defines the field, and two elements in the field:

$$F(t) = t^9 + t^5 + 1$$
$$A(t) = t^8 + t^7 + t^6 + t^5 + t^2 + 1$$
$$B(t) = t^8 + t^5 + t^3 + t + 1.$$

The corresponding half-size words with the prefixed elements are given in Table 1.

Table 1. Half-size parameters

Operand	Upper register	Lower register
$F(t)$	$F_1(t) = 10001$	$F_0(t) = 00001$
$A(t)$	$A_1(t) = 01111$	$A_0(t) = 00101$
$B(t)$	$B_1(t) = 01001$	$B_0(t) = 01011$

If we compute FIKO as usual, we would obtain all Karatsuba-Ofman products, bipartite terms, reduced terms, and final sum. However, since $F_0(t) = 1$, we can observe that the lower word of the Montgomery quotient is just the lower register of $C_0(t)$ since $a \pmod 1$ is always a. Similarly, we observe that the Blakely quotient for the entire product $C_2(t)$ is just the most significant $[k-2:r-1]$ bits of $C_2(t)$. Note that the Blakely quotient will always fit in less than r-bits. Because we are working with half-size words and we are only concerned with the upper register of the standard quotient $Q(t)$, we can easily see that this is just $C_2(t)[k-2:r-1]$. Because $A_1(t)$ and $B_1(t)$ are prefixed and their product is of degree at most $2(r-2)$, it can be observed that the $(k-1)-2(r-2)$ bits of $C_2(t)$ will always be zero and that the upper register of the quotient is found in the specified bits. However, the full upper register of $C_2(t)_{[k-1:r]}$ can be taken as $Q_1(t)$ if we multiply it by t or shift it left by one. This is illustrated in Fig. 3;

the left side shows RF-FIKO quotients taken directly from $C_0(t)$ and $C_2(t)$ and the right side shows the quotients we obtain after we apply bipartite reduction. They are the same.

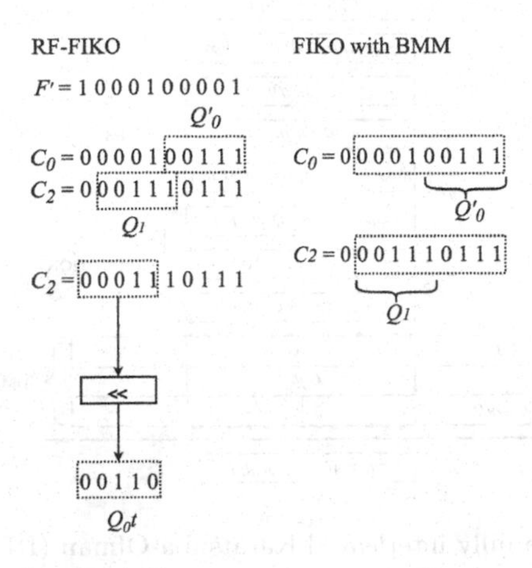

Fig. 3. RFFIKO and FIKO with a RFT.

There is no need for the Blakely and Montgomery reduction circuits, we can just take the quotients from the Karatsuba-Ofman terms $C_0(t)$ and $C_2(t)$. RF-FIKO, illustrated in Fig. 4, reduces to a total of three distinct multiplications, namely those to compute the Karatsuba-Ofman products $C_0(t)$, $C_1(t)$, and $C_2(t)$. The bipartite products and reduction terms are all computed with shifts and additions on the last clock cycle. In Example 1, our desired answer $C(t) = (C_2't^{2r} + C_0' + C_1't^r + C_0't^{2r} + C_2')t^{-r}$ is 1000010100. We can easily obtain the result in the standard domain by re-adding the r Montgomery zeroes (degrees we knocked down) to this bipartite residue and employing standard reduction. Our result is $C(t)t^{-r} \pmod{F(t)} \equiv A(t)B(t)(t^r)^{-1} \pmod{F(t)}$. Now, from closer inspection of Fig. 4, we can see that $Q_0'(t)F_0(t) = Q_0'(t)$ since $F_0(t) = 1$.

For an RFT, the cross term is simply computed as the sum of the quotients with multiplication by $(F_1(t) \oplus F_0(t))$ implemented as a shift left by $r-1$. Hence, we have a fully interleaved reduction-free modular multiplier with a total of three half-size word multiplications and six sums (two k-bit sums, three r-bit sums, and one 1-bit sum). The sums can be implemented differently, for example to reduce an r-bit sum to a 1-bit sum at the cost of space. In either case, the metrics are the same and we kept the original implementation with three k-bit sums and three r-bit sums.

7 Test Inputs

We consider finite fields of the form $GF(2^n)$ where the degree of the defining polynomial corresponds to an ECDSA binary field or a Mersenne exponent [6,13,

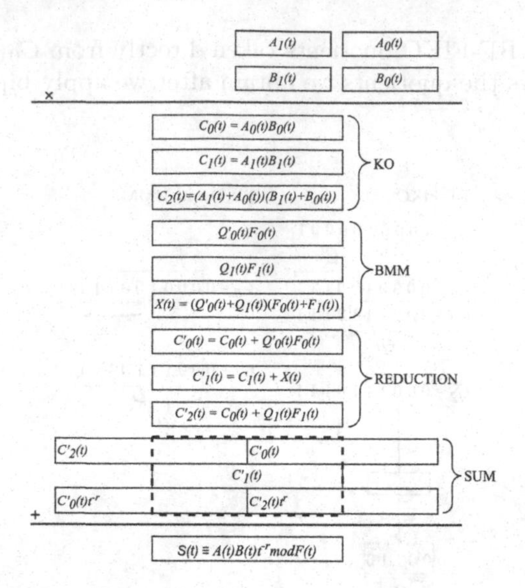

Fig. 4. Reduction-free fully interleaved Karatsuba-Ofman (RF-FIKO) algorithm.

29,38]. We consider three different types of trinomials, namely, RFTs of special form $x^n + x^{\lceil n/2 \rceil} + 1$, special trinomial type 1 of form $x^n + x^{\lfloor n/2 \rfloor} + 1$ and special trinomial type 2 of form $x^n + x + 1$.

Table 2. Test curves

Curve type	Exponents
ECDSA	163, 233, 283, 409, 571
Mersenne	107, 127, 521, 607, 1279, 2203, 2281, 3217

Table 2 shows the curves used and Table 3 shows our test field groups. For example, test group $F2$ from Table 3 consists of finite fields of the form $GF(2^n)$ where n varies over the ECDSA exponents listed in table Table 2 and the defining polynomial for all such fields is an RFT. More explicitly, the $F2$ group consists of fields: $GF(2^{163})$, $GF(2^{233})$, $GF(2^{283})$, $GF(2^{409})$, and $GF(2^{571})$ each of which is defined by a corresponding RFT $F(t) = t^{163} + t^{82} + 1$, $F(t) = t^{233} + t^{117} + 1$, $F(t) = t^{283} + t^{142} + 1$, $F(t) = t^{409} + t^{205} + 1$, and $F(t) = t^{571} + t^{286} + 1$, respectively. RF-FIKO is completely defined by an RFT and hence, we only test it with finite fields in groups $F1$ and $F2$. Blakely, Montgomery, FIKO, and BMM are independent of the defining polynomials and can be tested with all test groups.

Table 3. Test groups

Group	Curves	$F(t)$
$F1$	Mersenne	RFT
$F2$	ECDSA	RFT
$F3$	Mersenne	Type 1
$F3'$	ECDSA	Type 1
$F4$	Mersenne	Type 2
$F4'$	ECDSA	Type 2

8 Hardware Implementation

In this section, we present two bit-parallel hardware implementations in $GF(2^n)$. The algorithms implemented were Interleaved Montgomery, Interleaved Blakely, BMM, FIKO, and RF-FIKO. We implemented in Verilog and SystemVerilog and prototyped on a Virtex-7 FPGA. The ASIC design synthesis was done using TSMC 28-nm CMOS technology.

8.1 Non-recursive Decomposition

Figure 5 details the microarchitecture for RF-FIKO corresponding to our algorithm in Fig. 4. A direct and naïve implementation of Fig. 4 would yield a residue in k-CC with the KOA terms computed in r-CC followed by the bipartite reduction products in r-CC. However, a bit-parallel implementation allows for a fully interleaved implementation with all terms computed in parallel, one bit per clock cycle, in r-CC. Moreover, close inspection of Fig. 4 and the bipartite terms, allows us to be concerned only with the KOA products of $C_0(t)$, $C_1(t)$, and $C_2(t)$ since the bipartite and reduction terms can all be computed with shifts and additions. The critical delay path is then the multiplication of the three KOA terms for which there exist various techniques for improvement.

The fastest bit-parallel implementations are concerned with special trinomials and exploiting structure to compute the product $A(t) \times B(t) \mod F(t)$ while making efforts to simplify the reduction. We have explored the mathematical structure of the RFT together with noted observations so that we can solve the problem $A(t) \times B(t) \mod F(t)$ with only three multiplications, $C_0(t)$, $C_1(t)$, and $C_2(t)$, and obtain a true reduction-free residue. Because of the nature of the RFT, we obtain reduction-free quotients and compute the remaining terms with some shifts and adds.

This work sets forth an initial presentation of RF-FIKO concerned with the upper layer of recursion of Karatsuba-Ofman to compute the three multiplications using half-size operands. Further KOA layers imply more space and platform constraints. This decomposition was selected as an initial step to improve PIKO in time and space complexity and hence, obtain improvement over the

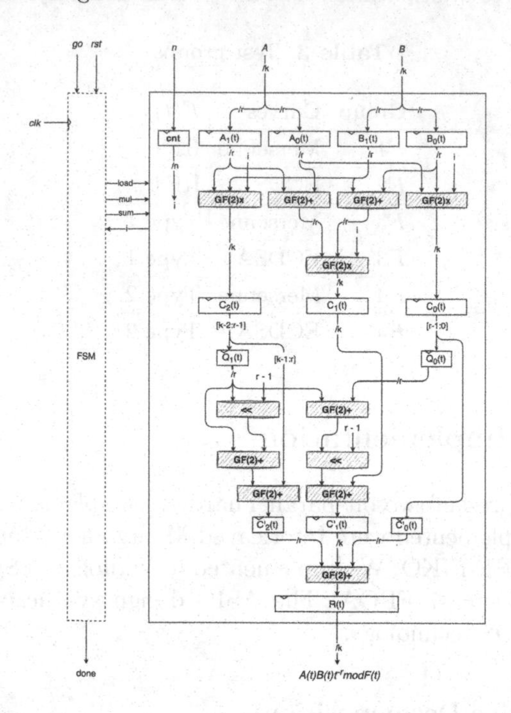

Fig. 5. RF-FIKO microarchitecture

bipartite reduction technique by completely removing it. Moreover, implementations that apply non-recursive KOA (only the upper layer), usually use 1/4 less space compared to the fastest implementations [8–10,15,22,35–37]. Further recursion and incorporation with other techniques is reserved for future work.

8.2 Register-Transfer Level Design

Close observation of Fig. 5 shows that the KOA terms are computed in a bit-parallel fashion. The data path is controlled by a four state Moore machine. In state zero all data registers are loaded with the value of zero and the down counter, *cnt*, is loaded with $r - 1$. In state two, the multiplication control signal controls the bit-parallel multiplication of $C_0(t)$, $C_1(t)$, and $C_2(t)$ to compute them one bit per clock cycle. Sketched rectangles highlight the three different arithmetic operations; the symbols $GF(2)+$, $GF(2)x$, and $\ll$ correspond to $GF(2)$ addition, $GF(2)$ multiplication, and shift left by $r - 1$. These operations are implemented as XORs, shift and add multiplication, and arithmetic left shift to multiply by a power of two. When *cnt* reaches zero, the multiplication is complete and the final *sum* and *done* signals are generated by the FSM. The remaining RTL details how the final sum is computed; all such registers are assigned inside a procedural block.

The r-bit quotients are assigned as $Q_1(t) = C_2(t)_{[k-2:r-1]}$ and $Q_0(t) = C_0(t)_{[r-1:0]}$. The reduced terms $C_0'(t)$ and $C_2'(t)$ are implemented as r-bits. The final sum only concerns the upper register of $C_0'(t)$ and hence, there is no need to compute the full $C_0'(t)$ term. Moreover $Q_0'(t)F_0(t)$ reduces to $Q_0'(t)$ for $F_0(t) = 1$ which does not affect the upper register of $C_0'(t)$. We let $C_0'(t) = C_0(t)_{[k-1:r]}$. Similarly for $C_2'(t)$, we implement $Q_1(t)F_1(t)$ as a left shift by $r-1$ added with Q_1, and take only the lower register of $C_2'(t) = C_2(t) \oplus (Q_1(t) \ll (r-1)) \oplus Q_1(t)$. For $C_1'(t)$ we require the full term, and implement the cross-term $X(t)$ as $(Q_1(t) \oplus Q_0(t)) \ll (r-1)$ since multiplication by $(F_1(t) \oplus F_0(t))$ reduces to a shift left by $r-1$. Finally, our sum of concatenated terms, is $\Sigma = C_0'(t)||C_2'(t) \oplus C_1'(t) \oplus C_2'(t)||C_0'(t)$ which corresponds to the final output in Fig. 5, $A(t)B(t)t^{-r} \pmod{F}(t)$.

9 Results

In this section we provide results for both FPGA and ASIC hardware implementations for sample ECDSA and Mersenne curves. In both implementations, Blakely and Montgomery form our baseline for comparison. Bipartite reduction is the fastest reduction technique prior to this work. Our main targets for comparison are BMM and FIKO for half-size words. A brief comparison with some of the fastest bit-parallel multipliers in the field is also provided along with estimates for FHE curves.

9.1 FPGA Results

Table 4 shows the FPGA results for sample curves in different groups–namely, the clock cycle (CC) count, LUT count, slices, frequency (MHz), and the latency in clock cycles × clock period (μs). For each sample field $GF(2^n)$, we list the defining polynomial $F(t)$ and the results for each algorithm. Blakely and Montgomery show some slight variance in all metrics. In comparison to the other three algorithms, these approximately double in the execution time but half the space used. FIKO outperforms BMM for $GF(2^{107})$ and $GF(2^{163})$ curves using several tens to hundreds more LUTs and slices. For the non-RFT curves, BMM and FIKO behave similar. As expected, RF-FIKO in turn outperforms FIKO which was our expected goal. For $GF(2^{107})$, RF-FIKO is 1.026 times faster than FIKO and 1.071 times faster than BMM using approximately half the LUTs. For $GF(2^{163})$ RF-FIKO computes the residue 1.064 times faster than FIKO and 1.076 times faster than BMM. This is more easily observed in Fig. 6 which shows the execution time for all algorithms for sample fields.

9.2 ASIC Results

Table 5 shows the results from our ASIC implementation. As expected, the results are significantly faster with the largest curve attaining a period of 0.41 ns for Montgomery. For 108-bit and 164-bit curves, FIKO outperforms BMM in

Table 4. FPGA sample results

Algorithms	*CC*	*LUT*	*Slices*	*f (MHz)*	*Latency (μs)*
$F1 : F(t) = t^{107} + t^{54} + 1$					
Blakely	107	179	64	299.4	0.357
Montgomery	107	177	61	277.8	0.385
BMM	54	344	114	327.9	0.165
FIKO	54	400	150	342.5	0.158
RF-FIKO	54	235	108	350.8	0.154
$F2 : F(t) = t^{163} + t^{82} + 1$					
Blakely	163	260	78	268.1	0.608
Montgomery	163	262	73	268.8	0.606
BMM	82	477	142	286.5	0.286
FIKO	82	768	259	289.9	0.283
RF-FIKO	82	512	212	308.6	0.266
$F_3 : F(t) = t^{521} + t^{260} + 1$					
Blakely	521	819	218	218.3	2.39
Montgomery	521	813	230	216.0	2.41
BMM	261	1492	435	219.3	1.19
FIKO	261	2,685	922	203.7	1.28
$F_4' : F(t) = t^{571} + t + 1$					
Blakely	571	893	238	215.1	2.66
Montgomery	571	895	264	207.9	2.75
BMM	286	1,632	491	215.1	1.33
FIKO	286	2,772	942	199.7	1.432

time but not space. For 108-bit and 164-bit curves RF-FIKO outperforms BMM in time and space. For 108-bit curves, BMM computes a residue in 54 CC × 0.36 ns × 10^{-3} = 0.01944 μs requiring 5,512 gates. RF-FIKO computes the residue in 54 CC × 0.32 ns × 10^{-3} = 0.01728 μs with 4,876 gates. RF-FIKO differs by 2.16 ns and is more optimal in space with 636 fewer gates. In terms of execution time, RF-FIKO is 19.44 ns / 17.28 ns = 1.125 times faster than BMM when working in $GF(2^{107})$. For 164-bit curves, BMM computes a residue in 0.03116 μs at the cost of 6551 gates and RF-FIKO in 0.02788 μs with 7059 gates. In this case RF-FIKO requires 508 more gates but is 31.16 ns / 27.88 ns = 1.118 times faster than BMM when working in $GF(2^{163})$. Figure 7 shows the latency for 108 and 164-bit curves in clock cycle count by clock cycle time in μs for all algorithms. For other curves, FIKO behaves similar to BMM.

Table 5. ASIC sample results

Algorithm	*CC*	*Time(ns)*	*Gate#*	*Area * time*
$F1 : F(t) = t^{107} + t^{54} + 1$				
Blakely	107	0.33	1,941	640.53
Montgomery	107	0.37	2,039	754.43
BMM	54	0.36	5,512	1,984.32
FIKO	54	0.34	7,812	2,656.08
RF-FIKO	54	0.32	4,876	1,560.32
$F2 : F(t) = t^{163} + t^{82} + 1$				
Blakely	163	0.34	3,115	1,059.1
Montgomery	163	0.38	3,835	1,457.3
BMM	82	0.38	6,551	2,489.38
FIKO	82	0.34	12,402	4,216.68
RF-FIKO	82	0.34	7,059	2,400.06
$F_3 : F(t) = t^{521} + t^{260} + 1$				
Blakely	521	0.4	9,240	3,696
Montgomery	521	0.41	11,785	4,831.85
BMM	261	0.39	20,690	8,069.1
FIKO	261	0.41	41,007	16,812.87
$F_4' : F(t) = t^{571} + t + 1$				
Blakely	571	0.4	11,031	4,412.4
Montgomery	571	0.41	12,399	5,083.59
BMM	286	0.39	22,370	8,724.3
FIKO	286	0.41	43,346	17,771.86

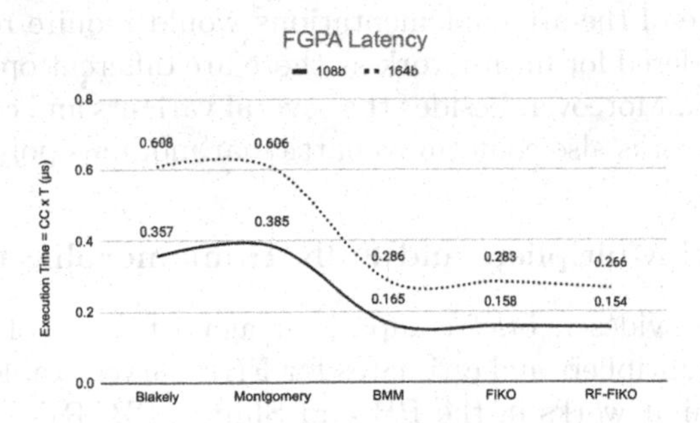

Fig. 6. FPGA execution time in CC × T (μs) for 108 and 164-bit curves

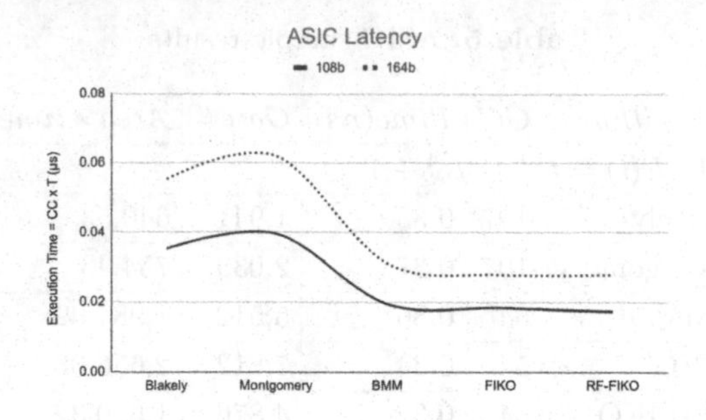

Fig. 7. ASIC execution time in CC × T (μs) for 108 and 164-bit curves

RF-FIKO removes the reduction part in modular multiplication simplifying the solution to the problem $A(t) \times B(t) \pmod{F(t)}$ to three multiplications followed by some shifts and additions on the last clock cycle; the results presented showed improvement in time and space complexity for specific curves. Further optimization can be obtained through experimentation and incorporation with other techniques such as, varying parameters, Mastrovito matrix for KOA terms, higher radices (e.g. 2^2, 2^3, etc.), refined and combined versions of KOA (e.g., with Toom-Cook), and pipelining.

Both FPGA and ASIC results show that for k-bit fields, Montgomery and Blakely can compute the residue in k CC, our main metric for time. BMM, FIKO, and RF-FIKO compute in $k/2$ CC. The focus of this work was to highlight the impact of our modulus polynomial. Hence, for comparison and additional consideration noted earlier, all algorithms were implemented in the same base and in a similar fashion. Results from state-of-the-art implementations, such as [16] and [17] for BMM, confirm similar results with respect to clock cycle count. For example in [17], for a radix-4 pipelined implementation, the residue is computed in $\frac{n}{2}+4$ CC for n-digit operands. A strict comparison in terms of time and space for state-of-the-art implementations would require reproducing such works and is considered for future work as there are different optimization levels that can be applied. Moreover, besides the several variants and considerations for comparison, such works also conform to particular modulus polynomials [21,23].

9.3 Bit-Parallel Multipliers and Fully Homomorphic Encryption

This subsection provides a brief comparison against some of the fastest bit-parallel modular multipliers and estimates for FHE curves. Table 6 lists the total gate count for similar works in the PB and Shifted PB. RF-FIKO space complexity consists of 7059 total gates for $GF(2^{163})$ which is six times less gates compared to [9,36,37] and four times less gates than [35]. The estimated time delay as a function of the signal propagation delay through total AND-gates (T_A) and XOR-gates (T_X) is expected to be significant.

Table 6. Bit-parallel multipliers in PB and SPB

$GF(2^{147})$		
Work	*Gate#*	*Time*
Sunar & Koç, Wu [36,37]	43,217	$T_A + 10T_X$
Elia et al. [8]	33,045	$T_A + 11T_X$
Fan et al. [9]	43,217	$T_A + 9T_X$
Negre [28]	49,000	$T_A + 8T_X$
Li et al. [35]	29,154	$T_A + 8T_X$

On the other hand, the degrees of defining polynomials for FHE are much larger than those of PQC, demanding more parallelism. Several functions with fully-homomorphic properties work with polynomial rings of the form $\mathbb{Z}_q[t]/(t^n + 1)$. These rings are applied to the FHE algorithms such as, CKKS and RLWE BVG [4,7]. For RLWE BGV, the moduli can take values in the intervals $q \in [2^{15}, 2^{500}]$ and $n \in [2^9, 2^{14}]$. CKKS works with polynomial rings of the form $\mathbb{Z}[t]/\phi(t)_m$ for which the defining polynomial $\phi(t)_m$ is the m-th cyclotomic polynomial for $m \in \mathbb{Z}^+$ and a power of 2. It also works with a ring $\mathbb{Z}_p[t]/\phi(x)_m$. Table 7 lists estimates for RF-FIKO for sample n for $F(t)$ that define rings for RLWE BVG, CKKS, and similar functions such as, FV and BFV [5,11]. The estimates for total gate count are based on the results for RF-FIKO for $GF(2^{163})$, as $\lceil n/163 \rceil \times 7059$. The CC is $n/2$ as noted previously. We can easily see that for $GF(2^{512})$, BVG would require 22,174 gates which is approximately half of the gate count for three of the fastest bit-parallel multipliers listed in Table 6 for a much smaller field.

10 Applications

Ring and finite field multiplication forms the fundamental operation in cryptographic schemes. Public-key cryptography, symmetric key cryptography, FHE,

Table 7. Estimates for FHE curves [4,5,7,11]

n	*CC*	*Gate#*
RLWE BVG		
2^9	256	22,174
2^{14}	8,192	709,538
CKKS and similar		
2^{15}	16,384	1,419,076
2^{16}	32,768	2,838,152
2^{17}	65,536	5,676,303

and primitives based on these, such as KEMs and HMACs, all employ modular arithmetic. These schemes are particularly interested is fast modular multiplication with polynomial rings and fields defined over different bases [18]. When working in higher radices, intermediate operations may be performed in $GF(2^n)$ for efficiency, such as eliminating carry propagations. In the modern regime, our work is applicable to schemes, such as RSA and ECC (e.g. ECDSA) [29]. In the quantum regime, NIST PQC Third Round Lattice-based finalists which share similar construction based on structured lattices, would benefit from our work. FHE uses higher degrees than PQC and requires a higher level of parallelism [4,26].

The arithmetic in Lattice-based schemes consists of matrix algebra and finite field and ring arithmetic. Cryptographic hashing (SHA-3 and XOFs), randomness generation, and ring multiplication are among the most expensive computations in these schemes [30]. The Number Theoretic Transform (NTT) is employed in all Lattice-based schemes for fast ring and field multiplication except in Saber and NTRU. Tables 8 and 9 list sizes for keys and moduli used in modern and PQC schemes. Table 10 lists rings used in different PQC schemes. Further details regarding the schemes and varying instances can be obtained by accessing the specification documents of each submission [31].

10.1 NTT-Unfriendly Rings

The polynomial arithmetic techniques applied to Lattice-based schemes can be grouped into three categories, namely, NTT-friendly, NTT-unfriendly, and combinations of Karatsuba and Toom-N. Karatsuba-Ofman multiplication is particularly suitable when the defining polynomial of the ring has degree above 16 and within 256. Variants of Karatsuba and Toom-N are more efficient when the degree is above 256. These variants are particularly suited for NTT-unfriendly rings where the moduli are a power of two, $\mathbb{Z}_{2^m}[t]$ (e.g. Saber and NTRU).

NTTs can be adapted for NTT-unfriendly rings to obtain significant speed-up through new implementations of the schemes and techniques (e.g., layering) [18]. However, improving NTTs by reducing the number of modular reductions is a sought venue for improvement [27]. Moreover, NTTs may not be applicable in all use cases (e.g. compression in Saber). Depending on the implementation, platform, and techniques applied, a speed-up may not be possible [31]. For example, in AVX2, a software speedup was not possible for $n = 509$ with NTT of length-1024 due to selected strategy and vector layout [18]. A fast hardware implementation uses schoolbook multiplication and highlights the difficulties of implementing recursive structures in hardware, such as Toom-N [32].

10.2 Number Theoretic Transforms

Kyber, Dilithium, and Falcon use NTT-friendly rings. NTRU-HRSS is flexible and the latest specification allows for variants that use a prime q allowing for security and size trade-offs not present when q is a power of two [31]. Falcon, based on NTRU lattices, uses a prime modulus $q = 12289$ of special form

$q = ((k \times 2n) + 1)$ that makes it suitable for NTTs [31]. Hardware acceleration of primitives through platform-specific ISA extensions and cryptographic processors such as, single-cycle multiplication and vectorized NTTs on target platforms, such as ARM Cortex-M4 and Intel AVX2 would improve all Lattice-based schemes.

Table 8. Sample modern schemes [29]

Modern schemes	Keys/Moduli (bits)
AES	128, 192, 256-bit keys
RSA	2048, 4096, 7680, 15,360-bit moduli
DH	2048-bit modulus
ECC	160–233, 224–255, 256–383, 384–511, 512+ moduli

The reduction-free property of RFTs is an extension from binomials. For simplicity, if we consider a small field of the form $\mathbb{Z}_2[t]/(x^8 + 1)$, we can easily see that with this binomial, elements can be split into even half-size words of d-bits or 8/2 bits. $F(t)$ can be split evenly by allowing the upper register to hold the d most significant bits and the lower register can be truncated to d-bits since it is 1 and reducing with $\pmod{00001}$ is equivalent to reducing with $\pmod{0001}$ or just 1. We can easily observe that the Blakely quotient will just be the $d-1$ most significant bits of C_2 and Montgomery is just the least significant d-bits of C_0. Modular multiplication in finite fields and rings can apply the reduction-free property when the defining polynomial is a binomial or trinomial that allows it.

When the coefficients of the polynomials are elements in a ring of field, such as $\mathbb{Z}_q$, such as a ring of the form $\mathbb{Z}_q[t]/(x^n + 1)$, the reduction-free property can be explored with respect to modulo q. For example, for $q = 8192_{10} = 10000000000000_2$ and intermediate multiplications and additions of elements in $\{0, 1, .., q-1\}$ may be done in the binary base and we may exploit reduction-free properties and split q into $q_1 = 1000000$ and $q_0 = 0000000$.

Table 9. Sample PQC keys [31]

PQC schemes	Key size (Bytes)	Security level
Kyber768	$sk = 2400(32^\dagger), pk = 1182$	3
FireSaber-KEM	$sk = 3040(1760^\ddagger), pk = 1312$	5
NTRUhrss701	$sk = 1452, pk = 1138$	$1, 3^*$
Dilithium5	$sk = 2592, pk = 4595$	5
Falcon-1024	$pk = 1793, \sigma = 1280$	5

† indicates option for only 32 bytes of randomness with trade-offs.
‡ indicates option to use compression to reduce the key size to 384 bytes. $*$ 1 for non-local models, 3 for local.

Table 10. Sample PQC fields and ring parameters [31]

Scheme	$\mathbb{Z}_q[t]/F(t)$
Saber	$\mathbb{Z}_{3329}[t]/(t^{256}+1)$
Kyber	$\mathbb{Z}_{3329}[t]/(t^{256}+1)$
NTRUhrss701	$\mathbb{Z}_{8192}[t]/(t^{701}+1)$
Dilithium	$\mathbb{Z}_{8380417}[t]/(t^{256}+1)$
Falcon-1024	$\mathbb{Z}_{12289}[t]/(t^{1024}+1)$

Improvements of Karatsuba and Toom-N that explore reduction-free $(\text{mod } q, F(t))$ is applicable for cases where NTT cannot be applied. If NTT can be adapted, there is no reduction $(\text{mod } F(t))$ and we are only concerned with reduction $(\text{mod } q)$ in which case, we can obtain reduction-free NTT if q can be expressed accordingly. The polynomial arithmetic and techniques applicable depend on the implementation type, software or hardware. In Saber, the reference software implementation notes that 50–70% of the time is spent on generating pseudorandomness [31]. Recent work also shows that optimization of polynomial multiplication in Lattice-based schemes, controls computation time to a large-scale [18]. Our work is specifically applicable to hardware implementations which are optimized through principled design. RF-FIKO can be designed in different bases, radices, incorporated with other techniques (pipelining, parallelism, refined KOA) and algorithms (e.g. Toom-N), and transformed into other domains (e.g. NTT) to obtain a significant speed up in modern, PQC, and FHE schemes.

Highly optimized software and specialized hardware implementations have paramount applications on the Internet and computing systems in general. These include embedded firmware (e.g. TPMs), cryptographic libraries (e.g. OpenSSL) to secure the cloud and VPNs through transport layer security (TLS) and IPSec implemented in the OS code on hosts and gateway routers, and devices in general such as, cryptographic cores and modules (e.g., secure enclaves on SoCs). Moreover, blockchain technology which is highly dependent on PKC, namely digital signatures, is faced with protecting against quantum attacks. To remain secure and practical in the quantum regime, blockchains must implement PQC schemes efficiently. General adaption of PQC must also be applied in a timely manner [26]. Being able to operate on encrypted data efficiently is also highly desired for FHE applications such as, zk-SNARKs.

11 Conclusions

Efficient implementations of modern PKC and lattice-based schemes are sought in both software and hardware. In this paper, we introduced two new algorithms (FIKO and RF-FIKO) which are based on fully interleaved bit-parallel Karatsuba-Ofman multipliers without the bipartite reduction circuits. Their

FPGA and ASIC implementations were faster than FIKO and BMM and showed promising results for PQC and FHE implementations. Moreover, further analysis of a complete system that applies cryptographic primitives must account for software and hardware attacks, such as side-channels. In this case, because RF-FIKO is reduction-free, it eliminates timing leakage via modular reductions without the need to recourse to alternative algorithms. This conforms to constant-time implementation requirements.

Further optimization of RF-FIKO through incorporation with other techniques or transforming NTT into reduction-free NTT, merits further research and consideration.

Acknowledgements. As the first author, I would like to thank all the beautiful people in my life.

References

1. Babbage, C.: Passages from the Life of a Philosopher. Longman, London (1864)
2. Bajard, J.C.: Useful Arithmetic for Cryptography, July 2013
3. Blakley, G.R.: A computer algorithm for the product AB modulo M. IEEE Trans. Comput. **32**(5), 497–500 (1983)
4. Brakerski, Z., Garg, S., Tsabary, R.: FHE-based bootstrapping of designated-prover NIZK. IACR ePrint Archive, Paper 2020/1168 (2020)
5. Brakerski, Z., Vaikuntanathan, V.: Fully homomorphic encryption from ring-LWE and security for key dependent messages. In: Rogaway, P. (ed.) CRYPTO 2011. LNCS, vol. 6841, pp. 505–524. Springer, Heidelberg (2011). https://doi.org/10.1007/978-3-642-22792-9_29
6. Brent, R.P., Zimmermann, P.: The great trinomial hunt. Not. Am. Math. Soc. **58**(2), 233–239 (2011)
7. Cheon, J.H., Kim, A., Kim, M., Song, Y.: Homomorphic encryption for arithmetic of approximate numbers. In: Takagi, T., Peyrin, T. (eds.) ASIACRYPT 2017. LNCS, vol. 10624, pp. 409–437. Springer, Cham (2017). https://doi.org/10.1007/978-3-319-70694-8_15
8. Elia, M., Leone, M., Visentin, C.: Low complexity bit-parallel multipliers for $GF(2^m)$ with generator polynomial $x^m + x^k + 1$. Electron. Lett. **35**, 551–552 (1999)
9. Fan, H., Dai, Y.: Fast bit-parallel $GF(2^n)$ multiplier for all trinomials. IEEE Trans. Circuits Syst. I Regul. Pap. **54**(4), 485–490 (2005)
10. Fan, H., Hasan, M.A.: Fast bit parallel-shifted polynomial basis multipliers in $GF(2^n)$. IEEE Trans. Circuits Syst. I Regul. Pap. **53**(12), 2606–2615 (2006)
11. Fan, J., Vercauteren, F.: Somewhat practical fully homomorphic encryption. IACR ePrint Archive (2012)
12. Fürer, M.: Faster Integer Multiplication, April 2007
13. Gallardo, L.H., Rahavandrainy, O.: On (unitary) perfect polynomials over $\mathbb{F}_2$ with only Mersenne primes as odd divisors (2019)
14. Halbutoğulları, A., Koç, Ç.K.: Mastrovito multiplier for general irreducible polynomials. IEEE Trans. Comput. **49**(5), 503–518 (2000)
15. Hariri, A., Reyhani-Masoleh, A.: Bit-serial and bit-parallel montgomery multiplication and squaring over $GF(2^m)$. IEEE Trans. Comput. **58**(10), 1332–1345 (2009)

16. Kaihara, M.E., Takagi, N.: Bipartite modular multiplication. In: Rao, J.R., Sunar, B. (eds.) CHES 2005. LNCS, vol. 3659, pp. 201–210. Springer, Heidelberg (2005). https://doi.org/10.1007/11545262_15
17. Kaihara, M.E., Takagi, N.: Bipartite modular multiplication method. IEEE Trans. Comput. **57**(2), 157–164 (2008)
18. Kannwischer, M.J.: Polynomial multiplication for post-quantum cryptography. Ph.D. thesis, Radboud Universiteit Nijmegen (2022)
19. Karatsuba, A., Ofman, Y.: Multiplication of multidigit numbers by automata. Soviet Physics-Doklady **7**, 595–596 (1963)
20. Koç, Ç.K.: High-speed RSA implementation. Technical report. TR 201, RSA Laboratories, 73 pages, November 1994
21. Li, Y., Chen, Y.: New bit-parallel Montgomery multiplier for trinomials using squaring operation. IACR ePrint Archive, Paper 2014/268 (2014)
22. Li, Y., Ma, X., Zhang, Y., Qi, C.: Mastrovito form of non-recursive Karatsuba multiplier for all trinomials. IEEE Trans. Comput. **66**(9), 1573–1584 (2017)
23. Li, Y., Zhang, Y.: An Efficient CRT-based Bit-parallel Multiplier for Special Pentanomials. IACR ePrint Archive (2020)
24. Mastrovito, E.: VLSI architectures for computation in Galois fields, Ph.D. dissertation. Ph.D. thesis, Linkoping University (1991)
25. Montgomery, P.L.: Modular multiplication without trial division. Math. Comput. **44**(170), 519–521 (1985)
26. Moody, D.: The beginning of the end: the first NIST PQC standards. Technical report, NIST (2022)
27. Navas, J.A., Dutertre, B., Mason, I.A.: Verification of an optimized NTT algorithm. In: Christakis, M., Polikarpova, N., Duggirala, P.S., Schrammel, P. (eds.) NSV/VSTTE -2020. LNCS, vol. 12549, pp. 144–160. Springer, Cham (2020). https://doi.org/10.1007/978-3-030-63618-0_9
28. Negre, C.: Efficient parallel multiplier in shifted polynomial basis. J. Syst. Architect. **53**, 109–116 (2007)
29. NIST: Digital Signature Standard (DSS). Technical report. Federal Information Processing Standards Publications (FIPS PUBS) 186-4, U.S. Department of Commerce, July 2013
30. NIST: SHA-3 standard: permutation-based hash and extendable-output functions. Technical report. Federal Information Processing Standards Publications (FIPS PUBS) 202, U.S. Department of Commerce, August 2015
31. NIST: Post-Quantum Cryptography, Round 3 Submissions (2020)
32. Roy, S.S., Basso, A.: High-speed instruction-set coprocessor for lattice-based key encapsulation mechanism: Saber in hardware. IACR Trans. Cryptogr. Hardw. Embed. Syst. **2020**(4), 443–466 (2020)
33. Saldamli, G., Baek, Y., Ertaul, L.: Partially interleaved modular Karatsuba-Ofman multiplication. IJCSNS **15**(5), 503–518 (2015)
34. Schönhage, A., Strassen, V.: Schnelle multiplikation großer zahlen. Computing **7**(3), 281–292 (1971)
35. Sun, J., Li, Y., Zhang, Y., Guo, X.: Efficient nonrecursive bit-parallel Karatsuba multiplier for a special class of trinomials. VLSI Des. (2018)
36. Sunar, B., Koç, C.: Mastrovito multiplier for all trinomials. IEEE Trans. Comput. **48**(5), 522–527 (1999)
37. Wu, H.: Bit-parallel finite field multiplier and squarer using polynomial basis. IEEE Trans. Comput. **51**(7), 750–758 (2002)
38. Zierler, N.: Primitive trinomials whose degree is a Mersenne exponent. Inf. Control **15**(1), 67–69 (1969)

Finite Field Arithmetic in Large Characteristic for Classical and Post-quantum Cryptography

Sylvain Duquesne(✉)

Univ Rennes, CNRS, IRMAR - UMR 6625, 35000 Rennes, France
sylvain.duquesne@univ-rennes1.fr
https://perso.univ-rennes1.fr/sylvain.duquesne/

Abstract. Both classical and post-quantum cryptography massively use large characteristic finite fields or rings. Consequently, basic arithmetic on these fields or rings (integer or polynomial multiplication, modular reduction) may significantly impact cryptographic devices' efficiency and power consumption. In this paper, we will present the most used and the less common methods, clarify their advantages and drawbacks and explain which ones are the more relevant depending on the implementation context and the chosen cryptographic primitive. We also explain why recent proposals such as RNS, PMNS or Montgomery-friendly primes may be a good alternative to classical methods depending on the context and suggest directions for further research to improve them.

Keywords: Finite field · Arithmetic · Cryptography · Modular reduction · Multi precision · Polynomial rings

1 Introduction

Most of the public key cryptosystems use large finite fields or rings as well as their polynomial extensions. The consequence of this massive usage is that basic arithmetic on these fields and rings may impact the efficiency and power consumption of cryptographic devices. Nevertheless, the sizes and degrees involved are very diverse. For example

- Discrete logarithm on the multiplicative subgroup of a finite field, as well as RSA, uses 1024 to 4096-bits integers.
- Elliptic curve cryptography uses 256 to 512-bits prime fields.
- Pairing-based cryptography uses 256 to 1024-bits prime fields together with small degrees extension fields.
- Isogeny-based post-quantum cryptography uses 400 to 1000-bits prime fields and quadratic extensions.

This work was supported in part by French project ANR-11-LABX-0020-01 "Centre Henri Lebesgue".

S. Mesnager and Z. Zhou (Eds.): WAIFI 2022, LNCS 13638, pp. 79–106, 2023.
https://doi.org/10.1007/978-3-031-22944-2_5

- Lattice-based post-quantum cryptography uses 13 to 60-bits prime fields and large-degree polynomial rings.

Because of this large spectrum, there is a wide collection of algorithms for performing finite field or ring arithmetic which have each their range of interest depending on the context. Many arithmetic operations (squarings, inversions, additions, Frobenius map, square roots, ...) are involved in various protocols. In this paper, we will concentrate on multiplications and modular reduction because there are the most impacting operations in practice.

We will present the most used methods (schoolbook, interpolation based, Montgomery) as well as less common ones (RNS, PMNS) and some that are specific to particular modules (Mersenne and pseudo-Mersenne primes, Montgomery friendly primes). We will discuss their advantages and drawbacks and explain which ones are the most relevant depending on the implementation context and the chosen cryptographic primitive. We will detail how they can be combined and in which direction they should or could be improved in further works. For some of them, we give some trails for such improvements. We also explain to what extent recent techniques (PMNS, Montgomery-friendly primes) are competitive with more classical ones and spotlight their advantages depending on the context.

The paper first recalls multi-precision arithmetic and then concentrates on modular reduction methods and polynomial operations.

2 Multiprecision and Large Integer Multiplication

Hardware devices have the native capacity to perform arithmetic operations on bounded inputs, usually the processor's word size (denoted w in this paper). This word size is classically between 8 and 128 bits; most common architectures use 32 or 64-bit words nowadays. However, classical cryptography and isogeny-based cryptography use much larger integers. Multiprecision is the way a device deals with such large integers. It consists in writing large integers in base β (where $\beta = 2^w$) and then using specific algorithms to get the result of a specific operation involving large inputs.

For example, to add 2 n-words integers $a < \beta^n$ and $b < \beta^n$, we have to add them word by word

$$a + b = \sum_{i=0}^{n-1} a_i \beta^i + \sum_{i=0}^{n-1} b_i \beta^i = \sum_{i=0}^{n-1} (a_i + b_i) \beta^i \tag{1}$$

Unfortunately, this is not so easy because of carries. Indeed, $a_i + b_i$ may be greater than or equal to β. The right formula is then

$$a + b = \sum_{i=0}^{n-1} a_i \beta^i + \sum_{i=0}^{n-1} b_i \beta^i = \sum_{i=0}^{n} (a_i + b_i + q_{i-1} \bmod \beta) \beta^i$$

where $q_{i-1} = 0$ or 1 is the quotient of the Euclidean division of $a_{i-1}+b_{i-1}+q_{i-2}$ by β. Contrary to Eq. (1), $a+b$ is now written in base β because its coefficients are less than β.

The carry is just 0 or 1, so it is not a big deal to add it to the next coefficient, and the complexity of multi-precision addition remains linear in n. Nevertheless, its potential propagation makes difficult the parallelization of this naive algorithm. Moreover, parallelization is very important today because multi-core devices are very current. It is possible to partially get around this problem using carry-save variants of this algorithm. Their concept is to delay and accumulate carries to the end of the computation [67].

2.1 Schoolbook Multiplication

We can follow the same process for multiplying 2 large integers. Let us consider a and b, two n-word integers written in base β. Then ab can be computed by a succession of basic word multiplications and additions in the way pupils learn multiplication in base 10 at school (which gives the name to the method)

$$\sum_{i=0}^{n-1} a_i\beta^i \sum_{i=0}^{n-1} b_i\beta^i = \sum_{i=0}^{2n-2} c_i\beta^i \text{ with } c_i = \sum_{k+l=i} a_k b_l \tag{2}$$

The complexity is n^2 word multiplications but again ab is not written in base β in formula (2) because c_i will generally overcome β. In order to take carries into account, we use Algorithm 1.

As was the case for multiprecision addition, the complexity is not affected (n^2 basic word multiplications) but managing carries makes this algorithm difficult to parallelize again. Note that if a specific squaring algorithm is used, it will be more efficient than using Algorithm 1 because if $a = b$, a_ib_j and a_jb_i are the same and then are computed only once.

2.2 Karatsuba Multiplication

There is a well-known way to reduce the number of basic multiplications required for multiplying large integers. It is due to Karatsuba [52]. To multiply 2-word integers with the schoolbook method, we do

$$(a_0 + a_1\beta)(b_0 + b_1\beta) = a_0b_0 + (a_0b_1 + a_1b_0)\beta + a_1b_1\beta^2$$

Karatsuba's method consists in computing the middle term as

$$a_0b_1 + a_1b_0 = (a_0 + a_1)(b_0 + b_1) - a_0b_0 - a_1b_1$$

This saves one multiplication (at the cost of 3 extra additions) because a_0b_0 and a_1b_1 are already computed as the constant term and the term in β^2, respectively. This process can be recursively applied to deal with larger integers (the well-known divide-and-conquer strategy). We then get the schoolbook method's

Algorithm 1: Schoolbook multiplication of large integers

Input: 2 integers $0 \le a, b < \beta^n$ written in base $\beta = 2^w$:

$$a = \sum_{i=0}^{n-1} a_i\beta^i, b = \sum_{i=0}^{n-1} b_i\beta^i, \text{ with } 0 \le a_i, b_i < \beta$$

Result: $r = ab < \beta^{2n}$ written in base β: $r = \sum_{i=0}^{2n-1} r_i\beta^i$, with $0 \le r_i < \beta$

$r_i \leftarrow 0 \ \forall i \in [0..2n-1]$
for $i = 0$ **to** $n-1$ **do**
  carry $\leftarrow 0$
  for $j = 0$ **to** $n-1$ **do**
    $t \leftarrow r_{i+j} + a_j b_i +$ carry
    carry $\leftarrow t$ quo β
    $r_{i+j} \leftarrow t \bmod \beta$
  end for
  $r_{i+j+1} \leftarrow$ carry
end for
return $\{r_i\}_{i=0..2n-1}$

complexity of $n^{\log_2 3}$ basic word multiplications instead of n^2. The main drawback is that the number of addition will grow much faster than the number of multiplications will decrease.

Then, depending on the context and the relative cost of additions and multiplications, the schoolbook or the Karatsuba method should be preferred. More precisely, the Karatsuba method can be used for the first steps of the divide-and-conquer strategy for large integers, but it is not necessarily adequate for the last steps. Indeed, operands are smaller so that the relative cost of additions is higher than those of multiplications. Then, in practice, both methods are combined. The threshold for switching from Karatsuba to a schoolbook depends on the implementation context. More details on this shallow Karatsuba approach and the choice of the threshold can be found in [69, 75]. For example, for the current ECC key sizes, Karatsuba approach is not competitive for software implementation [44].

Many other ways exist to improve the complexity of large integer multiplications at the cost of extra additions or other small operations. We do not present them here because there are not used in cryptography (their threshold of use is too far regarding the sizes of integers used in cryptography), but also because we will give them in Sect. 5 in the polynomial context, which is more straightforward (in particular, no carries are involved).

3 Modular Reduction in the General Case

The specificity of the finite field arithmetic is the reduction step modulo the characteristic p of the field. However, note that the algorithms for reducing modulo p presented in this section do not require p to be prime and are indeed not always used in cryptography for primes (e.g. for RSA ciphering).

3.1 Schoolbook Reduction

The reduction of an integer modulo p consists in finding the remainder of the Euclidean division of this integer by p. Again, the easiest way is to use the algorithm learned at school. Assume we want to reduce a m-word integer a modulo a n-word integer p (with of course $m \geq n$) written in base β:

$$a = \sum_{i=0}^{m-1} a_i\beta^i \text{ and } p = \sum_{i=0}^{n-1} p_i\beta^i$$

We assume that we have at our disposal a basic division of a 2-word integer by a 1-word integer, provided that the quotient is a 1-word integer. The principle of the schoolbook division is essentially to divide the main significant word a_{m-1} of a by the one of p (or the 2 main ones if $a_{m-1} < p_{n-1}$). The quotient q gives a first approximation of the result, and a is updated as $a - qp$ (up to some powers of β), and the process is iterated until a becomes less than p. This process is precise in Algorithm 2.

Algorithm 2: Schoolbook reduction

Input: $a = \sum_{i=0}^{m-1} a_i\beta^i$ and $p = \sum_{i=0}^{n-1} p_i\beta^i$
Result: a reduced modulo p
  Compute q_0, r_0 s.t. $a_{m-1} = p_{n-1}q_0 + r_0$ with $r_0 < p_{n-1}$
  $a \leftarrow a - q_0 p\beta^{m-n}$
  for $i = 1$ **to** $m - n$ **do**
    Compute q_i, r_i s.t. $a_{m-i}\beta + a_{m-i-1} = p_{n-1}q_i + r_i$ with $r_i < p_{n-1}$
    $s \leftarrow a - q_i p\beta^{m-n-i}$
    while $s < 0$ **do**
      $q_i \leftarrow q_i - 1$ $s \leftarrow s + p\beta^{m-n-i}$
    end while
    $a \leftarrow s$
  end for
  return a

The loop of this algorithm is $m - n + 1$-long and each step is, up to additions, made of one basic division (computing q_i and r_i) and n basic multiplications (multiplying the 1-word integer q_i by the n-word integer p). The overall complexity is then $m-n+1$ basic divisions and $n(m-n+1)$ basic multiplications. If a is the result of a multiplication in $\mathbb{F}_p$, we have $m = 2n$, and the complexity is then n divisions and n^2 multiplications. This is not so bad compared to multiplication. However, basic divisions are usually very expensive, so we want to avoid them. It can be, for example, done thanks to the Newton method applied to the function $f(x) = \frac{1}{x} - b$, which computes the inverse of b without any division. However, the best way to do it is to use the so-called Montgomery or Barrett reductions.

3.2 Montgomery Reduction

During the '80s, Peter Montgomery [58] and Paul Barrett [18] introduce new methods to perform a Euclidean division without basic divisions. Their approach are slightly different but the principle is the same: write $\frac{a}{p}$ as $\frac{a}{\beta^n}\frac{\beta^n}{p}$. If the second term is precomputed (it does not depend on a), the division by p is then replaced by a division by β^n, which is trivial in base β. Of course, this approach works for real divisions, not directly for Euclidean ones. Then, they need to consider correcting terms to get the right result. Barrett and Montgomery's approaches have their advantages and drawbacks. The main drawback of Montgomery reduction is that it requires changing the representation of numbers, but it is mitigated if several operations are performed successively. This is classically the case in cryptographic primitives, so the Montgomery reduction is the most popular in this domain, and we will detail it in this subsection.

To simplify, we assume we want to reduce modulo p the result a of multiplication of 2 integers less than p (which is the most common use case). The basic version of the Montgomery reduction is given by Algorithm 3.

Algorithm 3: Montgomery reduction modulo p

Input: $a < p\beta^n$, $\beta^{n-1} \leq p < \beta^n$
  the precomputed value $p' = -p^{-1} \bmod \beta^n$
Result: $r < 2p$ such that $r = a\beta^{-n} \bmod p$
  $q \leftarrow ap' \bmod \beta^n$
  $r \leftarrow (a + qp)/\beta^n$

It is easy to prove that the output $r < 2p$ and $r = a\beta^{-n} \bmod p$. But these two properties are not exactly the expected ones for the reduction of $a \bmod p$. For the first one, we can of course subtract p if necessary to get $r < p$. As an alternative, the output of Algorithm 3 can be used directly as input for the next step by adding a condition on p, specifically $4p < \beta^n$.

The fact that $r = a\beta^{-n}$ instead of a modulo p can be overcome by using the so-called Montgomery representation of numbers defined by $\overline{x} = x\,\beta^n \bmod p$. This representation is, of course, stable for the addition ($\overline{x} + \overline{y} = \overline{x+y}$), but it is also stable for the modular multiplication using Algorithm 3. Indeed, when we multiply x and y in Montgomery representation, we get a β^{2n} factor. But β^n is removed during the Montgomery reduction process, so we finally get the Montgomery representation of xy and we then have $\overline{x}\,\overline{y} = \overline{xy} \bmod p$. Thanks to this stability, we can chain several operations in Montgomery representation. Changing representation may only be necessary at the beginning and end of the complete cryptographic computation. Moreover, this change of representation can be easily obtained thanks to Algorithm 3. Indeed, the Montgomery representation of $x < p$ is obtained with $x\,(\beta^{2n} \bmod p)$ as an input. Similarly, we can recover x using Algorithm 3 with $\overline{x}$ as an input. Hence, we usually ignore the conversion cost from Montgomery to classic representation (and reciprocally) in cryptographic applications.

Concerning the complexity, 2 multiplications are involved in the algorithm, but there are incomplete ones. The first one (ap') is reduced mod β^n, so it is unnecessary to compute its most significant part. For the second one, we know in advance that $a+qp$ will be exactly divisible by β^n (by definition of q), so its least significant part is zero and may not need to be computed. The other operations in Algorithm 3 have negligible cost in base β, so the overall cost should be the one of only one full multiplication, say n^2. Nevertheless, this analysis does not take carries into account. For example, the least significant part of the second operation is zero but may produce some carries, so it must be computed anyway. For concrete applications and precise complexity analysis, we use Algorithm 4: a word-by-word version which requires $n^2 + n$ basic word multiplications [27,29].

Algorithm 4: Word version of the Montgomery reduction modulo p

---

Input: $a < p\beta^n$, $\beta^{n-1} \le p < \beta^n$
    the precomputed value $p' = -p^{-1} \bmod \beta$
Result: $r = a\beta^{-n} \bmod p$ and $0 \le r < p$
  $r \leftarrow a$;
  for $i = 0$ **to** $n-1$ **do**
    $r_0 \leftarrow r \bmod \beta$
    $q \leftarrow p' r_0 \bmod \beta$
    $r \leftarrow (r + qp)/\beta$
  end for
  $r' \leftarrow r - p + \beta^n$
  if $r' \ge \beta^n$ **then** // else $r < p$
    $r \leftarrow r' \bmod \beta^n$ // $r \leftarrow r - p$ if $r \ge p$
  return r

There also exists a version of Algorithm 4 that is interleaved with the multiplication steps [27,56] to get a complete modular multiplication algorithm. It has some advantages (smaller intermediate results) and drawbacks (only compatible with schoolbook multiplication). We do not give it here, but details can be found in Chap. 5 of [42], for example.

The Montgomery reduction has been used mainly in cryptography for decades as long as the module does not have a specific form allowing fast reduction. The only constraint is that p is odd so that it can be inverted modulo β. It is, for example, regularly used in RSA implementations, Elliptic Curve Cryptography, pairing-based protocols or, more recently, isogeny key exchanges or lattice-based cryptography using primes modules. For example, the reference and optimized implementation of the NIST candidates Frodo and NTRUprime for post-quantum standardization [61] are using Montgomery reduction. However, we still have carries issues with this technique. To avoid them, one can use the well-known Chinese reminder theorem.

3.3 The Residue Number Systems (RNS)

The principle of Residue Number Systems (RNS) is to represent an integer a by its residues $(a_1, a_2, \ldots, a_n)$ modulo a set of coprime numbers $\mathcal{B} = (m_1, m_2, \ldots, m_n)$. This set is called a RNS basis and we have $a_i = a \bmod m_i$ which we will also denote $|a|_{m_i}$ for clarity. We generally assume that $0 \leq a < M = \prod_{i=1}^{n} m_i$. The main interest of such a system is that it independently distributes large integer operations on the small residue values. In particular, large integer operations become linear relatively to their sizes, and there is no carry propagation. These systems were introduced and developed in [43,71,72]. A good introduction can be found in [55].

However, they cannot be used directly in the cryptographic context because we need to reduce modulo a prime p that cannot be factorized in small moduli by definition. For constructing an arithmetic over $\mathbb{F}_p$, we assume that $p < M = \prod_{i=1}^{n} m_i$ and we use a variant of the Montgomery reduction algorithm presented in Subsect. 3.2 to replace reductions modulo p by reductions modulo M instead of β^n [4,5,53,64]. Montgomery reduction algorithm cannot be used without adaptation. Otherwise, we would have to divide by M on the RNS basis. This is not possible because $M = \prod_{i=1}^{n} m_i$ is not invertible modulo m_i. It is then necessary to introduce an auxiliary RNS basis to handle the inverse of M. The consequence is that arithmetic operations (e.g. the initial multiplication) must be performed on the two bases. This doubles the cost of arithmetic operations, but this is not an important issue since this cost is now linear, contrary to classical arithmetic. The biggest issue is that some changes of basis become necessary before and after dividing by M.

As a derivative of the Montgomery reduction algorithm, this algorithm has the same drawbacks that can be solved in the same way. The Montgomery representation of a in this case is $\overline{a} = aM \mod p$ and is stable for Montgomery

Algorithm 5: RNS reduction based on Montgomery approach

Input: 2 coprime RNS basis $\mathcal{B}$ and $\mathcal{B}'$ such that $p < M < M'$
p given in basis $\mathcal{B}'$ and $-p^{-1}$ precomputed in basis $\mathcal{B}$
$a < Mp$ represented in both RNS basis

Result: $r = aM^{-1} \pmod p$ represented in both RNS basis, with $r < 2p$

1. $q \leftarrow a\times(-p^{-1})$ in $\mathcal{B}$
2. $[q \text{ in } \mathcal{B}] \longrightarrow [q \text{ in } \mathcal{B}']$ `// First basis extension`
3. $r \leftarrow (a + q\times p)\times M^{-1}$ in $\mathcal{B}'$
4. $[r \text{ in } \mathcal{B}\,] \longleftarrow [r \text{ in } \mathcal{B}']$ `// Second basis extension`

addition and multiplication. And again, changing representation is rare and easy to perform with a or $a(M^2 \bmod p)$ as input of Algorithm 5 [9].

Instructions 1 and 3 of Algorithm 5 are component-by-component operations performed independently for each basis element, so they are very efficient (linear complexity). On the other hand, the basis extensions of instructions 2 and 4 have quadratic complexity. To convert a RNS representation to another RNS basis, we usually use a Lagrange interpolation: if $(a_1, a_2, \ldots, a_n)$ is the RNS representation of a in the basis $\mathcal{B} = \{m_1, \ldots, m_n\}$, then,

$$a = \sum_{i=1}^{n} \left|a_i M_i^{-1}\right|_{m_i} M_i - \alpha M \text{ where } M_i = \frac{M}{m_i}$$

Furthermore, the main challenge is to compute α efficiently. There are several approaches in the literature. In [53], it is shown that the first conversion can be only an approximation because q is then multiplied by p. However, the second conversion needs to be exact and can be done if $w < 2p < (1 - \frac{1}{\rho})M'$ for some $\rho \geq 2$ and $\frac{c_i}{2^w} < \frac{1}{n}(1 - \frac{1}{\rho})$ assuming $m_i = 2^w - c_i$. The same idea is used in [12], but the conditions are relaxed. In [73], a binary tree construction is used with a logarithmic depth with a modulo reduction at each level. Finally, it is shown in [7] that the overall complexity of Algorithm 5 can be optimized up to $\frac{7}{5}n^2 + \frac{8}{5}n$ multiplications at the moduli level.

For efficiency reasons, the size of the elements of the RNS basis is related to the word size (usually 1 or 2 words), and they must be chosen carefully so that reductions modulo the m_i are cheap. Moduli of special form as in Sect. 4 are usually chosen so that the reduction modulo m_i is, in this case, obtained with few shifts and additions. For example, Kawamura et al. [53] are using moduli $m_i = 2^w - c_i$ with $c_i < 2^{w/2}$. In [12], a double Montgomery reduction is suggested to avoid internal modular reduction constraints and pseudo-Mersenne use. J. van der Hoven suggests in [73] to use *s-gentle moduli*, for example for $s = 2$, $m_i = 2^{2w} - \epsilon_i^2$ with $0 \leq \epsilon_i < 2^{(w-1)/2}$.

RNS arithmetic has many advantages compared to standard arithmetic. In particular, multiplication becomes a linear operation instead of a quadratic one. However, also, no more carriers are involved, and conjointly with the independence of the components, it makes RNS arithmetic very easy to implement and parallelize, especially in hardware, as shown in [45]. We can also introduce redundancy in the representation by adding new moduli. This allows for the introduction of randomization in the representation of elements that can be used as protection against differential side-channel attacks. Moreover, an RNS-based architecture is very flexible: with a given structure of n modular digit operators, it is possible to handle any value of $p < M$. Hence, the same architecture can be used for different cryptographic primitives, different levels of security and several base fields for each of these levels. The main drawback is the cost of the basis extensions, which makes the RNS reduction significantly more expensive than the classical Montgomery reduction. However, remember that RNS multiplication is much cheaper than classical multiprecision multiplication. So, there is a gap between the reduction and the multiplication cost, which does not occur in classical systems. We can take advantage of this gap by accumulating multiplications before reducing it. This method is called lazy reduction and is detailed in Sect. 6.

RNS systems have been successfully used in various cryptographic primitives, especially in hardware implementation (because it is necessary to have an efficient reduction modulo the m_i, which is challenging to achieve in software).

The first target was RSA (see [9] for example). The results are fascinating and competitive. However, the integers are very large in RSA, so a large basis are required for high-security levels, and it cannot always be realized with classical pseudo-Mersenne primes as in [53], especially if the word size is small. At this point, it would be very interesting to find other families of moduli allowing fast reduction, to develop variants of the RNS method, which is less demanding in terms of basis size (for example, by considering 2-words moduli). Some works already went in such direction recently [6,12,73] but for sure many improvements remain to be find and would be interesting not only for RSA.

RNS has also been successfully used in elliptic curve cryptography [7,8,45] and even allows to break of speed records in pairing-based cryptography [31]. Again, work remains to be done to adapt and optimize the elliptic curve and pairing formulas to exploit the complexity gap between the RNS multiplication and reduction steps more efficiently. RNS will also probably give interesting results in isogeny-based cryptography because, as in pairing-based cryptography, small degree extension fields are used in this context so that, as explained in Sect. 6, the lazy reduction can be systematically used, and this is advantaging RNS arithmetic.

More recently, an RNS-based implementation has been proposed for some protocols of Fully Homomorphic Encryption [49]. However, it generally has few interests in lattice-based cryptography because no large integers are used in this case, so we do not get the gain of fast multiplication.

Finally, RNS arithmetic may be helpful for constrained devices and hardware and when we plan to use the base field level parallelization.

3.4 PMNS

Another way to avoid carriers is to work with polynomials. The so-called PMNS (Polynomial Modular Number System) represents integers modulo p by a polynomial that gives the integers once evaluated in a given value $\gamma < p$: a polynomial P represents $P(\gamma) \bmod p$ [10]. For efficiency reasons, P must satisfy some constraints:

- The degree of P is (strictly) bounded degree by a given integer n.
- The coefficients of P are bounded by a given parameter ρ:

$$||P||_\infty < \rho \simeq \sqrt[n]{p}.$$

Note that ρ cannot be less than $(\sqrt[n]{p}+1)/2$ otherwise all the elements of $\mathbb{F}_p$ cannot be represented [63].

For example, if we choose $\gamma = 7$, then $-X-1$ represents 11 modulo 19. Note that X^2 also represents 11 so this representation is a redundant system [39,63].

Thanks to this representation, large integer arithmetic is replaced by polynomial arithmetic, which has many advantages (no carries, parallelization, independence of the components). The main problem is the growing of degrees and coefficients when operations are performed on the representative polynomials. Then degrees and coefficients may outreach the given constraints for a representative to be valid. So reduction techniques are necessary.

The easiest operation is to reduce the degree. This is called the external reduction, and this is done by reducing modulo a given polynomial E. This polynomial must be chosen as a monomial of degree n so that the degree of the reduced polynomial is less than $n-1$. It must also cancel γ modulo p so that the reduced polynomial represents the same element in $\mathbb{F}_p$. For example, if this polynomial is well chosen and sparse, this reduction step is very efficient. It should also be chosen with small coefficients for efficiency reasons but, above all, to minimize the growth of the coefficients induced by this step [37].

This is, for example, the case if we consider a polynomial of the form $X^n - \lambda$, ideally with λ very small [10] or a power of 2. In this case, these systems are called AMNS (Adapted Modular Number System) [10]. In any case, the external reduction cost is usually only some extra additions [63].

Reducing the coefficients is more complicated and costly. This step is called internal reduction. The idea is the same. Namely, we use a polynomial M that cancels in γ. We then hope that dividing by M will give a polynomial with smaller coefficients representing the same integer modulo p. Various methods have been proposed to do it efficiently [10,11,18,60]. The most conclusive one is based on the Montgomery reduction [60]: Algorithm 6 requires an auxiliary integer ϕ, which is usually a power of 2 and uses Montgomery's usual trick to replace expensive divisions by M with cheap divisions by ϕ:

Algorithm 6: Internal reduction using Montgomery reduction

Input: $R \in \mathbb{Z}_{n-1}[X]$ with $||R||_\infty \geq \rho$
the precomputed polynomial $M' = -M^{-1} \bmod (E, \phi)$
Result: $S \in \mathbb{Z}_{n-1}[X]$ s.t. $S(\gamma) = R(\gamma)\phi^{-1} \bmod p$
$Q \leftarrow R \times M' \bmod(E, \phi)$
$S \leftarrow (R + Q \times M)/\phi \bmod E$

Assuming that a Montgomery representation of elements is used ($\overline{R} = R\phi$), this algorithm outputs a polynomial S representing the same element as the input and whose coefficients are reduced if M and ϕ are satisfying the following conditions (for AMNS):

$$\rho > 2n|\lambda|\, ||M||_\infty \text{ and } \phi > 2n|\lambda|\rho$$

Contrary to the external reduction (which has linear complexity), the internal reduction is quadratic, as an instance of the Montgomery technique. As for the RNS reduction, we will not get any complexity gain compared to the classical Montgomery reduction of Sect. 3.2, but we get simpler arithmetic that can easily be parallelized.

Moreover, this representation is naturally redundant so that we can protect the arithmetic against differential side-channel attacks. For example, we can add a random polynomial that cancels in γ modulo p [37]. It is also shown in [38] that this representation allows base field arithmetic without conditional branching, which can be very interesting in the grey box context.

Until recently, the main drawback of this method was the parameter generation. We apply LLL to the lattice of polynomials cancelled by γ to get M. But it is not so trivial to ensure that the polynomial M' exists as an integer polynomial and that the parameter ϕ is a power of 2 to minimize the costs. A generator has been very recently given in [38].

In terms of cryptographic applications, there are still few implementations in the literature because of this parameter generation issue. However, it is very promising for ECC, pairing or isogeny context [22], especially for constrained devices and when parallelization is considered at the base field level. It may also be a good alternative if side-channel resistance is needed at the base field level, thanks to the redundancy of the representation.

4 Modular Reduction for Special Primes

The underlying base field can be chosen without restrictions in some cryptographic contexts. In this case, the module p can be taken in a specific form that allows much better complexity for reduction thanks to a dedicated algorithm. Usually, only several shifts and additions are needed for the modular reduction instead of $O(n^2)$ multiplications as in Sect. 3. The most known form is one of the generalized Mersenne primes we will describe now, but we will see that there are not the only ones. The main drawback of this strategy is that it requires an implementation/architecture dedicated to the specificity of p, which cannot be used for other base fields. Consequently, it is not always practical in either software or hardware implementation, and many users may prefer flexible products.

4.1 Mersenne and Pseudo-Mersenne Primes

The more efficient modules for reduction are the so-called Mersenne primes. They have the form $p = 2^e - 1$ and are historically used to break large prime numbers records. In this case, the reduction Algorithm 7 splits the input after e bits and uses that $2^e = 1 \bmod p$. Then adding the low and the high parts gives a reduced form of the input. It is not completely reduced since it is less than $2p$ instead of p, so an extra subtraction may be necessary to get the right result. Note that this subtraction is not directly done with p but with $2^e - 1$, which is mathematically equivalent but practically easier.

Algorithm 7: Reduction modulo a Mersenne prime $p = 2^e - 1$

Input: $0 \leq a < 2^e p$
Result: $r = a \bmod p$ and $0 \leq r < p$
  Write $a = a_1 2^e + a_0$
  $r \leftarrow a_0 + a_1$
  $r' \leftarrow r + 1$ // $r' \leftarrow r - p + 2^e$
  if $r' \geq 2^e$ **then** // if $r \geq p$
    $r \leftarrow r' \bmod 2^e$ // $r \leftarrow r - p$
return r

The complexity of this algorithm is only one (large) addition. This is of course much cheaper than all other reduction algorithms, especially if e is a multiple of the word-size. However, Mersenne primes are very rare in the cryptographic range. Only 3 of them can be encounter in the cryptographic literature:

- $2^{521} - 1$ introduced in [66] and used as a standard for elliptic curve cryptography at the 256-bits security level.
- $2^{127} - 1$ used for hyperelliptic curves at the 128-bit security level [20,65].
- $2^{31} - 1$ used as part of a RNS basis on 32-bit architecture.

In order to get more candidates, Mersenne primes can be generalized to any prime having the form $2^e - c$ for small values of c. Replacing 2^e by c in the writing of a large integer a in base 2^e will indeed reduce the size of a.

$$a = a_0 + a_1 2^e \rightarrow a = a_0 + a_1 c \bmod p$$

However, $a_1 c$ is not reduced enough, so this reduction step needs to be applied again to this term. It is easy to prove that if $|c|$ is less than $2^{\frac{e}{2}}$, then a third reduction step is not necessary, and we can then use Algorithm 8 involving only

2 reduction steps. In this case, the result is less than $3p$ (as the sum of 3 integers less than p), so 2 steps of subtraction of p may be necessary.

Algorithm 8: Reduction modulo $p = 2^e - c$, pseudo-Mersenne s.t. $|c| < 2^{\frac{e}{2}}$

```
Input: 0 ≤ a < 2^e p
Result: r = a mod p and 0 ≤ r < p
  Write a = a_0 + a_1 2^e
  b ← a_0 + a_1 c
  Write b = b_0 + b_1 2^e
  r = b_0 + b_1 c
  r' ← r + c;                                      // r' ← r − p + 2^e
  if r' ≥ 2^e then                                 // if r ≥ p
      r ← r' − 2^e                                 // r ← r − p
      r' ← r + c        // repeat the previous steps if r is still ≥ p
      if r' ≥ 2^e then
          r ← r' − 2^e
      end if
return r
```

The final cost of Algorithm 8 is then 2 multiplications by c (and some additions). Consequently, c must be chosen carefully so that multiplications by c are fast to compute. Of course, it can be chosen very small, as suggested by Crandall [35]. This is, for example, the case of the prime $2^{255} - 19$ introduced by Bernstein to define the famous elliptic curve Curve25519 [19]. However, it can also be chosen very sparse in the base of the machine word. This is, for example, the case of Solinas primes [32, 70], which are built as evaluation in the power of 2 (ideally in β) of sparse polynomials with very small coefficients. The polynomial should be irreducible [59]. For example, if we choose the polynomial $X^3 - X - 1$ and apply it to 2^{64}, we get a pseudo-Mersenne prime with $c = 2^{64} + 1$ so that multiplying by c just adds up to 64-bit architecture. This prime and other Solinas ones are suggested by the NIST [62] and included in many standards for elliptic curve cryptography.

Pseudo-Mersenne class is also the main provider of moduli (non necessarily primes) for RNS basis because of their efficient reduction algorithm and the fact that they can be freely chosen (as long as there are coprime). However, pseudo-Mersenne primes cannot be used in cryptographic contexts where modules have inherent constraints (e.g. RSA, pairings, Ring-lattices, isogenies).

4.2 Montgomery-Friendly Prime Numbers

In this subsection, we are interested in the so-called Montgomery-friendly primes [6, 25, 26, 50]. They are build in the form $p = 2^{e_2}\alpha \pm 1$ with e_2 larger than the word size w so that the Montgomery reduction Algorithm 4 is particularly efficient. Indeed, this algorithm involves multiplication by the inverse of p modulo β. In the case of Montgomery-friendly primes, p is ± 1 modulo β so that this multiplication is free. Then, the only operation to perform in the `for` loop is the computation of $(r + qp)/\beta$ where $q = r \bmod \beta$. As $p = \beta 2^{e_2-w}\alpha - 1$ and assuming that $r = \sum r_i \beta^i$ at the beginning of each step of the `for` loop, this operation can be computed as:

$$(r + qp)/\beta = \left(\sum r_i \beta^i + r_0 \beta 2^{e_2-w}\alpha - r_0\right)/\beta = \sum_{i \neq 0} r_i \beta^{i-1} + r_0 2^{e_2-w}\alpha$$

Algorithm 9: Word version of the Montgomery reduction if $p = 2^{e_2}\alpha - 1$

Input: $0 \leq a < p\beta^n$, $\beta^{n-1} \leq p < \beta^n$ with $\beta = 2^w$ and $w \leq e_2$
Result: $r = a\beta^{-n} \bmod p$ and $0 \leq r < p$

```
r ← a
for i = 0 to n − 1 do
    r0 ← r mod β
    r ← (r − r0)/β + r0 × α2^(e2−w)
end for
r' ← r + (β^n − p) ;
if r' ≥ β^n then
    r ← r' − β^n
return r
```

Algorithm 4 can then be rewritten as Algorithm 9 and the cost of each step of the for loop is reduced to one multiplication of $\alpha 2^{e_2 \bmod w}$ by the single word value r_0. Assuming $\alpha 2^{e_2 \bmod w}$ fits on n_α words, Algorithm 9 then requires nn_α word multiplications. Of course the case $p = 2^{e_2}\alpha + 1$ is very similar.

As detailed in [6], the choice of α and e_2 is crucial to get an efficient reduction algorithm.

- If α is a power of 2, the multiplication by α is very efficient, but we finally get a Mersenne prime.
- If $\alpha = 2^{e_2'} - c$ with c small, we get an equivalent of pseudo-Mersenne primes.
- If $\alpha = 2^{e_2'} - c$ with c sparse involving only coefficients of the word size, we get an equivalent of the Solinas primes.
- If e_2 (and e_2') is a multiple of the word size, this will reduce the word size n_α to the one of α.

If both α and e_2 are wisely chosen as above, the Montgomery reduction Algorithm 9 is very efficient by construction (each step of the for loop costs only one multiplication of a single word by c). Note that it will not give better efficiency than classical pseudo-Mersenne primes. It only provides more choices of good primes. This can be useful in many cryptographic contexts.

Such primes have been used in elliptic curve cryptography for years, such as

- $p_{192} = 2^{64}(2^{128} - 1) - 1$, which was already a Solinas prime used in most standards [62],

- $p_{448} = 2^{224}(2^{224} - 1) - 1$, used for the curve Ed448 [51],
- $p_{480} = 2^{448}(2^{32} - 1) - 1$ or $p_{512} = 2^{32}(2^{480} - 1) - 1$ for higher security levels.

Other Montgomery-friendly primes, which do not appear as pseudo-Mersenne primes, together with elliptic curves defined over the corresponding prime field, can be found in the recent cryptographic literature [26,28,50]. For example the primes $2^{240}(2^{16} - 88) - 1$ and $2^{240}(2^{14} - 127) - 1$ are recommended for the 128-bits security level. However, the parameters e_2 and e_2' were not multiples of the word size in these proposals. The most accomplished example is given in [6]: $p_{256} = 2^{192}(2^{64} - 4) - 1$ can be seen as a good alternative to the Bernstein prime $2^{255} - 19$ because reduction modulo p_{256} requires only a few additions while the Bernstein prime requires multiplication by 19. Moreover, it is possible to find an elliptic curve satisfying the security requirements given in [21] with a smaller curve coefficient than Curve25519 [6].

The most natural application is isogeny-based cryptography because primes involved have the form $2^{e_2}3^{e_3} - 1$ and are naturally Montgomery-friendly. It was even the initial motivation to study these primes [6]. For example, for the famous SIKE prime $p_{503} = 2^{250}3^{159} - 1$ given in the SIKE proposals [2,3] for the security category 2, we have $n = 8$ if $w = 64$ and $\alpha = 3^{159}2^{58}$ so that $n_\alpha = 5$. The overall cost of the reduction algorithm is then 40 word-multiplications (which fits with the reference implementation of SIKE). In order to follow the requirements on α and e_2 given above, it is not restrictive to consider primes p with a small cofactor f, as long as $p + 1$ is divisible by a sufficiently large power of 2 and 3 to ensure the existence of the isogenies [6]. This gives for example the prime $p_{512} = 31 \cdot 2^{256}\, 3^{158} - 1$ which ensures the same security than p_{503} but with $n_\alpha = 4$ so that the overall cost of the reduction algorithm is only 32 word-multiplications instead of 40. More such primes for various security levels are given in [6].

The main interest of these new primes is that we get new prime numbers with an efficient reduction algorithm that can be used when pseudo-Mersenne or Solinas primes are too rare. The situation is, in fact, even better because, contrary to pseudo-Mersenne primes, the value of c can be chosen even. Consequently, for the same size of constants c_i it is possible to find roughly twice many candidates. This can, for example, be very interesting for generating RNS bases handling larger values of p or involving smaller c_i values [6] (and then more efficient reduction algorithms). Consequently, RNS-based arithmetic can be considered for larger fields or rings.

Montgomery-friendly prime numbers form a large family and offer new (prime) numbers that can be used in many cryptographic contexts, such as elliptic curves, isogeny-based cryptography, or RNS arithmetic for large numbers.

Note that the prime used for the KYBER lattice-based primitive winner of the NIST post-quantum standardization process [1], $q = 13.2^8 - 1$, is Montgomery-friendly. However, the efficient reduction process could not be used at first sight (but on 8-bits devices which is quite rare today). For sure, some additional work has to be done to take advantage of the form of this prime in future implementations of KYBER. One could also use Montgomery-friendly prime numbers as RSA primes without loss of security while α is random and sufficiently large. Using these primes is recent in cryptography and requires further research.

5 Polynomial Rings and Extension Fields

Polynomials are used in various contexts in public key cryptography, mainly to build extension fields and their interest. For extension fields or polynomial rings, a reduction polynomial is necessary. In any case, it can be chosen sparse so that reducing is not really an efficiency problem, just like Mersenne primes. So we will concentrate in this section on multiplication or squaring algorithms or sometimes on both multiplication and reduction as one common operation. Polynomials in cryptography can be categorized into two main families.

- The first one is for small degree extension fields. For example, isogeny-based post-quantum cryptography is usually defined on finite quadratic fields, so it is important to compute with degree 1 polynomials efficiently. Pairing-based cryptography also massively involves extension fields of degrees 6 to 48. It is not so small, but in practice, the degrees are smooth, and the extension fields are built as a succession of very small degrees extensions. So we only need to deal with small degrees, principally 2 and 3 [42]. However, higher extension degrees were recently used in pairing-based cryptography [34, 46, 47] to overcome the Kim-Barbulescu attack, which takes great advantage of smooth embedding degrees [16, 54].
- The second family is made of very large degrees of polynomials. There are, for example, used in post-quantum cryptography based on Ring-lattices where polynomial rings of large degree (up to several thousand) are common [1, 30]

As was the case for integers multiplication, the reference algorithm for polynomial multiplication is the schoolbook one, given by the same formula with n^2 complexity to multiply 2 polynomials of degree $n-1$

$$\sum_{i=0}^{n-1} a_i X^i . \sum_{i=0}^{n-1} b_i X^i = \sum_{i=0}^{2n-2} c_i X^i \text{ with } c_i = \sum_{k+l=i} a_k b_l$$

It is even simpler than the integer situation because no carry is involved in the polynomial case. As in the case of integers, we can also be smarter to perform such multiplications with various methods depending on the context.

5.1 Interpolation Techniques for Small Degree Extensions

All the algorithms for efficient polynomial multiplication are based on the same principle. The general idea is to evaluate the 2 input polynomials of degree $n-1$ in $2n-1$ values. We then get $2n-1$ evaluations of the product polynomial, which has a degree $2n-2$. Finally, the product polynomial can be recovered using an interpolation algorithm. As long as the evaluation values are cleverly chosen, in the sense that evaluation (and interpolation) complexity is low, the multiplication cost becomes linear because only $2n-1$ base field multiplications are required instead of n^2 for the schoolbook method. Depending on the choice of these evaluation points, we obtain different algorithms that may be used in different contexts.

The most know method is undoubtedly the Karastuba multiplication (already seen in Subsect. 2.2 in the integer context). It evaluates degree 1 polynomials in $0, 1$ and ∞. If $A(X) = a_0 + a_1X$ and $B(X) = b_0 + b_1X$, we have

$$A(0) = a_0, A(1) = a_0 + a_1, A(\infty) = a_1 \text{ and}$$
$$B(0) = b_0, B(1) = b_0 + b_1, B(\infty) = b_1$$

Then AB evaluated in $0, 1$ and ∞ can be computed with only 3 multiplications.

$$AB(0) = a_0b_0, AB(1) = (a_0 + a_1)(b_0 + b_1), AB(\infty) = a_1b_1$$

So that AB can be easily and completely recovered with only few additions by classical interpolation.

$$AB(X) = a_0b_0 + ((a_0 + a_1)(b_0 + b_1) - a_0b_0 - a_1b_1)\,X + a_1b_1X^2$$

This method is very suitable for degree 2 finite fields extensions but can be generalized to any higher degree thanks to a "divide and conquer" procedure. It leads to a complexity of $O(n^{\log_3 2}) = O(n^{1.58})$. However, some other choices of evaluation points may be more appropriate for specific situations.

The Toom-Cook method allows, for example, to multiply degree 2 polynomials with only 5 multiplications instead of 9 using the evaluation points $0, 1, -1, 2$ and ∞ [55]. It is, of course, particularly well adapted for degree 3 extensions but, using the "divide and conquer" procedure, it asymptotically leads to complexity in $O(n^{\log_5 3}) = O(n^{1.46})$. However, it involves more additions than Karatsuba during the evaluation and interpolation steps and even some divisions by small numbers like 2 or 3 that must be treated very carefully and may invalidate the interest of the method depending on the context.

In the case of finite field extensions of small degrees (or towers of small degrees), the reducing polynomials have to be chosen carefully so that the reduction step is as cheap as possible. This step is usually included in the multiplication algorithm for efficiency reasons. The best choice is generally to choose an irreducible polynomial in the form $X^n - \mu$ such that multiplications by μ are easy, following the model of Mersenne primes. For example, the Karatsuba multiplication of degree 1 polynomials modulo $X^2 - \mu$ can be rewritten as

$$AB(X) = a_0b_0 + a_1b_1\mu + ((a_0 + a_1)(b_0 + b_1) - a_0b_0 - a_1b_1)\,X \tag{3}$$

5.2 Some Special Squarings: Complex and Chung-Hasan

The same analysis should be done for the specific operation of squaring. Of course, the same algorithms as for multiplication apply with a better complexity because there is only one input. For example, squaring a degree 1 polynomial modulo $X^2 - \mu$ can be done in 2 squarings and one multiplication, thanks to the schoolbook method.

$$(a_0 + a_1X)^2 = a_0^2 + \mu a_1^2 + 2a_0a_1X$$

while the Karatsuba method replaces $2a_0a_1$ by $(a_0 + a_1)^2 - a_0^2 - a_1^2$ and then one multiplication by one squaring in the base field (which yields to a global complexity of 3 squarings). But, we can alternatively use the so-called complex method that squares with only 2 multiplications in the base field

$$(a_0 + a_1X)^2 = (a_0 + \mu a_1)(a_0 + a_1) - (\mu + 1)a_0a_1 + 2a_0a_1X$$

This method is of course particularly efficient if $\mu = -1$, which explains its name.

In the case of degree 2 polynomials (and then degree 3 extension fields), the Schoolbook squaring modulo $X^3 - \mu$ requires 3 squarings and 3 multiplications in the base field

$$(a_0 + a_1X + a_2X^2)^2 = a_0^2 + \mathbf{2a_1a_2}\mu + \left[\mathbf{2a_0a_1} + \mu a_2^2\right] X + \left[a_1^2 + \mathbf{2a_0a_2}\right] X^2 \quad (4)$$

As usual, the Karatsuba method computes the bolded terms of (4) with only one squaring instead of one multiplication in the base field (and so a global complexity of 6 squarings). In this situation, we can also reduce the complexity to 5 operations in the base field using the Chung-Hasan method [33]. There are several variants but for example, the term in X^2 in (4) can be computed as

$$a_1^2 + 2a_0a_2 = (a_0 + a_1 + a_2)^2 - (2a_0a_1 + 2a_1a_2 + a_0^2 + a_2^2).$$

5.3 Application to Pairing Based Cryptography

Small degree extensions are used for isogenies and pairings. Determining which method is the best in a small degree extension is not immediate and not universal because it depends on the relative cost of multiplications, squarings and additions in the base field as well as multiplications by μ (if the reduction polynomial has the form $X^n - \mu$). But the base field may vary a lot depending on the targeted cryptosystem and even for a given use case. The situation is quite simple in isogeny-based cryptography because it is only using quadratic extensions.

Nevertheless, it is less trivial in the pairing context. Indeed extension degrees can be up to 48 and generally have the form 2^i3^j. This means that the extensions are built as a succession of degree 2 and degree 3 intermediate extensions. The first step is to choose this sequence and the reduction polynomials so that the global complexity is the best. It is, for example, shown in [36] choosing that even for a degree 12 extension in the pairing context, all the choices for building $\mathbb{F}_{p^{12}}$ are not equivalent and may result in significant differences in the final pairing cost (because the extension field arithmetic is predominant mainly in any pairing computation). Then, for each level in the extension tower, we have to choose the best algorithm for multiplying and squaring. As already mentioned, this choice depends on the relative cost of basic operations at the previous level. However, these relative costs are not always the same. They depend on the reduction polynomial used but also on the level itself: it is not the same at the ground level (for building $\mathbb{F}_{p^2}$ over $\mathbb{F}_p$, for example) and at the top level (for building $\mathbb{F}_{p^{12}}$ over $\mathbb{F}_{p^6}$ for example). Moreover, all these choices depend on the base field because an efficient reduction polynomial ($\mu = -1$, for example) will not always be irreducible. So there is no uniform answer to get the best possible arithmetic for pairings. New choices must be made for each case, and the final impact on the extension field arithmetic is important. It is also shown that it may influence the choice of the curve parameters themselves [41].

Of course, other operations on polynomials as additions, inversions or images of the Frobenius map may be considered for a complete study of the algorithmic complexity of the arithmetic of extension fields. This is, for example, done in [41] in the pairing context, but we did not go at this level of details here because multiplications and squarings have by far the largest impact.

5.4 Interpolation Techniques for Large Degrees

The situation is quite different for large degree polynomials/extension degrees. The first main reason is that the base field is finite, so the number of evaluation points is naturally limited. We then may not be able to have sufficiently many points (or efficient points) to set up an evaluation/interpolation technique. One elegant solution is to use Chudnovsky-type algorithms. These algorithms use points on an algebraic curve defined on the base field as evaluation values instead of base field elements [13,14,23]. It is interesting because, as the genus grows, we can have more points on the curve than elements in the base field. Their practical use in cryptography is still limited, so we do not give details here, but some recent publications on the subject make a good survey [15].

The second main reason is that large degrees make the famous Fast Fourier Transform (FFT) method competitive, and its complexity becomes theoretically quasi-linear in the degree. This method uses the $2n$-th roots of unity as evaluation points (of course, assuming they are lying in the base field). The problem, in this case, is, of course, to be able to evaluate the polynomials in these points efficiently. This can be done thanks to a famous divide-and-conquer algorithm from Cooley and Turkey [24]. This algorithm becomes computationally attractive only if n is large so that the evaluation complexity in the polynomial multiplication process is balanced by the gain on the number of base field multiplications.

This method was neglected until recently in cryptography because it becomes interesting only for very large integers or polynomials which were not used in this domain. The situation is different now because very large polynomial rings are used in lattice-based cryptography, especially in KYBER [1], which recently won the NIST post-quantum standardization process [61]. The FFT computes the Discrete Fourier Transform (DFT, when the base field is $\mathbb{C}$) or the Number Theoretic Transform (NTT, when the base field is finite, so DFT and NTT are just different names for the same thing depending on the context). Its general principle can be found in many algorithmic books because it is very old, famous, and widely used in many engineering, signal processing, mathematical and physical domains. We give it here in the general case to fix notations and explain how it applies to large polynomial multiplications and cryptographic applications. It is as follows: assume a primitive n-th roots of unity ω lies in the base field (for example, $\omega = e^{\frac{2i\pi}{n}}$ in $\mathbb{C}$) and let $A = \sum_{i=0}^{n-1} a_i X^i$ be a polynomial represented by the sequence of its coefficients $[a_0, a_1, \cdots, a_{n-1}]$. The Discrete Fourier Transform of A is defined as the evaluation of A in all the n-th root of unity:

$$\hat{A} = DFT(A) = [\hat{a}_0, \hat{a}_1, \cdots, \hat{a}_{n-1}] \text{ with } \hat{a}_k = \sum_{i=0}^{n-1} a_i \omega^{ik} = A(\omega^k)$$

We can also define the inverse transform which is nothing but the classical interpolation to recover A:

$$iDFT(\hat{A}) = [a'_0, a'_1, \cdots, a'_{n-1}] \text{ with } a'_j = \frac{1}{n} \sum_{k=0}^{n-1} \hat{a}_k \omega^{-kj} \tag{5}$$

And it is easy to prove that $a'_j = a_j$ so that the names are well chosen and

$$A = iDFT(DFT(A))$$

The application to degree $n-1$ polynomials multiplication immediately uses $2n$-th roots of unity (to get $2n-1$ evaluation points). The first step is to compute the DFT of inputs which can be asymptotically done in $O(n \log n)$ thanks to the Cooley-Turkey algorithm [24]. We then say that we are in the Fourier domain where multiplications can be done component by component (and then in $O(n)$). We finally return to the initial domain computing the inverse DFT (again in $O(n \log n)$) of the result. It is important to note that we can stay in the Fourier domain for consecutive operations, which may save many operations in practice. The inverse transformation needs to be applied only to the final result.

In the particular case of a polynomial ring defined by the specific reduction polynomial $X^n + 1$, it is possible to get an improved version of the multiplication algorithm based on the NTT that mimics a transform of size n instead of $2n$. Let ψ be a primitive $2n$-th root of unity in the base field.

- $\psi^2 = \omega$ is a n-th root of unity
- $\psi^n = -1$

There is an obvious correspondence between this second relation and the combination of the 2 operations we have to perform: reduce mod $X^n + 1$ and evaluate in ψ. This correspondence leads to the following writting for the product of 2 degree $n-1$ polynomials A and B modulo $X^n + 1$

$$C = AB = \sum_{j=0}^{2n-2} c_j X^j \Rightarrow C = \sum_{j=0}^{n-1} (c_j - c_{n+j}) X^j \bmod X^n + 1$$

and in the NTT context, we have

$$\begin{aligned} c_j - c_{n+j} &= \frac{1}{2n}\sum_{k=0}^{2n-1} \hat{c}_k \left(\psi^{-kj} - \psi^{-k(n+j)}\right) \\ &= \frac{1}{2n}\sum_{k=0}^{2n-1} \hat{c}_k \psi^{-kj}\left(1-(-1)^k\right) \text{ because } \psi^n = -1 \\ &= \frac{1}{n}\sum_{k=0}^{n-1} \hat{c}_{2k+1}\psi^{-j}\omega^{-kj} \end{aligned}$$

This expression is very similar to the inverse DFT of size n given by formula (5). There is only an extra factor ψ^{-j}. As a consequence, we can define a variant of the DFT/NTT of A by

$$\hat{a}_k = \sum_{i=0}^{n-1} a_i \psi^i \omega^{ik}$$

as well as its inverse. This transform is nothing but $A(\psi\omega^k)$. It has size n instead of $2n$ and allows polynomial multiplication in quasi-linear complexity with the modular reduction mod X^n+1 included. That is the reason why this polynomial ring is widely used in Ring/Module lattices cryptography [1,30].

6 Lazy Reduction

Lazy reduction is a technique commonly used in finite fields implementations [57, 68,74] but rarely presented in the literature [7,40] Its principle is to delay the reduction step when we have to compute several products that will be summed. For example, if a, b, c and d lie in $\mathbb{F}_p$, computing and reducing a pattern of the form $ab + cd$ is done with 2 multiplications (ab and cd), one (large) addition and one final modular reduction instead of 2 multiplications, 2 reductions and 1 (normal) addition. Of course, this implies that the reduction algorithm can take larger integers as input (less than $2p^2$ instead of less than p^2 in the above example), but it is not cumbersome in practice. There are 2 main contexts where using lazy reduction is particularly interesting.

The first is when the reduction step is costly compared to the multiplication step so factoring a reduction step for several other operations is particularly worthwhile. This is for example the case if RNS (see Sect. 3.3) or PMNS (see Sect. 3.4) arithmetic is used. It has been shown that lazy reduction was very profitable in this case in the elliptic curve context [7,8] as well as in the pairing context [31,40]. This was for RNS arithmetic because PMNS was not yet developed, but it is no doubt that the high cost of an internal reduction in PMNS would greatly advantage the lazy reduction technique for any cryptographic primitive. This may be an interesting trail to follow for people interested in PMNS implementations.

The second favourable context for lazy reduction is when patterns of the form $ab + cd$ naturally occur in a cryptographic protocol or can be revealed. This is, for example the case in elliptic curve cryptography where addition and doubling formulas can be adapted to involve more such patterns [7,8].

This is also the case in pairing-based cryptography and, more generally, in extension fields. Indeed, when a multiplication in an extension field $\mathbb{F}_{p^k}$ is performed, it is necessary to make only one reduction modulo p for each component of the result instead of one for each $\mathbb{F}_p$ multiplication involved [40]. This means that only k reduction steps are required instead of k^2 for a classical multiplication in $\mathbb{F}_{p^k}$. For example, if $\mathbb{F}_{p^2}$ is defined by $\mathbb{F}_p[X]/(X^2+1)$ and $A = a_0 + a_1X$ and $B = b_0 + b_1X \in \mathbb{F}_{p^2}$, we have $AB = a_0b_0 - a_1b_1 + (a_0b_1 + a_1b_0)X$.

We can see that $ab+cd$ patterns naturally occur which makes lazy reduction particularly relevant. Note that, as elements in $\mathbb{F}_{p^2}$ have 2 independent components, it is not possible to have less than 2 reductions in $\mathbb{F}_p$ in this case. The case of 6 and 12 extension degrees is done in detail in [40], showing a significant implementation improvement. These degrees were chosen because of their particular interest in the pairing context at that time, and especially because of the dominance of BN curves [17]. However, other degrees may now be considered (in the pairing context or in other contexts), and the detailed impact of a lazy reduction in these cases should be studied in the future.

Such patterns also naturally appear in lattice-based cryptography because most of the computations in this domain are inner products that are, by definition, well adapted to the lazy reduction trick. By the way, the reference implementations of Kyber [1] use lazy reduction. This situation is, in fact, even more, favourable in lattice-based cryptography because the reduction module involved is usually much less than a machine word (for example, Kyber prime is only 12-bits long). Consequently, many multiplications and additions may be performed and accumulated before reducing without overflowing a single machine-word [48]. One could even go further in this case by storing 2 integers on the same machine word. It would for sure be interesting to study in detail how this could be done in practice and what would be the expected gain.

7 Conclusion

Finite field arithmetic significantly impacts cryptographic performances but is very diverse and dependent on the cryptosystem. The efficient algorithms are not always the same and must be adapted to the context. They indeed depend on the size of the base field or ring, on the targeted device and its properties (possibilities of parallelism, size of machine words, software/hardware) but also on what is expected for the implementation in terms of performance and security (side-channel resistance ability for example).

Of course, this paper is far from exhaustive about algorithms for finite fields or ring arithmetic, and there are many missing improvements and variants or hidden things in what is presented. However, it gives a good overview of what can be done for efficient finite field arithmetic for cryptographic applications, and it gives some ideas to improve existing methods, extend their range of interest or even discover new ones.

References

1. Avanzi, R., et al.: CRYSTALS-Kyber (version 3.02) - Submission to round 3 of the NIST post-quantum project. https://pq-crystals.org/kyber/data/kyber-specification-round3-20210804.pdf
2. Azarderakhsh, R., et al.: Supersingular Isogeny Key Encapsulation, Submission to the NIST's post-quantum cryptography standardization process (2017). https://csrc.nist.gov/CSRC/media/Projects/Post-Quantum-Cryptography/documents/round-1/submissions/SIKE.zip
3. Azarderakhsh, R., et al.: Supersingular Isogeny Key Encapsulation, Submission to the NIST's post-quantum cryptography standardization process (2019). https://csrc.nist.gov/CSRC/media/Projects/Post-Quantum-Cryptography/documents/round-2/submissions/SIKE.zip
4. Bajard, J.C., Didier, L.S., Kornerup, P.: A RNS Montgomery's modular multiplication. IEEE Trans. Comput. **47**, 7 (1998)
5. Bajard, J.C., Didier, L.S., Kornerup, P.: Modular multiplication and base extension in residue number systems. In: 15th IEEE Symposium on Computer Arithmetic, pp. 59–65. IEEE Computer Society Press (2001)
6. Bajard, J.C., Duquesne, S.: Montgomery-friendly primes and applications to cryptography. J. Cryptogr. Eng. **11**, 399–415 (2021)
7. Bajard, J.C., Duquesne, S., Ercegovac, M.: Combining leak-resistant arithmetic for elliptic curves defined over Fp and RNS representation. Publications Mathématiques de Besancon **1**, 67–87 (2013)
8. Bajard, J.C., Duquesne, S., Ercegovac, M., Meloni, N.: Residue systems efficiency for modular products summation: application to Elliptic Curves Cryptography. SPIE **6313**, 631304 (2006)
9. Bajard, J.C., Imbert, L.: A full RNS implementation of RSA. IEEE Trans. Comput. **53**(6), 769–774 (2004)
10. Bajard, J.-C., Imbert, L., Plantard, T.: Modular number systems: beyond the Mersenne family. In: Handschuh, H., Hasan, M.A. (eds.) SAC 2004. LNCS, vol. 3357, pp. 159–169. Springer, Heidelberg (2004). https://doi.org/10.1007/978-3-540-30564-4_11
11. Bajard, J.C., Imbert, L., Plantard, T.: Arithmetic operations in the polynomial modular number system. In: 17th IEEE Symposium on Computer Arithmetic (ARITH 2017), pp. 206–213 (2005)
12. Bajard, J.-C., Merkiche, N.: Double level Montgomery Cox-Rower architecture, new bounds. In: Joye, M., Moradi, A. (eds.) CARDIS 2014. LNCS, vol. 8968, pp. 139–153. Springer, Cham (2015). https://doi.org/10.1007/978-3-319-16763-3_9
13. Ballet, S.: Curves with many points and multiplication complexity in any extension of Fq. Finite Fields Appl. **5**, 364–377 (1999)
14. Ballet, S., Rolland, R.: Multiplication algorithm in a finite field and tensor rank of the multiplication. J. Algebra **272**(1), 173–185 (2004)
15. Ballet, S., Chaumine, J., Pieltant, J., Rambaud, M., Randriambololona, H., Rolland, R.: On the tensor rank of multiplication in finite extensions of finite fields and related issues in algebraic geometry. Uspekhi Mathematichskikh Nauk **76:1**(457), 31–94 (2021)
16. Barbulescu, R., Duquesne, S.: Updating key size estimations for pairings. J. Cryptol. **32**(4), 1298–1336 (2019)
17. Barreto, P.S.L.M., Naehrig, M.: Pairing-friendly elliptic curves of prime order. In: Preneel, B., Tavares, S. (eds.) SAC 2005. LNCS, vol. 3897, pp. 319–331. Springer, Heidelberg (2006). https://doi.org/10.1007/11693383_22

18. Barrett, P.: Implementing the Rivest Shamir and Adleman public key encryption algorithm on a standard digital signal processor. In: Odlyzko, A.M. (ed.) CRYPTO 1986. LNCS, vol. 263, pp. 311–323. Springer, Heidelberg (1987). https://doi.org/10.1007/3-540-47721-7_24
19. Bernstein, D.J.: Curve25519: new Diffie-Hellman speed records. In: Yung, M., Dodis, Y., Kiayias, A., Malkin, T. (eds.) PKC 2006. LNCS, vol. 3958, pp. 207–228. Springer, Heidelberg (2006). https://doi.org/10.1007/11745853_14
20. Bernstein, D.J., Chuengsatiansup, C., Lange, T., Schwabe, P.: Kummer strikes back: new DH speed records. In: Sarkar, P., Iwata, T. (eds.) ASIACRYPT 2014. LNCS, vol. 8873, pp. 317–337. Springer, Heidelberg (2014). https://doi.org/10.1007/978-3-662-45611-8_17
21. Bernstein, D.J., Lange, T.: SafeCurves: choosing safe curves for elliptic-curve cryptography. http://safecurves.cr.yp.to
22. Bouvier, C., Imbert, L.: An alternative approach for SIDH arithmetic. In: Garay, J.A. (ed.) PKC 2021. LNCS, vol. 12710, pp. 27–44. Springer, Cham (2021). https://doi.org/10.1007/978-3-030-75245-3_2
23. Chudnovsky, D., Chudnovsky, G.: Algebraic complexities and algebraic curves over finite fields. J. Complex. **4**, 285–316 (1988)
24. Cooley, J.W., Tukey, J.W.: An algorithm for the machine calculation of complex Fourier series. Math. Comput. **19**(90), 290–301 (1965)
25. Bos, J.W., Costello, C., Hisil, H., Lauter, K.: Fast cryptography in genus 2. In: Johansson, T., Nguyen, P.Q. (eds.) EUROCRYPT 2013. LNCS, vol. 7881, pp. 194–210. Springer, Heidelberg (2013). https://doi.org/10.1007/978-3-642-38348-9_12
26. Bos, J., Costello, C., Longa, P., Naehrig, M.: Selecting elliptic curves for cryptography: an efficiency and security analysis. J. Cryptogr. Eng. **6**(4), 259–286 (2016)
27. Bos, J., Lenstra, A.: Topics in Computational Number Theory Inspired by Peter L. Montgomery. Cambridge University Press, Cambridge (2017)
28. Bos, J. Montgomery, P.L.W.: Topics in computational number theory inspired by Peter L. Montgomery. In: Bos, J., Lenstra, A. (eds.) Cambridge University Press (2017)
29. Bosselaers, A., Govaerts, R., Vandewalle, J.: Comparison of three modular reduction functions. In: Stinson, D.R. (ed.) CRYPTO 1993. LNCS, vol. 773, pp. 175–186. Springer, Heidelberg (1994). https://doi.org/10.1007/3-540-48329-2_16
30. Chen, C., et al.: NTRU - Submission to round 3 of the NIST post-quantum project. https://ntru.org/
31. Cheung, R.C.C., Duquesne, S., Fan, J., Guillermin, N., Verbauwhede, I., Yao, G.X.: FPGA implementation of pairings using residue number system and lazy reduction. In: Preneel, B., Takagi, T. (eds.) CHES 2011. LNCS, vol. 6917, pp. 421–441. Springer, Heidelberg (2011). https://doi.org/10.1007/978-3-642-23951-9_28
32. Chung, J., Hasan, A.: More generalized Mersenne numbers. In: Matsui, M., Zuccherato, R.J. (eds.) SAC 2003. LNCS, vol. 3006, pp. 335–347. Springer, Heidelberg (2004). https://doi.org/10.1007/978-3-540-24654-1_24
33. Chung, J., Hasan, A.: Asymmetric squaring formulae. In: 18th Symposium on Computer Arithmetic, pp. 113–122. IEEE Conference Publications (2017)
34. Clarisse, R., Duquesne, S., Sanders, O.: Curves with fast computations in the first pairing group. In: Krenn, S., Shulman, H., Vaudenay, S. (eds.) CANS 2020. LNCS, vol. 12579, pp. 280–298. Springer, Cham (2020). https://doi.org/10.1007/978-3-030-65411-5_14

35. Crandall, R.: Method and apparatus for public key exchange in a cryptographic system, U.S. Patent #5159632 (1992)
36. Devegili, A.J., O'Eigeartaigh, C., Scott, M., Dahab, R.: Multiplication and squaring on pairing-friendly fields, IACR Cryptology ePrint Archive 471 (2006). http://eprint.iacr.org/2006/471
37. Didier, L.-S., Dosso, F.-Y., El Mrabet, N., Marrez, J., Véron, P.: Randomization of arithmetic over polynomial modular number system. In: 2019 IEEE 26th Symposium on Computer Arithmetic (ARITH), pp. 199–206 (2019)
38. Didier, L.-S., Dosso, F.-Y., Véron, P.: Efficient modular operations using the Adapted Modular Number System. J. Cryptogr. Eng. **10**(2), 111–133 (2020)
39. Dosso, F.-Y.: Computer arithmetic contribution to side channel attacks resistant implementations. PhD, University of the South, Toulon-Var, France (2020)
40. Duquesne, S.: RNS arithmetic in $\mathbb{F}_{p^k}$ and application to fast pairing computation. J. Math. Cryptol. **5**(1), 51–88 (2011)
41. Duquesne, S., El Mrabet, N., Haloui, S., Rondepierre, F.: Choosing and generating parameters for pairing implementation on BN curves. Appl. Algebra Eng. Commun. Comput. **29**, 113–147 (2018)
42. El Mrabet, N., Joye, M.: Guide to Pairing Based Cryptography, Chapman & Hall/CRC Cryptography and Network Security (2016)
43. Garner, H.L.: The residue number system. IRE Trans. Electron. Comput. EL **8:6**, 140–147 (1959)
44. GNU MP. http://gmplib.org
45. Guillermin, N.: A high speed coprocessor for elliptic curve scalar multiplications over $\mathbb{F}_p$. In: Mangard, S., Standaert, F.-X. (eds.) CHES 2010. LNCS, vol. 6225, pp. 48–64. Springer, Heidelberg (2010). https://doi.org/10.1007/978-3-642-15031-9_4
46. Guillevic, A.: Comparing the Pairing Efficiency over Composite-Order and Prime-Order Elliptic Curves. In: Jacobson, M., Locasto, M., Mohassel, P., Safavi-Naini, R. (eds.) ACNS 2013. LNCS, vol. 7954, pp. 357–372. Springer, Heidelberg (2013). https://doi.org/10.1007/978-3-642-38980-1_22
47. Guillevic, A.: A short-list of pairing-friendly curves resistant to special TNFS at the 128-bit security level. In: Kiayias, A., Kohlweiss, M., Wallden, P., Zikas, V. (eds.) PKC 2020. LNCS, vol. 12111, pp. 535–564. Springer, Cham (2020). https://doi.org/10.1007/978-3-030-45388-6_19
48. Güneysu, T., Oder, T., Pöppelmann, T., Schwabe, P.: Software speed records for lattice-based signatures. In: Gaborit, P. (ed.) PQCrypto 2013. LNCS, vol. 7932, pp. 67–82. Springer, Heidelberg (2013). https://doi.org/10.1007/978-3-642-38616-9_5
49. Halevi, S., Polyakov, Y., Shoup, V.: An improved RNS variant of the BFV homomorphic encryption scheme. In: Matsui, M. (ed.) CT-RSA 2019. LNCS, vol. 11405, pp. 83–105. Springer, Cham (2019). https://doi.org/10.1007/978-3-030-12612-4_5
50. Hamburg, M.: Fast and compact elliptic-curve cryptography. IACR Cryptology ePrint Archive 309, http://eprint.iacr.org/2012/309 (2012)
51. Hamburg, M.: Ed448-Goldilocks, a new elliptic curve. IACR Cryptology ePrint Archive 625, https://eprint.iacr.org/2015/625 (2015)
52. Karatsuba, A., Ofman, Y.: Multiplication of multidigit numbers on automata, Sov Phys Dokl **7** (1963)
53. Kawamura, S., Koike, M., Sano, F., Shimbo, A.: Cox-Rower architecture for fast parallel montgomery multiplication. In: Preneel, B. (ed.) EUROCRYPT 2000. LNCS, vol. 1807, pp. 523–538. Springer, Heidelberg (2000). https://doi.org/10.1007/3-540-45539-6_37

54. Kim, T., Barbulescu, R.: Extended tower number field sieve: a new complexity for the medium prime case. In: Robshaw, M., Katz, J. (eds.) CRYPTO 2016. LNCS, vol. 9814, pp. 543–571. Springer, Heidelberg (2016). https://doi.org/10.1007/978-3-662-53018-4_20
55. Knuth, D.: Seminumerical Algorithms. The Art of Computer Programming 2. Addison-Wesley, Reading (1981)
56. Koç, Ç.K., Tolga, A., Burton, S.: Analysing and comparing Montgomery multiplication algorithms. IEEE Micro **16**(3), 26–33 (1996)
57. Lim, C.H., Hwang, H.S.: Fast implementation of elliptic curve arithmetic in GF(p^n). In: Imai, H., Zheng, Y. (eds.) PKC 2000. LNCS, vol. 1751, pp. 405–421. Springer, Heidelberg (2000). https://doi.org/10.1007/978-3-540-46588-1_27
58. Montgomery, P.: Modular multiplication without trial division. Math. Comput. **44**(170), 519–521 (1985)
59. Murty, R.: Prime numbers and irreducible polynomials. Am. Math. Mon. **109**(5), 452–458 (2002)
60. Negre, C., Plantard, T.: Efficient modular arithmetic in adapted modular number system using Lagrange representation. In: Information Security and Privacy, ACISP 2008, pp. 463–477 (2008)
61. NIST Post-Quantum Cryptography Standardization Process. https://csrc.nist.gov/Projects/post-quantum-cryptography
62. NIST ECC Standards. https://csrc.nist.gov/publications/detail/fips/186/4/final
63. Plantard, T.: Arithmétique modulaire pour la cryptographie. Ph.D. thesis, Montpellier 2 University, France (2005)
64. Posch, K.C., Posch, R.: Modulo reduction in residue number systems. IEEE Trans. Parallel Distrib. Syst. **6**(5), 449–454 (1995)
65. Renes, J., Schwabe, P., Smith, B., Batina, L.: μKummer: efficient hyperelliptic signatures and key exchange on microcontrollers. In: Gierlichs, B., Poschmann, A.Y. (eds.) CHES 2016. LNCS, vol. 9813, pp. 301–320. Springer, Heidelberg (2016). https://doi.org/10.1007/978-3-662-53140-2_15
66. Robinson, R.M.: Mersenne and Fermat numbers. Proc. Amer. Math. Soc. **5**, 842–846 (1954)
67. Savaş, E., Koç, Ç.K.: Finite field arithmetic for cryptography. IEEE Circuits Syst. Mag. **10**(2), 40–56 (2010)
68. Devegili, A.J., Scott, M., Dahab, R.: Implementing cryptographic pairings over Barreto-Naehrig curves. In: Takagi, T., Okamoto, T., Okamoto, E., Okamoto, T. (eds.) Pairing 2007. LNCS, vol. 4575, pp. 197–207. Springer, Heidelberg (2007). https://doi.org/10.1007/978-3-540-73489-5_10
69. Scott, M.: Missing a trick: Karatsuba variations. Cryptogr. Commun. **10**, 5–15 (2018)
70. Solinas, J. A.: Generalized Mersenne Numbers. Technical report Center for Applied Cryptographic Research, University of Waterloo (1999)
71. Svoboda, A., Valach, M.: Operational Circuits. Stroje na Zpracovani Informaci, Sbornik III, Nakl. CSAV, Prague, pp. 247–295 (1955)
72. Szabo, N.S., Tanaka, R.I.: Residue Arithmetic and its Applications to Computer Technology. McGraw-Hill, New York (1967)
73. Hoeven, J.: Fast Chinese remaindering in practice. In: Blömer, J., Kotsireas, I.S., Kutsia, T., Simos, D.E. (eds.) MACIS 2017. LNCS, vol. 10693, pp. 95–106. Springer, Cham (2017). https://doi.org/10.1007/978-3-319-72453-9_7
74. Weber, D., Denny, T.: The solution of McCurley's discrete log challenge. In: Krawczyk, H. (ed.) CRYPTO 1998. LNCS, vol. 1462, pp. 458–471. Springer, Heidelberg (1998). https://doi.org/10.1007/BFb0055747

75. Weimerskirch, A., Paar, C.: Generalizations of the Karatsuba Algorithm for Efficient Implementations, IACR Cryptol. ePrint Arch. 224 (2006). http://eprint.iacr.org/2006/224

Fast Enumeration of Superspecial Hyperelliptic Curves of Genus 4 with Automorphism Group V_4

Ryo Ohashi[1(✉)], Momonari Kudo[2], and Shushi Harashita[1]

[1] Graduate School of Environment and Information Sciences, Yokohama National University, Yokohama, Japan
ohashi-ryo-hg@ynu.jp, harasita@ynu.ac.jp

[2] Graduate School of Information Science and Technology, The University of Tokyo, Hongo 7-3-1, Bunkyo-ku, Tokyo 113-8656, Japan
kudo@mist.i.u-tokyo.ac.jp

Abstract. In arithmetic and algebraic geometry, *superspecial curves* have been studied as one of the most important objects, with practical applications to cryptography and coding theory. The enumeration of those curves is a central problem, but if $g \geq 4$ it is not even known whether a superspecial curve of genus g exists in general characteristic $p > 0$. In this paper, we propose an algorithm with complexity $O(p^3)$ to enumerate superspecial hyperelliptic curves of genus 4 with automorphism group V_4, where V_4 is the non-cyclic group of order 4. By executing the algorithm over Magma, we enumerate those curves over $\overline{\mathbb{F}_p}$ for p up to 200. We also succeeded in finding a superspecial hyperelliptic curve of genus 4 in every characteristic p with $19 \leq p \leq 6691$.

Keywords: Hyperelliptic curves · Superspecial curves · Genus-4 curves

1 Introduction

Throughout this paper, by a curve we mean a nonsingular projective variety of dimension one. Let K be a perfect field of characteristic $p > 0$, and $\overline{K}$ its algebraic closure. A curve C of genus g defined over K is said to be *superspecial* (s.sp. for short) if its Jacobian variety $\mathrm{Jac}(C)$ is isomorphic over $\overline{K}$ to the product of copies of a supersingular elliptic curve E, that is, $\mathrm{Jac}(C) \cong E^g$. S.sp. curves have been studied also for their applications to cryptography and coding theory since they or their forms have many rational points (with respect to genus) and their Jacobian varieties have large endomorphism rings.

For given g and p, it is very important to enumerate isomorphism classes of s.sp. curves of genus g in characteristic p. In the case of $g = 1$ (precisely the case of elliptic curves), the number of isomorphism classes of supersingular elliptic curves was studied by Deuring [5]. For $g = 2$ and 3, it follows from [15, Theorem 3.3] by Ibukiyama-Katsura-Oort that the number of isomorphism classes of s.sp. curves is determined by computing the class numbers of quaternion hermitian

S. Mesnager and Z. Zhou (Eds.): WAIFI 2022, LNCS 13638, pp. 107–124, 2023.
https://doi.org/10.1007/978-3-031-22944-2_6

lattices, which can be explicitly computed by Hashimoto-Ibukiyama [11] for $g = 2$ and Hashimoto [10] for $g = 3$ (cf. Brock's thesis [3] for explicit formulae). Ekedahl also proved in [6, Theorem 1.1] that if there exists a s.sp. curve of genus g in characteristic $p > 0$, then g and p must satisfy $2g \leq p^2 - p$ (and $2g \leq p - 1$ if the curve is hyperelliptic and $(g, p) \neq (1, 2)$).

The case of $g = 4$ is the next target, but in fact, it is still an open problem whether there exists a s.sp. curve of genus 4 for arbitrary $p > 7$. One of the mainstream approaches in recent years is the *computational* enumeration such as [20] and [22] (resp. [21]) for the non-hyperellliptic (resp. hyperelliptic) case. However, these algorithms might have exponential complexities, since they enumerate *all* s.sp. curves. On the other hand, in the non-hyperelliptic case, there is an algorithm [23] with complexity $\tilde{O}(p^4)$ restricting to a certain 4-dimensional family of non-hyperelliptic curves. In the following, we review in detail what we know currently about the case of $g = 4$:

- By Ekedahl's bound, there is no s.sp. curve of genus 4 for $p \leq 3$ (and for $p \leq 7$ in the hyperelliptic case). For $p = 5$, Fuhrmann-Garcia-Torres [7] proved that there exists a unique s.sp. curve of genus 4. Kudo-Harashita computationally enumerated genus-4 s.sp. non-hyperelliptic curves in [20] and [22] for $p \leq 11$, and a particular result in [20] is the non-existence of s.sp. curves for $p = 7$. In [21] at WAIFI2018, they also provided a computational enumeration in the hyperelliptic case for $p \leq 19$. The idea of Kudo-Harashita's approaches is to reduce the enumeration into solving multivariate systems over finite fields, where the number of variables is close to the whole moduli dimension ($3g - 3 = 9$ for the non-hyperelliptic case, and $2g - 1 = 7$ for the hyperelliptic case). Their algorithms are applicable to arbitrary p, but in practice infeasible for larger p. This is because the maximum total-degree of each multivariate system increases in $O(p)$, which might make the total complexities exponential with respect to p in worst case.
- To overcome the limitation of the enumeration in practical time, Kudo-Harashita-Howe recently proposed an alternative algorithm in [23] at ANTS-XIV for the *non-hyperelliptic* case. Their algorithm need not to solve *any* multivariate system, and we here briefly review details: They focused on a certain 4-dimensional family of curves of genus 4 which tend to be s.sp. except for several small p, that is, *Howe curves* (studied first in [12], and so named in [24]). A Howe curve D is the normalization of the fiber product over $\mathbb{P}^1$ of two genus-1 double covers $E_i \to \mathbb{P}^1$ which share exactly one ramified point. Note that the notion of Howe curves is generalized to higher genus, see [19] or Subsect. 2.3 below for a review. The authors of [23] proved that D is non-hyperelliptic [23, Lemma 2.1], and constructed an algorithm to enumerate s.sp. Howe curves in $\tilde{O}(p^4)$ arithmetic operations in $\mathbb{F}_{p^4}$, where Soft-O notation omits logarithmic factors. Their algorithm first produces all genus-2 s.sp. curves C by Richelot isogenies, and then constructs E_i by dividing the 6 ramified points of each C into two. By executing the algorithm on Magma, the authors of [23] enumerated s.sp. Howe curves for $p \leq 200$, and showed the existence of such a curve for every $5 \leq p \leq 20000$ with $p \neq 7$.

The aim of this paper is to provide a new efficient algorithm for the *hyperelliptic* case. For this, we consider a hyperelliptic analogue of the construction of a Howe curve, that is, the normalization H of the fiber product over $\mathbb{P}^1$ of two genus-2 curves C_1 and C_2 which share exactly 5 ramified points (we call such a pair (C_1, C_2) a *genus-4 hyperelliptic Howe pair*). It will be shown in Lemma 1 and Proposition 2 that such a curve H is a hyperelliptic curve of genus 4 with $\mathrm{Aut}(H) \supset V_4$, and vise versa, where V_4 is the non-cyclic group of order 4. Different from [21], our algorithm do not search the whole moduli, but nevertheless worthwhile; as in the case of Howe curves in [23], the dimension of the space of our curves is 4, which is more than a half of the whole moduli dimension 7.

Here, the first main result is as follows:

Theorem A. *There exists an algorithm (explicitly Algorithm 1 in Sect. 4) for enumerating s.sp. hyperelliptic curves of genus* 4 *with* V_4*-automorphism group. Under some assumption, it terminates in* $O(p^3)$ *arithmetic operations in* $\mathbb{F}_{p^4}$.

Note that we can modify Algorithm 1 so that it terminates once a single s.sp. curve is found. The modified algorithm has also the worst-case complexity $O(p^3)$.

Algorithm 1 consists of the following four parts. The first part is to enumerate s.sp. genus-2 curves by using *Richelot isogenies*, as in an algorithm of [23] (see the **Step 1**-part of Subsect. 5.1 for details). Second, we detect genus-4 hyperelliptic Howe pairs (C_1, C_2) of s.sp. genus-2 curves C_1 and C_2. Third, all pairs (C_1, C_2) with associated H satisfying $\mathrm{Aut}(H) = V_4$ are collected. Finally, we classify the isomorphism classes of collected H's as hyperelliptic curves.

The restriction $\mathrm{Aut}(H) = V_4$ is for the efficiency of the isomorphism classification; most of (C_1, C_2) satisfy $\mathrm{Aut}(H) = V_4$ as showed in Table 1, and thus this restriction would be practical. More precisely, since the number of produced (C_1, C_2)'s is expected to be $O(p^2)$ (Remark 4), that of required isomorphism tests is naively $O(p^4)$, but in fact $O(p^2)$ under the restriction $\mathrm{Aut}(H) = V_4$ due to our criteria (Lemma 4 and Proposition 5). Note that the algorithm also works even for the case where $\#\mathrm{Aut}(H) > 4$, but the complexity might become $O(p^4)$.

By executing the algorithm over Magma [2], we obtain the following results:

Theorem B. *(i) For every prime* p *with* $19 \leq p \leq 6691$, *there exists a s.sp. hyperelliptic curve* H *of genus* 4 *with* $\mathrm{Aut}(H) \supset V_4$.
(ii) For every prime $17 \leq p \leq 200$, *the number of isomorphism classes of s.sp. hyperelliptic curves* H *of genus* 4 *with* $\mathrm{Aut}(H) = V_4$ *is given in Table 1.*

The upper bounds on p in Theorem B are very larger than that of the enumeration result in [21], and it can be increased easily. For example, enumerating s.sp. hyperelliptic curves of genus 4 with automorphism group V_4 in characteristic 199 took 96,996 s ($\approx$ 27 h) by our environment stated in Subsect. 4.2. Among $17 \leq p \leq 6691$, the maximal time for finding a single s.sp. curve was about 7.5 h for $p = 4889$, while there are many large p for which such a curve was found in quite shorter time (e.g., 5 s for $p = 6529$). See [28] for the source codes and log files with explicit examples.

This paper is organized as follows. In Sect. 2, we will recall basic facts on hyperelliptic curves and generalized Howe curves. Section 3 studies hyperelliptic curves of genus 4 with automorphism group containing V_4. In Sect. 4, we propose our main algorithm, and state computational results obtained by our implementation. Section 5 is devoted to the complexity analysis of the algorithm. Finally we give a concluding remark in Sect. 6. All the complexities are measured by the number of arithmetic operations in $\mathbb{F}_{p^4}$ (cf. Lemma 3).

2 Preliminaries

In this section, we will review some facts on hyperelliptic curves, and their isomorphisms and automorphisms. The notion of generalized Howe curves and their fundamental properties will be also described.

2.1 Hyperelliptic Curves and Their Isomorphisms

Let K be a field of characteristic not equal to 2. In general, a hyperelliptic curve H of genus g over K is realized as the normalization of the projective closure of the affine plane curve $y^2 = f(x)$, where $f(x) \in K[x]$ is a separable polynomial of degree $2g+1$ or $2g+2$. The (affine) model $y^2 = f(x)$ is said to be imaginary or real if $\deg(f) = 2g+1$ or $2g+2$, respectively.

Let $H_1 : y^2 = f_1(x)$ and $H_2 : y^2 = f_2(x)$ be hyperelliptic curves of genus $g > 1$ over K, where f_1 and f_2 are polynomials over K of degree $2g+1$ or $2g+2$ with no multiple root. It is well-known (see, e.g., [21, Lemma 1]) that any isomorphism over K between H_1 and H_2 is represented by a pair (h, λ) of $h := \alpha E_{11} + \beta E_{12} + \gamma E_{21} + \delta E_{22} \in \mathrm{GL}_2(K)$ with $\alpha, \beta, \gamma, \delta \in K$ and $\lambda \in K^*$, where each E_{ij} denotes the 2×2-matrix with 1 in the (i, j) entry and 0's elsewhere. The isomorphism associated with (h, λ) is given by $(x, y) \mapsto \left(\frac{\alpha x+\beta}{\gamma x+\delta}, \frac{\lambda y}{(\gamma x+\delta)^{g+1}}\right)$, or by considering the homogenization $y^2 z^{2g} = F_i(x, z) := z^{2g+2} f(x/z)$, it is given by $(x, z, y) \mapsto (\alpha x + \beta z, \gamma x + \delta z, \lambda y)$. The representation (h, λ) is unique up to the equivalence $(h, \lambda) \sim (\mu h, \mu^{g+1} \lambda)$ for some $\mu \in K^*$. If K is algebraically closed, then we can take a representative with $\lambda = 1$ in each equivalence class.

Let a_i (resp. a_i') be the x^i-coefficient of f_1 (resp. f_2), and assume that K is algebraically closed. A method to determine if H_1 and H_2 are isomorphic over K or not is to test if there exists a root over K of a multivariate system with respect to α, β, γ and δ obtained by comparing coefficients in $F_1(\alpha x + \beta z, \gamma x + \delta z)$ and those in $F_2(x, z)$. More explicitly, consider the following (homogeneous) system:

$$\begin{cases} \displaystyle\sum_{i=0}^{2g+2} a_i \sum_{k=0}^{i} \binom{i}{k} \alpha^k \beta^{i-k} \binom{2g+2-i}{\ell-k} \gamma^{\ell-k} \delta^{2g+2-i-(\ell-k)} = a_\ell' \\ (\alpha\delta - \beta\gamma)\mu = 1 \end{cases} \tag{2.1.1}$$

with $0 \leq \ell \leq 2g+2$ and an extra variable μ, and test if a Gröbner basis for the ideal in $K[\alpha, \beta, \gamma, \delta, \mu]$ associated to (2.1.1) contains a unit. There are only finite elements in $\mathrm{PGL}_2(K)$ translating the ramified points of H_2 to those of H_1, and thus the ideal associated to the homogeneous system (2.1.1) is zero-dimensional. Since the system also has the maximum total-degree $\leq 2g+2$ and a constant number of variables, the Gröbner basis is computed in constant time with respect to p for a fixed g.

In the case of genus two (resp. three), the isomorphism test using *Igusa* (resp. *Shioda*) invariants is much more effective than the method described above. We here briefly recall Igusa invariants which will be used in our main algorithm. Let $C : y^2 = f(x)$ be a genus-2 hyperelliptic curve over K, where $f(x)$ is a sextic over K with mutually distinct roots $a_1, \ldots, a_6$ and leading coefficient c_6. Then the *Igusa invariants* of C are defined as certain symmetric functions J_2, J_4, J_6, J_8 and J_{10} of the roots, e.g., $J_{10} = 2^{-12} c_6^{10} \prod_{i<j}(a_i - a_j)^2$, and the *absolute invariant* of C (cf. [16, p. 641]), which we denote by $j_{\mathrm{abs}}(C)$, is defined as a tuple of ten quotients of the Igusa invariants. Two genus-2 curves C_1 and C_2 are isomorphic over $\overline{K}$ if and only if $j_{\mathrm{abs}}(C_1) = j_{\mathrm{abs}}(C_2)$. Since J_2, J_4, J_6, J_8 and J_{10} are represented as polynomials in the coefficients of f with degree bounded by a constant, $j_{\mathrm{abs}}(C)$ can be computed in constant time. Several computer algebra systems have functions for computing the Igusa and absolute invariants, e.g., AbsoluteInvariants in Magma [2] for computing $j_{\mathrm{abs}}(C)$.

2.2 Automorphism Groups of Hyperelliptic Curves

It is well-known that the automorphism group over $\overline{K}$ of a curve of genus g over K is finite, and has size $\leq 16g^4$ unless it is a Hermitian curve [27]. In particular, we here consider the case of hyperelliptic curves, say, $C : y^2 = f(x)$ a hyperelliptic curve with $\deg(f) = 2g+2$ and $\mathrm{Aut}(C)$ its automorphism group over $\overline{K}$. Putting $F(x, z) := z^{2g+2} f(x/z)$, we have the following group isomorphism:

$$\mathrm{Aut}(C) \simeq \left\{ h \in \mathrm{GL}_2(\overline{K}) \,\middle|\, F((x, z) \cdot {}^t h)) = F(x, z) \right\} / \boldsymbol{\mu}_{g+1}, \tag{2.2.1}$$

where $\boldsymbol{\mu}_{g+1} = \{\mu I_2 : \mu \in \overline{K}^\times,\ \mu^{g+1} = 1\}$ with the 2×2-identity matrix I_2. Each automorphism is represented by a root of (2.1.1) with $a_i = a_i'$ for $0 \leq i \leq 2g+2$. Since both $|\mathrm{Aut}(C)|$ and $|\boldsymbol{\mu}_{g+1}|$ are finite, the number $N(C)$ of roots of (2.1.1) is also finite, and $|\mathrm{Aut}(C)|$ is computed as $N(C)/|\boldsymbol{\mu}_{g+1}|$. The number $N(C)$ is equal to the dimension of the coordinate ring of the ideal associated to (2.1.1) as a K-vector space, and thus it can be computed by the Gröbner basis computation. By the same reason for the case of isomorphisms in Subsect. 2.1, computing $|\mathrm{Aut}(C)|$ is done in constant time with respect to p, when g is fixed.

This paper focuses on the case where $g = 4$ and $\mathrm{Aut}(C) \supset V_4$. In this case, Paulhus [26, Theorem 4] classified possible finite groups isomorphic to $\mathrm{Aut}(C)$ in characteristic zero, whereas unfortunately we could not find any reference stating such a classification in positive characteristic case. This paper does not rely on the classification of $\mathrm{Aut}(C)$, and discusses only the case where $\mathrm{Aut}(C) \supset V_4$.

2.3 Generalized Howe Curves

This subsection reviews the notion of *generalized Howe curves*, which are very useful to produce superspecial (or supersingular) curves (e.g., see [23] and [24] for genus-4 non-hyperelliptic case). Let C_1 and C_2 be two hyperelliptic curves of genera g_1 and g_2 such that C_1 and C_2 share precisely r ramified points, say

$$C_1 : y^2 = (x - a_1) \cdots (x - a_r)(x - b_1) \cdots (x - b_{2g_1+2-r}),$$
$$C_2 : y^2 = (x - a_1) \cdots (x - a_r)(x - c_1) \cdots (x - c_{2g_2+2-r}),$$

where a_i, b_i and c_i are all distinct. Let $\pi_i : C_i \to \mathbb{P}^1$ be double covers. Note that π_i is unique if the genus of C_i is greater than or equal to 2, see [9, Section IV.5]. The normalization H of the fiber product $C_1 \times_{\mathbb{P}^1} C_2$ is called *a generalized Howe curve* as defined in [19]. When $g_1 = g_2 = 1$ and $r = 1$ with $g(H) = 4$, it is called simply *a Howe curve*, which had already been treated in [23] and [24]. Let σ_i be the involution on H whose quotient map is $H \to C_i$. We then have a diagram

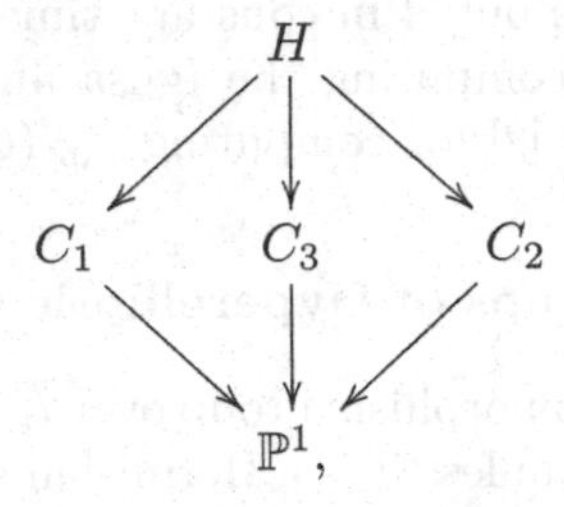

where C_3 is the quotient by the other involution $\sigma_3 := \sigma_1\sigma_2$ on H:

$$C_3 : y^2 = (x - b_1) \cdots (x - b_{2g_1+2-r})(x - c_1) \cdots (x - c_{2g_2+2-r}).$$

Recall a result by Katsura-Takashima [19, Section 2] on the genus of H and on a criterion for H to be hyperelliptic:

Proposition 1. *Let C_1, C_2 and H be as above. The genus of H is equal to $2g_1 + 2g_2 + 1 - r$. If the genus of H is greater than or equal to 4, then H is hyperelliptic if and only if $r = g_1 + g_2 + 1$, that is, the curve C_3 is of genus 0.*

3 Construction of Our Curves and Their Superspeciality

In this section, we study hyperelliptic curves of genus 4 with automorphism group containing V_4. In particular, it will be shown in Subsect. 3.1 below that such a curve D is realized as a hyperelliptic generalized Howe curve of genus 4, that is, the normalization H of the fiber product of two genus-2 hyperelliptic curves C_1 and C_2 which share exactly five ramified points. We also show in Subsect. 3.2 that the superspeciality of H is reduced into that of C_1 and C_2.

3.1 Construction

Let K be a field of characteristic $p \geq 3$. Let C_1 and C_2 be hyperelliptic curves of genus 2 over K defined by

$$C_i : y^2 = (x - a_1)(x - a_2)(x - a_3)(x - a_4)(x - a_5)(x - b_i) \tag{3.1.1}$$

for $i = 1, 2$ with distinct elements $a_1, a_2, a_3, a_4, a_5, b_1, b_2$ of $\overline{K} \cup \{\infty\}$, where by that an element α of $\{a_1, a_2, a_3, a_4, b_1, b_2\}$ is ∞ we mean the factor $(x - \alpha)$ is excluded from (3.1.1). Let $\pi_1 : C_1 \to \mathbb{P}^1$ and $\pi_2 : C_2 \to \mathbb{P}^1$ be the usual double covers, where f_i is ramified over S_i with

$$S_1 := \{a_1, a_2, a_3, a_4, a_5, b_1\}, \quad S_2 := \{a_1, a_2, a_3, a_4, a_5, b_2\}.$$

We consider the normalization H of the fiber product $C_1 \times_{\mathbb{P}^1} C_2$. Note that H is a generalized Howe curve defined in Subsect. 2.3, and we have the following:

Lemma 1. *H is a hyperelliptic curve of genus 4, and* $\mathrm{Aut}(H)$ *contains* V_4.

Proof. It follows from Proposition 1 that H is a genus-4 hyperelliptic curve. The hyperelliptic involutions of C_1 and C_2 lift to automorphisms of H, and thus the automorphism group of H contains a subgroup isomorphic to V_4. □

Example 1. Consider the case where $b_1 = 0$ and $b_2 = \infty$, say $C_1 : y^2 = xf(x)$ and $C_2 : y^2 = f(x)$ with $f(x) = (x - a_1)(x - a_2)(x - a_3)(x - a_4)(x - a_5)$. Then H is isomorphic to the normalization of the curve defined by

$$D : Y^2 = (X^2 - a_1)(X^2 - a_2)(X^2 - a_3)(X^2 - a_4)(X^2 - a_5).$$

Indeed, there exist the morphism $D \to C_1$ defined by $x \mapsto X^2$, $y \mapsto XY$ and the morphism $D \to C_2$ defined by $x \mapsto X^2$, $y \mapsto Y$. These morphisms are of degree 2 and make a commutative diagram with π_1 and π_2, and hence the normalization of $C_1 \times_{\mathbb{P}^1} C_2$ is isomorphic to D by the universality of the fiber product.

Example 2. Consider the case where $\{a_1, a_2, a_3\} = \{0, \infty, 1\}$, say

$$C_i : \quad y^2 = x(x - 1)(x - a_4)(x - a_5)(x - b_i)$$

for $i = 1, 2$. The map $x \mapsto \frac{x - b_1}{x - b_2}$ transforms C_1 and C_2 into $C_1' : y^2 = xf(x)$ and $C_2' : y^2 = f(x)$ with $f(x) = (x - 1)(x - c_2)(x - c_3)(x - c_4)(x - c_5)$, where

$$c_2 = \frac{b_1}{b_2}, \quad c_3 = \frac{1 - b_1}{1 - b_2}, \quad c_4 = \frac{a_4 - b_1}{a_4 - b_2}, \quad c_5 = \frac{a_5 - b_1}{a_5 - b_2}.$$

Hence H is isomorphic to the normalization of the curve defined by

$$D : Y^2 = (X^2 - 1)(X^2 - c_2)(X^2 - c_3)(X^2 - c_4)(X^2 - c_5) \tag{3.1.2}$$

as in Example 1.

We also prove the converse of Lemma 1:

Proposition 2. *Any hyperelliptic genus-4 curve D with* $\mathrm{Aut}(D) \supset V_4$ *is isomorphic to the normalization of* $C_1 \times_{\mathbb{P}^1} C_2$ *for some* C_1 *and* C_2 *as in* (3.1.1).

Proof. Let D be a hyperelliptic curve of genus 4 such that $\mathrm{Aut}(D) \supset G$ with $G \cong V_4$. Let σ be the hyperelliptic involution (which is unique) of D. We claim that there exists a V_4-subgroup G' of $\mathrm{Aut}(D)$ such that $\sigma \in G'$. Indeed, if $\sigma \in G$, there is nothing to prove. If not, then we choose an involution $\tau \in G$ and let G' be the subgroup generated by σ and τ. Since σ and τ are commutative (otherwise we have another hyperelliptic involution $\tau^{-1}\sigma\tau$), we have $G' \cong V_4$.

We choose two involutions ι_1, ι_2 in G' with $\iota_1, \iota_2 \neq \sigma$ and set the curves $C_1 := D/\langle\iota_1\rangle$, $C_2 := D/\langle\iota_2\rangle$ and $P := D/\langle\iota_1, \iota_2\rangle$. We have the following diagram:

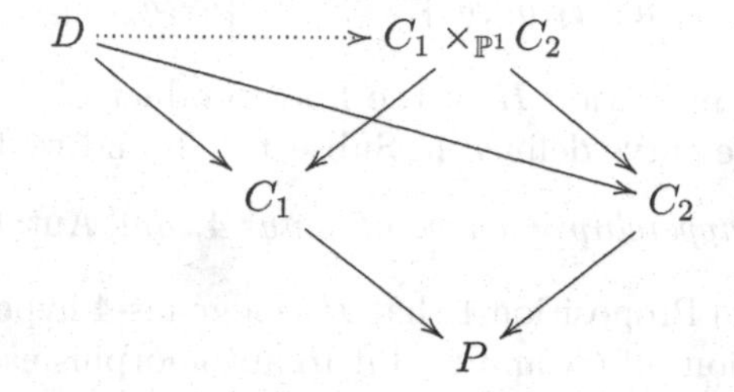

Here the curve P is isomorphic to $\mathbb{P}^1$ as this is a quotient of $\mathbb{P}^1 \simeq D/\langle\sigma\rangle$ with $\sigma = \iota_1\iota_2$, and therefore there exists the morphism $D \to C_1 \times_{\mathbb{P}^1} C_2$ by the universality of the fiber product. Since both the morphisms $C_1 \times_{\mathbb{P}^1} C_2 \to C_i$ and $D \to C_i$ are of degree 2, the morphism $D \to C_1 \times_{\mathbb{P}^1} C_2$ is of degree one. Hence $D \to C_1 \times_{\mathbb{P}^1} C_2$ gives the normalization map.

To complete the proof, it suffices to show that C_1 and C_2 are of genus 2 and share exactly 5 ramified points. Let $g_i \geq 1$ be the genus of C_i and r the number of common ramified points. It follows from Proposition 1 that $2(g_1+g_2)+1-r = 4$ since D is of genus 4, and $r = g_1 + g_2 + 1$ since D is hyperelliptic. Thus we have $g_1 + g_2 = 4$. On the other hand, there is no dominant morphism $D \to C_i$ where C_i is of genus 3 by Hurwitz's formula, and hence we have $g_i \leq 2$. Therefore, we obtain $g_1 = g_2 = 2$ and $r = 5$. □

Remark 1. We have another proof of Proposition 2 as follows. A hyperelliptic curve D of genus 4 with $\mathrm{Aut}(D) \supset V_4$ is isomorphic to the normalization of the curve defined by $y^2 = f(x^2)$, where $f(x)$ is a monic polynomial whose degree is five by [8, Lemma 2.2]. It follows from Example 1 that D is isomorphic to the normalization of $C_1 \times_{\mathbb{P}^1} C_2$ for $C_1 : y^2 = xf(x)$ and $C_2 : y^2 = f(x)$.

3.2 Superspeciality

Let K be a field of characteristic $p \geq 3$. Let H be the normalization of $C_1 \times_{\mathbb{P}^1} C_2$ as in Subsect. 3.1, where C_1 and C_2 are hyperelliptic curves of genus 2 over K which share exactly five ramification points. Then we have the following criterion on the superspeciality of H:

Lemma 2. *H is superspecial if and only if C_1 and C_2 is superspecial.*

Proof. Recall the fact that there exists an isogeny of Jacobian varieties from $\mathrm{Jac}(C_1) \times \mathrm{Jac}(C_2)$ to $\mathrm{Jac}(H)$ of 2-power degree (cf. [18, Theorem C]). By $p \neq 2$, we have an isomorphism between their p-kernels

$$\mathrm{Jac}(C_1)[p] \times \mathrm{Jac}(C_2)[p] \cong \mathrm{Jac}(H)[p]. \tag{3.2.1}$$

For a curve D over K, its *a-number* $a(D)$ is defined to be $\dim \mathrm{Hom}_{\overline{K}}(\alpha_p, \mathrm{Jac}(D))$, where α_p is the kernel of the Frobenius map on the one-dimensional additive group $\mathbb{G}_a$. Note that $a(D)$ is at most the genus of D. Since the p-multiplication on α_p is zero, the a-number $a(D)$ is equal to $\dim \mathrm{Hom}_{\overline{K}}(\alpha_p, \mathrm{Jac}(D)[p])$. Hence we have $a(H) = a(C_1) + a(C_2)$ by (3.2.1). The lemma follows from tha fact (cf. [25, 1.6]) that a curve is s.sp. if and only if its a-number is equal to its genus. □

By Lemma 2, we have an idea to find or enumerate s.sp. hyperelliptic curves D of genus 4 with $\mathrm{Aut}(D) = V_4$. First, we enumerate s.sp. genus-2 curves in characteristic p. As it is shown in [23], this is done by using *Richelot isogenies*, see the **Step 1**-part of Subsect. 5.1 below for details. Second, we detect pairs (C_1, C_2) of s.sp. genus-2 curves C_1 and C_2 which share exactly 5 ramified points (we call such a pair a *genus-4 hyperelliptic Howe pair*). Third, all pairs (C_1, C_2) with associated D satisfying $\#\mathrm{Aut}(D) = 4$ are collected. Finally, we classify the isomorphism classes of collected D's as hyperelliptic curves.

4 Algorithm and Computational Results

This section presents an algorithm to enumerate s.sp. hyperelliptic curves of genus 4 with V_4-automorphism group, based on the idea described in the last paragraph of Subsect. 3.2. Moreover, computational results obtained by our implementation of the algorithm on Magma are summarized. Note that all the computations are done in $\mathbb{F}_{p^4}$, see Lemma 3 below. We fix a primitive element $\zeta \in \mathbb{F}_{p^4}$ for each p, and order elements $a = \zeta^i$ and $b = \zeta^j$ in $\mathbb{F}_{p^4}$ by $a \leq b$ if $i \leq j$.

4.1 Main Algorithm

Algorithm 1 below is our main algorithm. The four steps in Algorithm 1 correspond to those in the idea described in the last paragraph of Subsect. 3.2.

Algorithm 1 (Main algorithm).

Input: A rational prime $p > 2$.
Output: A list of the isomorphism classes of s.sp. genus-4 hyperelliptic curves over the algebraic closure of $\mathbb{F}_p$ with V_4-automorphism group.

Step 1. Enumerate all s.sp. curves C of genus 2 in characteristic p, by a method in Subsect. 5.1. Let $\mathsf{SSp}_2(p)$ be a list of enumerated genus-2 s.sp. curves.

Step 2. Collect genus-4 hyperelliptic Howe pairs as follows: For each s.sp. curve C of genus 2, produce its imaginary models $y^2 = x(x-1)(x-a_4)(x-a_5)(x-b)$ for some $a_4, a_5, b \in \mathbb{F}_{p^4} \smallsetminus \{0,1\}$, by translating 3 points among 6 ramified points of $C \to \mathbb{P}^1$ to $\{0, 1, \infty\}$. For each imaginary model, and we generate three triples of the form (a_4, a_5, b) with $a_4 < a_5$. Detect all pairs (C_1, C_2) of imaginary models C_1 and C_2 constructed as above which share exactly 5 ramified points, namely pairs of (a_4, a_5, b_1) and (a_4', a_5', b_2) with $a_4 = a_4'$, $a_5 = a_5'$ and $b_1 \neq b_2$. Let HowePairsList be a list of (C_1, C_2) found as above.
Step 3. Among the genus-4 hyperelliptic Howe pairs collected in **Step 2**, detect all (C_1, C_2) such that the hyperelliptic curve D obtained as $C_1 \times_{\mathbb{P}^1} C_2$ satisfies $\# \mathrm{Aut}(D) = 4$. Let HowePairsV4 be a list of D's satisfying $\# \mathrm{Aut}(D) = 4$.
Step 4. Classify the isomorphism classes of D's in HowePairsV4 as follows:
  (4-1) Sorting (C_1, C_2)'s in a total order on the set of pairs of absolute invariants of C_1 and C_2, divide them into groups to have the same absolute invariants pair. Namely, if two pairs (C_1, C_2) and (C_1', C_2') belong to the same group, then $C_1 \cong C_1'$ and $C_2 \cong C_2'$ as hyperelliptic curves. We denote by SameInv(C_1, C_2) the group represented by (C_1, C_2).
  (4-2) Classify the isomorphism classes of s.sp. hyperelliptic curves in each of the groups constructed in **(4-1)**. Finally, output a list AllHypHoweCurves of the computed isomorphism classes.

Remark 2. In Sect. 5 below, we will estimate the complexity of Algorithm 1 as $O(p^3)$, assuming that we use a hash table to collect pairs of (a_4, a_5, b_1) and (a_4', a_5', b_2) in Step 2. Using a hash table is suggested by one of the reviewers. However, we have not succeeded in implementing the generation of such a hash table yet, and thus our current implementation collects those pairs by simply sorting (a_4, a_5, b_1)'s. Therefore, the practical behavior would follow $\tilde{O}(p^3)$.

4.2 Computational Results

We implemented the algorithm and its variant on Magma [2] V2.25-3 in its 64bit version, where the variant terminates as soon as a single s.sp. hyperelliptic curve D of genus 4 with $\mathrm{Aut}(D) \supset V_4$ is found. The codes and log files for our implementation are available at [28]. Executing the codes on a PC with Windows 10 Pro OS at 1.80 GHz CPU (Intel Core i7-8565U) and 16.0 GB memory, Theorem B is proved. Table 1 shows computational results for Theorem B (ii). We see from Table 1 that the time acts almost as $\tilde{O}(p^3)$ estimated in Remark 2.

For $17 \leq p \leq 200$ (except for several small p), most of the time is spent at Step 2, which is one of the dominant steps (cf. the proof of Proposition 3 below). From Table 1, we can also confirm the validity of our assumption in Proposition 3

Table 1. The number N_p of s.sp. hyperelliptic curves of genus 4 with automorphism group V_4 (namely N_p = #AllHypHoweCurves), and the time for executing Algorithm 1 for each p with $17 \leq p \leq 200$. The notation HowePairsList, HowePairsV4 and SameInv(C_1, C_2)) are same as in Steps 3 and 4 of Algorithm 1.

p	#HowePairsList	#HowePairsV4	max(#SameInv(C_1, C_2))	N_p	Time (sec.)
17	0	0	0	0	< 1.0
19	96	0	0	0	4.9
23	120	60	60	1	7.0
29	96	0	0	0	6.1
31	450	300	120	5	35.2
37	480	240	60	4	38.4
41	510	420	60	7	63.5
43	600	480	120	8	75.2
47	1110	840	60	14	145.8
53	1140	780	60	13	148.9
59	1236	1020	120	17	183.8
61	1140	900	120	15	173.6
67	1200	960	60	16	178.0
71	4020	3660	120	61	576.2
73	2610	2280	120	38	397.0
79	4266	3780	120	63	662.3
83	4320	3720	120	62	675.3
89	3966	3600	120	60	751.7
97	4440	4200	120	70	1029.5
101	6000	5580	180	93	1411.7
103	5700	5100	180	85	1927.2
107	6720	6060	120	101	2535.4
109	6336	5940	120	99	2604.2
113	6000	5220	120	87	2595.8
127	11970	11160	180	186	5114.2
131	10260	9540	180	159	4805.0
137	8730	8100	120	135	6882.4
139	10356	9240	120	154	7907.9
149	12216	11340	120	189	11821.1
151	18240	17640	180	294	15865.3
157	13140	12180	120	203	20647.5
163	14580	14100	120	235	25176.4
167	24300	23160	180	386	30838.6
173	19020	18360	120	306	32910.1
179	20616	19260	120	321	47790.8
181	16620	15780	180	263	61735.1
191	32250	30060	120	501	104923.2
193	18420	17340	180	289	64025.0
197	21360	20640	120	344	82069.3
199	34896	33360	180	556	96996.3

and the correctness of Lemma 5: The number of pairs (C_1, C_2) of s.sp. curves of genus 2 such that they share 5 ramified points up to $\mathrm{Aut}(\mathbb{P}^1)$ is expected to be $O(p^2)$, and $\#\mathsf{SameInv}(C_1, C_2)$ is bounded by a constant.

As an example, for $p = 199$, we found a s.sp. genus-4 hyperelliptic curve $D : y^2 = (x^2-1)(x^2-9^2)(x^2-77^2)(x^2-79^2)(x^2-58^2)$ over $\mathbb{F}_p$ with $\mathrm{Aut}(D) = V_4$, with $C_i : y^2 = x(x-1)(x-70)(x-23)(x-b_i)$ for $(b_1, b_2) = (25, 180)$. See log files at [28] for further examples.

Remark 3. We also examined with Magma that all of the obtained s.sp. hyperelliptic curves D with $\mathrm{Aut}(D) = V_4$ for $19 \leq p \leq 200$ are $\mathbb{F}_{p^2}$-maximal or $\mathbb{F}_{p^2}$-minimal, where a genus-g curve C over a finite field $\mathbb{F}_q$ is said to be $\mathbb{F}_q$-maximal (resp. $\mathbb{F}_q$-minimal) if the number of its $\mathbb{F}_q$-rational points attains the Hasse-Weil upper (resp. lower) bound $q+1+2g\sqrt{q}$ (resp. $q+1-2g\sqrt{q}$). Maximal curves are useful to construct good algebraic-geometric codes (cf. [14]), and their concrete examples are given at `manypoint.org`. (Unfortunately, the site does not cover the case $q = p^2$ for $p > 19$.)

The smallest case $p = 19$ among ours is covered by the site, but in this case there is no s.sp. D with $\mathrm{Aut}(D) = V_4$. On the other hand, we also found an $\mathbb{F}_{19^2}$-maximal curve $D : \zeta y^2 = x^{10} + \zeta^{127}x^8 + \zeta^{173}x^6 + \zeta^{225}x^4 + \zeta^{283}x^2 + 3$ with $\mathrm{Aut}(D) = D_4 \supsetneq V_4$, where ζ is a primitive element of $\mathbb{F}_{p^2}$. This is a new example of $\mathbb{F}_{19^2}$-maximal curves of genus 4 since one can check by Magma that it is not $\mathbb{F}_{19^2}$-isomorphic to the examples by Rovi and Fischer shown at `manypoint.org`.

5 Complexity Analysis

The main purpose of this section is to analyze the complexity of the main algorithm (Algorithm 1) presented in Subsect. 4.1. For this, we first explain details of each step concretely, in Subsect. 5.1 below. After that, we shall prove in Subsect. 5.2 below that the complexity of Algorithm 1 is bounded by $O(p^3)$.

5.1 Concrete Description of Each Step

We here precisely describe each of **Steps 1-4** in Algorithm 1. The notation is same as in Sect. 4, unless otherwise noted. The absolute invariants of genus-2 hyperelliptic curves will be ordered by the lexicographic order on $(\mathbb{F}_{p^4})^{\oplus 10}$ derived from the order on $\mathbb{F}_{p^4}$ defined at the beginning of Sect. 4.

Step 1: Enumeration of s.sp. Curves of Genus Two. This step enumerates all the isomorphism classes of s.sp. curves of genus 2 in characteristic p. For this, we apply a method given in [23, Section 5A] (see also Steps (1)–(3) of [23, Algorithm 5.1] for concrete procedures). In this method, we first produce $\mathbb{F}_{p^2}$-maximal s.sp. curves of genus 2 over $\mathbb{F}_{p^2}$ by gluing supersingular elliptic curves together along their 2-torsion subgroups [13, Proposition 4], and then construct more such curves by applying Richelot isogenies to the curves already produced. This construction from curves newly obtained is repeated until no new isomorphism class is produced. We use the absolute invariant defined in Subsect.

2.1 to test if two genus-2 curves are isomorphic or not. The termination of this method is assured by a fact that there are only finitely many s.sp. curves of genus 2, and the correctness follows from [17, Theorem 43] by Jordan-Zaytman, where they showed the connectedness of superspecial isogeny graphs in arbitrary dimension. Note that the complexity of this method is bounded by the number of s.sp. curves of genus 2 in characteristic p, which is asymptotically equal to $O(p^3)$, as $p \to \infty$.

Lemma 3. *Let $y^2 = f(x)$ be any hyperelliptic curve of genus 2 appearing in* **Step 1** *as above. Then all the roots of $f(x)$ belong to $\mathbb{F}_{p^4}$. Hence all the computations in Algorithm 1 are done in $\mathbb{F}_{p^4}$.*

Proof. Any supersingular elliptic curve has a model of the form $y^2 = g(x)$, where all roots of $g(x)$ belong to $\mathbb{F}_{p^2}$ (cf. [1, Proposition 2.2]). The genus-2 s.sp. curve constructed from two such models by the method in [13, Proposition 4] is of the form $y^2 = G_1(x)G_2(x)G_3(x)$ with quadrics $G_i(x)$ over $\mathbb{F}_{p^2}$. According to [4, Proposition 1], the curve obtained by any Richelot isogeny from a genus-2 curve $y^2 = G_1(x)G_2(x)G_3(X)$ as above is also of the form $y^2 = H_1(x)H_2(x)H_3(x)$ with quadrics $H_i(x)$ over $\mathbb{F}_{p^2}$. Thus we have the required result. □

Step 2: Detection of Genus-4 Hyperelliptic Howe Pairs. As in Algorithm 1, let $\mathsf{SSp}_2(p)$ denote a list of s.sp. genus-2 curves enumerated in **Step 1**. This step computes imaginary models of each s.sp. curve $C \in \mathsf{SSp}_2(p)$ so that the model has 0, 1 and ∞ as the x-coordinates of ramification points. More concretely, it follows from Lemma 3 that we can write C as $C : y^2 = f(x)$ and let $x_i \in \mathbb{F}_{p^4}$ be the roots of $f(x)$ with $1 \le i \le 6$. Choosing three roots x_1, x_2 and x_3, we produce an imaginary model of C given by

$$C_{a_4,a_5,b} : y^2 = x(x-1)(x-a_4)(x-a_5)(x-b) \tag{5.1.1}$$

obtained by the map $x \mapsto \frac{(x-x_1)(x_3-x_2)}{(x-x_2)(x_3-x_1)}$ from C, where a_4, a_5 and b are its images of x_4, x_5 and x_6, respectively. The map also sends x_1, x_2 and x_3 into 0, ∞ and 1, respectively. Since the number of choices of ordered three points (x_1, x_2, x_3) is $6 \times 5 \times 4 = 120$, each $C \in \mathsf{SSp}_2(p)$ produces 120 imaginary models in total.

We collect all pairs (C_1, C_2) of imaginary models C_1 and C_2 produced as above which share exactly five ramification points. For this, we generate three triples of the form (a_4, a_5, b_1) with $a_4, a_5, b_1 \in \mathbb{F}_{p^4}$ and $a_4 < a_5$ from each imaginary model as in (5.1.1), and collect pairs of triples (a_4, a_5, b_1) and (a_4', a_5', b_2) with $(a_4, a_5) = (a_4', a_5')$ and $b_1 \neq b_2$. Computing the hash $h = H(a_4, a_5)$ of the first two elements of each triple, this can be efficiently done by preparing a hash table of size $p^3 \times 120 \times 3 = O(p^3)$. We store each found pair (C_1, C_2) as a sextuple $(a_4, a_5, b_1, a_4', a_5', b_2)$, and let $\mathsf{HowePairsList}$ be a list of collected sextuples. Note that we expect $\#\mathsf{HowePairsList} = O(p^2)$ by Remark 4.

Step 3: Detection of s.sp. Hyperelliptic Curves with Automorphism Group V_4. Let $\mathsf{HowePairsV4}$ be an empty list. For each $(C_1, C_2) \in \mathsf{HowePairsList}$ (stored as a sextuple $(a_4, a_5, b_1, a_4', a_5', b_2)$), we set c_2, c_3, c_4 and c_5 as in

Example 1, and then the genus-4 hyperelliptic curve D defined by (3.1.2) is isomorphic to the normalization H of $C_1 \times_{\mathbb{P}^1} C_2$.

We compute the number of automorphisms of D by a method described in Subsect. 2.2, and add D to the list $\mathsf{HowePairsV4}$ if $\#\mathrm{Aut}(D) = 4$. Note that $\#\mathsf{HowePairsV4} = O(p^2)$, assuming $\#\mathsf{HowePairsList} = O(p^2)$ by Remark 4.

We store each D as a quadruple $(j_{\mathrm{abs}}(C_1), j_{\mathrm{abs}}(C_2), (a_4, a_5, b_1), (a_4', a_5', b_2))$ with $a_4 = a_4'$, $a_5 = a_5'$ and $b_1 \neq b_2'$, where $j_{\mathrm{abs}}(C_1)$ and $j_{\mathrm{abs}}(C_2)$ are the absolute invariants defined in Subsect. 2.1 with $j_{\mathrm{abs}}(C_1) \leq j_{\mathrm{abs}}(C_2)$, respectively.

Step (4-1): Dividing Into Groups to Have the Same Absolute Invariants Pair. This step divides elements in $\mathsf{HowePairsV4}$ into groups so that any distinct two elements (C_1, C_2) and (C_1', C_2') in a group satisfy $C_1 \cong C_1'$ and $C_2 \cong C_2'$. This is because, for isomorphism classification in the next step, it suffices from Lemma 4 to classify elements in the same group.

To divide the elements, we sort $\mathsf{HowePairsV4}$ with respect to the lexicographic order on $(j_{\mathrm{abs}}(C_1), j_{\mathrm{abs}}(C_2))$, and add (C_1, C_2) and (C_1', C_2') into the same group if $(j_{\mathrm{abs}}(C_1), j_{\mathrm{abs}}(C_2)) = (j_{\mathrm{abs}}(C_1'), j_{\mathrm{abs}}(C_2'))$. Denoting by $\mathsf{SameInv}(C_1, C_2)$ the group whose representative is (C_1, C_2), we have $\mathsf{HowePairsV4} = \bigsqcup \mathsf{SameInv}(C_1, C_2)$.

Step (4-2): Isomorphism Classification of Collected Hyperelliptic Curves. Let $\mathsf{AllHypHoweCurves}$ be an empty list. In this step, we classify the isomorphism classes of Howe curves in each of the groups constructed in **Step (4-1)**. Concretely, we write each $\mathsf{SameInv}(C_1, C_2)$ as $\mathsf{SameInv}(C_1, C_2) = \{D_1, \ldots, D_k\}$, and then conduct the following: For i from 1 to k by 1, test whether D_i is isomorphic to D_j for some $j < i$ or not. If D_i is not isomorphic to D_j for any $j < i$, then add D_i into $\mathsf{AllHypHoweCurves}$ as a representative of an isomorphism class. Each isomorphism test for hyperelliptic curves is done by a method described in Subsect. 2.1. Finally, we return the list $\mathsf{AllHypHoweCurves}$.

5.2 Criterion for Efficiency and Complexity Analysis

In this subsection, we shall prove some criteria for making isomorphism classification in Algorithm 1 efficient, and then determine the total complexity of Algorithm 1. Since **Step 3** of Algorithm 1 is expected to produce $O(p^2)$ s.sp. genus-4 hyperelliptic curves in practice (cf. Remark 4 below), the number of required isomorphism tests is $O(p^4)$ naively. However, in fact, it suffices to do $O(p^2)$ isomorphism tests, due to our criteria below.

The first criterion is Lemma 4, which implies that it suffices to do isomorphism classification on (C_1, C_2)'s with the same invariants pair $(j_{\mathrm{abs}}(C_1), j_{\mathrm{abs}}(C_2))$.

Lemma 4. *If $D^{(1)}$ and $D^{(2)}$ are isomorphic, then $\{C_1^{(1)}, C_2^{(1)}\}$ are isomorphic to $\{C_1^{(2)}, C_2^{(2)}\}$, where $D^{(i)}$ is the hyperelliptic curve with $\mathrm{Aut}(D^{(i)}) \cong V_4$ constructed from $C_1^{(i)}$ and $C_2^{(i)}$ as in the equation* (3.1.1) *for $i = 1$ and 2.*

Proof. There are exactly three order-2 subgroups of V_4 and one of them is the hyperelliptic involution on each of $D^{(1)}$ and $D^{(2)}$. The other two make the double covers $D^{(i)} \to C_1^{(i)}$ and $D^{(i)} \to C_2^{(i)}$ for each i. Thus the lemma follows. $\square$

The second criterion is Lemma 5. This lemma shows that the number of (C_1, C_2)'s which have the same absolute invariants pair does not depend on p.

Lemma 5. *Let C_1 and C_2 be curves of genus 2. Then the cardinality of the set $\{(a_4, a_5, b_1, b_2) \mid C_{a_4,a_5,b_1} \cong C_1,\ C_{a_4,a_5,b_2} \cong C_2\}$ is bounded by a constant independent of p, where $C_{a_4,a_5,b}$ is as in* (5.1.1).

Proof. It suffices to show that for a given curve C of genus 2, the number of (a_4, a_5, b) such that $C_{a_4,a_5,b} \cong C$ is bounded by a constant independent of p. It is bounded by the number (explicitly 720) of arrangements of the ramification points of the double cover $C \to \mathbb{P}^1$. Indeed for each arrangement, say $P_1, \ldots, P_6$, we obtain a $C_{a_4,a_5,b}$ by sending the first three ramification points P_1, P_2 and P_3 to 0, 1 and ∞ respectively in this order by an element h of $\mathrm{Aut}(\mathbb{P}^1)$ and by setting $a_i := h(P_i)$ for $i = 4$ and 5, and $b := h(P_6)$. □

Remark 4. Let $M(p)$ be the number of pairs (C_1, C_2) of s.sp. genus-2 curves which share 5 ramified points up to $\mathrm{Aut}(\mathbb{P}^1)$, namely $M(p) = $ #HowePairsList. Then $M(p) = O(p^2)$ would be expected. Indeed the number of (C_1, C_2) of isomorphism classes of s.sp. genus-2 curves is $O(p^6)$ and the probability that (a_4, a_5)'s of C_1 and C_2 coincide can be expected to be $1/p^4$. Moreover, it can be expected that the genus-4 curves constructed by "almost all" pairs have the automorphism group V_4 (see also #HowePairsV4 in Table 1 in practice), and there is at most constant contribution from taking account of isomorphisms of curves of genus 4.

Now we determine the complexity of the main algorithm. All the complexities are measured by the number of arithmetic operations in $\mathbb{F}_{p^4}$.

Proposition 3. *Under an assumption that $M(p) = O(p^2)$, the total complexity of Algorithm 1 is bounded by $O(p^3)$.*

Proof. **Step 1** is done in $O(p^3)$, as noted in Subsect. 5.1. Note that the number of generated s.sp. genus-2 curves is estimated as $O(p^3)$ by [15, Theorem 3.3], and thus the number of iterations in **Step (2)** is also $O(p^3)$.

In **Step (2)**, we produce $6 \times 5 \times 4 = 120$ imaginary models for each s.sp. curve of genus 2, and for each imaginary model, we also construct 3 ordered triples $(a_4, a_5, b) \in (\mathbb{F}_{p^4})^3$ with $a_4 \leq a_5$. Since the cost of computing an imaginary model and a triple is constant in p, the complexity of **Step (2)** is estimated as the size of a generated hash table (see Subsect. 5.1 for details), say $O(p^3)$.

Note that the number of constructed genus-4 hypperelliptic Howe pairs (C_1, C_2) (stored as $(a_4, a_5, b_1, a'_4, a'_5, b_2)$ in HowePairsList) is $O(p^2)$ by our assumption, which comes from our heuristic (Remark 4).

Step 3 constructs a s.sp. hyperelliptic curve D of genus 4 for each pair of (a_4, a_5, b_1) and (a_4, a_5, b_2), and computes the number of its automorphisms. The construction of each D requires a constant number of operations in $\mathbb{F}_{p^4}$. Recall from Subsect. 2.2 that the computation of automorphisms on D is done in constant time with respect to p. Therefore, the complexity of collecting (C_1, C_2) such that the corresponding D satisfies $\mathrm{Aut}(D) = V_4$ is $O(p^2)$. Since the absolute invariants $j_{\rm abs}(C_1)$ and $j_{\rm abs}(C_2)$ are computed in constant time with respect to p, the complexity of this step is estimated as $O(p^2)$.

Step (4-1) sorts the list HowePairsV4 which has $O(p^2)$ elements, and thus its complexity is $O(p^2 \log(p^2)) = \tilde{O}(p^2)$. It follows from Lemma 5 that the size of each group is bounded by a constant, and thus the number of groups is $O(p^2)$.

In **(4-2)**, we classify the isomorphism classes of s.sp. hyperelliptic curves in each of the groups constructed in **(4-1)**. Recall from Subsect. 2.1 that each isomorphism test is done in constant time with respect to p, and hence the complexity of **(4-2)** is $O(p^2)$. □

6 Concluding Remarks

In this paper, we proposed an algorithm for enumerating (or finding) s.sp. hyperelliptic curves of genus 4 with V_4-automorphism group (Theorem A with Algorithm 1). We characterized such curves as generalized Howe curves, and constructed the algorithm as a hyperellptic analogue of an algorithm in [23] together with our criteria for efficient isomorphism classification. Unlike [21], our algorithm need not to solve any multivariate system to collect our s.sp. curves. The complexity of the algorithm was shown to be $O(p^3)$ (under some practical assumption), which is asymptotically faster than the algorithm of [21], and than the most efficient existing algorithm [23] with complexity $\tilde{O}(p^4)$ for the non-hyperelliptic case. By our implementation of the algorithm over Magma, we enumerated the isomorphism classes of s.sp. hyperelliptic curves of genus 4 with automorphism group V_4 in characteristic $17 \leq p \leq 200$ (Theorem B (ii)). We also succeeded in finding a s.sp. hyperelliptic curve of genus 4 in every characteristic p with $19 \leq p \leq 6691$ (Theorem B (i)). From our computational results on the existence, we also expect that the following conjecture would be true:

Conjecture 1. There exists a s.sp. hyperelliptic curve of genus 4 with automorphism group containing V_4 for arbitrary characteristic $p \geq 19$.

A theoretical proof of this conjecture and that of the $\mathbb{F}_{p^2}$-maximality or minimality of obtained s.sp. curves (cf. Remark 3) are future works.

Acknowledgements. The authors thank the referees for valuable comments and suggestions. This work was supported by JSPS Grant-in-Aid for Young Scientists 20K14301, and JSPS Grant-in-Aid for Scientific Research (C) 21K03159.

References

1. Auer, R., Top, J.: Legendre elliptic curves over finite fields. J. Number Theory **95**(2), 303–312 (2002)
2. Bosma, W., Cannon, J., Playoust, C.: The Magma algebra system. I. The user language. J. Symb. Comput. **24**, 235–265 (1997)
3. Brock, B.W.: Superspecial curves of genera two and three. Thesis (Ph.D.)-Princeton University, 69 pp (1993)

4. Castryck, W., Decru, T., Benjamin, S.: Hash functions from superspecial genus-2 curves using Richelot isogenies. J. Math. Cryptol. **14**, 268–292 (2020)
5. Deuring, M.: Die Typen der Multiplikatorenringe elliptischer Funktionenkörper. Abh. Math. Sem. Univ. Hamburg **14**(1), 197–272 (1941)
6. Ekedahl, T.: On supersingular curves and abelian varieties. Math. Scand. **60**, 151–178 (1987)
7. Fuhrmann, R., Garcia, A., Torres, F.: On maximal curves. J. Number Theory **67**, 29–51 (1997)
8. Gutierrez, J., Shaska, T.: Hyperelliptic curves with extra involutions. LMS J. Comput. Math. **8**, 102–115 (2005)
9. Hartshorne, R.: Algebraic Geometry. GTM, vol. 52. Springer, New York (1977). https://doi.org/10.1007/978-1-4757-3849-0
10. Hashimoto, H.: Class numbers of positive definite ternary quaternion Hermitian forms. Proc. Japan Acad. Ser. A Math. Sci. **59**(10), 490–493 (1983)
11. Hashimoto, K., Ibukiyama, T.: On class numbers of positive definite binary quaternion Hermitian forms II. J. Fac. Sci. Univ. Tokyo Sect. IA Math. **28**(3), 695–699 (1982)
12. Howe, E.W.: Quickly constructing curves of genus 4 with many points. In: Kohel, D., Shparlinski, I. (eds.) Frobenius Distributions: Sato-Tate and Lang-Trotter conjectures Contemporary Mathematics, vol. 663, pp. 149–173. American Mathematical Society, Providence (2016)
13. Howe, E.W., Leprévost, F., Poonen, B.: Large torsion subgroups of split Jacobians of curves of genus two or three. Forum Math. **12**(3), 315–364 (2000)
14. Hurt, N.E.: Many Rational Points: Coding Theory and Algebraic Geometry. Kluwer Academic Publishers, Dordrecht (2003)
15. Ibukiyama, T., Katsura, T., Oort, F.: Supersingular curves of genus two and class numbers. Compositio Math. **57**(2), 127–152 (1986)
16. Igusa, J.-I.: Arithmetic variety of moduli for genus two. Ann. Math. **72**, 612–649 (1960)
17. Jordan, B.W., Zaytman, Y.: Isogeny graphs of superspecial abelian varieties and generalized Brandt matrices arXiv: 2005.09031v4
18. Kani, E., Rosen, M.: Idempotent relations and factors of Jacobians. Math. Ann. **284**, 307–327 (1989)
19. Katsura, T., Takashima, K.: Decomposed Richelot isogenies of Jacobian varieties of hyperelliptic curves and generalized Howe curves arXiv: 2108.06936
20. Kudo, M., Harashita, S.: Superspecial curves of genus 4 in small characteristic. Finite Fields Appl. **45**, 131–169 (2017)
21. Kudo, M., Harashita, S.: Superspecial hyperelliptic curves of genus 4 over small finite fields. In: Budaghyan, L., Rodríguez-Henríquez, F. (eds.) WAIFI 2018. LNCS, vol. 11321, pp. 58–73. Springer, Cham (2018). https://doi.org/10.1007/978-3-030-05153-2_3
22. Kudo, M., Harashita, S.: Computational approach to enumerate non-hyperelliptic superspecial curves of genus 4. Tokyo J. Math. **43**(1), 259–278 (2020). https://doi.org/10.3836/tjm/1502179310
23. Kudo, M., Harashita, S., Howe, E.W.: Algorithms to enumerate superspecial Howe curves of genus 4. In: Proceedings of Fourteenth Algorithmic Number Theory Symposium (ANTS-XIV). Open Book Series, vol. 4 no. 1, pp. 301–316 (2020)
24. Kudo, M., Harashita, S., Senda, H.: The existence of supersingular curves of genus 4 in arbitrary characteristic. Res. Number Theory **6** (2020). Article Number 44

25. Li, K.-Z., Oort, F.: Moduli of Supersingular Abelian Varieties. Lecture Notes in Mathematics, vol. 1680. Springer, Heidelberg (1998). https://doi.org/10.1007/BFb0095931
26. Paulhus, J.: Decomposing Jacobians of curves with extra automorphisms. Acta Arith **132**, 231–244 (2008)
27. Stichtenoth, H.: Über die Automorphismengruppe eines algebraischen Funktionenkörpers von Primzahlcharakteristik. I. Eine Abschätzung der Ordnung der Automorphismengruppe. Arch. Math. **24**, 527–544 (1973)
28. https://sites.google.com/view/m-kudo-official-website/english/code/genus4hypv4

Coding Theory

Two Classes of Constacyclic Codes with Variable Parameters

Cunsheng Ding[1(✉)], Zhonghua Sun[2], and Xiaoqiang Wang[3]

[1] Department of Computer Science and Engineering, The Hong Kong University of Science and Technology, Clear Water Bay, Kowloon, Hong Kong, China
cding@ust.hk

[2] School of Mathematics, Hefei University of Technology, Hefei 230601, Anhui, China

[3] Hubei Key Laboratory of Applied Mathematics, Faculty of Mathematics and Statistics, Hubei University, Wuhan 430062, China

Abstract. Constacyclic codes over finite fields are a family of linear codes and contain cyclic codes as a subclass. Constacyclic codes are closely related to many areas of mathematics and outperform cyclic codes in several aspects. Hence, constacyclic codes are of theoretical importance. On the other hand, constacyclic codes are important in practice, as they have rich algebraic structures and may have efficient decoding algorithms. In this extended abstract, two classes of constacyclic codes are constructed using a general construction of constacyclic codes with cyclic codes. The first class of constacyclic codes is motivated by the punctured Dilix cyclic codes, and the second class is motivated by the punctured generalised Reed-Muller codes. The two classes of constacyclic codes contain optimal linear codes. The parameters of the two classes of constacyclic codes are analysed, and some open problems are presented in this extended abstract.

Keywords: Constacyclic code · Cyclic code · Dilix code · Generalised Reed-Muller code · Projective Reed-Muller code

1 Introduction and Motivations

1.1 Constacyclic Codes and Cyclic Codes

Let $\mathrm{GF}(q)$ be the finite field with q elements, and let $\mathrm{GF}(q)^*$ denote the multiplicative group of $\mathrm{GF}(q)$. By an $[n, k, d]$ code $\mathcal{C}$ over $\mathrm{GF}(q)$ we mean a k-dimensional linear subspace of $\mathrm{GF}(q)^n$ with minimum distance d. Let $\lambda \in \mathrm{GF}(q)^*$. A linear code $\mathcal{C}$ of length n is said to be λ-*constacyclic* if $(c_0, c_1, \ldots, c_{n-1})$

C. Ding's research was supported by The Hong Kong Research Grants Council, Proj. No. 16301522, Z. Sun's research was supported by The National Natural Science Foundation of China under Grant Number 62002093. X. Wang's research was supported by The National Natural Science Foundation of China under Grant Number 12001175.

S. Mesnager and Z. Zhou (Eds.): WAIFI 2022, LNCS 13638, pp. 127–141, 2023.
https://doi.org/10.1007/978-3-031-22944-2_7

$\in \mathcal{C}$ implies $(\lambda c_{n-1}, c_0, c_1, \ldots, c_{n-2}) \in \mathcal{C}$. Let Φ be the mapping from $\mathrm{GF}(q)^n$ to the quotient ring $\mathrm{GF}(q)[x]/\langle x^n - \lambda \rangle$ defined by

$$\Phi((c_0, c_1, \ldots, c_{n-1})) = \sum_{i=0}^{n-1} c_i x^i.$$

It is well known that every ideal of the ring $\mathrm{GF}(q)[x]/\langle x^n - \lambda \rangle$ is *principal* and a linear code $\mathcal{C} \subseteq \mathrm{GF}(q)^n$ is λ-constacyclic if and only if $\Phi(\mathcal{C})$ is an ideal of $\mathrm{GF}(q)[x]/\langle x^n - \lambda \rangle$. Consequently, we will identify $\mathcal{C}$ with $\Phi(\mathcal{C})$ for any λ-constacyclic code $\mathcal{C}$. Let $\mathcal{C} = \langle g(x) \rangle$ be a λ-constacyclic code over $\mathrm{GF}(q)$, where $g(x)$ is monic and has the smallest degree. Then $g(x)$ is called the *generator polynomial* and $h(x) = (x^n - \lambda)/g(x)$ is referred to as the *check polynomial* of $\mathcal{C}$. The dual code $\mathcal{C}^\perp$ of $\mathcal{C}$ is a λ^{-1}-constacyclic code generated by the reciprocal polynomial of the check polynomial $h(x)$ of $\mathcal{C}$. By definition, 1-constacyclic codes are the classical cyclic codes. Hence, cyclic codes form a subclass of constacyclic codes. In other words, constacyclic codes are a generalisation of the classical cyclic codes. For more information on constacyclic codes over finite fields, the reader is referred to [5–9,13,18,20,21,24,25,27,30–32,35] and the references therein.

1.2 Motivations and Objectives

By definition, cyclic codes are a proper subclass of constacyclic codes and constacyclic codes are a proper subclass of linear codes. Clearly, cyclic codes have a better algebraic structure than λ-constacyclic codes with $\lambda \neq 1$ and constacyclic codes have a better algebraic structure than other linear codes. A better algebraic structure may mean a better decoding algorithm. Then the following two questions are interesting and good motivations for studying constacyclic codes.

Question 1. Is a given linear code over $\mathrm{GF}(q)$ monomially-equivalent to a cyclic code over $\mathrm{GF}(q)$?

Question 2. Is a given linear code over $\mathrm{GF}(q)$ monomially-equivalent to a λ-constacyclic code over $\mathrm{GF}(q)$ with $\lambda \neq 1$?

For example, the Hamming code of length $(q^m - 1)/(q - 1)$ over $\mathrm{GF}(q)$ is monomially-equivalent to a cyclic code over $\mathrm{GF}(q)$ when $\gcd(m, q-1) = 1$, and is always monomially-equivalent to a contacyclic code over $\mathrm{GF}(q)$. This shows that the Hamming code is attractive. Notice that the two questions are open for most linear codes.

Recall that cyclic codes have a better algebraic structure. Then one would ask why we would study constacyclic codes. Below is a list of motivations for studying λ-constacyclic codes with $\lambda \neq 1$:

- There does not exist a cyclic code over $\mathrm{GF}(q)$ with parameters $[n, k, d]$ for certain q, n, k and d; but there is a λ-constacyclic codes over $\mathrm{GF}(q)$ with parameters $[n, k, d]$.

- The best $[n, k]$ constacyclic code over $\mathrm{GF}(q)$ has a better error-correcting capability than the best $[n, k]$ cyclic code over $\mathrm{GF}(q)$ for certain q, n and k.
- Constacyclic codes can do many things that cyclic codes cannot do. For example, the Hamming code of length $(q^m - 1)/(q - 1)$ can always be constructed as a constacyclic code, but cannot be constructed as a cyclic code when $\gcd(q - 1, m) \neq 1$.

The original binary Reed-Muller codes were introduced by Reed and Muller in 1954 [26,29]. They are called geometric codes, as all the minimum weight codewords of the r-th order Reed-Muller code $\mathcal{R}_2(r, m)$ are the incidence vectors of all the $(m - r)$-flats in the affine geometry $\mathrm{AG}(m, \mathrm{GF}(2))$ and they generate $\mathcal{R}_2(r, m)$ [2]. The automorphism group of $\mathcal{R}_2(r, m)$ is known to be the general affine group $\mathrm{GA}_m(\mathrm{GF}(2))$, which is triply transitive on $\mathrm{GF}(2)^m$. Hence, the binary Reed-Muller codes support 3-designs. It was later discovered that the binary Reed-Muller codes become cyclic codes if they are punctured in a special coordinate position. These properties show that the original Reed-Muller codes are very interesting in theory. Binary Reed-Muller codes are also interesting in practice as they have efficient decoding algorithms [29]. The binary Reed-Muller codes and their punctured codes were later generalised to codes over $\mathrm{GF}(q)$ for general q. In 2018, the binary Reed-Muller codes were generalised into another type of linear codes [10], which were called *Dilix codes* for the purpose of distinguishing the two types of generalisations [11, Chapter 6]. The Dilix codes have also interesting properties and are extended cyclic codes by definition. In other words, if the Dilix codes are punctured in the last coordinate, the punctured Dilix codes are cyclic. Motivated by the punctured generalized Reed-Muller codes and punctured Dilix codes, the objective of this extended abstract is to construct and analyse two classes of constacyclic codes.

2 Preliminaries

Throughout this extended abstract, we fix the following notation, unless it is stated otherwise:

- q is a prime power.
- $m \geq 2$ is an integer.
- r is a positive divisor of $q - 1$.
- $N = q^m - 1$.

For a linear code $\mathcal{C}$, we use $\dim(\mathcal{C})$ and $d(\mathcal{C})$ to denote its dimension and minimum Hamming distance, respectively. For a linear code $\mathcal{C} \subset \mathrm{GF}(q)^n$, let A_i denote the number of codewords with Hamming weight i in $\mathcal{C}$. The *weight enumerator* of $\mathcal{C}$ is defined as $1 + A_1 z + \cdots + A_n z^n$. The sequence $(1, A_1, \ldots, A_n)$ is called the *weight distribution* of $\mathcal{C}$. If the number of nonzero A_i in the sequence $(A_1, A_2, \ldots, A_n)$ equals t, then $\mathcal{C}$ is called a t-weight code.

2.1 The Hamming Weight and q-weight of Nonnegative Integers

Let $N = q^m - 1$. For each $i \in Z_N$, let the q-adic expansion of i be $i = \sum_{j=0}^{m-1} i_j q^j$, where $0 \le i_j \le q-1$. The *Hamming weight* of i, denoted by $\mathtt{wt}(i)$, is defined to be the Hamming weight of the vector $(i_0, i_1, \ldots, i_{m-1})$. The *$q$-weight* of i, denoted by $\mathtt{wt}_q(i)$, is defined to be $\sum_{j=0}^{m-1} i_j$.

2.2 Cyclotomic Cosets

Let q be a prime power, n be a positive integer with $\gcd(q, n) = 1$, r be a positive divisor of $q - 1$, and let λ be an element of GF(q) with order r. To deal with λ-constacyclic codes of length n over GF(q), we have to study the factorization of $x^n - \lambda$ over GF(q). To this end, we need to introduce q-cyclotomic cosets modulo rn.

Let $\mathbb{Z}_{rn} = \{0, 1, 2, \cdots, rn-1\}$ be the ring of integers modulo rn. For any $i \in \mathbb{Z}_{rn}$, the *q-cyclotomic coset of i modulo rn* is defined by

$$C_i^{(q,rn)} = \{i, iq, iq^2, \cdots, iq^{\ell_i - 1}\} \bmod rn \subseteq \mathbb{Z}_{rn},$$

where ℓ_i is the smallest positive integer such that $i \equiv iq^{\ell_i} \pmod{rn}$, and is the *size* of the q-cyclotomic coset $C_i^{(q,rn)}$. The smallest integer in $C_i^{(q,rn)}$ is called the *coset leader* of $C_i^{(q,rn)}$. Let $\Gamma_{(q,rn)}$ be the set of all the coset leaders. We have then $C_i^{(q,rn)} \cap C_j^{(q,rn)} = \emptyset$ for any two distinct elements i and j in $\Gamma_{(q,rn)}$, and

$$\bigcup_{i \in \Gamma_{(q,rn)}} C_i^{(q,rn)} = \mathbb{Z}_{rn}.$$

Let $m = \mathrm{ord}_{rn}(q)$. It is easily seen that there is a primitive element α of GF(q^m) such that $\beta = \alpha^{(q^m-1)/rn}$ and $\beta^n = \lambda$. Then β is a primitive rn-th root of unity in GF(q^m). The *minimal polynomial* $\mathbb{M}_{\beta^i}(x)$ of β^i over GF(q) is the monic polynomial of the smallest degree over GF(q) with β^i as a zero. We have $\mathbb{M}_{\beta^i}(x) = \prod_{j \in C_i^{(q,rn)}} (x - \beta^j) \in \mathrm{GF}(q)[x]$, which is irreducible over GF(q). It then follows that $x^{rn} - 1 = x^{rn} - \lambda^r = \prod_{i \in \Gamma_{(q,rn)}} \mathbb{M}_{\beta^i}(x)$. Define

$$\Gamma_{(q,rn,r)}^{(1)} = \{i : i \in \Gamma_{(q,rn)},\ i \equiv 1 \pmod{r}\}.$$

Then $x^n - \lambda = \prod_{i \in \Gamma_{(q,rn,r)}^{(1)}} \mathbb{M}_{\beta^i}(x)$.

2.3 Automorphism Groups and Equivalence of Linear Codes

Two linear codes $\mathcal{C}_1$ and $\mathcal{C}_2$ are said to be *permutation-equivalent* if there is a permutation of coordinates which sends $\mathcal{C}_1$ to $\mathcal{C}_2$. This permutation could be described by employing a *permutation matrix*, which is a square matrix with exactly one 1 in each row and column and 0s elsewhere. The set of coordinate permutations that map a code $\mathcal{C}$ to itself forms a group, which is referred to as the *permutation automorphism group* of $\mathcal{C}$ and denoted by $\mathrm{PAut}(\mathcal{C})$.

A *monomial matrix* over $\mathrm{GF}(q)$ is a square matrix having exactly one nonzero element of $\mathrm{GF}(q)$ in each row and column. A monomial matrix M can be written either in the form DP or the form PD_1, where D and D_1 are diagonal matrices and P is a permutation matrix.

Let $\mathcal{C}_1$ and $\mathcal{C}_2$ be two linear codes of the same length over $\mathrm{GF}(q)$. Then $\mathcal{C}_1$ and $\mathcal{C}_2$ are said to be *monomially-equivalent* if there is a monomial matrix over $\mathrm{GF}(q)$ such that $\mathcal{C}_2 = \mathcal{C}_1 M$. Monomial equivalence and permutation equivalence are precisely the same for binary codes. If $\mathcal{C}_1$ and $\mathcal{C}_2$ are monomially-equivalent, then they have the same weight distribution. The set of monomial matrices that map $\mathcal{C}$ to itself forms the group $\mathrm{MAut}(\mathcal{C})$, which is called the *monomial automorphism group* of $\mathcal{C}$. By definition, we have $\mathrm{PAut}(\mathcal{C}) \subseteq \mathrm{MAut}(\mathcal{C})$. Two linear codes $\mathcal{C}_1$ and $\mathcal{C}_2$ of the same length over $\mathrm{GF}(q)$ are said to be *scalar-equivalent* if there is an invertible diagonal matrix D over $\mathrm{GF}(q)$ such that $\mathcal{C}_2 = \mathcal{C}_1 D$.

Two codes $\mathcal{C}_1$ and $\mathcal{C}_2$ are said to be *equivalent* if there is a monomial matrix M and an automorphism γ of $\mathrm{GF}(q)$ such that $\mathcal{C}_1 = \mathcal{C}_2 M\gamma$. All three are the same if the codes are binary; monomial equivalence and equivalence are the same if the field considered has a prime number of elements.

The *automorphism group* of $\mathcal{C}$, denoted by $\mathrm{Aut}(\mathcal{C})$, is the set of maps of the form $M\gamma$, where M is a monomial matrix and γ is a field automorphism, that map $\mathcal{C}$ to itself. In the binary case, $\mathrm{PAut}(\mathcal{C})$, $\mathrm{MAut}(\mathcal{C})$ and $\mathrm{Aut}(\mathcal{C})$ are the same. If q is a prime, $\mathrm{MAut}(\mathcal{C})$ and $\mathrm{Aut}(\mathcal{C})$ are identical. In general, we have

$$\mathrm{PAut}(\mathcal{C}) \subseteq \mathrm{MAut}(\mathcal{C}) \subseteq \mathrm{Aut}(\mathcal{C}).$$

In this extended abstract, we consider the monomial equivalence of linear codes. Two monomially-equivalent codes have the same parameters and weight distribution. If a linear code $\mathcal{C}$ is monomially-equivalent to a constacyclic code $\mathcal{C}_1$, we prefer $\mathcal{C}_1$ to $\mathcal{C}$ as constacyclic codes have a better algebraic structure than general linear codes.

2.4 The Projective Reed-Muller Codes

Let q be a power of a prime p and let $m \geq 2$. A point of the projective geometry $\mathrm{PG}(m-1, \mathrm{GF}(q))$ is given in homogeneous coordinates by $(x_1, x_2, \ldots, x_m)$ where all x_i are in $\mathrm{GF}(q)$ and are not all zero. Each point of $\mathrm{PG}(m-1, \mathrm{GF}(q))$ has $q-1$ coordinate representations, as $(ax_1, x_2, ..., ax_m)$ and $(x_1, x_2, ..., x_m)$ generate the same 1-dimensional subspace of $\mathrm{GF}(q)^m$ for any nonzero $a \in \mathrm{GF}(q)$.

Let $\mathrm{GF}(q)[x_1, x_2, \ldots, x_m]$ be the set of polynomials in m indeterminates over $\mathrm{GF}(q)$, which is a linear space over $\mathrm{GF}(q)$. Let $A(q, m, h)$ be the subspace of $\mathrm{GF}(q)[x_1, x_2, \ldots, x_m]$ generated by all the homogeneous polynomials of degree h. Let $\{\mathbf{x}^1, \mathbf{x}^2, \cdots, \mathbf{x}^n\}$ be a set of projective points in $\mathrm{PG}(m-1, \mathrm{GF}(q))$, where $n = (q^m-1)/(q-1)$. Then, the *h-th order projective Reed-Muller code* $\mathrm{PRM}(q, m, h)$ of length n is defined by

$$\mathrm{PRM}(q, m, h) = \left\{ \left(f(\mathbf{x}^1), f(\mathbf{x}^2), \ldots, f(\mathbf{x}^n) \right) : f \in A(q, m, h) \right\}.$$

The code $\mathrm{PRM}(q,m,h)$ depends on the choice of the set $\{\mathbf{x}^1, \mathbf{x}^2, \cdots, \mathbf{x}^n\}$ of coordinate representatives of the point set in $\mathrm{PG}(m-1, \mathrm{GF}(q))$, but is unique up to monomial equivalence (in fact, up to scalar equivalence). The parameters of $\mathrm{PRM}(q,m,h)$ are known and documented in the following theorem [3,19,34].

Theorem 1. *Let $m \geq 2$ and $1 \leq h \leq (m-1)(q-1)$. Then the linear code $\mathrm{PRM}(q,m,h)$ has length $n = (q^m-1)/(q-1)$ and minimum distance $(q-v)q^{m-2-u}$, where $h-1 = u(q-1)+v$ and $0 \leq v < q-1$. Furthermore,*

$$\dim(\mathrm{PRM}(q,m,h)) = \sum_{\substack{t \equiv h \pmod{q-1} \\ 0<t\leq h}} \left(\sum_{j=0}^{m} (-1)^j \binom{m}{j} \binom{t-jq+m-1}{t-jq} \right).$$

By Theorem 1 and definition, $\mathrm{PRM}(q,m,1)$ is monomially-equivalent to the Simplex code. The weight distribution of $\mathrm{PRM}(q,m,2)$ was settled in [22]. It was pointed out in [4] that the code $\mathrm{PRM}(q,m,h)$ is not cyclic in general, but could be cyclic or quasi-cyclic under special conditions. Later in this extended abstract, we will compare some newly constructed constacyclic codes with the projective Reed-Muller codes. This explains why we introduced the projective Reed-Muller codes here.

2.5 Projective Generalized Reed-Muller Codes

For an integer $h \geq 0$, let $\mathrm{PP}(q,m,h)$ be the linear subspace of $\mathrm{GF}(q)[x_1, x_2, \ldots, x_m]$, which is spanned by all monomials $x_1^{i_1} x_2^{i_2} \cdots x_m^{i_m}$ satisfying the following two conditions:

1. $\sum_{j=1}^{m} i_j \equiv 0 \pmod{q-1}$,
2. $\sum_{j=1}^{m} i_j \leq h(q-1)$.

Each $a \in \mathrm{GF}(q)$ is viewed as the constant function $f_a(x_1, x_2, \ldots, x_m) \equiv a$.

Let $\{\mathbf{x}^1, \mathbf{x}^2, \ldots, \mathbf{x}^n\}$ be the set of projective points in $\mathrm{PG}(m-1, \mathrm{GF}(q))$, where $n = \frac{q^m-1}{q-1}$. Then, the *h-th order projective generalized Reed-Muller code* $\mathrm{PGRM}(q,m,h)$ of length n is defined by

$$\mathrm{PGRM}(q,m,h) = \left\{ \left(f(\mathbf{x}^1), f(\mathbf{x}^2), \ldots, f(\mathbf{x}^n)\right) : f \in \mathrm{PP}(q,m,h) \right\}.$$

Theorem 2. *Let $0 \leq h \leq m-1$. Then, the minimum weight of* $\mathrm{PGRM}(q,m,h)$ *is $\frac{q^{m-h}-1}{q-1}$ and*

$$\dim(\mathrm{PGRM}(q,m,h)) = \left| \left\{ 0 \leq j \leq \frac{q^m-1}{q-1} : \mathtt{wt}_q(j(q-1)) \leq h(q-1) \right\} \right|. \quad (1)$$

Note that the minimum distance of $\mathrm{PGRM}(q,m,h)$ is known to be $\frac{q^{m-h}-1}{q-1}$. But the expression in (1) is not specific, and there is no known specific formula for $\dim(\mathrm{PGRM}(q,m,h))$. Later, we will compare the codes $\mathrm{PGRM}(q,m,h)$ with the constacyclic codes presented in this extended abstract. To this end, we present the following example.

Example 1. The parameters of the codes PGRM(3, 4, h) for $0 \le h \le 3$ are given below.

$$[40, 1, 40],\ [40, 11, 13],\ [40, 30, 4],\ [40, 40, 1].$$

2.6 The Punctured Dilix Codes

In this subsection, we outline a type of cyclic codes, called *punctured Dilix codes* [10]. Let m be a positive integer and let $N = q^m - 1$, where q is a prime power. Let β be a primitive element of GF(q^m). For any $1 \le h \le m$, we define a polynomial

$$\omega_{(q,m,h)}(x) = \prod_{\substack{1 \le i \le N-1 \\ 1 \le \mathtt{wt}(i) \le h}} (x - \beta^i).$$

Since $\mathtt{wt}(i)$ is a constant function on each q-cyclotomic coset modulo N, $\omega_{(q,m,h)}(x)$ is a polynomial over GF(q). By definition, $\omega_{(q,m,h)}(x)$ is a divisor of $x^N - 1$. Let $\Omega(q, m, h)$ denote the cyclic code over GF(q) with length N and generator polynomial $\omega_{(m,q,h)}(x)$.

Theorem 3. *[10] Let $m \ge 2$ and $1 \le h \le m-1$. Then $\Omega(q, m, h)$ has parameters $[N, k, d \ge (q^{h+1} - 1)/(q - 1)]$, where*

$$k = q^m - \sum_{i=0}^{h} \binom{m}{i} (q - 1)^i.$$

Later, we will use the codes $\Omega(q, m, h)$ to construct some constacyclic codes. This explains why we introduced the punctured Dilix codes $\Omega(q, m, h)$ here.

3 A General Construction of Constacyclic Codes of Length $\frac{q^m-1}{r}$ with Cyclic Codes of Length $q^m - 1$

In this section, we present a general construction of constacyclic codes of length $\frac{q^m-1}{r}$ with cyclic codes of length $q^m - 1$ over GF(q), where r is a positive divisor of $q - 1$. Throughout this section, let $n = \frac{q^m-1}{r}$, where m is an integer with $m \ge 2$. Define $N = rn = q^m - 1$. Let β be a primitive element of GF(q^m) and $\lambda = \beta^n$. Then λ is an element of GF(q)* with order r.

Let $\mathcal{C}$ be a cyclic code of length N over GF(q) with generator polynomial

$$g(x) = \prod_{i \in D(\mathcal{C})} (x - \beta^i),$$

where $D(\mathcal{C})$ is the union of some q-cyclotomic cosets modulo N and is called the *defining set* of $\mathcal{C}$ with respect to the primitive element β of GF(q^m). Put

$$\underline{D}(\mathcal{C}) = \{i \in D(\mathcal{C}) : i \equiv 1 \pmod{r}\}.$$

If $\underline{D}(\mathcal{C}) = \emptyset$, then define $\underline{g}(x) = 1$. If $\underline{D}(\mathcal{C}) \neq \emptyset$, then define

$$\underline{g}(x) = \prod_{i \in \underline{D}(\mathcal{C})} (x - \beta^i).$$

Then the following hold:

1. $\underline{g}(x)$ is a polynomial over GF(q).
2. $\underline{g}(x) = \gcd(g(x), x^n - \lambda)$.

Let $\underline{\mathcal{C}}$ denote the λ-constacyclic code of length n over GF(q) with generator polynomial $\underline{g}(x)$. By definition, $\underline{\mathcal{C}}$ is constructed from the given cyclic code $\mathcal{C}$.

By definition,

$$\dim(\mathcal{C}) = N - \deg(g) = N - |D(\mathcal{C})|$$

and

$$\dim(\underline{\mathcal{C}}) = n - \deg(\underline{g}) = n - |\underline{D}(\mathcal{C})|.$$

Hence, it may not be easy to determine $\dim(\underline{\mathcal{C}})$ even if $\dim(\mathcal{C})$ is known. However, this may be possible in some special cases. By definition, there is no clear connection between $d(\underline{\mathcal{C}})$ and $d(\mathcal{C})$ in general.

Later in this extended abstract, we will use this general construction to obtain two classes of λ-constacyclic codes of length $(q^m - 1)/(q - 1)$ over GF(q). Specifically, we will consider only the special case $r = q - 1$ in this extended abstract.

Example 2. Let $q > 2$ be a prime power and let $m \geq 2$ and $r = q - 1$. Let β be a primitive element of GF(q^m) and $\lambda = \beta^{(q^m-1)/(q-1)}$. Let $\mathcal{C}$ denote the cyclic code of length $N = q^m - 1$ with generator polynomial $g(x) = \mathbb{M}_\beta(x)\mathbb{M}_{\beta^{q+1}}(x)$. It is easily seen that $\underline{D}(\mathcal{C}) = C_1^{(q,N)}$ and $\underline{g}(x) = \mathbb{M}_\beta(x)$. Then the λ-constacyclic code $\underline{\mathcal{C}}$ is the Hamming code and $\underline{\mathcal{C}}^\perp$ is the Simplex code.

4 The First Class of Constacyclic Codes

We follow the previous notation. Throughout this section, let $n = \frac{q^m-1}{q-1}$, where m is an integer with $m \geq 2$. Define $N = rn = q^m - 1$, where $r = q - 1$. Then it is easily seen that $\mathrm{ord}_n(q) = \mathrm{ord}_N(q) = m$. Let $\Gamma_{(q,N)}$ be the set of q-cyclotomic coset leaders modulo N and let

$$\Gamma^{(1)}_{(q,N,q-1)} = \{i : i \in \Gamma_{(q,N)}, \ i \equiv 1 \pmod{q-1}\}.$$

Let β be a primitive element of GF(q^m) and let $\lambda = \beta^{(q^m-1)/(q-1)}$. Then λ is a primitive element of GF(q). Let ℓ be a positive integer with $1 \leq \ell \leq m$. Define

$$g'_{(q,m,\ell)}(x) = \prod_{\substack{i \in \Gamma^{(1)}_{(q,N,q-1)} \\ 1 \leq \mathrm{wt}(i) \leq \ell}} \mathbb{M}_{\beta^i}(x).$$

Let

$$D'_{(q,m,\ell)} = \bigcup_{\substack{i \in \Gamma^{(1)}_{(q,N,q-1)} \\ 1 \le \mathtt{wt}(i) \le \ell}} C_i^{(q,N)}.$$

Then $\{\beta^i : i \in D'_{(q,m,\ell)}\}$ is the set of all zeros of $g'_{(q,m,\ell)}(x)$. It is easily verified that $D'_{(q,m,\ell)}$ is invariant under the permutation $qy \mod N$ of $\mathbb{Z}_N$. Consequently, $g'_{(q,m,\ell)}(x)$ is over $\mathrm{GF}(q)$ and is a divisor of $x^n - \lambda$. Let $\mathcal{C}'(q,m,\ell)$ denote the λ-constacyclic code of length n over $\mathrm{GF}(q)$ with generator polynomial $g'_{(q,m,\ell)}(x)$. By definition, $g'_{(q,m,m)}(x) = x^n - \lambda$ and the code $\mathcal{C}(q,m,m)$ is the zero code and $\mathcal{C}'(q,m,m)^{\perp}$ is the $[n,n,1]$ code $\mathrm{GF}(q)^n$ over $\mathrm{GF}(q)$. Hence, we will consider the code $\mathcal{C}'(q,m,\ell)$ only for $1 \le \ell \le m-1$, and call $D'_{(q,m,\ell)}$ the *defining set* of $\mathcal{C}'(q,m,\ell)$ with respect to the primitive element β of $\mathrm{GF}(q^m)$.

Theorem 4. *Let $1 \le \ell \le m-1$. Then*

$$\dim(\mathcal{C}'(q,m,\ell)) = \frac{q^m - \sum_{i=0}^{\ell} \binom{m}{i}(q-1)^i}{q-1}$$

and

$$d(\mathcal{C}'(q,m,\ell)) \ge \left\lfloor \frac{q^{\ell+1} - 1 - 2(q-1)}{(q-1)^2} \right\rfloor + 2. \quad (2)$$

Theorem 5. *Let $1 \le \ell \le m-1$ and $q \ge 3$. Then*

$$\dim(\mathcal{C}'(q,m,\ell)^{\perp}) = \sum_{i=1}^{\ell} \binom{m}{i}(q-1)^{i-1}$$

and

$$d(\mathcal{C}'(q,m,\ell)^{\perp}) \ge q^{m-\ell}. \quad (3)$$

Corollary 1. *Let $m \ge 2$. Then the constacyclic code $\mathcal{C}'(q,m,1)$ has parameters*

$$[(q^m-1)/(q-1), (q^m-1)/(q-1) - m, 3]$$

and is monomially-equivalent to the Hamming code. In addition, $\mathcal{C}'(q,m,1)^{\perp}$ has parameters $[(q^m-1)/(q-1), m, q^{m-1}]$ and is monomially-equivalent to the Simplex code.

Let $\Omega(q,m,\ell)$ denote the punctured Dilix code constructed in [10] (see also Sect. 2.6). Theorem 4 tells us that

$$\dim(\Omega(q,m,\ell)) = (q-1)\dim(\mathcal{C}'(q,m,\ell)).$$

Experimental data indicates that the lower bound in (2) is good in general. But the following problem is worth of investigation.

Open Problem 6. *Determine the minimum distance of* $\mathcal{C}'(q,m,\ell)$ *or improve the lower in (2) for* $2 \le \ell \le m-1$.

Experimental data shows that the lower bound in (3) is quite away from the true minimum distance.

Open Problem 7. *Determine the minimum distance of* $\mathcal{C}'(q,m,\ell)^\perp$ *or improve the lower bound in (3) for* $2 \le \ell \le m-1$.

Example 3. Let $(q,m,\ell) = (3,4,1)$. Let β be the primitive element of $\mathrm{GF}(3^4)$ with $\beta^4 + 2\beta^3 + 2 = 0$. Then the code $\mathcal{C}'(q,m,\ell)$ has parameters $[40,36,3]$ and $\mathcal{C}'(q,m,\ell)^\perp$ has parameters $[40,4,27]$. The former is a perfect code and the latter meets the Griesmer bound.

Example 4. Let $(q,m,\ell) = (3,4,2)$. Let β be the primitive element of $\mathrm{GF}(3^4)$ with $\beta^4 + 2\beta^3 + 2 = 0$. Then the code $\mathcal{C}'(q,m,\ell)$ has parameters $[40,24,8]$ and $\mathcal{C}'(q,m,\ell)^\perp$ has parameters $[40,16,12]$. The best ternary code known of length 40 and dimension 24 has minimum distance 9 [16].

Example 5. Let $(q,m,\ell) = (3,4,3)$. Let β be the primitive element of $\mathrm{GF}(3^4)$ with $\beta^4 + 2\beta^3 + 2 = 0$. Then the code $\mathcal{C}'(q,m,\ell)$ has parameters $[40,8,21]$ and has the best parameters known [16], and $\mathcal{C}'(q,m,\ell)^\perp$ has parameters $[40,32,4]$.

The forgoing examples demonstrate that the code $\mathcal{C}'(q,m,\ell)$ and its dual $\mathcal{C}'(q,m,\ell)^\perp$ may be optimal or have the best parameters known some times. Below we explain some connection and difference among the code $\mathcal{C}'(q,m,\ell)$, the projective Reed-Muller codes and the projective generalised Reed-Muller codes.

By Corollary 1, $\mathcal{C}'(q,m,1)^\perp$ is monomially-equivalent to $\mathrm{PRM}(q,m,1)$, as both codes are monomially-equivalent to the Simplex code. This is one connection between the codes $\mathcal{C}'(q,m,\ell)$ and the projective Reed-Muller codes. Consider now all the projective codes $\mathrm{PRM}(3,4,\ell)$ for all ℓ with $1 \le \ell \le 6$. It follows from Theorem 1 that

$$\begin{aligned} d(\mathrm{PRM}(3,4,1)) &= 27, \\ d(\mathrm{PRM}(3,4,2)) &= 18, \\ d(\mathrm{PRM}(3,4,3)) &= 9, \\ d(\mathrm{PRM}(3,4,4)) &= 6, \\ d(\mathrm{PRM}(3,4,5)) &= 3, \\ d(\mathrm{PRM}(3,4,6)) &= 2. \end{aligned}$$

By Example 4, $d(\mathcal{C}'(3,4,2)) = 8$ and $d(\mathcal{C}'(3,4,2)^\perp) = 12$. This means that both $\mathcal{C}'(3,4,2)$ and $\mathcal{C}(3,4,2)^\perp$ cannot be monomially-equivalent to a code $\mathrm{PRM}(3,4,\ell)$ for all ℓ with $1 \le \ell \le 6$. Hence, the two families of codes $\mathcal{C}'(q,m,\ell)$ and $\mathrm{PRM}(q,m,\ell)$ are different in general. Notice that $\mathcal{C}'(2,m,\ell)$ and the punctured Dilix code $\Omega(2,m,\ell)$ are identical. But $\mathcal{C}'(q,m,\ell)$ and the punctured Dilix code

$\Omega(q, m, \ell)$ are not monomially-equivalent when $q > 2$, as they have different lengths.

Compared with parameters of the codes PGRM$(3, 4, \ell)$ in Example 1, both $\mathcal{C}'(3, 4, 2)$ and $\mathcal{C}'(3, 4, 2)^{\perp}$ cannot be monomially-equivalent to a code PGRM$(3, 4, \ell)$ for all ℓ with $0 \leq \ell \leq 3$. Hence, the class of codes $\mathcal{C}'(q, m, \ell)$ and the class of codes PGRM(q, m, ℓ) are different.

5 The Second Class of Constacyclic Codes

We follow the previous notation. Throughout this section, let $n = \frac{q^m-1}{q-1}$, where m is an integer with $m \geq 2$. Define $N = rn = q^m - 1$, where $r = q - 1$. Then it is easily seen that $\mathrm{ord}_n(q) = \mathrm{ord}_N(q) = m$. Let $\Gamma_{(q,N)}$ be the set of q-cyclotomic coset leaders modulo N and let

$$\Gamma^{(1)}_{(q,N,q-1)} = \{i : i \in \Gamma_{(q,N)}, \, i \equiv 1 \pmod{q-1}\}.$$

5.1 Definition and Basic Properties of the Constacyclic Codes

Let β be a primitive element of GF(q^m) and let $\lambda = \beta^{(q^m-1)/(q-1)}$. Then λ is a primitive element of GF(q). Let ℓ be a positive integer with $0 \leq \ell < (q-1)m-1$. Define

$$g_{(q,m,\ell)}(x) = \prod_{\substack{i \in \Gamma^{(1)}_{(q,N,q-1)} \\ \mathtt{wt}_q(i) < (q-1)m-\ell}} \mathrm{M}_{\beta^i}(x).$$

Let

$$D_{(q,m,\ell)} = \bigcup_{\substack{i \in \Gamma^{(1)}_{(q,N,q-1)} \\ \mathtt{wt}_q(i) < (q-1)m-\ell}} C_i^{(q,N)}.$$

Then $\{\beta^i : i \in D_{(q,m,\ell)}\}$ is the set of all zeros of $g_{(q,m,\ell)}(x)$. It is easily verified that $D_{(q,m,\ell)}$ is invariant under the permutation $qy \mod N$ of $\mathbb{Z}_N$. Consequently, $g_{(q,m,\ell)}(x)$ is over GF(q) and is a divisor of $x^n - \lambda$. Let $\mathcal{C}(q, m, \ell)$ denote the λ-constacyclic code of length n over GF(q) with generator polynomial $g_{(q,m,\ell)}(x)$. We call $D_{(q,m,\ell)}$ the *defining set* of $\mathcal{C}(q, m, \ell)$ with respect to the primitive element β of GF(q^m).

Lemma 1. *Let $m \geq 2$ and $q \geq 3$. Then $\mathcal{C}(q, m, \ell) = \{\mathbf{0}\}$ for all ℓ with $1 \leq \ell \leq q-3$. Furthermore, $\mathcal{C}(q, m, (q-1)u + q - 2) = \mathcal{C}(q, m, (q-1)(u+1) + v)$ for all $0 \leq u \leq m-2$ and $0 \leq v \leq q-3$.*

This lemma shows that this class of constacyclic codes $\mathcal{C}(q, m, \ell)$ contain only the following distinct nonzero codes:

$$\mathcal{C}(q, m, (q-1)u + q - 2), \quad 0 \leq u \leq m-2.$$

Theorem 8. *Let $m \geq 2$, $q \geq 3$ and $0 \leq u \leq m-2$. Then $\mathcal{C}(q,m,(q-1)u+q-2)$ is monomially-equivalent to $\mathrm{PRM}(q,m,(q-1)u+q-2)$. Consequently,*

$$\dim(\mathcal{C}(q,m,(q-1)u+q-2)) = \sum_{\substack{t \equiv q-2 \pmod{q-1} \\ 0<t\leq(q-1)u+q-2}} \left(\sum_{j=0}^{m} (-1)^j \binom{m}{j} \binom{t-jq+m-1}{t-jq} \right),$$

and

$$d(\mathcal{C}(q,m,(q-1)u+q-2)) = 3q^{m-2-u}.$$

The main contribution of this section is Theorem 8, which shows that the projective Reed-Muller code $\mathrm{PRM}(q,m,\ell)$ has a constacyclic code construction when $\ell = (q-1)u+q-2$ for any u with $0 \leq u \leq m-2$ up to monomial equivalence. However, the following question is still open.

Open Problem 9. *Is $\mathrm{PRM}(q,m,\ell)$ monomially-equivalent to a constacyclic code when $\ell \not\equiv q-2 \pmod{q-1}$ and $q-2 \leq \ell \leq (m-1)(q-1)$?*

5.2 Some Special Cases of the Code $\mathcal{C}(q,m,\ell)$

In this subsection, we study the code $\mathcal{C}(q,m,\ell)$ in some special cases. The code is very interesting in some special cases.

Corollary 2. *Let $q \geq 3$ and $m \geq 2$. Then $\mathcal{C}(q,m,(q-1)(m-2)+q-2)$ has parameters*

$$\left[\frac{q^m-1}{q-1}, \frac{q^m-1}{q-1}-m, 3\right]$$

and is monomially-equivalent to the Hamming code. Hence, $\mathcal{C}(q,m,(q-1)(m-2)+q-2)^\perp$ has parameters

$$\left[\frac{q^m-1}{q-1}, m, q^{m-1}\right]$$

and is monomially-equivalent to the Simplex code.

Corollary 3. *Let $m \geq 2$. Then $\mathcal{C}(4,m,2)$ has parameters*

$$\left[\frac{4^m-1}{3}, \frac{m(m+1)}{2}, 3\times 4^{m-2}\right]$$

and $\mathcal{C}(4,m,2)^\perp$ has parameters

$$\left[\frac{4^m-1}{3}, \frac{4^m-1}{3}-\frac{m(m+1)}{2}, 4\right].$$

The following four examples show that $\mathcal{C}(4,m,2)$ is a $(m+1)$-weight code for even m and m-weight code for odd m.

Example 6. Let $(q, m, \ell) = (4, 2, 2)$. Let β be the primitive element of $\mathrm{GF}(4^2)$ with $\beta^4+\beta+1 = 0$. Then $\mathcal{C}(4, 2, 2)$ has parameters $[5, 3, 3]$ and weight enumerator $1 + 30z^3 + 15z^4 + 18z^5$. Furthermore, $\mathcal{C}(4, 2, 2)^\perp$ has parameters $[5, 2, 4]$. Both codes are MDS and optimal.

Example 7. Let $(q, m, \ell) = (4, 3, 2)$. Let β be the primitive element of $\mathrm{GF}(4^3)$ with $\beta^6 + \beta^4 + \beta^3 + \beta + 1 = 0$. Then $\mathcal{C}(4, 3, 2)$ has parameters $[21, 6, 12]$ and weight enumerator

$$1 + 630z^{12} + 3087z^{16} + 378z^{20}.$$

Notice that the code $\mathcal{C}(4, 3, 2)$ is distance-optimal [16]. Furthermore, $\mathcal{C}(4, 3, 2)^\perp$ has parameters $[21, 15, 4]$ and is almost-distance optimal [16].

Example 8. Let $(q, m, \ell) = (4, 3, 5)$. Let β be the primitive element of $\mathrm{GF}(4^3)$ with $\beta^6 + \beta^4 + \beta^3 + \beta + 1 = 0$. Then the code $\mathcal{C}(q, m, \ell)$ has parameters $[21, 18, 3]$ and is distance-optimal [16], and $\mathcal{C}(q, m, \ell)^\perp$ has parameters $[21, 3, 16]$ and is distance-optimal [16].

Example 9. Let $(q, m, \ell) = (4, 3, 4)$. Let β be the primitive element of $\mathrm{GF}(4^3)$ with $\beta^6 + \beta^4 + \beta^3 + \beta + 1 = 0$. Then the code $\mathcal{C}(q, m, \ell)$ has parameters $[21, 6, 12]$ and is distance-optimal [16], and $\mathcal{C}(q, m, \ell)^\perp$ has parameters $[21, 15, 4]$.

These examples above show that the code $\mathcal{C}(q, m, \ell)$ could be optimal in some cases. Thus, the code $\mathcal{C}(q, m, \ell)$ is interesting in terms of its error-correcting capability.

5.3 The Difference Between the Codes $\mathcal{C}(q, m, \ell)$ and the Codes PGRM(q, m, h)

According to Theorem 2, $\mathcal{C}(3, 4, 5)$ has minimum distance 3. By Example 1, none of the codes PGRM$(3, 4, h)$ for $0 \le h \le 3$ has minimum distance 3. Consequently, the class of codes $\mathcal{C}(q, m, \ell)$ and the class of codes PGRM(q, m, h) are different.

6 Summary and Concluding Remarks

The main contributions of this extended abstract are the constructions and analyses of the two classes of constacyclic codes $\mathcal{C}'(q, m, \ell)$ and $\mathcal{C}(q, m, \ell)$. The codes are interesting in theory as they contain optimal codes and codes with best-known parameters and they are constacyclic. In addition, the codes $\mathcal{C}'(q, m, \ell)$ are new, and the codes $\mathcal{C}(q, m, \ell)$ give a costacyclic-code construction of some projective Reed-Muller codes. It would be very interesting to settle the open problems presented in this extended abstract and determine the automorphism groups of the first class of constacyclic codes $\mathcal{C}'(q, m, \ell)$.

Acknowledgements. The first author thanks Sihem Mesnager and Zhengchun Zhou for inviting him to present the talk at WAIFI 2022.

References

1. Abdukhalikov, K., Ho, D.: Extended cyclic codes, maximal arcs and ovoids. Des. Codes Crypt. **89**(10), 2283–2294 (2021). https://doi.org/10.1007/s10623-021-00915-2
2. Assmus, E.F., Jr., Key, J.D.: Polynomial codes and finite geometries. In: Pless, V.S., Huffman, W.C. (eds.) Handbook of Coding Theory, vol. II, pp. 1269–1343. Elsevier, Amsterdam (1998)
3. Ballet, S., Rolland, R.: On low weight codewords of generalized affine and projective Reed-Muller codes. Des. Codes Cryptogr. **73**, 271–297 (2014)
4. Berger, T.P., de Maximy, L.: Cyclic projective reed-muller codes. In: Boztaş, S., Shparlinski, I.E. (eds.) AAECC 2001. LNCS, vol. 2227, pp. 77–81. Springer, Heidelberg (2001). https://doi.org/10.1007/3-540-45624-4_8
5. Blackford, E.R.: Negacyclic codes for the Lee metric. In: Proceedings of the Conference on Combinatorial Mathematics and its Applications, pp. 298–316. Chapel Hill, NC (1968)
6. Chen, B., Dinh, H.Q., Fan, Y., Ling, S.: Polyadic constacyclic codes. IEEE Trans. Inf. Theory **61**(9), 4895–4904 (2015)
7. Chen, B., Fan, Y., Lin, L., Liu, H.: Constacyclic codes over finite fields. Finite Fields Appl. **18**, 1217–1231 (2012)
8. Dahl, C., Pedersen, J.P.: Cyclic and pseudo-cyclic MDS codes of length $q+1$. J. Comb. Theory Ser. A **59**, 130–133 (1992)
9. Danev, D., Dodunekov, S., Radkova, D.: A family of constacyclic ternary quasi-perfect codes with covering radius 3. Des. Codes Cryptogr. **59**, 111–118 (2011)
10. Ding, C., Li, C., Xia, Y.: Another generalization of the Reed-Muller codes. Finite Fields Appl. **53**, 147–174 (2018)
11. Ding, C., Tang, C.: Designs from Linear Codes, 2nd edn. World Scientific, Singapore (2022)
12. Ding, P., Key, J.D.: Subcodes of the projective generalized Reed-Muller codes spanned by minimum-weight vectors. Des. Codes Cryptogr. **26**, 197–211 (2002)
13. Dong, X., Yin, S.: The trace representation of λ-constacyclic codes over $\mathbb{F}_q$. J. Liaoning Normal Univ. (Nat. Sci. ed.) **33**, 129–131 (2010)
14. Fang, W., Wen, J., Fu, F.: A q-polynomial approach to constacyclic codes. Finite Fields Appl. **47**, 161–182 (2017)
15. Georgiades, J.: Cyclic $(q+1,k)$-codes of odd order q and even dimension k are not optimal. Atti Sent. Mat. Fis. Univ. Modena **30**, 284–285 (1982)
16. Grassl, M.: Bounds on the minimum distance of linear codes and quantum codes. http://www.codetables.de
17. Huffman, W.C., Pless, V.: Fundamentals of Error Correcting Codes. Cambridge University Press, Cambridge (2003)
18. Krishna, A., Sarwate, D.V.: Pseudocyclic maximum-distance-separable codes. IEEE Trans. Inf. Theory **36**(4), 880–884 (1990)
19. Lachaud, G.: Projective reed-muller codes. In: Cohen, G., Godlewski, P. (eds.) Coding Theory 1986. LNCS, vol. 311, pp. 125–129. Springer, Heidelberg (1988). https://doi.org/10.1007/3-540-19368-5_13
20. Li, F., Yue, Q., Liu, F.: The weight distribution of constacyclic codes. Adv. Math. Commun. **11**(3), 471–480 (2017)
21. Li, F., Yue, Q.: The primitive idempotents and weight distributions of irreducible constacyclic codes. Des. Codes Cryptogr. **86**, 771–784 (2018)

22. Li, S.: On the weight distribution of second order Reed-Muller codes and their relatives. Des. Codes Cryptogr. **87**, 2447–2460 (2019)
23. Lidl, R., Niederreiter, H.: Finite Fields. Addison-Wesly, New York (1983)
24. Liu, Y., Li, R., Lv, L., Ma, Y.: A class of constacyclic BCH codes and new quantum codes. Quantum Inf. Process. **16**(3), 1–16 (2017). https://doi.org/10.1007/s11128-017-1533-y
25. Mi, J., Cao, X.: Constructing MDS Galois self-dual constacyclic codes over finite fields. Discret. Math. **334**(6), 1–15 (2021)
26. Muller, D.E.: Application of boolean algebra to switching circuit design and to error detection. IEEE Trans. Comput. **3**, 6–12 (1954)
27. Pedersen, J.P., Dahl, C.: Classification of pseudo-cyclic MDS codes. IEEE Trans. Inf. Theory **37**(2), 365–370 (1991)
28. Peterson, W.W., Weldon, E.J., Jr.: Error-Correcting Codes, 2nd edn. MIT Press, Cambridge (1972)
29. Reed, I.S.: A class of multiple-error-correcting codes and the decoding scheme. IRE Trans. Inf. Theory **4**, 38–49 (1954)
30. Wang, L., Sun, Z., Zhu, S.: Hermitian dual-containing narrow-sense constacyclic BCH codes and quantum codes. Quantum Inf. Process. **18**(10), 1–40 (2019). https://doi.org/10.1007/s11128-019-2440-1
31. Wolfmann, J.: Projective two-weight irreducible cyclic and constacyclic codes. Finite Fields Appl. **14**(2), 351–360 (2008)
32. Sharma, A., Rani, S.: Trace description and Hamming weights of irreducible constacyclic codes. Adv. Math. Commun. **12**(1), 123–141 (2018)
33. Shi, Z., Fu, F.: The primitive idempotents of irreducible constacyclic codes and LCD cyclic codes. Cryptogr. Commun. **12**, 29–52 (2020)
34. Sørensen, A.: Projective Reed-Muller codes. IEEE Trans. Inf. Theory **37**(6), 1567–1576 (1991)
35. Sun, S., Zhu, S., Wang, L.: A class of constacyclic BCH codes. Cryptogr. Commun. **12**, 265–284 (2020)
36. Zhu, S., Sun, Z., Li, P.: A class of negacyclic BCH codes and its application to quantum codes. Des. Codes Cryptogr. **86**(10), 2139–2165 (2018)

Near MDS Codes with Dimension 4 and Their Application in Locally Recoverable Codes

Ziling Heng and Xiaoru Li(✉)

School of Science, Chang'an University, Xi'an 710064, China
zilingheng@chd.edu.cn, lx_lixiaoru@163.com

Abstract. In this paper, several infinite families of near MDS codes with dimension four are constructed with special matrixes and oval polynomials. The weight enumerators of them are explicitly determined. As an application, the duals of these near NMDS codes are proved to be both distance-optimal and dimension-optimal locally recoverable codes.

Keywords: Linear code · Near MDS code · Locally repairable code

1 Introduction

Denote by $\mathbb{F}_q$ the finite field with q elements for a prime power q. Let $\mathbb{F}_q^* := \mathbb{F}_q \setminus \{0\}$.

1.1 Near MDS Codes

For a positive integer n, a subset $\mathcal{C}$ of $\mathbb{F}_q^n$ is called an $[n,k,d]$ linear code over $\mathbb{F}_q$ if it is a k-dimensional linear subspace $\mathbb{F}_q^n$, where d denotes the minimum (Hamming) distance of $\mathcal{C}$. It is well known that the minimum distance of a linear code $\mathcal{C}$ equals the minimum Hamming weight of nonzero codewords in $\mathcal{C}$. The dual of an $[n,k,d]$ linear code over $\mathbb{F}_q$ is defined by $\mathcal{C}^\perp = \{\mathbf{c}' \in \mathbb{F}_q^n : \langle \mathbf{c}', \mathbf{c} \rangle = 0 \, \forall\, \mathbf{c} \in \mathcal{C}\}$, where $\langle \mathbf{c}', \mathbf{c} \rangle$ represents the Euclidean inner product of $\mathbf{c}'$ and $\mathbf{c}$. Clearly, $\mathcal{C}^\perp$ is an $[n, n-k]$ linear code over $\mathbb{F}_q$. Denote by $A_i = |\{\mathbf{c} \in \mathcal{C} : \mathrm{wt}(\mathbf{c}) = i\}|$, $0 \le i \le n$, where $\mathrm{wt}(\mathbf{c})$ denotes the Hamming weight of $\mathbf{c}$. Then the sequence $(1, A_1, \cdots, A_n)$ is referred to as the weight distribution of the linear code $\mathcal{C}$. The weight enumerator of $\mathcal{C}$ is defined as the following polynomial: $A(z) = 1 + A_1 z + A_2 z^2 + \cdots + A_n z^n$. The weight distributions of many special linear codes were determined in the literature [3,4,9,11,20,21].

In coding theory, we hope to construct an $[n,k,d]$ linear code with both large code rate k/n and large minimum distance d. Nevertheless, there exists a tradeoff among the parameters n,k and d. The well-known Singleton bound on an $[n,k,d]$ linear code $\mathcal{C}$ is given by $n \ge k + d - 1$. A linear code achieving this bound is called an MDS (maximum

This work was supported by Natural Science Foundation of China under Grant 11901049, in part by the Young Talent Fund of University Association for Science and Technology in Shaanxi, China, under Grant 20200505 and in part by the Fundamental Research Funds for the Central Universities, CHD, under Grant 300102122202.

S. Mesnager and Z. Zhou (Eds.): WAIFI 2022, LNCS 13638, pp. 142–158, 2023.
https://doi.org/10.1007/978-3-031-22944-2_8

distance separable) code, i.e. it has parameters $[n,k,n-k+1]$. The dual of an MDS code is also an MDS code. MDS codes have many nice applications and were investigated in [10, 18]. However, according to the well-known MDS conjecture, the lengths of a nontrivial $[n,k]$ MDS codes over $\mathbb{F}_q$ are limited by the size of the field, i.e. $n \leq q+2$ if q is even and $k=3$ or $k=q-1$, and $n \leq q+1$ for other cases.

It is interesting to study linear codes nearly achieving the Singleton bound with larger lengths. Linear codes with parameters $[n,k,n-k]$ are called almost MDS codes. In general, the dual of an almost MDS code may not be almost MDS. If both a code and its dual are almost MDS, this code is called an near MDS code (NMDS code for short). Denote by $n(k,q)$ the largest possible length of an NMDS code over $\mathbb{F}_q$ with dimension k. In [5], it was proved that $n(k,q) \leq 2q+k$. For $k > q$, there indeed exist NMDS codes with parameters $[2q+k,k,2q]$ constructed in [1]. Hence, NMDS codes can have larger lengths than those of MDS codes. Near MDS codes have nice applications in combinatorics, finite geometry, cryptography and data storage [3–5, 19–21].

1.2 Locally Recoverable Codes

For a block code, if any symbol in the encoding is a function of r other symbols, then this code is called a locally recoverable code (LRC for short) with locality r. In this paper, we only consider linear LRCs. We denote an $[n,k,d]$ linear code over $\mathbb{F}_q$ with locality r by $(n,k,d,q;r)$-LRC. There exist some tradeoffs among the parameters of LRCs. For an $(n,k,d,q;r)$-LRC, the Singleton-like bound (see [2]) is given by

$$d \leq n-k-\left\lceil \frac{k}{r} \right\rceil +2. \tag{1}$$

If a LRC achieves this bound, then it is said to be distance-optimal. For an $(n,k,d,q;r)$-LRC, the Cadambe-Mazumdar bound (see [7]) is given as

$$k \leq \min_{t\in\mathbb{Z}^+}[rt + k_{opt}^{(q)}(n-t(r+1),d)], \tag{2}$$

where $k_{opt}^{(q)}(n,d)$ denotes the minimum dimension of a linear code of length n, minimum distance d over $\mathbb{F}_q$ and $\mathbb{Z}^+$ represents the set of all positive integers. If a LRC achieves the Cadambe-Mazumdar bound, then it is said to be dimension-optimal. In [15], the authors constructed optimal locally repairable codes of distance 3 and 4 from cyclic codes. In [19], the minimum linear locality of general linear codes was investigated and many families of optimal LRCs were derived with certain families of linear codes. In [8], optimal LRCs were constructed with a family of almost MDS codes. In [12, 13], several families of distance-optimal or dimension-optimal LRCs were constructed with NMDS codes with dimension 3.

1.3 The Objectives of This Paper

Since many infinite families of NMDS codes with dimension 3 were constructed with oval polynomials in [12, 13], a natural question arises: whether can we construct infinite families of NMDS codes with bigger dimensions from oval polynomials? This paper will tackle this question. The first objective of this paper is to construct several infinite families of NMDS codes with dimension 4 from some special matrixes and oval polynomials. The weight enumerators of the NMDS codes will also be determined. The second objective is to construct dimension-optimal and distance-optimal LRCs via the duals of these NMDS codes.

2 Preliminaries

For a linear code $\mathcal{C}$ of length n, let $(1, A_1, \cdots, A_n)$ and $(1, A_1^{\perp}, \cdots, A_n^{\perp})$ represent the weight distributions of it and its dual, respectively. The following lemma provides crucial information on the weight distributions of NMDS codes and their duals.

Lemma 1 *[5]. For an $[n,k]$ NMDS code $\mathcal{C}$ over $\mathbb{F}_q$, its weight distribution satisfies the following recurrence relation:*

$$A_{n-k+s} = \binom{n}{k-s} \sum_{j=0}^{s-1} (-1)^j \binom{n-k+s}{j} (q^{s-j}-1) + (-1)^s \binom{k}{s} A_{n-k}$$

for $s \in \{1, 2, \ldots, k\}$. The weight distribution of $\mathcal{C}^{\perp}$ satisfies the following recurrence relation:

$$A_{k+s}^{\perp} = \binom{n}{k+s} \sum_{j=0}^{s-1} (-1)^j \binom{k+s}{j} (q^{s-j}-1) + (-1)^s \binom{n-k}{s} A_k^{\perp}$$

for $s \in \{1, 2, \ldots, n-k\}$.

The following lemma provides a useful property on the relationship between the support of the minimum weight codewords of an NMDS code and that of its dual.

Lemma 2 *[6]. For an NMDS code $\mathcal{C}$ of length n, define the support of $\mathbf{c} = (c_1, \ldots, c_n) \in \mathcal{C}$ by $\mathrm{suppt}(\mathbf{c}) = \{1 \leq i \leq n : c_i \neq 0\}$. For any minimum weight codeword $\mathbf{c}$ in $\mathcal{C}$, then there exists, up to a multiple, a unique minimum weight codeword $\mathbf{c}^{\perp}$ in $\mathcal{C}^{\perp}$ such that $\mathrm{suppt}(\mathbf{c}) \cap \mathrm{suppt}(\mathbf{c}^{\perp}) = \emptyset$. Besides, the number of minimum weight codewords in $\mathcal{C}$ is equal to that of minimum weight codewords in $\mathcal{C}^{\perp}$.*

In this paper, we will use oval polynomials to construct NMDS codes. The definition of an oval polynomial is as follows.

Definition 1 *[14]. Let $q = 2^m$ for $m \geq 2$. A polynomial $f \in \mathbb{F}_q[x]$ is called an oval polynomial if it satisfies the following two conditions:*

- *f is a permutation polynomial of $\mathbb{F}_q$ such that $\deg(f) < q$ and $f(0) = 0$, $f(1) = 1$;*
- *for each $a \in \mathbb{F}_q$, $g_a(x) := (f(x+a) + f(a))x^{q-2}$ is also a permutation polynomial of $\mathbb{F}_q$.*

Some known infinite families of oval polynomials are listed as follows.

Theorem 1 *[16]. Let $m \geq 2$ be an integer and $q = 2^m$. Then the following polynomials are all oval polynomials over $\mathbb{F}_q$.*

- $f(x) = x^{2^h}$ *with* $\gcd(h,m) = 1$.
- $f(x) = x^6$ *for odd m.*
- $f(x) = x^{3\times 2^{(m+1)/2}+4}$ *for odd m.*

Some useful properties on oval polynomials are also presented as follows.

Lemma 3 *[17]. Let f be a polynomial over $\mathbb{F}_q$ such that $f(0) = 0$. Let $f_u := f(x) + ux$ with $u \in \mathbb{F}_q^*$. Then it is an oval polynomial if and only if f_u is 2-to-1 for each $u \in \mathbb{F}_q^*$.*

Lemma 4. *Let f be a polynomial over $\mathbb{F}_q$. Then f is an oval polynomial over $\mathbb{F}_q$ if and only if it satisfies the following two conditions:*

1. *f is a permutation polynomial of $\mathbb{F}_q$ such that* $\deg(f) < q$ *and* $f(0) = 0$, $f(1) = 1$;
2. *if $x, y, z \in \mathbb{F}_q$ are pairwise distinct, then*

$$\frac{f(x)+f(y)}{x+y} \neq \frac{f(x)+f(z)}{x+z}.$$

It is easy to prove Lemma 4 by the definition of oval polynomials.

Lemma 5 *[21]. Let m be an odd integer such that $m > 2$. Let $f(x)$ be an oval polynomial over $\mathbb{F}_q$ such that its coefficients are in $\mathbb{F}_2$. Then $f(x) + x + 1 \neq 0$ for all $x \in \mathbb{F}_q$.*

3 Several Infinite Families of NMDS Codes

In this section, let $q = 2^m$, where m is an odd integer with $m \geq 3$. For convenience, let $\dim(C)$ and $d(C)$ represent the dimension and minimum distance of a linear code C, respectively. Let $\mathbb{F}_q = \{\alpha_0 = 0, \alpha_1 = 1, \alpha_2, \cdots, \alpha_{q-1}\}$. By Theorem 1, both $f_1(x) = x^2$ and $f_2(x) = x^4$ are oval polynomials for odd m.

Lemma 6. *Let $x_1, x_2, \cdots, x_n$ be pairwise distinct elements of $\mathbb{F}_q$, then*

$$D = \begin{vmatrix} 1 & 1 & \cdots & 1 \\ x_1 & x_2 & \cdots & x_n \\ x_1^2 & x_2^2 & \cdots & x_n^2 \\ \vdots & \vdots & \ddots & \vdots \\ x_1^{n-2} & x_2^{n-2} & \cdots & x_n^{n-2} \\ x_1^n & x_2^n & \cdots & x_n^n \end{vmatrix} = -(x_1 + x_2 + \cdots + x_n) \prod_{0 \leq i < j \leq n} (x_j - x_i).$$

Proof. The proof is easy and omitted.

3.1 NMDS Codes with Parameters $[q+3,4,q-1]$

Define a 4 by $q+3$ matrix over $\mathbb{F}_q$ by

$$G_1 = \begin{bmatrix} 1 & 1 & \cdots & 1 & 1 & 0\,0\,0 \\ \alpha_1 & \alpha_2 & \cdots & \alpha_{q-1} & \alpha_0 & 1\,0\,0 \\ \alpha_1^2 & \alpha_2^2 & \cdots & \alpha_{q-1}^2 & \alpha_0 & 0\,1\,0 \\ \alpha_1^4 & \alpha_2^4 & \cdots & \alpha_{q-1}^4 & \alpha_0 & 0\,0\,1 \end{bmatrix} = \begin{bmatrix} 1 & 1 & \cdots & 1 & 1\,0\,0\,0 \\ \alpha_1 & \alpha_2 & \cdots & \alpha_{q-1} & 0\,1\,0\,0 \\ \alpha_1^2 & \alpha_2^2 & \cdots & \alpha_{q-1}^2 & 0\,0\,1\,0 \\ \alpha_1^4 & \alpha_2^4 & \cdots & \alpha_{q-1}^4 & 0\,0\,0\,1 \end{bmatrix}.$$

Then G_1 generates a linear code C over $\mathbb{F}_q$. In the following, we study the linear code C.

Theorem 2. *Let m be an odd integer with $m \geq 3$ and $q = 2^m$. Then C is a $[q+3,4,q-1]$ NMDS code over $\mathbb{F}_q$ whose weight enumerator is given by*

$$A(z) = 1 + \frac{q(q-1)^2(q-2)}{24} z^{q-1} + \frac{(q-1)(3q^2+3q+2)}{2} z^q + \frac{q(q-1)^2(q-2)}{4} z^{q+1} + \frac{q(q-1)^2(2q+11)}{6} z^{q+2} + \frac{3q(q-1)^2(q-2)}{8} z^{q+3}.$$

Proof. Note that the q-th, $q+1$-th, $q+2$-th and $q+3$-th columns of the generator matrix G_1 are linearly independent. Then $\dim(C) = 4$ and $\dim(C^\perp) = q-1$.

We next prove that $C^\perp$ has parameters $[q+3,q-1,4]$. Firstly, we prove $d(C^\perp) > 3$ by considering the following cases.

Case 1.1: Selecting any three columns from the first q columns of G_1, we obtain a submatrix

$$M_{1,1} = \begin{bmatrix} 1 & 1 & 1 \\ x & y & z \\ x^2 & y^2 & z^2 \\ x^4 & y^4 & z^4 \end{bmatrix},$$

where x,y,z are three pairwise distinct elements in $\mathbb{F}_q$. To determine the rank of $M_{1,1}$, we consider a submatrix of $M_{1,1}$ as

$$M_{1,1,1} = \begin{bmatrix} 1 & 1 & 1 \\ x & y & z \\ x^2 & y^2 & z^2 \end{bmatrix}.$$

Note that $|M_{1,1,1}| = (x+y)(x^2+z^2) + (x+z)(x^2+y^2) \neq 0$ by Lemma 4 as $f_1(x) = x^2$ is an oval polynomial. Then the rank of $M_{1,1}$ is 3, i.e. any three columns in the first q columns of G_1 are linearly independent.

Case 1.2: Now we consider the following submatrixes whose first two columns are chosen from the first q columns of G_1 and third column is chosen from the last three columns of G_1. These submatrixes is denoted by

$$M_{1,2} = \begin{bmatrix} 1 & 1 & 0 \\ x & y & 1 \\ x^2 & y^2 & 0 \\ x^4 & y^4 & 0 \end{bmatrix}, M_{1,3} = \begin{bmatrix} 1 & 1 & 0 \\ x & y & 0 \\ x^2 & y^2 & 1 \\ x^4 & y^4 & 0 \end{bmatrix}, M_{1,4} = \begin{bmatrix} 1 & 1 & 0 \\ x & y & 0 \\ x^2 & y^2 & 0 \\ x^4 & y^4 & 1 \end{bmatrix},$$

where x, y are distinct elements in $\mathbb{F}_q$. Choose a 3×3 submatrix $M_{1,2,1}$ of $M_{1,2}$ as

$$M_{1,2,1} = \begin{bmatrix} 1 & 1 & 0 \\ x & y & 1 \\ x^2 & y^2 & 0 \end{bmatrix}.$$

It is easy to deduce that $|M_{1,2,1}| \neq 0$ as $x \neq y$. Then the rank of $M_{1,2}$ is 3. Similarly, we can prove that both $M_{1,3}$ and $M_{1,4}$ have rank 3.

Case 1.3: We then consider the following submatrixes whose first column is chosen from the first q columns of G_1 and last two columns are chosen from the last three columns of G_1. These submatrixes are denoted by

$$M_{1,5} = \begin{bmatrix} 1 & 0 & 0 \\ x & 1 & 0 \\ x^2 & 0 & 1 \\ x^4 & 0 & 0 \end{bmatrix}, M_{1,6} = \begin{bmatrix} 1 & 0 & 0 \\ x & 1 & 0 \\ x^2 & 0 & 0 \\ x^4 & 0 & 1 \end{bmatrix}, M_{1,7} = \begin{bmatrix} 1 & 0 & 0 \\ x & 0 & 0 \\ x^2 & 1 & 0 \\ x^4 & 0 & 1 \end{bmatrix},$$

where $x \in \mathbb{F}_q$. Choose a 3×3 submatrix $M_{1,5,1}$ of $M_{1,5}$,

$$M_{1,5,1} = \begin{bmatrix} 1 & 0 & 0 \\ x & 1 & 0 \\ x^2 & 0 & 1 \end{bmatrix}.$$

It is easy to deduce that $|M_{1,5,1}| = 1$. Then the rank of $M_{1,5}$ is 3. Similarly, we can see both that $M_{1,6}$ and $M_{1,7}$ have rank 3.

Case 1.4: Consider the following submatrix whose columns are chosen from the last three columns of G_1. The submatrix is given by

$$M_{1,8} = \begin{bmatrix} 0 & 0 & 0 \\ 1 & 0 & 0 \\ 0 & 1 & 0 \\ 0 & 0 & 1 \end{bmatrix}.$$

Then the rank of $M_{1,8}$ is 3.

In conclusion, any three columns of G_1 are $\mathbb{F}_q$-linearly independent. Then $d(C^\perp) > 3$. To prove $d(C^\perp) = 4$ and calculate the total number of codewords of weight 4 in $C^\perp$, we consider the following cases.

Case 2.1: Let x, y, z, w be four pairwise different elements in $\mathbb{F}_q$. Consider the following submatrix as

$$M_{2,1} = \begin{bmatrix} 1 & 1 & 1 & 1 \\ x & y & z & w \\ x^2 & y^2 & z^2 & w^2 \\ x^4 & y^4 & z^4 & w^4 \end{bmatrix}.$$

Then $|M_{2,1}| = (x+y+w+z)(w-z)(w-y)(w-x)(z-y)(z-x)(y-x)$ by Lemma 6. It is easy to deduce that $|M_{2,1}| = 0$ if and only if $x+y+z+w = 0$. Then we have the number of pair (x, y, z, w) is $\frac{q(q-1)(q-2)}{4!}$. Besides, $\mathrm{rank}(M_{2,1}) = 3$ if $x+y+z+w = 0$.

Hence, the number of codewords of weight 4 in $C^{\perp}$ whose nonzero coordinates are at the first q locations is equal to $\frac{q(q-1)^2(q-2)}{24}$.

Case 2.2: Let x, y, z be three pairwise different elements in $\mathbb{F}_q$. Consider the following submatrix as

$$M_{2,2} = \begin{bmatrix} 1 & 1 & 1 & 0 \\ x & y & z & 1 \\ x^2 & y^2 & z^2 & 0 \\ x^4 & y^4 & z^4 & 0 \end{bmatrix}.$$

Then $|M_{2,2}| = (x^4+z^4)(x^2+y^2)+(x^4+y^4)(x^2+z^2)$. Since $f(x)=x^2$ is an oval polynomial over $\mathbb{F}_q$, then x^2, y^2, z^2 are also three pairwise different elements in $\mathbb{F}_q$. By Lemma 4, $|M_{2,2}| \neq 0$. Hence, there is no codeword of weight 4 in $C^{\perp}$ such that its first three nonzero coordinates are at the first q locations and the rest is at the $q+1$-th location.

Case 2.3: Let x, y, z be three pairwise different elements in $\mathbb{F}_q$. Consider the following submatrix as

$$M_{2,3} = \begin{bmatrix} 1 & 1 & 1 & 0 \\ x & y & z & 0 \\ x^2 & y^2 & z^2 & 1 \\ x^4 & y^4 & z^4 & 0 \end{bmatrix}.$$

Then $|M_{2,3}| = (x^4+z^4)(x+y)+(x^4+y^4)(x+z)$. By Lemma 1, we know $f(x)=x^4$ is an oval polynomial over $\mathbb{F}_q$ when m is an odd integer with $m \geq 3$. Then by Lemma 4, we have $|M_{2,3}| \neq 0$. Hence, there is no codeword of weight 4 in $C^{\perp}$ such that its first three nonzero coordinates are at the first q locations and the rest is at the $q+2$-th location.

Case 2.4: Let x, y, z be three pairwise different elements in $\mathbb{F}_q$. Consider the following submatrix as

$$M_{2,4} = \begin{bmatrix} 1 & 1 & 1 & 0 \\ x & y & z & 0 \\ x^2 & y^2 & z^2 & 0 \\ x^4 & y^4 & z^4 & 1 \end{bmatrix}.$$

Then $|M_{2,4}| = (x^2+z^2)(x+y)+(x^2+y^2)(x+z)$. By Lemma 4, $|M_{2,4}| \neq 0$. Hence, there is no codeword of weight 4 in $C^{\perp}$ such that its first three nonzero coordinates are at the first q locations and the rest is at the $q+3$-th location.

Case 2.5: Let x, y be two different elements in $\mathbb{F}_q$. Consider the following submatrix as

$$M_{2,5} = \begin{bmatrix} 1 & 1 & 0 & 0 \\ x & y & 1 & 0 \\ x^2 & y^2 & 0 & 1 \\ x^4 & y^4 & 0 & 0 \end{bmatrix}.$$

We have $|M_{2,5}| = x^4+y^4 = (x+y)^4 \neq 0$ as $x \neq y$. Hence, there is no codeword of weight 4 in $C^{\perp}$ such that its first two nonzero coordinates are at the first q locations and the others are at $q+1$-th and $q+2$-th locations.

Case 2.6: Let x, y be two different elements in $\mathbb{F}_q$. Consider the following submatrix as

$$M_{2,6} = \begin{bmatrix} 1 & 1 & 0 & 0 \\ x & y & 1 & 0 \\ x^2 & y^2 & 0 & 0 \\ x^4 & y^4 & 0 & 1 \end{bmatrix}.$$

We have $|M_{2,6}| = x^2 + y^2 = (x+y)^2 \neq 0$ as $x \neq y$. Hence, there is no codeword of weight 4 in $\mathcal{C}^{\perp}$ such that its first two nonzero coordinates are at the first q locations and the others are at $q+1$-th and $q+3$-th locations.

Case 2.7: Let x, y be two different elements in $\mathbb{F}_q$. Consider the following submatrix

$$M_{2,7} = \begin{bmatrix} 1 & 1 & 0 & 0 \\ x & y & 0 & 0 \\ x^2 & y^2 & 1 & 0 \\ x^4 & y^4 & 0 & 1 \end{bmatrix}.$$

We have $|M_{2,7}| = x + y \neq 0$ as $x \neq y$. Hence, there is no codeword of weight 4 in $\mathcal{C}^{\perp}$ such that its first two nonzero coordinates are at the first q locations and the others are at $q+2$-th and $q+3$-th locations.

Case 2.8: Let $x \in \mathbb{F}_q$. Consider the following submatrix as

$$M_{2,8} = \begin{bmatrix} 1 & 0 & 0 & 0 \\ x & 1 & 0 & 0 \\ x^2 & 0 & 1 & 0 \\ x^4 & 0 & 0 & 1 \end{bmatrix}.$$

We have $|M_{2,8}| = 1$. Hence, there is no codeword of weight 4 in $\mathcal{C}^{\perp}$ such that its first nonzero coordinate is at the first q locations and the others are at $q+1$-th, $q+2$-th and $q+3$-th locations.

Consequently, the total number of codewords of weight 4 in $\mathcal{C}^{\perp}$ is $\frac{q(q-1)^2(q-2)}{24}$ from the discussions above.

We finally prove that the minimum distance of $\mathcal{C}$ is $q-1$. Assume that $d(\mathcal{C}) \leq q-2 = q+3-5$ and let $\mathbf{c} = a\mathbf{g}_1 + b\mathbf{g}_2 + c\mathbf{g}_3 + d\mathbf{g}_4$ be a codeword with the minimum weight in $\mathcal{C}$, where $\mathbf{g}_1$, $\mathbf{g}_2$, $\mathbf{g}_3$ and $\mathbf{g}_4$ respectively denote the first, second, third and fourth rows of G_1. Thus, there at least five coordinates in $\mathbf{c}$ are zero. We now consider the following cases.

Case 3.1: Assume that none of the last three coordinates in $\mathbf{c}$ are zero. Then there exist five pairwise distinct elements x_1, x_2, x_3, x_4, x_5 in $\mathbb{F}_q$ such that

$$\begin{cases} a + bx_1 + cx_1^2 + dx_1^4 = 0, \\ a + bx_2 + cx_2^2 + dx_2^4 = 0, \\ a + bx_3 + cx_3^2 + dx_3^4 = 0, \\ a + bx_4 + cx_4^2 + dx_4^4 = 0, \\ a + bx_5 + cx_5^2 + dx_5^4 = 0. \end{cases}$$

Let $f(x) = a + bx + cx^2 + dx^4$. Then x_1, x_2, x_3, x_4, x_5 are five pairwise different roots of $f(x)$. This contradicts with the fact that the degree of $f(x)$ is 4 and it has at most four roots in $\mathbb{F}_q$ as $(a,b,c,d) \neq (0,0,0,0)$.

Case 3.2: Assume that one of the last three coordinates in $\mathbf{c}$ is zero. Then there exist four elements x_1, x_2, x_3, x_4 in $\mathbb{F}_q$ such that

$$\begin{cases} a+bx_1+cx_1^2+dx_1^4=0, \\ a+bx_2+cx_2^2+dx_2^4=0, \\ a+bx_3+cx_3^2+dx_3^4=0, \\ a+bx_4+cx_4^2+dx_4^4=0, \\ b=0, \end{cases} \text{or} \begin{cases} a+bx_1+cx_1^2+dx_1^4=0, \\ a+bx_2+cx_2^2+dx_2^4=0, \\ a+bx_3+cx_3^2+dx_3^4=0, \\ a+bx_4+cx_4^2+dx_4^4=0, \\ c=0, \end{cases} \text{or} \begin{cases} a+bx_1+cx_1^2+dx_1^4=0, \\ a+bx_2+cx_2^2+dx_2^4=0, \\ a+bx_3+cx_3^2+dx_3^4=0, \\ a+bx_4+cx_4^2+dx_4^4=0, \\ d=0. \end{cases}$$

Since $f(x) = x^2$ and $f(x) = x^4$ are oval polynomials over $\mathbb{F}_q$ for odd $m \geq 3$, we can deduce that $a = b = c = d = 0$ and $\mathbf{c} = 0$ by Lemma 5. This contradicts with the fact that $\mathbf{c}$ is a minimum weight codeword in $\mathcal{C}$.

Case 3.3: Assume that two of the last three coordinates in $\mathbf{c}$ are zero. Then there exist three pairwise distinct elements x_1, x_2, x_3 in $\mathbb{F}_q$ such that

$$\begin{cases} a+bx_1+cx_1^2+dx_1^4=0, \\ a+bx_2+cx_2^2+dx_2^4=0, \\ a+bx_3+cx_3^2+dx_3^4=0, \\ b=0, \\ c=0, \end{cases} \text{or} \begin{cases} a+bx_1+cx_1^2+dx_1^4=0, \\ a+bx_2+cx_2^2+dx_2^4=0, \\ a+bx_3+cx_3^2+dx_3^4=0, \\ b=0, \\ d=0, \end{cases} \text{or} \begin{cases} a+bx_1+cx_1^2+dx_1^4=0, \\ a+bx_2+cx_2^2+dx_2^4=0, \\ a+bx_3+cx_3^2+dx_3^4=0, \\ c=0, \\ d=0. \end{cases}$$

Since $f(x) = x^2$ and $f(x) = x^4$ are permutation polynomials over $\mathbb{F}_q$ for odd $m \geq 3$ and x_1, x_2, x_3 are pairwise distinct elements in $\mathbb{F}_q$, then it is easy to deduce that $a = b = c = d = 0$ and $\mathbf{c} = 0$. This contradicts with the fact that $\mathbf{c}$ is a minimum weight codeword in $\mathcal{C}$.

Case 3.4: Assume that the last three coordinates in $\mathbf{c}$ are all zero. Then there exist two different elements x_1, x_2 in $\mathbb{F}_q$ such that

$$\begin{cases} a+bx_1+cx_1^2+dx_1^4=0, \\ a+bx_2+cx_2^2+dx_2^4=0, \\ b=c=d=0. \end{cases}$$

Obviously, we have $a = b = c = d = 0$. This contradicts with the fact that $\mathbf{c}$ is a minimum weight codeword in $\mathcal{C}$.

As a result, we conclude that $d(\mathcal{C}) \geq q-1$. Besides, $d(\mathcal{C}) \leq q$ by the Singleton bound. If $d(\mathcal{C}) = q$, then $\mathcal{C}$ is an $[q+3,4,q]$ MDS code and $\mathcal{C}^\perp$ is also an MDS code with parameters $[q+3,q-1,5]$. This is contrary to $d(\mathcal{C}^\perp) = 4$. So $\mathcal{C}$ is a $[q+3,4,q-1]$ AMDS code. Then $\mathcal{C}$ is an NMDS code as both $\mathcal{C}$ and $\mathcal{C}^\perp$ are AMDS. By Lemma 2, the total number A_{q-1} of the minimum weight codewords in $\mathcal{C}$ is equal to the total number of the codewords of weight 4 in $\mathcal{C}^\perp$. Hence $A_{q-1} = \frac{q(q-1)^2(q-2)}{24}$. By Lemma 1, the desired conclusion follows.

3.2 NMDS Codes with Parameters $[q+2, 4, q-2]$

Define the following $4 \times (q+2)$ matrix over $\mathbb{F}_q$ as

$$G_2 = \begin{bmatrix} 1 & 1 & \cdots & 1 & 1 & 0 & 0 \\ \alpha_1 & \alpha_2 & \cdots & \alpha_{q-1} & \alpha_0 & 1 & 0 \\ \alpha_1^2 & \alpha_2^2 & \cdots & \alpha_{q-1}^2 & \alpha_0 & 0 & 1 \\ \alpha_1^4 & \alpha_2^4 & \cdots & \alpha_{q-1}^4 & \alpha_0 & 0 & 0 \end{bmatrix} = \begin{bmatrix} 1 & 1 & \cdots & 1 & 1 & 0 & 0 \\ \alpha_1 & \alpha_2 & \cdots & \alpha_{q-1} & 0 & 1 & 0 \\ \alpha_1^2 & \alpha_2^2 & \cdots & \alpha_{q-1}^2 & 0 & 0 & 1 \\ \alpha_1^4 & \alpha_2^4 & \cdots & \alpha_{q-1}^4 & 0 & 0 & 0 \end{bmatrix}$$

Then G_2 generates a linear code over $\mathbb{F}_q$ denoted by $\mathcal{D}$. In the following, we study the linear code $\mathcal{D}$.

Theorem 3. *Let m be an odd integer with $m \geq 3$. Then $\mathcal{D}$ is a $[q+2, 4, q-2]$ NMDS code over $\mathbb{F}_q$ with weight enumerator*

$$A(z) = 1 + \frac{q(q-1)^2(q-2)}{24} z^{q-2} + q^2(q-1)z^{q-1} + \frac{(q-1)(q^3-q^2+8q+4)}{4} z^q + \frac{q(q-1)^2(q+4)}{3} z^{q+1} + \frac{q(3q-2)(q-1)^2}{8} z^{q+2}.$$

Proof. Since the first, q-th, $q+1$-th and $q+2$-th columns of G_2 are linearly independent, we have $\dim(\mathcal{D}) = 4$ and $\dim(\mathcal{D}^\perp) = (q+2) - 4 = q-2$.

We next prove that $\mathcal{D}^\perp$ has parameters $[q+2, q-2, 4]$. To prove $d(\mathcal{D}^\perp) > 3$, we consider the following cases.

Case 1.1: Selecting any three columns from the first q columns of generator matrix G_2, we obtain the submatrix

$$M_{1,1} = \begin{bmatrix} 1 & 1 & 1 \\ x & y & z \\ x^2 & y^2 & z^2 \\ x^4 & y^4 & z^4 \end{bmatrix},$$

where x, y, z are the three pairwise different elements in $\mathbb{F}_q$. Consider the 3×3 submatrix of $M_{1,1}$ as

$$M_{1,1,1} = \begin{bmatrix} 1 & 1 & 1 \\ x & y & z \\ x^2 & y^2 & z^2 \end{bmatrix}.$$

Then $|M_{1,1,1}| \neq 0$ by Lemma 4. Then we have the rank of $M_{1,1}$ is 3, which means that any three columns in the first q columns of G_2 are linearly independent.

Case 1.2: Now we consider the following submatrixes whose first two columns are chosen from the first q columns of G_2 and third column is chosen from the last two columns of G_2:

$$M_{1,2} = \begin{bmatrix} 1 & 1 & 0 \\ x & y & 1 \\ x^2 & y^2 & 0 \\ x^4 & y^4 & 0 \end{bmatrix}, M_{1,3} = \begin{bmatrix} 1 & 1 & 0 \\ x & y & 0 \\ x^2 & y^2 & 1 \\ x^4 & y^4 & 0 \end{bmatrix},$$

where x, y are the distinct elements in $\mathbb{F}_q$. Note that the submatrix $M_{1,2,1}$ of $M_{1,2}$ satisfying $|M_{1,2,1}| \neq 0$ as $x \neq y$, where

$$M_{1,2,1} = \begin{bmatrix} 1 & 1 & 0 \\ x & y & 1 \\ x^2 & y^2 & 0 \end{bmatrix}.$$

Then the rank of $M_{1,2}$ is 3. Similarly, we can prove that $M_{1,3}$ has rank 3.

Case 1.3: We now consider the submatrix whose first column is chosen from the first q columns of G_2 and last two columns are given as the last two columns of G_2, where the submatrix is given by

$$M_{1,4} = \begin{bmatrix} 1 & 0 & 0 \\ x & 1 & 0 \\ x^2 & 0 & 1 \\ x^4 & 0 & 0 \end{bmatrix}, \ x \in \mathbb{F}_q.$$

Note that there is a submatrix $M_{1,4,1}$ of $M_{1,4}$ such that $|M_{1,4,1}| = 1$, where

$$M_{1,4,1} = \begin{bmatrix} 1 & 0 & 0 \\ x & 1 & 0 \\ x^2 & 0 & 1 \end{bmatrix}.$$

Then the rank of $M_{1,4}$ is 3.

Consequently, any three columns of G_2 are linearly independent over $\mathbb{F}_q$, which implies $d(\mathcal{D}^\perp) > 3$. To prove $d(\mathcal{D}^\perp) = 4$ and calculate the total number of codewords of weight 4 in $\mathcal{D}^\perp$, we consider the following cases.

Case 2.1: Let x, y, z, w be four pairwise different elements in $\mathbb{F}_q$. Consider the submatrix

$$M_{2,1} = \begin{bmatrix} 1 & 1 & 1 & 1 \\ x & y & z & w \\ x^2 & y^2 & z^2 & w^2 \\ x^4 & y^4 & z^4 & w^4 \end{bmatrix}.$$

Then $|M_{2,1}| = (x+y+w+z)(w-z)(w-y)(w-x)(z-y)(z-x)(y-x)$ by Lemma 6. Then $|M_{2,1}| = 0$ if and only if $x+y+z+w = 0$. Then we have the number of pair (x, y, z, w) is $\frac{q(q-1)(q-2)}{4!}$. Hence, the number of codewords of weight 4 in $\mathcal{D}^\perp$ whose nonzero coordinates are at the first q locations is equal to $\frac{q(q-1)^2(q-2)}{24}$.

Case 2.2: Let x, y, z be three pairwise different elements in $\mathbb{F}_q$. Consider the submatrix

$$M_{2,2} = \begin{bmatrix} 1 & 1 & 1 & 0 \\ x & y & z & 1 \\ x^2 & y^2 & z^2 & 0 \\ x^4 & y^4 & z^4 & 0 \end{bmatrix}.$$

Then $|M_{2,2}| = (x^4+z^4)(x^2+y^2) + (x^4+y^4)(x^2+z^2)$. Since $f(x) = x^2$ is an oval polynomial over $\mathbb{F}_q$, then $|M_{2,2}| \neq 0$ by Lemma 4. Hence, there is no codeword of weight

4 in $\mathcal{D}^{\perp}$ such that its first three nonzero coordinates are at the first q locations and the rest is at the $q+1$-th location.

Case 2.3: Let x, y, z be three pairwise different elements in $\mathbb{F}_q$. Consider the submatrix

$$M_{2,3} = \begin{bmatrix} 1 & 1 & 1 & 0 \\ x & y & z & 0 \\ x^2 & y^2 & z^2 & 1 \\ x^4 & y^4 & z^4 & 0 \end{bmatrix}.$$

Then $|M_{2,3}| = (x^4+z^4)(x+y)+(x^4+y^4)(x+z)$. By Lemma 1, $f(x) = x^4$ is an oval polynomial over $\mathbb{F}_q$ when m is an odd integer with $m \geq 3$. Then by Lemma 4, we have $|M_{2,3}| \neq 0$. Hence, there is no codeword of weight 4 in $\mathcal{D}^{\perp}$ such that its first three nonzero coordinates are at the first q locations and the rest is at the $q+2$-th location.

Case 2.4: Let x, y be two different elements in $\mathbb{F}_q$. Consider the submatrix

$$M_{2,4} = \begin{bmatrix} 1 & 1 & 0 & 0 \\ x & y & 1 & 0 \\ x^2 & y^2 & 0 & 1 \\ x^4 & y^4 & 0 & 0 \end{bmatrix}.$$

Then $|M_{2,4}| = x^4+y^4 = (x+y)^4$. $|M_{2,4}| \neq 0$ as $x \neq y$. Hence, there is no codeword of weight 4 in $\mathcal{D}^{\perp}$ such that its first two nonzero coordinates are at the first q locations and the others are at $q+1$-th and $q+2$-th locations.

Based on these discussions, we deduce that the total number of codewords of weight 4 in $\mathcal{D}^{\perp}$ is $\frac{q(q-1)^2(q-2)}{24}$.

We finally prove that the minimum distance of $\mathcal{D}$ is $q-2$. Assume that $d(\mathcal{D}) \leq q-3 = q+2-5$ and let $\mathbf{c} = a\mathbf{g}_1 + b\mathbf{g}_2 + c\mathbf{g}_3 + d\mathbf{g}_4$ be a codeword with the minimum weight in $\mathcal{D}$, where $\mathbf{g}_1$, $\mathbf{g}_2$, $\mathbf{g}_3$ and $\mathbf{g}_4$ respectively represent the first, second, third and fourth rows of G_2. Hence there are at least five coordinates in $\mathbf{c}$ are zero. Consider the following cases.

Case 3.1: Assume that none of the last two coordinates in $\mathbf{c}$ are zero. Then there exist five pairwise distinct elements x_1, x_2, x_3, x_4, x_5 in $\mathbb{F}_q$ such that

$$\begin{cases} a+bx_1+cx_1^2+dx_1^4 = 0, \\ a+bx_2+cx_2^2+dx_2^4 = 0, \\ a+bx_3+cx_3^2+dx_3^4 = 0, \\ a+bx_4+cx_4^2+dx_4^4 = 0, \\ a+bx_5+cx_5^2+dx_5^4 = 0. \end{cases}$$

Let $f(x) = a+bx+cx^2+dx^4$. Note that x_1, x_2, x_3, x_4, x_5 are the roots of $f(x)$. This contradicts with the fact that $f(x)$ has at most four roots in $\mathbb{F}_q$ as $(a,b,c,d) \neq (0,0,0,0)$.

Case 3.2: Assume that one of the last two coordinates in $\mathbf{c}$ is zero. Then there exist four elements x_1,x_2,x_3,x_4 in $\mathbb{F}_q$ such that

$$\begin{cases} a+bx_1+cx_1^2+dx_1^4=0, \\ a+bx_2+cx_2^2+dx_2^4=0, \\ a+bx_3+cx_3^2+dx_3^4=0, \\ a+bx_4+cx_4^2+dx_4^4=0, \\ b=0, \end{cases} \text{or} \begin{cases} a+bx_1+cx_1^2+dx_1^4=0, \\ a+bx_2+cx_2^2+dx_2^4=0, \\ a+bx_3+cx_3^2+dx_3^4=0, \\ a+bx_4+cx_4^2+dx_4^4=0, \\ c=0. \end{cases}$$

Since $f(x)=x^2$ and $f(x)=x^4$ are oval polynomials over $\mathbb{F}_q$ for odd $m\geq 3$, it is easy to deduce that $a=b=c=d=0$ and $\mathbf{c}=0$ by Lemma 5. This contradicts with the fact that $\mathbf{c}$ is a minimum weight codeword in $\mathcal{D}$.

Case 3.3: Assume that the last two coordinates in $\mathbf{c}$ are zero. Then there exist three pairwise distinct elements x_1,x_2,x_3 in $\mathbb{F}_q$ such that

$$\begin{cases} a+bx_1+cx_1^2+dx_1^4=0, \\ a+bx_2+cx_2^2+dx_2^4=0, \\ a+bx_3+cx_3^2+dx_3^4=0, \\ b=0, \\ c=0. \end{cases}$$

Since $f(x)=x^4$ is an oval polynomial of $\mathbb{F}_q$ for odd $m\geq 3$, then x_1^4,x_2^4,x_3^4 are pairwise distinct elements in $\mathbb{F}_q$. So we have $a=b=c=d=0$ and $\mathbf{c}=0$. This contradicts with the fact that $\mathbf{c}$ is a minimum weight codeword in $\mathcal{D}$.

As a result, we then have $d(\mathcal{D})\geq q-2$. Besides, $d(\mathcal{D})\leq q-1$ by the Singleton bound. If $d(\mathcal{D})=q-1$, then $\mathcal{D}$ is a $[q+2,4,q-1]$ MDS code and $\mathcal{D}^\perp$ is also an MDS code with parameters $[q+2,q-2,5]$. This is contrary to $d(\mathcal{D}^\perp)=4$. So $\mathcal{D}$ is an AMDS code with parameters $[q+2,4,q-2]$ and $\mathcal{D}^\perp$ is also an AMDS code with parameters $[q+2,q-2,4]$. Then $\mathcal{D}$ is an NMDS code. By Lemma 2, the total number A_{q-2} of the minimum weight codewords in $\mathcal{D}$ is equal to the total number of the codewords of weight 4 in $\mathcal{D}^\perp$. Hence $A_{q-2}=\frac{q(q-1)^2(q-2)}{24}$. By Lemma 1, the weight enumerator of $\mathcal{D}$ follows.

It is well known that any $[n,k,n-k+1]$ MDS code over $\mathbb{F}_q$ must have a unique weight enumerator. But this fact is not true for NMDS codes. We give another construction of NMDS code with the same parameters $[q+2,4,q-2]$ but different weight enumerators in the following. Define

$$G_{2,1}=\begin{bmatrix} 1 & 1 & \cdots & 1 & 0\,0\,0 \\ \alpha_1 & \alpha_2 & \cdots & \alpha_{q-1} & 1\,0\,0 \\ \alpha_1^2 & \alpha_2^2 & \cdots & \alpha_{q-1}^2 & 0\,1\,0 \\ \alpha_1^4 & \alpha_2^4 & \cdots & \alpha_{q-1}^4 & 0\,0\,1 \end{bmatrix}.$$

Then $G_{2,1}$ is a 4 by $q+2$ matrix over $\mathbb{F}_q$. Let $G_{2,1}$ generate a linear code $\mathcal{D}_1$ over $\mathbb{F}_q$. With a similar proof to that of Theorem 3, we can obtain the parameters and weight enumerator of $\mathcal{D}_1$ in the following theorem.

Theorem 4. *Let m be an odd integer with $m \geq 3$. Then $\mathcal{D}_1$ is a $[q+2,4,q-2]$ NMDS code over $\mathbb{F}_q$ with weight enumerator*

$$A(z) = 1 + \frac{(q-1)^2(q-2)(q-4)}{24}z^{q-2} + \frac{(q-1)(5q^2-6q+4)}{3}z^{q-1} + \frac{(q-1)(q^3-5q^2+20q-4)}{4}z^{q} + \frac{(q-1)^2(q^2+6q-4)}{3}z^{q+1} + \frac{(q-1)^2(9q^2-10q+8)}{24}z^{q+2}.$$

3.3 NMDS Codes with Parameters $[q+1,4,q-3]$

Define

$$G_3 = \begin{bmatrix} 1 & 1 & \cdots & 1 & 1 & 0 \\ \alpha_1 & \alpha_2 & \cdots & \alpha_{q-1} & \alpha_0 & 1 \\ \alpha_1^2 & \alpha_2^2 & \cdots & \alpha_{q-1}^2 & \alpha_0 & 0 \\ \alpha_1^4 & \alpha_2^4 & \cdots & \alpha_{q-1}^4 & \alpha_0 & 0 \end{bmatrix} = \begin{bmatrix} 1 & 1 & \cdots & 1 & 1 & 0 \\ \alpha_1 & \alpha_2 & \cdots & \alpha_{q-1} & 0 & 1 \\ \alpha_1^2 & \alpha_2^2 & \cdots & \alpha_{q-1}^2 & 0 & 0 \\ \alpha_1^4 & \alpha_2^4 & \cdots & \alpha_{q-1}^4 & 0 & 0 \end{bmatrix}.$$

Then G_3 is a 4 by $q+1$ matrix over $\mathbb{F}_q$. Let G_3 generate a linear code $\mathcal{F}$ over $\mathbb{F}_q$. The parameters and weight enumerator of $\mathcal{F}$ are determined in the following theorem.

Theorem 5. *Let m be an odd integer with $m \geq 3$. Then $\mathcal{F}$ is a $[q+1,4,q-3]$ NMDS code over $\mathbb{F}_q$ with weight enumerator*

$$A(z) = 1 + \frac{q(q-1)^2(q-2)}{24}z^{q-3} + \frac{q(q-1)^2}{2}z^{q-2} + \frac{q(q^3+5q-6)}{4}z^{q-1} + \frac{(q-1)(q+2)(2q^2-q+3)}{6}z^{q} + \frac{q(3q+2)(q-1)^2}{8}z^{q+1}.$$

Proof. By the proof of Theorem 3, we can similarly prove that $\mathcal{F}$ has parameters $[q+1,4,q-3]$ and $\mathcal{F}^{\perp}$ has parameters $[q+1,q-3,4]$. In addition, the total number of codewords of weight 4 in $\mathcal{F}^{\perp}$ is $\frac{q(q-1)^2(q-2)}{24}$ and $A_{q-3} = \frac{q(q-1)^2(q-2)}{24}$ by Lemma 2. By Lemma 1, the weight enumerator of $\mathcal{F}$ are proved.

Below we give another construction of NMDS code with the same parameters $[q+1,4,q-3]$ but different weight enumerators. Define

$$G_{3,1} = \begin{bmatrix} 1 & 1 & \cdots & 1 & 0 & 0 \\ \alpha_1 & \alpha_2 & \cdots & \alpha_{q-1} & 1 & 0 \\ \alpha_1^2 & \alpha_2^2 & \cdots & \alpha_{q-1}^2 & 0 & 1 \\ \alpha_1^4 & \alpha_2^4 & \cdots & \alpha_{q-1}^4 & 0 & 0 \end{bmatrix}.$$

Then $G_{3,1}$ is a 4 by $q+1$ matrix over $\mathbb{F}_q$. Let $G_{3,1}$ be a generator matrix of the linear code $\mathcal{F}_1$ over $\mathbb{F}_q$. With a similar proof to that of Theorem 3, the parameters and weight enumerator of $\mathcal{F}_1$ can be determined in the following theorem.

Theorem 6. *Let m be an odd integer with $m \geq 3$. Then $\mathcal{F}_1$ is a $[q+1,4,q-3]$ NMDS code over $\mathbb{F}_q$ with weight enumerator*

$$A(z) = 1 + \frac{(q-1)^2(q-2)(q-4)}{24} z^{q-3} + \frac{(7q-8)(q-1)^2}{6} z^{q-2} + \frac{(q-1)(q^3-3q^2+18q-8)}{4} z^{q-1} + \frac{(q-1)(2q^3+7q^2-11q+14)}{6} z^{q} + \frac{(q-1)^2(9q^2+2q+8)}{24} z^{q+1}.$$

Note that the NMDS codes in [19, Theorems 37, 38], Theorems 3 and 4 have the same parameters. But they are inequivalent to each other as they have different enumerators.

4 Optimal Locally Recoverable Codes

In this section, we prove that the duals of NMDS codes constructed in Sect. 3 are optimal LRCs.

Lemma 7 *[19]. Let C be an NMDS code. Denote by $d^{\perp} = d(C^{\perp})$. If $\bigcap_{S \in \mathcal{B}_{d^{\perp}}(C^{\perp})} S = \emptyset$, then the minimum linear locality of $C^{\perp}$ is equal to $d(C)-1$, where $\mathcal{B}_{d^{\perp}}(C^{\perp})$ is the set of the supports of all codewords with weight $d^{\perp}$ in $C^{\perp}$.*

Theorem 7. *The dual of the NMDS code C in Theorem 2 is a $(q+3, q-1, 4, q; q-2)-$ LRC. Besides, $C^{\perp}$ is both distance-optimal and dimension-optimal.*

Proof. By the proof of Theorem 2, $\bigcap_{S \in \mathcal{B}_4(C^{\perp})} S = \emptyset$. Then by Lemma 7, the minimum linear locality of $C^{\perp}$ is $d(C)-1 = q-2$. Now we prove $C^{\perp}$ is an optimal LRC. Putting the parameters of the $(q+3, q-1, 4, q; q-2)$-LRC into the right-hand side of the Singleton-like bound in (1), we have

$$n-k-\left\lceil \frac{k}{r} \right\rceil + 2 = q+3-(q-1)-\left\lceil \frac{q-1}{q-2} \right\rceil + 2 = 4.$$

Hence $C^{\perp}$ is a distance-optimal LRC. Putting $t=1$ and the parameters of the $(q+3, q-1, 4, q; q-2)$-LRC into the right-hand side of the Cadambe-Mazumdar bound in (2), we have

$$k \leq r + k_{opt}^{(q)}(n-(r+1), d) = q-2+k_{opt}^{(q)}(4,4) = q-1,$$

where $k_{opt}^{(q)}(4,4) = 1$ by the classical Singleton bound. Hence $C^{\perp}$ is a dimension-optimal LRC. The proof is completed.

Theorem 8. *The dual of the NMDS code $\mathcal{D}$ in Theorem 3 is a$(q+2, q-2, 4, q; q-3)-$ LRC. Besides, $\mathcal{D}^{\perp}$ is both distance-optimal and dimension-optimal.*

Proof. It is easy to deduce that $\bigcap_{S \in \mathcal{B}_4(\mathcal{D}^{\perp})} S = \emptyset$ by the proof of Theorem 3. The rest of this proof is similar to that of Theorem 7.

Theorem 9. *The dual of the NMDS code $\mathcal{D}_1$ in Theorem 4 is a $(q+2, q-2, 4, q; q-3)-LRC$. Besides, $\mathcal{D}_1^{\perp}$ is both distance-optimal and dimension-optimal.*

Proof. Similarly to the proof of Theorem 3, it is easy to deduce that $\bigcap_{S \in \mathcal{B}_4(\mathcal{D}_1^{\perp})} S = \emptyset$. The rest of this proof is similar to that of Theorem 7.

Theorem 10. *The dual of the NMDS code $\mathcal{F}$ in Theorem 5 is a $(q+1, q-3, 4, q; q-4)-LRC$. Besides, $\mathcal{F}^{\perp}$ is both distance-optimal and dimension-optimal.*

Proof. Similarly to the proof of Theorem 3, it is easy to deduce that $\bigcap_{S \in \mathcal{B}_4(\mathcal{F}^{\perp})} S = \emptyset$. The rest of this proof is similar to that of Theorem 7.

Theorem 11. *The dual of the NMDS code $\mathcal{F}_1$ in Theorem 6 is a $(q+1, q-3, 4, q; q-4)-LRC$. Besides, $\mathcal{F}_1^{\perp}$ is both distance-optimal and dimension-optimal.*

Proof. Similarly to the proof of Theorem 3, we have $\bigcap_{S \in \mathcal{B}_4(\mathcal{F}_1^{\perp})} S = \emptyset$. The rest of this proof is similar to that of Theorem 7.

5 Concluding Remarks

In this paper, we constructed several families of NMDS codes and explicitly determined their weight enumerators. The duals of these NMDS codes were proved to be optimal locally recoverable codes. We remark that the optimal locally repairable codes of distance 4 in this paper are not contained in [15].

References

1. de Boer, M.A.: Almost MDS codes. Des. Codes Cryptogr. **9**, 143–155 (1996). https://doi.org/10.1023/A:1018014013461
2. Cadambe, V., Mazumdar, A.: An upper bound on the size of locally recoverable codes. In: International Symposium on Network Coding, pp. 1–5 (2013). https://doi.org/10.1109/NetCod.2013.6570829
3. Ding, C.: Designs from Linear Codes. World Scientific, Singapore (2019)
4. Ding, C., Tang, C.: Infinite families of near MDS codes holding t-designs. IEEE Trans. Inform. Theory **66**(9), 5419–5428 (2020). https://doi.org/10.1109/TIT.2020.2990396
5. Dodunekov, S., Landgev, I.: On near-MDS codes. J. Geometry **54**, 30–43 (1995). https://doi.org/10.1007/BF01222850
6. Faldum, A., Willems, W.: Codes of small defect. Des. Codes Cryptogr. **10**, 341–350 (1997). https://doi.org/10.1023/A:1008247720662
7. Gopalan, P., Huang, C., Simitci, H., Yekhanin, S.: On the locality of codeword symbols. IEEE Trans. Inform. Throry **58**(11), 6925–6934 (2012). https://doi.org/10.1109/TIT.2012.2208937
8. Geng, X., Yang, M., Zhang, J., Zhou, Z.. A class of almost MDS codes. Finite Fields Appl. **79**, 101996 (2022). https://doi.org/10.1016/j.ffa.2022.101996
9. Heng, Z., Ding, C., Zhou, Z.: Minimal linear codes over finite fields. Finite Fields Appl. **54**, 176–196 (2018). https://doi.org/10.1016/j.ffa.2018.08.010

10. Huang, D., Yue, Q., Niu, Y., Li, X.: MDS or NMDS self-dual codes from twisted generalized Reed-Solomon codes. Designs Codes Cryptogr. **89**(9), 2195–2209 (2021). https://doi.org/10.1007/s12095-022-00564-9
11. Li, C., Wu, P., Liu, F.: On two classes of primitive BCH Codes and some related codes. IEEE Trans. Inform. Theory **65**(6), 3830–3840 (2019). https://doi.org/10.1109/TIT.2018.2883615
12. Li, X., Heng, Z.: A construction of optimal locally recoverable codes. ArXiv:2204.12034 (2022). https://doi.org/10.48550/arXiv.2204.12034
13. Li, X., Heng, Z.: Constructions of near MDS codes which are optimal locally recoverable codes. ArXiv:2204.11208 (2022). https://doi.org/10.48550/arXiv.2204.11208
14. Lidl, R., Niederreiter, H.: Finite Fields. Cambridge University Press, Cambridge (1997). https://doi.org/10.1016/S1570-7954(96)80013-1
15. Luo, Y., Xing, C., Chen, Y.: Optimal locally repairable codes of distance 3 and 4 via cyclic codes. IEEE Trans. Inform. Theory **65**(2), 1048–1053 (2018). https://doi.org/10.1109/TIT.2018.2854717
16. Mesnager, S.: Bent vectorial functions and linear codes from o-polynomials. Designs Codes Cryptography **77**(1), 99–116 (2014). https://doi.org/10.1007/s10623-014-9989-6
17. Maschietti, A.: Difference sets and hyperovals. Des. Codes Cryptogr. **14**(1), 89–98(1998). https://doi.org/10.1023/A:1008264606494
18. Shi, X., Yue, Q., Wu, Y.: New quantum MDS codes with large minimum distance and short length from generalized Reed-Solomon codes. Dis. Math. **342**(7), 1989–2001(2019). https://doi.org/10.1016/j.disc.2019.03.019
19. Tan, P., Fan, C., Ding, C., Tang, C., Zhou, Z.: The minimum locality of linear codes. Des. Codes Cryptogr. 1–32 (2022). https://doi.org/10.1007/s10623-022-01099-z
20. Tang, C., Ding, C.: An infinite family of linear codes supporting 4-designs. IEEE Trans. Inform. Theory **67**(1), 244–254 (2020). https://doi.org/10.1109/TIT.2020.3032600
21. Wang, Q., Heng, Z.: Near MDS codes from oval polynomials. Discrete Math. **344**(4), 112277 (2021). https://doi.org/10.1016/j.disc.2020.112277

Optimal Possibly Nonlinear 3-PIR Codes of Small Size

Henk D.L. Hollmann[1(✉)] and Urmas Luhaäär[2]

[1] Institute of Computer Science, University of Tartu, 50409 Tartu, Estonia
henk.hollmann@ut.ee

[2] Institute of Mathematics and Statistics, University of Tartu, 51009 Tartu, Estonia
urmas.luhaaar@ut.ee

Abstract. First, we state a generalization of the minimum-distance bound for PIR codes. Then we describe a construction for linear PIR codes using packing designs and use it to construct some new 5-PIR codes. Finally, we show that no encoder (linear or nonlinear) for the binary r-th order Hamming code produces a 3-PIR code except when $r = 2$. We use these results to determine the smallest length of a binary (possibly nonlinear) 3-PIR code of combinatorial dimension up to 6. A binary 3-PIR code of length 11 and size 2^7 is necessarily nonlinear (as a PIR code), and we pose the existence of such a code as an open problem.

Keywords: Batch codes · PIR codes · Nonlinear code · Hamming code · Packing design

1 Introduction

Private Information Retrieval (PIR) schemes enable a user to extract a bit of information from a database, stored in encoded form on a multi-server distributed data storage system, without leaking information to the servers in which particular bit the user was interested in, see, e.g., [3].

A (binary) t-PIR code of length n and size 2^k is an encoder that encodes k data bits one-to-one into n encoded bits in such a way that each data bit has t mutually disjoint recovery sets. If the encoder employs only linear operations, then we speak of a *linear* PIR code. Linear t-PIR codes can be used to implement a classical (linear) t-server PIR scheme [3] with less storage overhead than the original scheme, by using the PIR code to emulate the t servers [5,6]; see also [17] for another explanation of how this magic is worked.

A *batch code* is a special type of PIR code where for *any* batch of t data symbols, there exist t mutually disjoint recovery sets. Batch codes were initially introduced in [9] as a method to improve load-balancing in distributed data storage systems. Later, so-called *switch codes* (a special case of batch codes) were proposed in [19] as a method to increase the throughput rate in network

The research of the first author was supported by the Estonian Research Council grant PRG49.

S. Mesnager and Z. Zhou (Eds.): WAIFI 2022, LNCS 13638, pp. 159–168, 2023.
https://doi.org/10.1007/978-3-031-22944-2_9

switches. In such applications, there is no need for the batch code to be linear. We remark that a PIR or batch code can be nonlinear because the associated code is nonlinear, or because it consists of a nonlinear encoder onto a linear code.

For an overview of PIR- and batch-type codes and other similar codes, we refer to [16]. In this paper, all PIR codes are multiset primitive [16], and we will mostly consider only binary codes. Precise definitions will be given in the next section.

For linear PIR-codes, much work has been done to find bounds on the smallest n for which a linear t-PIR code of dimension k and length n exists, see for example [10] for a recent overview. For linear batch codes, the situation is similar. Nonlinear PIR-codes are interesting combinatorial objects in their own right, but in contrast, virtually nothing is known about their possible parameters. In fact, we do not know a single example of an "interesting" nonlinear PIR code, that is, with parameters for which no linear PIR-code exists. One of our aims in this paper is to at least identify some parameters for which such an interesting nonlinear code could exist, were we concentrate on 3-PIR codes since there are linear optimal t-PIR codes for $t = 1, 2$ (see, e.g., [10, p. 560]).

The contents of this paper are as follows. In Sect. 2, we define the notion of a t-PIR code and various other notions that we will need. Our results strongly depend on a simple bound on the minimum distance of a (linear or nonlinear) t-PIR code. In Sect. 3, we derive a generalization of this lower bound for a broad class of (not necessarily linear) PIR-like codes. For linear 3-PIR and 3-batch codes, the optimal codes are known. Bounds and constructions for linear 3-PIR codes and some generalizations of these constructions are discussed in Sect. 4. In Sect. 5 we prove one of our main results, stating that no encoder for a binary length $2^r - 1$ Hamming code with $r \geq 3$ is a 3-PIR code. We use this result to determine the optimal length of 3-PIR codes of size 2^k for $1 \leq k \leq 6$ in Sect. 6, and we pose the question of the existence of a (necessarily nonlinear) 3-PIR code of length 11 and size 2^7 as an open problem. We end with some conclusions and further questions in Sect. 7.

2 Preliminaries

Let q be a positive integer. We use Σ to denote an alphabet with q symbols; if q is a prime-power, we identify these symbols with the q elements of the finite field $\mathbb{F}_q$ of size q. For a positive integer n, we let $[n]$ denote the set $\{1, \ldots, n\}$, and we use this set to index the positions in code words of length n.

Informally, PIR- and batch-type codes are characterized by the property that given the encoded data, certain *simultaneous* requests for specific data symbols can each be handled by reading and decoding data from a set of positions called a *recovery set*, where these sets are supposed to be of bounded size, with limited overlap between the sets. We now introduce some useful terminology to make this precise.

Definition 1. A *k-to-n encoder* over an alphabet Σ is a one-to-one map $\epsilon : \Sigma^k \to \Sigma^n$; the image $C = C_\epsilon$ of ϵ is referred to as the *associated code* of ϵ. By definition, such an encoder ϵ has a *decoder* $\delta : C \to \Sigma^k$ with the property that if $c = \epsilon(a)$, then $\delta(c) = a$. We refer to ϵ as a q-ary encoder if $|\Sigma| = q$.

Let $I = \{i_1, \ldots, i_s\} \subseteq [n]$ with $i_1 < \ldots < i_s$. Given a code word $c \in \Sigma^n$, the *restriction* c_I of c to I is the word $c_I = (c_{i_1}, \ldots, c_{i_s})$.

Definition 2. We say that I is a *recovery set* of the j-th data symbol for a k-to-n encoder ϵ over Σ if for every $a \in \Sigma^k$, when $c = \epsilon(a)$, the restriction c_I of c to I uniquely determines a_j; it is called *minimal* if no proper subset of I has this property.

A *query* of ϵ is a sequence $i_1, \ldots, i_t$ of (not necessarily distinct) elements of $[n]$. Given a code word $c = \epsilon(a)$, the query $i_1, \ldots, i_t$ should be considered as a request to obtain the data symbols $a_{i_1}, \ldots, a_{i_t}$. We will say that the sets $I_1, \ldots, I_t \subseteq [n]$ *serve* the query of ϵ if for every $j \in [t]$, the set I_j is a recovery set of ϵ for the i_j-th data symbol. We say that $I_1, \ldots, I_t$ serve the query with *width* w and *multiplicity* μ if $|I_j| \leq w$ $(j = 1, \ldots, t)$ and if every position $i \in [n]$ occurs in at most μ of the sets $I_1, \ldots, I_t$.

Now we are ready for a definition of batch-type codes.

Definition 3. Let ϵ be a k-to-n encoder over Σ, let w, μ be positive integers, and let $\mathcal{Q}$ be a collection of queries of ϵ. We say that ϵ is a $(\mathcal{Q}, w, \mu)$*-batch code* if ϵ can serve every query in $\mathcal{Q}$ with width at most w and multiplicity at most μ. The encoder ϵ is a (t, w, μ)-PIR code if ϵ is a $(\mathcal{Q}, w, \mu)$-batch code with $\mathcal{Q}$ consisting of all queries of the form $i, i, \ldots, i$ (t times) with $i \in [k]$; a $(t, \infty, 1)$-PIR code is called a *t-PIR code*. The encoder ϵ is a *t-batch code* if ϵ is a $(\mathcal{Q}, \infty, 1)$-batch code with $\mathcal{Q}$ consisting of all queries of the form $i_1, \ldots, i_t$ with $i_1, \ldots, i_t \in [k]$.

More informally, a recovery set for a data symbol allows the recovery of a certain data symbol by inspecting only the code word symbols in the positions of the recovery set. Then a t-PIR code has the property that every encoded data symbol has t mutually disjoint recovery sets, while for a t-batch code we can find t mutually disjoint recovery sets for every batch of t data symbols.

We remark that what we call here a batch code is referred to by some authors as a primitive (multiset) batch code, see, e.g., [16].

A *linear* k-to-n encoder over a q-ary alphabet is an $\mathbb{F}_q$-linear map $\epsilon : \mathbb{F}_q^k \to \mathbb{F}_q^n$, which can thus be represented by a $k \times n$ matrix G over $\mathbb{F}_q$; here G is the generator matrix of the associated linear code $C = \epsilon(\mathbb{F}_q^k)$. In this case, a set $I \subseteq [n]$ is a recovery set for the j-th data symbol if and only if some $\mathbb{F}_q$- linear combination of the columns of G indexed by I sum up to e_j, the j-th unit vector in $\mathbb{F}_q^k$, for a proof see [13, Theorem 1].

In this paper, we are mainly interested in "optimal" binary t-PIR and t-batch codes with $1 \leq t \leq 4$.

Definition 4. Let k and t be positive integers. We let $P(k, t)$, $PL(k, t)$, $B(k, t)$, and $BL(k, t)$ denote the smallest length n of a binary possibly nonlinear t-PIR code, a binary linear t-PIR code, a binary possibly nonlinear t-batch code, or a binary linear t-batch code, of size 2^k, respectively.

We will refer to a code of the above types with an optimal, minimal length as an *optimal* code for that type.

3 The Minimum-Distance Bound for Batch-Type Codes

Let Σ denote an alphabet of size q. An $(n, M, d)_q$-code C is a subset of Σ^n, of size M, where any two distinct code words in C have (Hamming) distance at least d. Here, the (Hamming) distance between two words $v, w \in \Sigma^n$ is the number of positions in which v and w differ. An $[n, k, d]_q$ code is a *linear* code of length n and dimension k over $\mathbb{F}_q$, with minimum distance d. One of the very few known lower bounds for the length of a t-PIR code of a given size results from the observation that a t-PIR code must have minimum distance at least t. This was first stated for binary linear batch codes in [13] and for non-linear batch codes over general alphabets in [22]. See also [16,21], and [11,12] where the result was stated for PIR codes. Here we present a slight generalization of these results.

Theorem 5. Let C be an $(n, q^k, d)_q$-code over an alphabet Σ, and suppose that C has an encoder $\epsilon : \Sigma^k \to C$ that is a (t, ∞, μ)-PIR code. Then $\lceil t/\mu \rceil \leq d$.

Proof. Let $\delta : C \to \Sigma^k$ be the corresponding decoder. Let $c^{(1)}, c^{(2)}$ be distinct code words from C. Then there is an s such that $\delta(c^{(1)})_s \neq \delta(c^{(2)})_s$. By our assumption on C, there are sets $I_1, \ldots, I_t$ that serve the query $s, s, \ldots, s$ (t times) with multiplicity at most μ. So for every position set I_j, the restrictions $c^{(1)}_{I_j}$ and $c^{(2)}_{I_j}$ determine distinct data symbols, hence I_j must contain a position i_j for which $c^{(1)}_{i_j} \neq c^{(1)}_{i_j}$. By the multiplicity condition there must be at least $\lceil t/\mu \rceil$ distinct positions among $i_1, \ldots, i_t$, so as a consequence, $c^{(1)}$ and $c^{(2)}$ differ in at least $\lceil t/\mu \rceil$ positions. Since the code words were arbitrary, we conclude that $d \geq \lceil t/\mu \rceil$.

We will refer to a code that attains the bound in Theorem 5 as *distance-optimal*.

4 Some Bounds and Constructions

For later use, we first state the following simple result.

Theorem 6. If $P(k, 2t - 1) = PL(k, 2t - 1)$, then $P(k, 2t) = PL(k, 2t) = P(k, 2t - 1) + 1$.

Proof. Suppose that the condition in the theorem holds, and let C be a linear $(2t - 1)$-PIR code of dimension k and length $n = P(k, 2t - 1)$. Then by a well-known argument (see [6]), the extended code $\overline{C}$ (adding an overall parity-check bit) is a $(2t)$-PIR code, hence $P(k, 2t) \leq P(k, 2t - 1) + 1$. On the other hand, if C' is any s-PIR code of size 2^k and length n, then the code obtained from C' by deleting a position is obviously an $(s - 1)$-PIR code. By taking $s = 2t$, we conclude that $P(k, 2t - 1) \leq P(k, 2t) - 1$. Combining these inequalities shows that $P(k, 2t) = P(k, 2t-1)+1$, and since $\overline{C}$ has length $P(k, 2t-1)+1 = P(k, 2t)$, we also have that $P(k, 2t) = PL(k, 2t)$.

As a consequence of Theorem 6, we can restrict our search for binary nonlinear t-PIR codes to the cases where t is odd. We obviously have $P(k,1) = PL(k,1) = k$ and $P(k,2) = PL(k,2) = k+1$, where the optimal codes are the entire k-dimensional space and the even-weight vectors in a $(k+1)$-dimensional space, respectively (see, e.g., [10, p. 560]). This leads us to consider the case where $t = 3$.

In [15], it was shown that a linear 3-PIR code with dimension k and length n, so with redundancy $r = n - k$, satisfies the bound $r(r-1)/2 \geq k$. Moreover, this bound is attained by the codes with generator matrix of the form $(I_k P)$, where P is the $k \times r$ matrix that has rows consisting of distinct binary vectors of weight 2 (note that such a matrix exists by the condition on k and r). We even have the following.

Theorem 7. *Let $k \geq 1$ be integer. The code C with generator matrix $(I_k P)$ as defined above is 3-batch, and the extended code is 4-batch. Hence both are optimal linear codes, $BL(k,3) = PL(k,3)$, and $BL(k,4) = PL(k,4)$. Both the code C and its extension are also distance-optimal.*

Proof. The batch properties of the two codes can easily be proved directly, but also follow from [18, Lemma 3, 4, 5] since the matrices of the form $(I_k P)$ as defined above are systematic. Since $PL(k,4) = PL(k,3)+1$ (see [6]), both codes must be optimal both as PIR and as batch codes. Since the code C has code words of weight 3 in its generator, by the minimum distance bound Theorem 5, it has distance 3, and the extension has minimum distance 4.

In fact, the above code construction can be generalized. To this end, we need a special type of combinatorial structure. Let $v \geq k \geq t$. A $t-(v,k,\lambda)$ *packing design* or, more briefly, a *packing*, consists of a collection $\mathcal{B}$ of subsets of $[v]$, each of size k, with the property that any subset of $[v]$ of size t occurs in at most λ sets in $\mathcal{B}$. We will refer to the elements of $[v]$ as *points* and to the elements of $\mathcal{B}$ as *blocks*. We write $D_\lambda(v,k,t)$ to denote the *packing number*, the largest possible number of blocks in a $t-(v,k,\lambda)$ packing; in the case where $\lambda = 1$, we denote the packing number by $D(v,k,t)$. For a general overview of packing designs, we refer to [4, Part IV, Sect. 40].

Here, we will be interested in the case $t = 2$ and $\lambda = 1$. Note that in this case, any two blocks of the design intersect in at most one point (indeed, otherwise a pair of points from the intersection would be contained in at least two blocks). We now have the following generalization of Theorem 7.

Theorem 8. *Let r, t be positive integers with $r \geq t-1$, and let k be a positive integer such that $k \leq D(r, t-1, 2)$. Let P be a $k \times r$ matrix whose rows are the incidence vectors of k pairwise distinct blocks from a $2-(r,t-1,1)$ packing design with at least k blocks (note that this is possible by the condition on k). Then the matrix $(I_k P)$ is the generator matrix of a t-PIR code. As a consequence, we have that $PL(k,t) \leq k+r$, where r is the smallest integer for which $k \leq D(r,t-1,2)$.*

Proof. By the properties of a packing design, this follows immediately from [6, Lemma 7] or [5, Lemma 7].

Strictly speaking, the above result is not new. But the authors of [5] did not explicitly make the connection with packing designs, so they did not quantify their result except for the case of Steiner systems.

Note that this theorem indeed generalizes Theorem 7 since in the case where $t = 3$, a $2-(r,2,1)$ packing design is simply a collection of pairs from $[r]$, so that $D(r,2,2) = r(r-1)/2$. Since, as remarked before, $PL(k,4) = PL(k,3)+1$, the next interesting case of the above theorem is when $t = 5$. Interestingly, the packing numbers $D(r,4,2)$ are completely known.

Theorem 9 (See [2]**).** Let

$$U(r,4,2) = \left\lfloor \frac{r}{4} \left\lfloor \frac{r-1}{3} \right\rfloor \right\rfloor,$$

and write

$$J(r,4,2) = \begin{cases} U(r,4,2) - 1, & \text{for } r \equiv 7 \text{ or } 10 \pmod{12}; \\ U(r,4,2), & \text{otherwise.} \end{cases} \tag{1}$$

Then $D(r,4,2) = J(2,4,r)$ if $r \notin \{8,9,10,11,17,19\}$ and $D(r,4,2) = J(r,4,2) - \epsilon$ with $\epsilon = 1$ for $r \in \{9,10,17\}$ and $\epsilon = 2$ for $r \in \{8,11,19\}$.

In the next example, we discuss some applications of Theorem 8 and Theorem 9.

Example 10. We mention some improvements of [6, Table III].
(i) First, $D(12,4,2) = 9$, so $P(9,5) \leq 9 + 12 = 21$ and $P(9,6) \leq 22$, which improves the known value by 1, but loses against the more recent [10, Table 1].
(ii) We have $D(15,4,2) = 15$ and $D(16,4,2) = 20$. So $P(15,5) \leq 30$, hence $P(15,6) \leq 31$, improving the value in [6, Table III] by 3, and $P(16+i,5) \leq 32+i$, hence $P(16+i,6) \leq 33+i$, for $i = 0,\ldots,4$, improving the values in [6, Table III] by 4. After completion of this work, we learned that these results are similar to those in [7] (unpublished), see also [8] in these proceedings. □

5 The Hamming Codes as PIR-codes

For an integer $r \geq 2$, the binary r-th order Hamming code is a linear code of length $n = 2^r - 1$ and dimension $k = 2^r - 1 - r$, with the $k \times n$ parity-check matrix H_k whose columns are the nonzero binary vectors of length r. Obviously, these codes have minimum Hamming distance 3. We will now prove the following.

Theorem 11. For $r \geq 2$, the all-one word $\mathbf{1}$ is in the r-th order Hamming code. Moreover, let $r \geq 3$ and suppose that for some encoder for the r-th order Hamming code, the position subsets I_1, I_2, I_3 are three mutually disjoint, minimal recovery sets for a particular data bit. Then for every code word c, both c and its complement $\mathbf{1} + c$ decode to the same value of that data bit.

Proof. It is natural to label the positions with the nonzero binary vectors of length r. In what follows, we will not distinguish between a set $S \subseteq \mathbb{F}_2^r \setminus \{0\}$ and its characteristic vector χ_S of length $2^r - 1$ that has a 1 in the positions of S and a 0 in the other positions. Note that with this convention, a set $S = \{v, w, v+w\}$ corresponds to a word of (minimal) weight 3 in the Hamming code, so the minimum weight vectors in the Hamming code correspond to the lines in the projective geometry $\mathrm{PG}(r-1,2)$. Note also that every point in $\mathrm{PG}(r-1,2)$ is on $(2^r-2)/2 = 2^{r-1}-1$ lines, so for $r \geq 2$ the all-one vector $\mathbf{1}$ is contained in the code. In what follows, we associate the points of $\mathrm{PG}(r-1,2)$ with the nonzero vectors in $\mathbb{F}_2^r$.

First, we claim that a line L intersecting two of the sets I_1, I_2, I_3 also intersects the third one. Indeed, if not, we may assume without loss of generality that L intersects I_1 only in P and does not intersect I_3. Let ℓ be the code word corresponding to the line L. Then for every code word c, the code words c and $c+\ell$ have the same restriction to I_3, so decode to the same value for the data bit, while their restrictions to I_1 differ exactly in position P. As a consequence, the restriction of c to $I_1 \setminus \{P\}$ already contains sufficient information to decode, contradicting the minimality of I_1.

Next, we claim that none of I_1, I_2, I_3 contains a line. Indeed, suppose that I_1 contains the line $L = \{P_1, P_2, P_3\}$. Let R be a point in I_3. Then L and R together span a $\mathrm{PG}(2,2)$. Now consider the lines L_i though P_i and R ($i = 1,2,3$). By the first claim, the third point Q_i on the line containing R and P_i is in I_2. Then the third line through P_1 in this $\mathrm{PG}(2,2)$ is $\{P_1, Q_1, Q_2\}$, intersecting I_1 in one point and I_2 in two points, contradicting the first claim.

Finally, as a consequence of the above two claims, if P, Q are two points in some I_i, then the third point R on the line L through P and Q is outside I_i and by the first claim R is outside $I_1 \cup I_2 \cup I_3$. Consider any code word c. If ℓ is the code word corresponding to the line L, then since c and $c+\ell$ have the same restriction to the sets I_j with $j \neq i$, they decode to the same value of the data bit. Since the two points and the set I_i are arbitrary, it follows that on each of I_1, I_2, I_3, the restrictions that have even weight all decode to the same value of the data bit, and the restrictions that have odd weight all decode to the complement of that value.

Since the all-one word is contained in the code, it follows from the above that to prove the theorem, we are done if we can show that each of the sets I_1, I_2, I_3 has even size. To this end, let H consist of the all-zero vector 0 together with all the nonzero vectors associated with the points outside $I_1 \cup I_2 \cup I_3$. By the minimality of the I_i's, no line containing two points from $H \setminus \{0\}$ can have its third point outside H, hence H is a subspace of $\mathbb{F}_2^r$. Moreover, for every i, the line through two points on I_i has its third point on H, hence I_i is contained in a coset of H. Moreover, by our first claim, each of these cosets are distinct, and since H and the I_i together partition $\mathbb{F}_2^r$, we conclude that $|H| = |I_1| = |I_2| = |I_3| = 2^{r-2}$. As a consequence, for every i, the set I_i indeed has even size provided that $r \geq 3$.

Obviously, since the all-one vector is a code word, Theorem 11 implies that no encoder for the r-th order Hamming code with $r \geq 3$ can be a 3-PIR code. Since the second order Hamming code is just the repetition code of length 3, which is easily seen to be a linear 3-PIR code, we have proved the following.

Corollary 12. *The r-th order Hamming code ($r \geq 2$) has a (linear or nonlinear) 3-PIR encoder if and only if $r = 2$.*

6 Optimal (not Necessarily Linear) 3-PIR Codes

Earlier, we have already remarked that the best q-ary (not necessarily linear) 1-PIR code of size q^k has length $n = k$ and consists of all words of length k, and the best 2-PIR code of size q^k has length $n = k + 1$ and consists of all words $c = (c_0, \ldots, c_{n-1})$ for which $\sum c_i = 0$ (in the binary case, this is the even-weight code).

In Theorem 7 we have seen that a binary linear 3-PIR code of length n and dimension k, so with a linear encoder and completely described by a $k \times n$ generator matrix, has a redundancy $r = n - k$ satisfying $r(r-1)/2 \geq k$. We also saw that codes satisfying this bound exist: they have a generator matrix of the form $G = (I_k P)$ where P is a $k \times r$ matrix that has distinct weight-two vectors as its rows. In Table 1 below, we list the optimal length of a binary linear k-dimensional 3-PIR code of this form, for various values of k.

Table 1. Optimal (smallest) length of binary linear k-dimensional 3-PIR codes

k	1	2	3	4	5	6	7	8
n	3	5	6	8	9	10	12	13

A priory, it is possible that there exist shorter non-linear codes. By the minimum-distance bound in Theorem 5, any 3-PIR code has minimum distance $d \geq 3$. In Table 2 we list the values of $A_2(n, 3)$, the maximum number M of code words in a binary code of length n and distance 3, see [1]. Inspection of Table 2 shows that there are no shorter binary codes of length n and minimum distance 3 than those in Table 1 for $k = 1, 2, 3, 5, 6$. For $k = 4$, there is a unique code of length 7, size 16, and minimum distance 3 (see [20]), which is the Hamming code of that length. We have shown that there is no encoder (linear or nonlinear) that turns that code into a 3-PIR code. For $k = 7$, there are 7398 inequivalent binary codes of length 11, size 144, and minimum distance 3 (see [14]). As a consequence, there are many nonlinear binary codes of length 11,

Table 2. Maximum size $A_2(n, 3)$ of a binary code of length n and minimum distance 3

n	3	4	5	6	7	8	9	10	11	12
M	2	2	4	8	16	20	40	72	144	256

size 2^7 and minimum distance 3. We do not know if there exist a (nonlinear) 3-PIR code with these parameters.

Problem. Does there exist a (nonlinear) binary 3-PIR code of length 11 and size 2^7?
In fact, we believe that the answer is no. (Note that the underlying code could be linear, for example a shortened Hamming code, but in view of Theorem 7, the encoder is necessarily non-linear.) Indeed, we suspect that $P(k,3) = PL(k,3)$, that is, for every $k \geq 1$, there are no nonlinear codes of size 2^k with a shorter length than the linear 3-PIR codes of size 2^k in Theorem 7, but presently we have neither a proof nor a counterexample.

7 Conclusions

First, we have shown how packing designs can be used to construct new PIR codes. Then, we have shown that for $r \geq 2$, the r-th order Hamming code has a (linear or nonlinear) 3-PIR encoder if and only if $r = 2$. Using the fact that a (linear or nonlinear) t-PIR code has minimum Hamming distance at least t, this result has allowed us to determine $P(k,3)$, the shortest length of a (not necessarily linear) 3-PIR code of size 2^k, for $k \leq 6$. We posed the existence of a (necessarily nonlinear) 3-PIR code of length 11 and size 2^7 as an open problem.

Acknowledgments. The research of the first author was supported by the Estonian Research Council grant PRG49. It is a great pleasure to thank our colleagues Vitaly Skachek, Karan Khathuria, Ago-Erik Riet, and Ludo Tolhuizen for their help in preparing this paper.

References

1. Brouwer, A.: Table of general binary codes. http://www.win.tue.nl/aeb/codes/binary-1.html. Accessed 6 Oct 2021
2. Brouwer, A.: Optimal packings of K_4's into a K_n. J. Combin. Theory, Ser. A **26**, 278–297 (1979)
3. Chor, B., Kushilevitz, E., Goldreich, O., Sudan, M.: Private information retrieval. In: Proceedings of the 36-th IEEE Symposium on Foundations of Computer Science (FOCS), pp. 41–50 (1995)
4. Colbourn, C., Dinitz, J. (eds.): Handbook of Combinatorial Designs, 2 edn. CRC Press. Boca Raton (2007)
5. Fazeli, A., Vardy, A., Yaakobi, E.: Codes for distributed PIR with low storage overhead. In: Proceedings of the IEEE Symposium Information Theory (ISIT), pp. 2852–2856. Hong Kong (2015)
6. Fazeli, A., Vardy, A., Yaakobi, E.: PIR with low storage overhead: coding instead of replication (2015). https://arxiv.org/abs/1505.06241
7. Giulietti, M., Sabatini, A., Timpanella, M.: PIR codes from combinatorial structures (2021). https://arxiv.org/abs/2107.01169
8. Giulietti, M., Sabatini, A., Timpanella, M.: PIR codes from combinatorial structures. In: Mesnager, S., Zhou, Z. (eds.) WAIFI 2022. LNCS, vol. 13638, pp. 169–182. Springer, Cham (2022)

9. Ishai, Y., Kushilevitz, E., Ostrovsky, R., Sahai, A.: Batch codes and their applications. In: Proceedings of the 36th ACM Symposium on Theory of Computing (STOC), pp. 1057–1061. Chicago (2004)
10. Kurz, S., Yaakobi, E.: PIR codes with short block length, Des. Codes, Cryptogr. **89** 559–587 (2021)
11. Lin, H.-Y., Rosnes, E.: Lengthening and extending binary private information retrieval codes. In: Proceedings of the International Zurich Seminar on Information and Communication (IZS), pp. 113–117. ETH Zurich (2018)
12. Lin, H.-Y., Rosnes, E.: Lengthening and extending binary private information retrieval codes (2018). https://arxiv.org/abs/1707.03495
13. Lipmaa, H., Skachek, V.: Linear batch codes. In: Proceedings of the 4th International Castle Meeting on Coding Theory and Applications (ICMCTA), pp. 245–253. Palmela, Portugal (2014)
14. Östergård, P.R., Baicheva, T., Kolev, E.: Optimal binary one-error-correcting codes of length 10 have 72 codewords. IEEE Trans. Inform. Theory **45**(4), 1229–1231 (1999)
15. Rao, S., Vardy, A.: Lower bound on the redundancy of PIR codes (2017). http://arxiv.org/abs/1605.01869
16. Skachek, V.: Batch and PIR codes and their connections to locally repairable codes. In: Greferath, M., Pavčević, M.O., Silberstein, N., Vázquez-Castro, M.Á. (eds.) Network Coding and Subspace Designs. SCT, pp. 427–442. Springer, Cham (2018). https://doi.org/10.1007/978-3-319-70293-3_16
17. Vardy, A.: Private Information Retrieval: coding instead of Replication. Talk at the Institate Henri Poincaré (2016). https://www.youtube.com/watch?v=WU2-6Da8IyE&t=934s
18. Vardy, A., Yaakobi, E.: Constructions of batch codes with near-optimal redundancy. In: Proceedings of the 2016 IEEE International Symposium on Information Theory, pp. 1197–1201. Barcelona (2016)
19. Wang, Z., Shaked, O., Cassuto, Y., Bruck, J.: Codes for network switches. In: Proceedings of the 2013 IEEE International Symposium on Information Theory (ISIT), pp. 1057–1061. Istanbul (2013)
20. Zaremba, S.: Covering problems concerning abelian groups. J. London Math. Soc. **27**, 242–246 (1952)
21. Zhang, H., Skachek, V.: Bounds for batch codes with restricted query size. In: Proceedings of the 2016 IEEE International Symposium on Information Theory, pp. 1192–1196. Barcelona (2016)
22. Zumbrägel, J., Skachek, V.: Talk: On bounds for batch codes, Algebraic Combinatorics and Applications (ALCOMA) (2015)

PIR Codes from Combinatorial Structures

Massimo Giulietti, Arianna Sabatini, and Marco Timpanella(✉)

Università degli Studi di Perugia, Perugia, Italy
{massimo.giulietti,marco.timpanella}@unipg.it,
arianna.sabatini@studenti.unipg.it

Abstract. A k-server Private Information Retrieval (PIR) code is a binary linear $[m, s]$-code admitting a generator matrix such that for every integer i with $1 \leq i \leq s$ there exist k disjoint subsets of columns (called recovery sets) that add up to the vector of weight one, with the single 1 in position i. As shown in [8], a k-server PIR code is useful to reduce the storage overhead of a traditional k-server PIR protocol. Finding k-server PIR codes with a small blocklength for a given dimension has recently become an important research challenge. In this work, we propose new constructions of PIR codes from combinatorial structures, introducing the notion of k-partial packing. Several bounds over the existing literature are improved.

Keywords: Privacy information retrieval · PIR codes · Configurations · Packings

1 Introduction

A Distributed Storage System (DSSs) consists of a set of hard drives (disks), or nodes, and it is used to store data in a distributed manner. DSSs are an integral part of modern data centers which support large scale computing applications. Reasons why one may want to store data in a distributed manner (rather than on a single disk) include ease of scale and reliability. To achieve reliability, redundancy is needed. Instead of using replication of the nodes, more advanced coding techniques are implemented because of storage efficiency.

Fazeli, Vardy and Yaacobi [8] proposed the definition of a k-server PIR code as an important ingredient in the construction of coded PIR protocols. PIR codes are one of the classes of linear codes that received more attention for their applications to DSSs. A k-server PIR code is a binary linear $[m, s]$-code admitting a generator matrix such that for every integer i with $1 \leq i \leq s$ there exist k disjoint subsets of columns (called recovery sets) that add up to the vector of weight one, with the single 1 in position i. Here m is the total number of bits stored on all the servers and s is the number of bits in the database. Clearly, for given k and s the optimal m is the minimal one. Given k and s, let $P(s, k)$ denote the least integer m for which a k-server PIR $[m, s]$-code exists. The storage overhead of a k-server PIR $[m, s]$-code is the ratio m/s.

S. Mesnager and Z. Zhou (Eds.): WAIFI 2022, LNCS 13638, pp. 169–182, 2023.
https://doi.org/10.1007/978-3-031-22944-2_10

Already in [8] it was noted that notions and tools from incidence geometry and design theory could be useful to construct good PIR codes. In particular, Lemma 7 in [8] states that a collection $S_1, \ldots, S_r$ of subsets of a finite set X such that every element of X belongs to at least $k-1$ subsets and two distinct subsets meet in at most one element give rise to a k-server $[r+s, s]$-code. This result motivates the following definition.

Definition 1.1. *Let X be a finite set of size s. A k-partial packing of X is a set of $k-1$ partitions of X such that*

(i) each subset in any partition has size at least two;
(ii) two subsets from two distinct partitions meet in at most one point.

The order r *of a k-partial packing is the total number of subsets of X belonging to its partitions. A k-partial packing is* uniform *if all the subsets from any partition have the same size.*

It is clear that any k-partial packing of order r of a set of size s gives rise to a k-server PIR $[r+s, s]$-code, with storage overhead $1 + \frac{r}{s}$.

A k-partial packing $\mathfrak{P}$ of a set X clearly defines h-partial packings of X for every $h < k$. We will call them *partial subpackings of* $\mathfrak{P}$. It is known that from a k_1-server PIR $[m_1, s]$-code and a k_2-server PIR $[m_2, s]$-code one can construct a (k_1+k_2)-server PIR $[m_1+m_2, s]$-code; see e.g. [7, Theorem 2]. Here, it is interesting to note that if $h_1 + h_2 \leq k+1$, then we can construct partial subpackings of $\mathfrak{P}$ giving rise to an h_1-server PIR $[m_1, s]$-code, an h_2-server PIR $[m_2, s]$-code, and an (h_1+h_2-1)-server PIR $[m_3, s]$-code with

$$m_3 = m_1 + m_2 - s < m_1 + m_2.$$

This provides a strong motivation for searching k-partial packings with large k with respect to s.

Table 1. New upper bounds on $P(s, k)$.

s	k	$P(s,k) \leq$
$a_1 \cdot a_2 \cdots a_c$	$\leq c+1$	$s(1 + \frac{1}{a_1} + \cdots + \frac{1}{a_{k-1}})$
$2^{N+1} - 1$, N odd	$\leq 2^N$	$s(1 + \frac{k-1}{3})$
$\frac{q^{N+1}-1}{q-1}$, $N = 2^{i+1} - 1$	$\leq 1 + \frac{q^N - 1}{q-1}$	$s(1 + \frac{k-1}{q+1})$
q^N	$\leq 1 + \frac{q^N - 1}{q-1}$	$s(1 + \frac{k-1}{q})$
$2^{n+n'} - 2^n + 2^{n'}, 0 \leq n' \leq n$	$\leq 2^n + 2$	$s(1 + \frac{k-1}{2^{n'}})$
$q^3 + 1$	$\leq q^2 + 1$	$s(1 + \frac{k-1}{q+1})$
$\frac{q^2-q}{2}$	$\leq q+1$	$s + (k-1)q$
$\equiv 3 \pmod 6$	$\leq 1 + \frac{s-1}{2}$	$s(1 + \frac{k-1}{3})$
$\equiv 4 \pmod{12}$	$\leq 1 + \frac{s-1}{3}$	$s(1 + \frac{k-1}{4})$
$\equiv 5 \pmod{20}, \neq 45, 345, 465, 645$	$\leq 1 + \frac{s-1}{4}$	$s(1 + \frac{k-1}{5})$
$\equiv 7 \pmod{42} > 294427$	$\leq 1 + \frac{s-1}{6}$	$s(1 + \frac{k-1}{7})$
$\equiv 8 \pmod{56}, > 24480$	$\leq 1 + \frac{s-1}{7}$	$s(1 + \frac{k-1}{8})$
sh multiple of $k+1$, sufficiently large	arbitrary	$s(1 + \frac{h}{k+1})$
$s \geq 13$	3	$2s$

Other combinatorial objects which provide k-server PIR codes are the so-called configurations; see [5, Chapter VI, Sect. 7].

Definition 1.2. **i)** *A (v_t, b_z)-configuration is an incidence structure of v points and b lines, such that each line contains z points, each point lies on t lines, and two distinct points are connected by at most one line.*
ii) *If $v = b$, and hence $t = z$, the configuration is* symmetric, *and it is denoted by v_z.*

It is straightforward to check that a (v_t, b_z)-configuration produces a $(t+1)$-server PIR code with $s = v$ and storage overhead $1 + \frac{b}{v}$. In particular, any symmetric configuration defines a PIR code with storage overhead equal to 2. The dual incidence structure of a configuration is still a configuration, which defines a $(z+1)$-server PIR code with $s = b$ and storage overhead $1 + \frac{v}{b}$.

We remark that a uniform k-partial packing of a set X, together with its partial subpackings, naturally define configurations.

The aim of this paper is to obtain new upper bounds on $P(s, k)$ through the notions of k-partial packings and configurations. Our constructions provide

Table 2. Best known bounds for $P(s, k)$ for small values of s and k.

k \ s	2		3		4		5		6		7	
2	3*	1.50	5*	2.50	6*	3.00	8*	4.00	9*	4.50	11*	5.50
3	4*	1.33	6*	2.00	7*	2.33	10*	3.33	11*	3.67	13*	4.33
4	5*	1.25	8	2.00	9	2.25	11	2.75	12*	3.00	14	3.50
5	6*	1.20	9	1.80	10	2.00	12	2.40	13	2.60	17	3.40
6	7*	1.17	10	1.67	11	1.83	13	2.17	14	2.33	18	3.00
7	8*	1.14	12	1.71	13	1.86	14	2.00	15	2.14	20	2.86
8	9*	1.13	13	1.63	14	1.75	17	2.13	18	2.25	22	2.75
9	10*	1.11	14	1.56	15	1.67	19	2.11	20	2.22	24	2.67
10	11*	1.10	15	1.50	16	1.60	20	2.00	21	2.10	25	2.50
11	12*	1.09	**17**^{PR}	**1.55**	**18**^{PR}	**1.64**	24	2.18	25	2.27	36	3.27
12	13*	1.08	**18**$^{T.4.2}$	**1.50**	**20**^{PR}	**1.67**	25	2.08	26	2.17	38	3.17
13	14*	1.08	21	1.62	22	1.69	26	2.00	27	2.08	**39**$^{T.4.1}$	**3.00**
14	15*	1.07	22	1.57	23	1.64	28	2.00	29	2.07	42	3.00
15	16*	1.07	23	1.53	24	1.60	**30**$^{S.4.1}$	**2.00**	**31**^{PR}	**2.07**	43	2.87
16	17*	1.06	24	1.50	25	1.56	**32**$^{T.3.2}$	**2.00**	**33**^{PR}	**2.06**	44	2.75
17	18*	1.06	**26**^{PR}	**1.53**	**27**^{PR}	**1.59**	**33**^{PR}	**1.94**	**34**^{PR}	**2.00**	45	2.65
18	19*	1.06	**27**$^{T.2.2}$	**1.50**	**28**^{PR}	**1.56**	**34**^{PR}	**1.89**	**35**^{PR}	**1.94**	46	2.56
19	20*	1.05	**28**^{PR}	**1.47**	**29**^{PR}	**1.53**	**35**^{PR}	**1.84**	**36**^{PR}	**1.89**	47	2.47
20	21*	1.05	**29**$^{T.2.2}$	**1.45**	**30**^{PR}	**1.50**	**36**$^{T.4.3}$	**1.80**	**37**^{PR}	**1.85**	48	2.40
21	22*	1.05	31	1.48	32	1.52	41	1.95	42	2.00	49	2.33
22	23*	1.05	32	1.45	33	1.50	**42**^{PR}	**1.91**	**43**^{PR}	**1.95**	50	2.27
23	24*	1.04	33	1.43	34	1.48	**43**^{PR}	**1.87**	**44**^{PR}	**1.91**	51	2.22
24	25*	1.04	34	1.42	35	1.46	**44**^{PR}	**1.83**	**45**^{PR}	**1.88**	52	2.17
25	26*	1.04	35	1.40	36	1.44	**45**$^{S.3}$	**1.80**	**46**^{PR}	**1.84**	53	2.22
26	27*	1.04	**37**^{PR}	**1.42**	**38**^{PR}	**1.46**	**46**^{PR}	**1.77**	**47**^{PR}	**1.81**	54	2.08
27	28*	1.04	**38**^{PR}	**1.41**	**39**^{PR}	**1.44**	**47**^{PR}	**1.74**	**48**^{PR}	**1.78**	55	2.04
28	29*	1.04	**39**$^{T.2.2}$	**1.39**	**40**^{PR}	**1.43**	**48**^{PR}	**1.71**	**49**^{PR}	**1.75**	56	2.00
29	30*	1.03	**40**^{PR}	**1.38**	**41**^{PR}	**1.41**	**49**^{PR}	**1.69**	**50**^{PR}	**1.72**	57	1.97
30	31*	1.03	**41**$^{T.2.2}$	**1.37**	**42**^{PR}	**1.40**	**50**$^{T.4.2}$	**1.67**	**51**^{PR}	**1.70**	58	1.93

both families of PIR codes whose storage overhead is asymptotically optimal (see Table 1), and PIR codes that provide improvements over the existing literature for small values of s and k (see Table 2).

We also recall that the PIR codes obtained in this paper are systematic. Then, by [16, Corollary 1], they also produce families of locally recoverable codes.

2 Families of k-partial Packings

2.1 Direct Product Construction

Assume that s can be written as the product of $k-1$ integers greater than 2, that is

$$s = a_1 \cdot a_2 \cdots a_{k-1}, \qquad \text{with } a_i \geq 2.$$

For an integer $a \geq 2$, let C_a denote the cyclic group of order a. Let

$$G = C_{a_1} \times C_{a_2} \times \cdots \times C_{a_{k-1}}$$

be the direct product of the groups C_{a_i} for $i = 1, \ldots, k-1$.

Finally, let $\mathcal{P}_i$ be the partition induced by the cosets of the subgroup C_{a_i}, naturally embedded in G.

Proposition 2.1. *For each $w \leq k-1$,*

$$\mathfrak{P} = \{\mathcal{P}_1, \ldots, \mathcal{P}_w\}$$

is a $(w+1)$-partial packing of G of order $\frac{s}{a_1} + \ldots + \frac{s}{a_w}$.

Proof. As $a_i \geq 2$ for $i = 1, \ldots, w$, property (i) of Definition 1.1 holds. Also, for any two distinct indices i, j, the intersection of a coset in $\mathcal{P}_i$ and a coset in $\mathcal{P}_j$ clearly contains at most one element, and hence (ii) holds. Finally, observe that $|\mathcal{P}_i| = \frac{s}{a_i}$ for any $i = 1, \ldots, w$. □

The following result is a straightforward corollary.

Theorem 2.2. *Let*

$$s = a_1 \cdot a_2 \cdots a_{k-1}, \qquad \textit{with } a_i \geq 2.$$

Then for each $w \leq k-1$ there exists a $(w+1)$-server PIR $[m, s]$-code with

$$m = s + \frac{s}{a_1} + \ldots + \frac{s}{a_w}$$

and storage overhead $1 + \sum_{i=1}^{w} \frac{1}{a_i}$. In particular, if $s = h^{k-1}$, for each $w \leq k-1$ there exists a $(w+1)$-server PIR $[s + w\frac{s}{h}, s]$-code with storage overhead $1 + \frac{w}{h}$.

2.2 Uniform Partial Packings from Projective Geometry

For q a prime power, let $\mathrm{PG}(N, q)$ be the projective space of dimension N over the finite field with q elements $\mathbb{F}_q$. We recall that the size of $\mathrm{PG}(N, q)$ is

$$s(N, q) = \frac{q^{N+1} - 1}{q - 1} = q^N + q^{N-1} + \ldots + q + 1,$$

and the total number of lines is

$$L(N, q) = \frac{(q^{N+1} - 1)(q^N - 1)}{(q^2 - 1)(q - 1)}.$$

Also, a line in $\mathrm{PG}(N, q)$ consists of $q + 1$ points, and two distinct lines meet in at most one point.

A *resolution class* (also called a spread of lines) of $\mathrm{PG}(N, q)$ is a set of lines which partition the point set. A *packing* (or *resolution*) of the lines of $\mathrm{PG}(N, q)$ is a partition of the lines into resolution classes. Clearly, any $k - 1$ resolution classes from a packing are a k-partial packing of $\mathrm{PG}(N, q)$.

Sufficient conditions on N and q for a packing to exist are known since the seventies.

Proposition 2.3. *[1, 2] A packing of the lines of* $\mathrm{PG}(N, q)$ *exists if*

(a) $N = 2z + 1$, $q = 2$, $z \geq 1$;
(b) $N = 2^{i+1} - 1$, $i \geq 1$, *q a prime power.*

Then the following holds.

Theorem 2.4. *Let N and q be as in (a) or (b) of Proposition 2.3. Then for $s = s(N, q)$ and any $k \leq 1 + (q^{N-1} + \ldots + q + 1)$, there exists a k-server PIR $[m, s]$-code with*

$$m = s + \frac{(k-1)s}{q+1}$$

and storage overhead $1 + \frac{k-1}{q+1}$.

Proof. Note that there are $\ell(N, q) = \frac{s}{q+1}$ lines in any resolution class of $\mathrm{PG}(N, q)$, and a packing of the lines of $\mathrm{PG}(N, q)$ comprises $\frac{L(N,q)}{\ell(N,q)} = \frac{q^N - 1}{q - 1} = q^{N-1} + \ldots + q + 1$ resolution classes. Then the k-partial packing of $\mathrm{PG}(N, q)$ obtained taking any $k - 1$ resolution classes gives rise to a k-server PIR as in the claim. □

2.3 Uniform Partial Packings from Affine Geometry

In $\mathrm{AG}(N, q)$ a resolution is easily obtained for any N and q. Here a resolution class is just a parallelism class. Taking into account that every line contains q points, and that the number of parallelism classes is $s(N - 1, q)$, the following result is easily obtained.

Theorem 2.5. *Let q be a prime power and N an integer with $N \geq 2$. Then for $s = q^N$ and any $k \leq 1 + s(N-1, q)$ there exists a k-server PIR $[m, s]$-code with*

$$m = s + \frac{(k-1)s}{q}$$

and storage overhead $1 + \frac{k-1}{q}$.

Now we consider subsets E of $\mathrm{AG}(N, q)$ of size hq^{N-1} consisting of $h \leq q$ parallel hyperplanes. There are q^{N-1} directions not determined by these hyperplanes and each line with such directions meets E in precisely h points. Then the following holds.

Theorem 2.6. *Let q be a prime power and N an integer with $N \geq 2$. Then for $s = hq^{N-1}$, $h \leq q$, and any $k \leq 1 + q^{N-1}$ there exists a k-server PIR $[m, s]$-code with*

$$m = (h + k - 1)q^{N-1} = s + (k-1)q^{N-1}$$

and storage overhead $1 + \frac{k-1}{h}$.

2.4 Partial Packings from Other Geometrical Objects

2.4.1 Maximal Arcs

In a projective plane $\mathrm{PG}(2, q)$, a maximal arc is a set of v points $\mathcal{K}$ such that every line of $\mathrm{PG}(2, q)$ is either disjoint from $\mathcal{K}$ or meets $\mathcal{K}$ in the same number z of points. If this happens $\mathcal{K}$ is said to be a $\{v; z\}$-maximal arc.

The existence problem for maximal arcs of given size is completely solved; see [11].

Theorem 2.7. *A $\{v; z\}$-maximal arc of $\mathrm{PG}(2, q)$ exists if and only if there exist $0 \leq n' \leq n$ such that*

$$q = 2^n, \qquad z = 2^{n'}, \qquad v = zq - q + z.$$

For a point P not in $\mathcal{K}$, the lines through P that are not disjoint from $\mathcal{K}$ give rise to a partition of $\mathcal{K}$ in subsets of size z. Also, joining $k - 1$ partitions corresponding to $q + 1$ collinear points gives rise to a k-partial packing of $\mathcal{K}$. Then the following holds.

Corollary 2.8. *Let s be an integer of the form $s = 2^{n+n'} - 2^n + 2^{n'}$, for some $1 \leq n' \leq n$. Then for each $k \leq 2^n + 2$ there exists a k-server PIR $[m, s]$-code with*

$$m = s + \frac{(k-1)s}{2^{n'}}$$

and storage overhead $1 + \frac{k-1}{2^{n'}}$.

2.4.2 Classical Unitals

A classical unital U in $\mathrm{PG}(2,q^2)$ is the set of points whose homogeneous coordinates (x_0,x_1,x_2) satisfy the equation $x_0^{q+1}+x_1^{q+1}+x_2^{q+1}=0$, up to projectivities. It is well known that U consists of q^3+1 points in $\mathrm{PG}(2,q^2)$, and that through a point $P\in \mathrm{PG}(2,q^2)\setminus U$, there are q^2-q secant lines intersecting U in $q+1$ points, and $q+1$ tangent lines intersecting U in one point. Also, the $q+1$ points lying on the tangent lines through P are collinear; see [14, Section 7.3]. Therefore, the following result holds.

Theorem 2.9. *The set U consists of q^3+1 points, and each point in $\mathrm{PG}(2,q^2)\setminus U$ defines a partition of U in q^2-q+1 subsets of $q+1$ collinear points.*

If we consider a line l meeting U in precisely one point P, then the q^2 points on l distinct from P define disjoint partitions. Then the following holds.

Corollary 2.10. *Let s be an integer of the form $s=q^3+1$, for some prime power q. Then for each $k\le q^2+1$ there exists a k-server PIR $[m,s]$-code with*

$$m=s+\frac{(k-1)s}{q+1}$$

and storage overhead $1+\frac{k-1}{q+1}$.

2.4.3 Internal Points to a Conic

Let $\mathcal{C}$ be an irreducible conic in $\mathrm{PG}(2,q)$, with q an odd prime power. A point $P\in \mathrm{PG}(2,q)\setminus\mathcal{C}$ is external if it lies on a tangent line to C, and internal otherwise.

There exist precisely $(q^2-q)/2$ internal points. Also, a secant line of $\mathrm{PG}(2,q)$ contains $(q-1)/2$ internal points of $\mathcal{C}$, while an external line contains $(q+1)/2$ internal points of $\mathcal{C}$. Then, clearly the lines through an external point P, distinct from the tangent lines at P, define a partition of the set of internal points of $\mathcal{C}$ in $q-1$ subsets of collinear points of cardinalities $(q-1)/2$ and $(q+1)/2$; see [12].

If $q>3$, taking $k-1$ distinct external points lying on a same tangent line to $\mathcal{C}$, we obtain a k-partial packing of the set of internal points of $\mathcal{C}$. Note that, unlike the other partial packings from geometrical objects, this construction provides a non-uniform k-partial packings. The following result then holds.

Corollary 2.11. *Let s be an integer of the form $s=(q^2-q)/2$, for some odd prime power $q>3$. Then for each $k\le q+1$ there exists a k-server PIR $[m,s]$-code with*

$$m=s+(k-1)(q-1),$$

and storage overhead $1+\frac{2(k-1)}{q}$.

3 Uniform Partial Packings from Resolvable Configurations and BIBDs

Recently, in [10], the notion of resolvable configuration has been introduced. A parallel class in a configuration $\mathcal{C}$ is a set of lines which partition the set of points; a *resolution* of $\mathcal{C}$ is a partition of the set of lines into parallel classes. A configuration $\mathcal{C}$ is said to be *resolvable* if it admits a resolution. A resolution of a (v_t, b_z) resolvable configuration consists of t parallel classes, each of which has size $\frac{v}{z}$. Therefore, if a (v_t, b_z)-configuration is resolvable, then a k-partial packing of the set of its v points can be defined for each $k \leq 1+t$.

Theorem 3.1. *Let (v_t, b_z) be a resolvable configuration. Then for any $k \leq 1+t$ there exists a k-server PIR $[m, v]$-code with*

$$m = v + (k-1)\frac{v}{z},$$

and storage overhead $1 + \frac{k-1}{z}$.

Existence results for symmetric resolvable configurations were investigated in [4]. Here we list the parameters for which a v_z resolvable configuration exists.

- $3 \leq z \leq 5$, $v = wz$, $w \geq z$, see [4, Theorem 3.2];
- $6 \leq z \leq 13$, $v = wz$, $w \geq z$, with the following possible exceptions

$$(z,w) \in \{(9,10),(10,12),(11,12),(11,14),(12,12),(12,14),(12,15),(13,14),(13,15)\},$$

  see [4, Theorem 4.7];
- $z \geq 3$, $w \geq z^2$, $v = wz$, see [4, Corollary 4.6];
- q a prime power, $z \leq q$, $v = zq$, see [4, Corollary 3.4].

If a (v_t, b_z)-configuration is such that any two distinct points are connected by *exactly* one line, then $\mathcal{C}$ is called a Balanced Incomplete Block Design (BIBD), or a Steiner system. In [8] it was noticed that one can construct a PIR code from a given Steiner system; see also [15]. Here we focus on resolvable Steiner systems, since they give rise to uniform partial packings and hence to a large number of distinct PIR codes, each one with a different number of servers. By a counting argument it is easy to see that the number of parallel classes in a resolution of a BIBD is $\frac{v-1}{z-1}$. Therefore, the following result holds.

Theorem 3.2. *Let (v_t, b_z) be a resolvable BIBD. Then for any $k \leq 1 + \frac{v-1}{z-1}$ there exists a k-server PIR $[m, v]$-code with*

$$m = v + (k-1)\frac{v}{z}.$$

We list here some families of parameters for which there exists a (v_t, b_z)-configuration which is also a resolvable BIBD; see [5, Chapter II, Sect. 7] and [13].

- $z = 3$, v such that $v \equiv 3 \pmod{6}$;
- $z = 4$, v such that $v \equiv 4 \pmod{12}$;
- $z = 5$, $v \equiv 5 \pmod{20}$, $v \neq 45, 345, 465, 645$;
- $z = 7$, $v \equiv 7 \pmod{42}$, $v > 294427$;
- $z = 8$, $v \equiv 8 \pmod{56}$, $v > 24480$.

Also, the following general result holds.

Theorem 3.3. *[5, Chapter II, Theorem 7.10] If v and z are both powers of the same prime, and $z-1$ divides $v-1$, then a (v_t, b_z) resolvable BIBD exists.*

4 Families of Configurations

4.1 Symmetric Configurations

As already pointed out, any symmetric configuration v_z defines a $(z+1)$-server PIR $[2v, v]$-code with storage overhead equal to 2. In this section we provide a list of infinite families of symmetric configurations that are known to exist, see [5,6]. In the following, q is a prime power and p is any prime number.

v	z	Conditions
v	4	$v \geq 13$
q^2-1	q	none
p^2-p	$p-1$	none
q^2-qs	$q-s$	$q > s \geq 0$
$q^2-(q-1)s-1$	$q-s$	$q > s \geq 0$
$c(q+\sqrt{q}+1)$	$\sqrt{q}+c$	q square, $c = 2, 3, \ldots, q-\sqrt{q}$
$2p^2$	$p+s$	$p+s > 0$, $0 < s \leq q+1$, $q^2+q+1 \leq p$
$c(q-1)$	$c-\delta$	$\delta \geq 0$, $c = \delta, \ldots, b$, $b = q$ if $\delta \geq 1$, $b = \lceil \frac{q}{2} \rceil$ if $\delta = 0$
$\frac{q(q-1)}{2}$	$\frac{q+1}{2}$	q odd
$\frac{q(q+1)}{2}$	$\frac{q-1}{2}$	q odd
$q^2+q-q\sqrt{q}$	$q-\sqrt{q}$	q square
q^2-rq-1	$q-r$	$q-3 \geq r \geq 0$
q^2-q-2	$q-1$	$q-3 \geq r \geq 0$
$rq-1$	r	$r > 0$, $q > r \geq 3$
$rq-2$	r	$r > 0$, $q > r \geq 3$

For small values of v and z, more symmetric configurations are known; see [5, Table 7.13].

- $v \in \{21, 23, 24, 25, 26, 27, 28\}$ and $z = 5$;
- $v \in \{31, 34, 35, 36, 37, 38\}$ and $z = 6$;

- $v \in \{45, 48, 49, 50\}$ and $z = 7$;
- $v \in \{57, 63, 64\}$ and $z = 8$;
- $v \in \{73, 78, 80\}$ and $z = 9$;
- $v \in \{91, 98\}$ and $z = 10$;
- $v \in \{133, 135\}$ and $z = 12$.

4.2 Non-symmetric Configurations

Non-symmetric configurations allow to obtain PIR codes with storage overhead smaller than 2. Indeed, let (v_t, b_z) be a configuration with $v \neq b$. Then, up to taking the dual configuration, we can assume $b < v$ and hence this configuration produces a $(t+1)$-server PIR $[v+b, v]$-code, with storage overhead $1 + \frac{b}{v} < 2$. The existence problem of configurations with $z = 3$ is completely solved; see [13, Theorem 3.1].

Theorem 4.1. *A (v_t, b_3) configuration exists if and only if $vt = 3b$ and $v \geq 2t+1$.*

For $z = 4, 5$, the following results hold; see [13, Sections 3.2 and 3.4].

Theorem 4.2. *In the following cases, a configuration (v_t, b_4) exists.*

- $v \equiv 4 \pmod{12}$, $v > 3t+1$ *and* $vt = 4b$;
- $v \equiv 0 \pmod{12}$, $v \geq 3t+1$, $vt = 4b$, *and* $v \notin E$, *where*

$$E = \{84, 120, 132, 180, 216, 264, 312, 324, 372, 456, 552, 648, 660, 804, 852, 888\};$$

- $v \equiv 0 \pmod{12}$, $v = 3t+3$ *and* $vt = 4b$;
- $t = 4s$, $v \geq 3t+1$, $vt = 4b$, *and* $1 \leq s \leq 15$, *except possibly* $s = 3$ *and* $v = 38$;
- $t = 6$, $v \geq 20$ *even*, $b = \frac{3v}{2}$.

Theorem 4.3. *In the following cases, a configuration (v_t, b_5) exists.*

- $v = 4t+4$, $v \equiv 0 \pmod{20}$, *and* $vt = 5b$;
- $v \equiv 5 \pmod{20}$, $v \geq 4t+1$, $vt = 5b$, *and* $v \geq 7865$;
- $t = 5s$, $v \geq 4t+1$, $vt = 5b$, *and* $1 \leq s \leq 10$, *except possibly for the cases* $(t, v) \in E$, *where*

$$E = \{(1,22), (2,42), (2,43), (3,62)(3,63)(4,82), (5,102), (7,142)(9,182), \\ (9,183), (9,185), (9,186), (9,187), (9,188), (9,189), (9,190), (9,191), (9,192)\}.$$

4.3 Asymptotic Results

It was proven in [3] that for fixed t and z there exist integers v_0, b_0 such that for every $v \geq v_0$ and $b \geq b_0$ with $vt = bz$, there exists a (v_t, b_z)-configuration.

This means that if we fix the number of servers k and an arbitrary fraction of $\frac{r}{k+1}$ then for s sufficiently large and such that sr is a multiple of $k+1$, there exists a k-server PIR $[m, s]$-code with $m = s(1 + \frac{r}{k+1})$ and storage overhead $1 + \frac{r}{k+1}$.

4.4 Dual Configurations from Partial Packings

In the direct product construction, if $G = C_h^\ell$ we obtain a uniform partial packing. Since it defines a configuration, we can also consider the dual configuration. Therefore, k-server PIR $[m, s]$-codes with the following parameters are obtained:

- any h, any ℓ:

$$k = h + 1, \qquad s = vh^{\ell-1} \text{ with } 2 \leq v \leq \ell, \qquad m = s + h^\ell,$$

  and storage overhead $1 + \frac{h}{v}$.

The same approach can be used for the other constructions that provide uniform partial packings. Therefore, we obtain PIR codes with the following parameters:

- Projective case (q and N as in (a) or (b) of Proposition 2.3):

$$k = q+2, \qquad s = v\frac{s(N,q)}{q+1} \text{ with } 2 \leq v \leq (q+1)\frac{L(N,q)}{s(N,q)}, \qquad m = s+s(N,q),$$

  and storage overhead $1 + \frac{q+1}{v}$.
- Affine case, Theorem 2.6 (any q prime power, $N \geq 2$):

$$k \leq q + 1, \qquad s = vq^{N-1} \text{ with } 2 \leq v \leq q^{N-1}, \qquad m = s + (k-1)q^{N-1},$$

  and storage overhead $1 + \frac{k-1}{v}$.
- Maximal arcs case (maximal arcs of size $2^{n+n'} - 2^n + 2^{n'}$, for some $0 \leq n' \leq n$):

$$k = 2^{n'} + 1, \quad s = h(2^n - 2^{n-n'} + 1) \text{ with } 2 \leq h \leq 2^n + 1, \quad m = s + 2^{n+n'} - 2^n + 2^{n'}$$

  and storage overhead $1 + \frac{2^{n'}}{h}$.
- Classical unitals case:

$$k = q + 2, \qquad s = h(q^2 - q + 1) \text{ with } 2 \leq h \leq q^2, \qquad m = s + q^3 + 1$$

  and storage overhead $1 + \frac{q+1}{h}$.
- Resolvable BIBD case: if a (v_t, b_z)-configuration which is also a resolvable BIBD exists, then the dual construction provide k-server PIR $[m, s]$-codes with

$$k = z + 1, \qquad s = h\frac{v}{z} \text{ with } h \leq \frac{v-1}{z-1}, \qquad m = s + v$$

  and storage overhead $1 + \frac{z}{h}$.

5 General Constructions of k-server PIR Codes

In the previous sections we constructed PIR codes whose lengths had a specific form. Here we explicitly construct PIR codes of arbitrary length.

The proof of the following statement is straightforward.

Proposition 5.1. *Let $\mathfrak{P} = \{\mathcal{P}_1, \dots, \mathcal{P}_{k-1}\}$ be a k-partial packing of a set X. Let Y be a subset of X and for each $i = 1, \dots, k-1$ let $\mathcal{P}_i^Y$ be the partition of Y induced by $\mathcal{P}_i$. Then $\mathfrak{P}^Y = \{\mathcal{P}_1^Y, \dots, \mathcal{P}_{k-1}^Y\}$ is a k-partial packing of Y if and only if for each i no subset of $\mathcal{P}_i$ meets Y in precisely one element. In this case, the order of $\mathfrak{P}^Y$ is less than or equal to that of $\mathfrak{P}$.*

As an illustration, we apply Proposition 5.1 to the partial packings described in Sect. 2.3.

Let q^N be the least prime power such that $k \leq 1 + s(N-1, q) - q^{N-1}$ and $s \leq q^N$. The condition on k allows to construct a k-partial packing $\mathfrak{P}$ according to Theorem 2.5, in which the parallelism classes of the lines belonging to a fixed hyperplane H are avoided.

If in addition $s \geq 2q^{N-1}$, then one can fix a subset Y of $\mathrm{AG}(N, q)$ with size s that contains two hyperplanes parallel to H. Then clearly every line belonging to the partitions of $\mathfrak{P}$ meets Y in at least two points, and $\mathfrak{P}^Y$ is a k-partial packing.

Theorem 5.2. *For integers k and s, let q^N be the least prime power such that $k \leq 1 + s(N-1, q) - q^{N-1}$ and $2q^{N-1} \leq s \leq q^N$. Then there exists a k-server PIR $[m, s]$-code with*

$$m = s + (k-1)q^{N-1}$$

and storage overhead $1 + \frac{(k-1)q^{N-1}}{s}$.

The best case is clearly when s is close to a prime power. However, something very general can be stated.

Corollary 5.3. *For integers k and s, let q^N be the least prime power such that $k \leq 1 + s(N-1, q) - q^{N-1}$ and $2q^{N-1} \leq s \leq q^N$. Then there exists a k-server PIR $[m, s]$-code with storage overhead O with*

$$1 + \frac{k-1}{q} \leq O \leq 1 + \frac{k-1}{2}.$$

6 Conclusions

In recent years, finding k-server PIR codes with a small blocklength for a given dimension has become an important research challenge. Let $P(s, k)$ denote the minimum value of m for which a k-server PIR $[m, s]$-code exists.

In this paper several upper bounds on $P(s, k)$ have been obtained through the notions of k-partial packings and configurations. Here we summarize our result on $P(s, k)$, taking into account that the function P is strictly increasing in both variables s and k, as the following propagation rules show.

Proposition 6.1. *[9, Lemmas 13 and 14]*

(i) $P(s,k) \leq P(s,k+1) - 1$;
(ii) if k *is odd, then* $P(s,k) = P(s,k+1) - 1$;
(iii) $P(s,k) \leq P(s+1,k) - 1$.

In the next table q denotes a prime power, whereas N and a_i any integer greater than 1. The integer k is always assumed to be greater than 2.

Finally, in the next table we report the best known bounds for $P(s,k)$ for small values of s and k. In particular, the improvements over the existing literature that are provided by our constructions are printed in bold. In these cases, we state the Section (briefly S), or Theorem (briefly T) from which the improvement is obtained. Also, we use PR to denote the improvements that are obtained using the constructions of this paper together with the above-mentioned propagation rules. We don't know if some of the codes that we obtained are actually optimal. In this direction, it would be interesting to find new lower bounds on the parameter $P(s,k)$, at least for small values of $s \geq 3$.

Acknowledgments. This research was partially supported by the Italian National Group for Algebraic and Geometric Structures and their Applications (GNSAGA - INdAM). The first author is funded by the project "Strutture Geometriche, Combinatoria e loro Applicazioni" (Fondo Ricerca di Base, 2019, University of Perugia). The third author is funded by the project "Metodi matematici per la firma digitale ed il cloud computing" (Programma Operativo Nazionale (PON) "Ricerca e Innovazione" 2014–2020, University of Perugia). The authors would like to thank Marco Buratti for his helpful suggestions.

References

1. Baker, R.D.: Partitioning the planes of $AG_{2m}(2)$ into 2-designs. Discret. Math. **15**, 205–211 (1976)
2. Beutelspacher, A.: On parallelisms in finite projective spaces. Geom. Dedicata. **3**, 35–40 (1974)
3. Bras-Amorós, M., Stokes, K.: The semigroup of combinatorial configurations. Semigroup Forum **84**, 91–96 (2012)
4. Buratti, M., Stinson, D.R.: On resolvable Golomb rulers, symmetric configurations and progressive dinner parties. J. Algebraic Combin. **55**, 141–156 (2022)
5. Colbourn, C.J., Dinitz, J.H.: Handbook of Combinatorial Designs, Discrete Mathematics and Its Applications, Second Edition, Chapman & Hall/CRC (2007)
6. Davydov, A.A., Faina, G., Giulietti, M., Marcugini, S., Pambianco, F.: On constructions and parameters of symmetric configurations v_k. Des. Codes Cryptogr. **80**, 125–147 (2016)
7. Kurz, S., Yaakobi, E.: PIR codes with short block length. Des. Codes Crypt. **89**(3), 559–587 (2021). https://doi.org/10.1007/s10623-020-00828-6
8. Fazeli, A., Vardy, A., Yaakobi, E.: Codes for distributed PIR with low storage overhead. In: 2015 IEEE International Symposium on Information Theory (ISIT), pp. 2852–2856 (2015)
9. Fazeli, A., Vardy, A., Yaakobi, E.: PIR with low storage overhead: coding instead of replication. arXiv:1505.06241 (2015)

10. Gévay, G.: Resolvable configurations. Discret. Appl. Math. **266**, 319–330 (2019)
11. Gezek, M., Mathon, R., Tonchev, V.D.: Maximal arcs, codes, and new links between projective planes of order 16. Electron. J. Combinat. **27** (2020)
12. Giulietti, M.: Line partitions of internal points to a conic in PG$(2, q)$. Combinatorica **29**(1), 19–25 (2009)
13. Gropp, H.: Non-symmetric configurations with natural index. Discrete Math. **124**, 87–98 (1994)
14. Hirschfeld, J.W.P.: Projective Geometries Over Finite Fields, 2nd edn. Oxford Univ. Press, Oxford (1998)
15. Lin, H.Y., Rosnes, E.: Lengthening and extending binary private information retrieval codes. In: International Zurich Seminar on Information and Communication, pp. 113–117 (2018)
16. Skachek, V.: Batch and PIR codes and their connections to locally repairable codes, Network Coding and Subspace Designs, Cham, Switzerland, pp. 427–442 (2018)

The Projective General Linear Group PGL(2, 5^m) and Linear Codes of Length $5^m + 1$

Yue Zhao[1], Chunming Tang[2(✉)], and Yanfeng Qi[1,2]

[1] School of Mathematics and Information, China West Normal University, Nanchong 637002, Sichuan, China

[2] School of Science, Hangzhou Dianzi University, Hangzhou 310018, China

tangchunmingmath@163.com

Abstract. The importance of interactions between groups, linear codes and t-designs has been well recognized for decades. Linear codes that are invariant under groups acting on the set of code coordinates have found important applications for the construction of combinatorial t-designs. Examples of such codes are the Golay codes, the quadratic-residue codes, and the affine-invariant codes. Let $q = 5^m$. The projective general linear group PGL$(2, q)$ acts as a 3-transitive permutation group on the set of points of the projective line. This paper is to present two infinite families of cyclic codes over GF(5^m) such that the set of the supports of all codewords of any fixed nonzero weight is invariant under PGL$(2, q)$, therefore, the codewords of any nonzero weight support a 3-design. A code from the first family has parameters $[q + 1, 4, q - 5]_q$, where $q = 5^m$, and $m \geq 2$. A code from the second family has parameters $[q + 1, q - 3, 4]_q$, $q = 5^m$, $m \geq 2$. This paper also points out that the set of the support of all codewords of these two kinds of codes with any nonzero weight is invariant under $\mathrm{Stab}_{U_{q+1}}$, thus the corresponding incidence structure supports 3-design.

Keywords: Linear code · Cyclic code · t-design · Projective general linear group · Automorphism group

1 Introduction

A $t - (\nu, k, \lambda)$ design is an incidence structure $(X, \mathcal{B})$, where X is a set of ν points and $\mathcal{B}$ a set of b k-subsets of X called blocks, such that any t points are contained in exactly λ blocks, where $\lambda > 0$. A t-design is a $t - (\nu, k, \lambda)$ design for some parameters ν, k, λ.

The incidence matrix $A = (a_{i,j})$ of a design $\mathbb{D}$ is a (0,1)-matrix with rows indexed by the blocks, and columns indexed by the points of $\mathbb{D}$, where $a_{i,j} = 1$ if the jth point belongs to the ith block, and $a_{i,j} = 0$ otherwise.

It is known that groups, linear codes and t-designs are closely related. Linear codes that are invariant under groups acting on the set of code coordinates have found important applications for the construction of combinatorial t-designs. Recently, an infinity

Tang's research was supported by The National Natural Science Foundation of China (Grant No. 11871058) and China West Normal University (14E013, CXTD2014-4 and the Meritocracy Research Funds). Qi was supported by Zhejiang provincial Natural Science Foundation of China (No. LY21A010013).

S. Mesnager and Z. Zhou (Eds.): WAIFI 2022, LNCS 13638, pp. 183–193, 2023.
https://doi.org/10.1007/978-3-031-22944-2_11

family of linear codes holding 4-designs was constructed by Tang and Ding in [3]. It remains an interesting open problem if there exists an infinite family of linear codes holding an infinite family of t-designs for $t \geq 5$. In fact, only a few infinite families of cyclic codes holding an infinite family of 3-designs are reported in the literature. For more results on linear codes and t-designs, we refer the reader to [4–7, 10]. Now, we will consider a class of cyclic codes

$$\mathcal{C}_m = \{(\mathrm{Tr}(a_2 u^2 + a_3 u^3))_{u \in U_{q+1}} : a_i \in \mathrm{GF}(q^2)\}$$

over GF(q) and its dual, where $q = 5^m$ with $m \geq 2$ being an integer, Tr is the trace function from GF(q^2) to GF(q) and U_{q+1} is the set of all $(q+1)$-th roots of unity in GF(q^2).

The objective of this paper is to present two infinite families of cyclic codes over GF(5^m) such that the set of the supports of all codewords of any fixed nonzero weight is invariant under PGL(2, q), therefore, the codewords of any nonzero weight support a 3-design.

2 Preliminaries

2.1 Linear Codes and Cyclic Codes

Let GF(r) be the finite field with r elements, where r is a power of a prime p. For a positive integer n, a linear code of length n over GF(r) is defined to be a subspace of GF(r)^n. A linear code $\mathcal{C}$ of length n and dimension k over GF(r) is called an $[n, k]_r$ linear code over GF(r). The vectors in $\mathcal{C}$ are called codewords. If it has minimum distance d it is also called an $[n, k, d]_r$ code. An $[n, k]_r$ code $\mathcal{C}$ is called cyclic if for every codeword $(c_0, c_1, ..., c_{n-1}) \in \mathcal{C}$, its cyclic shift $(c_{n-1}, c_0, ..., c_{n-2})$ is also in $\mathcal{C}$.

The dual code $\mathcal{C}^\perp$ of a linear code $\mathcal{C}$ over GF(r) of length n is defined to be the set

$$\mathcal{C}^\perp = \{\boldsymbol{x} \in \mathrm{GF}(r)^n : \langle \boldsymbol{x}, \boldsymbol{c} \rangle = 0, \text{for all } \mathbf{c} \in \mathcal{C}\},$$

where $\langle \boldsymbol{x}, \boldsymbol{c} \rangle$ is the usual Euclidean inner product of $\mathbf{c}$ and $\boldsymbol{x}$. Let $\boldsymbol{a} = (a_0, \ldots, a_{n-1}) \in (\mathrm{GF}(r)^*)^n$, $\boldsymbol{a} \cdot \mathcal{C}$ stands for the linear code $\{(a_0 c_0, \ldots, a_{n-1} c_{n-1}) : (c_0, \ldots, c_{n-1}) \in \mathcal{C}\}$. It is a simple matter to check that

$$(\boldsymbol{a} \cdot \mathcal{C})^\perp = \boldsymbol{a}^{-1} \cdot \mathcal{C}^\perp, \tag{1}$$

where $\boldsymbol{a}^{-1} = (a_0^{-1}, \ldots, a_{n-1}^{-1})$ (see [8]).

There are two classical ways to construct a code over GF(r) from a given code over GF(r^h). Let $\mathcal{C}$ be a code of length n over GF(r^h). Then the *subfield subcode* $\mathcal{C}|_{\mathrm{GF}(r)}$ equals $\mathcal{C} \cap \mathrm{GF}(r)^n$, the set of those codewords of $\mathcal{C}$ all of whose coordinate entries belonging to the subfield GF(r). The *trace code* of $\mathcal{C}$ is given by

$$\mathrm{Tr}_{r^h/r}(\mathcal{C}) = \left\{ \left(\mathrm{Tr}_{r^h/r}(c_0), \ldots, \mathrm{Tr}_{r^h/r}(c_{n-1})\right) : (c_0, \ldots, c_{n-1}) \in \mathcal{C} \right\},$$

where $\mathrm{Tr}_{r^h/r}$ denotes the trace function from GF(r^h) to GF(r). A celebrated result of Delsarte [6] states that the subfield code $\mathcal{C}^\perp|_{\mathrm{GF}(r)}$ and the trace code $\mathrm{Tr}_{r^h/r}(\mathcal{C})$ are duals of each other, namely,

$$\left(\mathrm{Tr}_{r^h/r}(\mathcal{C})\right)^\perp = \mathcal{C}^\perp|_{\mathrm{GF}(r)} \tag{2}$$

Conversely, given a linear code $\mathcal{C}$ of length n and dimension k over GF(r), we define a linear code GF(r^h) $\otimes\, \mathcal{C}$ over GF(r^h) by

$$\mathrm{GF}(r^h) \otimes \mathcal{C} = \left\{ \sum_{i=1}^{k} a_i \mathbf{c}_i : (a_1, \ldots, a_k) \in \mathrm{GF}(r^h)^k \right\},$$

where $\mathbf{c}_1, \mathbf{c}_2, \ldots, \mathbf{c}_k$ is a basis of $\mathcal{C}$ over GF(r). This code is independent of the choice of the basis $\mathbf{c}_1, \mathbf{c}_2, \ldots, \mathbf{c}_k$ of $\mathcal{C}$, is called the *lifted code* of $\mathcal{C}$ to GF(r^h). Clearly, GF(r^h) $\otimes$ $\mathcal{C}$ and $\mathcal{C}$ have the same length, dimension and minimum distance, but different weight distributions. A trivial verification shows that if $(c_0, \ldots, c_{n-1}) \in \mathrm{GF}(r^h) \otimes \mathcal{C}$, then $(c_0^r, \ldots, c_{n-1}^r) \in \mathrm{GF}(r^h) \otimes \mathcal{C}$. Applying Lemma 7 in [9], one has

$$\mathrm{Tr}_{r^h/r}\left(\mathrm{GF}(r^h) \otimes \mathcal{C}\right) = \left(\mathrm{GF}(r^h) \otimes \mathcal{C}\right)|_{\mathrm{GF}(r)}$$

Let n be a positive integer with gcd(n, r) = 1 and $h = \mathrm{ord}_n(r)$. Let U_n be the cyclic multiplicative group of all n-th roots of unity in GF(r^h). By polynomial interpolation, every function f from U_n to GF(r) has a unique univariate polynomial expansion of the form

$$f(u) = \sum_{i=0}^{n-1} a_i u^i,$$

where $a_i \in \mathrm{GF}(r^h)$, $u \in U_n$.

2.2 Group Actions and t-designs

A *permutation group* is a subgroup of the *symmetric group* Sym(X), where X is a finite set. An *action* σ of a finite group G on a set X is a homomorphism σ from G to Sym(X). We denote the image $\sigma(g)(x)$ of $x \in X$ under $g \in G$ by $g(x)$ when no confusion can arise. The $G-$ *orbit* of $x \in X$ is $\mathrm{Orb}_x = \{g(x) : g \in G\}$. The *stabilizer* of x is $\mathrm{Stab}_x = \{g \in G : g(x) = x\}$.

Given a t-homogeneous group G on a finite set X with $|X| = v$ and a subset B of X with $|B| = k > t$, the pair (X, Orb_B) is a $t-(\nu, k, \lambda)$ design, where Orb_B is the set of images of B under the group G, $\lambda = \frac{\binom{k}{t}|G|}{\binom{v}{t}|\mathrm{Stab}_B|}$ and Orb_B is the setwise stabilizer of B in X. Let $\binom{X}{k}$ be the set of subsets of X consisting of k elements. A nonempty subset $\mathcal{B}$ of $\binom{X}{k}$ is called *invariant* under G if $\mathrm{Orb}_B \subseteq \mathcal{B}$ for any $B \in \mathcal{B}$.

2.3 Projective General Linear Groups

Let PGL(2,q) be the projective general linear group acting as a permutation group on the set of points of the projective line PG(1,q) over a finite field GF(q) with q elements. Every vector in the $(q+1)$-dimensional vector space $\mathrm{GF}(r)^{q+1}$ can be written as $(c_x)_{x \in \mathrm{PG}(1,q)}$, where $c_x \in \mathrm{GF}(r)$ and r is a prime power. In other words, the coordinates of the vectors in $\mathrm{GF}(r)^{q+1}$ can be indexed by the points in PG(1,q). Consider the induced action of PGL(2,q) on $\mathrm{GF}(r)^{q+1}$ by the left translation:

$$\pi : (c_x)_{x \in \mathrm{PG}(1,q)} \mapsto \left(c_{\pi(x)}\right)_{x \in \mathrm{PG}(1,q)},$$

where $(c_x)_{x\in \mathrm{PG}(1,q)} \in \mathrm{GF}(r)^{q+1}$ and $\pi \in \mathrm{PGL}(2,q)$. Let $\mathcal{C}$ be a linear code of length $q+1$ over GF(r). We say that $\mathcal{C}$ is *invariant under* PGL(2,q) if each element of PGL(2,q) carries each codeword of $\mathcal{C}$ into a codeword of $\mathcal{C}$. For a codeword $\mathbf{c} = (c_x)_{x\in \mathrm{PG}(1,q)}$ in $\mathcal{C}$, the support of $\mathbf{c}$ is defined as

$$\mathrm{Supp}(\mathbf{c}) = \{x \in \mathrm{PG}(1,q) : c_x \neq 0\}.$$

Let $A_w(\mathcal{C}) = |\{\mathbf{c} \in \mathcal{C} : wt(\mathbf{c}) = w\}|$ and $\mathcal{B}_w(\mathcal{C}) = \{\mathrm{Supp}(\mathbf{c}) : wt(\mathbf{c}) = w \; and \; \mathbf{c} \in \mathcal{C}\}$, where wt($\mathbf{c}$) denotes the Hamming weight of $\mathbf{c}$. $\mathcal{B}_w(\mathcal{C})$ is said to be invariant under PGL(2,q) if the support $\mathrm{Supp}\left(\left(c_{\pi(x)}\right)_{x\in \mathrm{PG}(1,q)}\right)$ belongs to $\mathcal{B}_w(\mathcal{C})$ for every $\pi \in \mathrm{PGL}(2,q)$ and any codeword $(c_x)_{x\in \mathrm{PG}(1,q)}$ of weight w in $\mathcal{C}$. It is easily seen that if $\mathcal{C}$ is invariant under PGL(2,q), then so is $\mathcal{B}_w(\mathcal{C})$ for each w. Moreover, if $\mathcal{B}_w(\mathcal{C})$ is invariant under PGL(2,q), then $(\mathrm{PG}(1,q), \mathcal{B}_w(\mathcal{C}))$ holds a 3-design provided $A_w(\mathcal{C}) \neq 0$, since the action of PGL(2,q) on PG(1,q) is 3-transitive (see [2]).

3 Linear Codes of Length $5^m + 1$ and 3-designs

Let U_{q+1} be the subset of the projective line $\mathrm{PG}(1,q^2) = \mathrm{GF}(q^2) \cup \{\infty\}$ consisting of all the $(q+1)$-th roots of unity. Denote by $\mathrm{Stab}_{U_{q+1}}$ the setwise stabilizer of U_{q+1} under the action of $\mathrm{PGL}_2(\mathrm{GF}(q^2))$ on $\mathrm{PG}(1,q^2)$.

Lemma 1. *Let* $q = 5^m$. *Then the setwise stabilizer* $\mathrm{Stab}_{U_{q+1}}$ *of* U_{q+1} *is generated by the following three types of linear fractional transformations:*

1. $u \mapsto u_0 u$, *where* $u_0 \in U_{q+1}$;
2. $u \mapsto u^{-1}$;
3. $u \mapsto \frac{u+c^q}{cu+1}$, *where* $c \in \mathrm{GF}(q^2)^* \setminus U_{q+1}$.

Similar proofs are shown in [1].

Let $q = 5^m$ and $U_{q+1} = \{u : u \in \mathrm{GF}(q^2), u^{q+1} = 1\}$. Let $\mathcal{C}_{\{2,3\}}$ be the linear code defined by

$$\mathcal{C}_{\{2,3\}} = \left\{(a_2 u^2 + a_{q-1} u^{q-1} + a_3 u^3 + a_{q-2} u^{q-2})_{u\in U_{q+1}} : a_2, a_{q-1}, a_3, a_{q-2} \in \mathrm{GF}(q^2)\right\}. \tag{3}$$

We index the coordinates of the codewords in $\mathcal{C}_{\{2,3\}}$ and related codes with the elements in U_{q+1}. The dual of $\mathcal{C}_{\{2,3\}}$ is given as

$$\mathcal{C}_{\{2,3\}}{}^{\perp} = \left\{(c_u)_{u\in U_{q+1}} \in \mathrm{GF}(q^2)^{q+1} : \sum_{u\in U_{q+1}} c_u \boldsymbol{h}_u = \mathbf{0}\right\}, \tag{4}$$

where $\boldsymbol{h}_u$ is the transpose of the row vector $(u^{-3}, u^{-2}, u^2, u^3)$.

It is obvious that if $(c_u)_{u\in U_{q+1}} \in \mathcal{C}_{\{2,3\}}$ (resp., $(c_u)_{u\in U_{q+1}} \in \mathcal{C}_{\{2,3\}}{}^{\perp}$), then $(c_u^q)_{u\in U_{q+1}} \in \mathcal{C}_{\{2,3\}}$ (resp., $(c_u^q)_{u\in U_{q+1}} \in \mathcal{C}_{\{2,3\}}{}^{\perp}$). In fact, $\mathcal{C}_{\{2,3\}}$ is the lifted code of $\mathrm{Tr}_{q^2/q}(\mathcal{C}_{\{2,3\}})$ to $\mathrm{GF}(q^2)$ and has cyclicity-defining set $\{2, 3, q-1, q-2\}$. Similarly, $\mathcal{C}_{\{2,3\}}{}^{\perp}$ is the lifted code of $\mathrm{Tr}_{q^2/q}\left(\mathcal{C}_{\{2,3\}}{}^{\perp}\right)$ to $\mathrm{GF}(q^2)$.

From Lemma 7 in [9] we deduce that

$$\mathrm{Tr}_{q^2/q}(\mathcal{C}_{\{2,3\}}) = \mathcal{C}_{\{2,3\}}|_{\mathrm{GF}(q)}, \tag{5}$$

and

$$\mathrm{Tr}_{q^2/q}\left(\mathcal{C}_{\{2,3\}}{}^{\perp}\right) = \mathcal{C}_{\{2,3\}}{}^{\perp}|_{\mathrm{GF}(q)}. \tag{6}$$

In order to describe the supports of the codewords of $\mathrm{Tr}_{q^2/q}(\mathcal{C}_{\{2,3\}})$ and $\mathcal{C}_{\{2,3\}}{}^{\perp}|_{\mathrm{GF}(q)}$, we need to employ symmetric polynomials and elementary symmetric polynomials. A polynomial f is said to be symmetric if it is invariant under any permutation of its variables. The *elementary symmetric polynomial*(ESP) of degree ℓ in k variables $u_1, u_2, \ldots, u_k$, written $\sigma_{k,\ell}$, is defined by

$$\sigma_{k,\ell}(u_1, \ldots, u_k) = \sum_{I \subseteq [k], |I| = \ell} \prod_{j \in I} u_j, \tag{7}$$

where $[k] = \{1, 2, ..., k\}$.

For any k-variable symmetric polynomial f with coefficients in GF(q^2), write

$$\mathcal{B}_{f,q+1} = \left\{ \{u_1, \ldots, u_k\} \in \binom{U_{q+1}}{k} : f(u_1, \ldots, u_k) = 0 \right\}.$$

To determine the parameters of $\mathrm{Tr}_{q^2/q}(\mathcal{C}_{\{2,3\}})$ and $\mathcal{C}_{\{2,3\}}{}^{\perp}|_{\mathrm{GF}(q)}$, we prove several lemmas below. To simplify notation and expressions below, we use $\sigma_{k,\ell}$ to denote $\sigma_{k,\ell}(u_1, \ldots, u_k)$ for any $\{u_1, \ldots, u_k\} \in \binom{U_{q+1}}{k}$ whenever $\{u_1, \ldots, u_k\}$ is specified.

Lemma 2. *Let $q = 5^m$, where $m \geq 2$ is a positive integer. For any $\{u_1, u_2, u_3\} \in \binom{U_{q+1}}{3}$, we have the following results.*

1. $u_1 + u_2 + 3u_3 \neq 0$;
2. $u_1 + 2u_2 + 2u_3 \neq 0$;
3. $u_1 + 3u_2 + u_3 \neq 0$.

Proof. *We only give the proof of the first conclusion and the proofs of other conclusions are similar to the first conclusion.*

Assume that $u_1 + u_2 + 3u_3 = 0$, this is equivalent to $u_1 + u_2 - 2u_3 = 0$, then

$$(u_1 + u_2 - 2u_3)^q = \frac{1}{u_1} + \frac{1}{u_2} - \frac{2}{u_3} = \frac{1}{u_1} + \frac{1}{u_2} - \frac{4}{2u_3} = 0.$$

It follows from $u_1 + u_2 = 2u_3$ that

$$\frac{1}{u_1} + \frac{1}{u_2} - \frac{4}{u_1 + u_2} = 0,$$

Multiplying both sides of the previous equation by $u_1 u_2 (u_1 + u_2)$ yields

$$(u_1 + u_2)^2 - 4u_1 u_2 = 0,$$

which means that $(u_1 - u_2)^2 = 0$. This is contrary to our assumption that u_1, u_2, u_3 are pairwise distinct. Thus, $u_1 + u_2 + 3u_3 \neq 0$. This completes the proof.

For a positive integer $\ell \leq q+1$, define a $4 \times \ell$ matrix M_ℓ by

$$\begin{bmatrix} u_1^{-3} & u_2^{-3} & \dots & u_\ell^{-3} \\ u_1^{-2} & u_2^{-2} & \dots & u_\ell^{-2} \\ u_1^{+2} & u_2^{+2} & \dots & u_\ell^{+2} \\ u_1^{+3} & u_2^{+3} & \dots & u_\ell^{+3} \end{bmatrix}, \tag{8}$$

where $u_1, \dots, u_\ell \in U_{q+1}$. For $r_1, \dots, r_i \in \{\pm 2, \pm 3\}$, let $M_\ell[r_1, \dots, r_i]$ denote the submatrix of M_ℓ obtained by deleting the rows $(u_1^{r_1}, u_2^{r_1}, \dots, u_\ell^{r_1}), \dots, (u_1^{r_i}, u_2^{r_i}, \dots, u_\ell^{r_i})$ of the matrix M_ℓ, where $1 \leq i \leq 4$.

Lemma 3. *Let M_ℓ be the matrix given by (8) with $\{u_1, \dots, u_\ell\} \in \binom{U_{q+1}}{\ell}$. Consider the system of homogeneous linear equations defined by*

$$M_\ell (x_1, \dots, x_\ell)^T = 0 \tag{9}$$

Then (9) has a nonzero solution $(x_1, \dots, x_\ell)$ in GF$(q)^\ell$ if and only if rank$(M_\ell) < \ell$, where rank (M_ℓ) denotes the rank of the matrix M_ℓ.

Lemma 4. *Let $m \geq 2$ be a positive integer, $q = 5^m$ and M_3 be the matrix given by (8) with $\{u_1, u_2, u_3\} \in \binom{U_{q+1}}{3}$. Then rank (M_3) = 3.*

Proof. *Suppose that $(M_3) < 3$. Then det $(M_3[3]) = \frac{\prod_{1 \leq i < j \leq 3}(u_i + 4u_j)}{\sigma_{3,3}^3}(\sigma_{3,1}^3 + 3\sigma_{3,1}\sigma_{3,2} + \sigma_{3,3}) = (u_1 + u_2 + 3u_3)(u_1 + 2u_2 + 2u_3)(u_1 + 3u_2 + u_3) = 0$, which is contrary to Lemma 2. This completes the proof.*

Lemma 5. *Let $m \geq 2$ be a positive integer, $q = 5^m$ and M_4 be the matrix given by (8) with $\{u_1, u_2, u_3, u_4\} \in \binom{U_{q+1}}{4}$. Then rank $(M_4) = 3$ if and only if $\sigma_{4,2}^3 + 3\sigma_{4,3}\sigma_{4,2}\sigma_{4,1} + \sigma_{4,3}^2 + \sigma_{4,1}^2\sigma_{4,4} + 4\sigma_{4,2}\sigma_{4,4} = 0$.*

Proof. *Note that*

$$\det(M_4) = \frac{\prod_{1 \leq i < j \leq 4}(u_i + 4u_j)}{\sigma_{4,4}^3}(\sigma_{4,2}^3 + 3\sigma_{4,3}\sigma_{4,2}\sigma_{4,1} + \sigma_{4,3}^2 + \sigma_{4,1}^2\sigma_{4,4} + 4\sigma_{4,2}\sigma_{4,4}),$$

which completes the proof.

In this subsection, we will determine the parameters of the cyclic code $\mathcal{C}_{\{2,3\}}$ and its dual $\mathcal{C}_{\{2,3\}}^\perp$.

Theorem 1. *Let $q = 5^m$ with $m \geq 2$ being a positive integer. Then the subfield subcode $\mathcal{C}_{\{2,3\}}^\perp$ over GF(q) has the parameters $[q+1, q-3, 4]_q$.*

Proof. *It follows from definitions that the code $\mathcal{C}_{\{2,3\}}^{\perp}|_{\mathrm{GF}(q)}$ has length $q+1$. Let α be a generator of the multiplicative group GF(q^2)* and define $\beta = \alpha^{q-1}$. Then $\beta \in U_{q+1}$ is a $(q+1)$-th primitive root of unity in the field GF(q^2). Let $g_i(x)$ denote the minimal polynomial of β^i over GF(q), where $i \in \{2,3\}$. Note that $g_i(x)$ has only the roots β^i and β^{-i}. We then deduce that $g_2(x)$ and $g_3(x)$ are pairwise distinct irreducible polynomials of degree 2. By definition, the generator polynomial of $\mathcal{C}_{\{2,3\}}^{\perp}|_{\mathrm{GF}(q)}$ is $g_2(x)g_3(x)$ with degree 4. Thus, $\mathcal{C}_{\{2,3\}}^{\perp}|_{\mathrm{GF}(q)}$ has dimension $(q+1)-4 = q-3$.*

By Lemma 4, we have the minimum distance d of $\mathcal{C}_{\{2,3\}}^{\perp}|_{\mathrm{GF}(q)}$ is at least 4. Next we will prove that $d = 4$.

Let $\{u_1, u_2, u_3, u_4\} \in \binom{U_{q+1}}{4}$, Without the loss of generality, we assume that

$$u_1 = u_{i_1}, u_2 = u_{i_2}, u_3 = u_{i_3}, u_4 = u_{i_4},$$

where $1 \leq i_1 < i_2 < i_3 < i_4 \leq q+1$. Since $d \geq 4$, the rank of $M(u_1, u_2, u_3, u_4)$ equals 3, where $M(u_1, u_2, u_3, u_4)$ was defined by (9). Let $(x_{i_1}, x_{i_2}, x_{i_3}, x_{i_4}) \in \mathrm{GF}(q)^4$ denote a nonzero solution of

$$\begin{bmatrix} 1 & 1 & 1 & 1 \\ u_1 & u_2 & u_3 & u_4 \\ u_1^5 & u_2^5 & u_3^5 & u_4^5 \\ u_1^6 & u_2^6 & u_3^6 & u_4^6 \end{bmatrix} \begin{bmatrix} x_{i_1} \\ x_{i_2} \\ x_{i_3} \\ x_{i_4} \end{bmatrix} = 0\,. \tag{10}$$

Since the rank of the matrix $M(u_1, u_2, u_3, u_4)$ is 3, all these $x_{i_j} \neq 0$. Define a vector $\mathbf{c} = (c_0, c_1, \ldots, c_n) \in \mathrm{GF}(q)^{n+1}$, where $c_{i_j} = x_{i_j}$ for $j \in \{1,2,3,4\}$ and $c_h = 0$, for all $h \in \{0, 1, \ldots, n\} \setminus \{i_1, i_2, i_3, i_4\}$. It is easily observed that $\mathbf{c}$ is a codeword with Hamming weight 4 in $\mathcal{C}_{\{2,3\}}^{\perp}|_{\mathrm{GF}(q)}$. The set $\{a\mathbf{c} : a \in \mathrm{GF}(q)^*\}$ consists of all such codewords of Hamming weight 4 with nonzero coordinates in $\{i_1, i_2, i_3, i_4\}$. Hence, the code $\mathcal{C}_{\{2,3\}}{}^{\perp}|_{\mathrm{GF}(q)}$ has minimum distance $d = 4$. Meanwhile, every codeword of Hamming weight 4 in $\mathcal{C}_{\{2,3\}}{}^{\perp}|_{\mathrm{GF}(q)}$ with nonzero coordinates in $\{i_1, i_2, i_3, i_4\}$ must correspond to the set $\{u_1, u_2, u_3, u_4\}$. This completes the proof.

The minimum-weight codewords in $\mathrm{Tr}_{q^2/q}(\mathcal{C}_{\{2,3\}})$ are described in the following lemma.

Lemma 6. *Let $f(u) = \mathrm{Tr}_{q^2/q}(au^2 + bu^3)$ where $(a,b) \in \mathrm{GF}(q^2)^2 \setminus \{\mathbf{0}\}$. Define*

$$\mathrm{zero}(f) = \{u \in U_{q+1} : f(u) = 0\},$$

Then $|\mathrm{zero}(f)| \leq 6$. Moreover, $|\mathrm{zero}(f)| = 6$ if and only if $a = -\sigma_{6,1}\frac{\tau}{\sqrt{\sigma_{6,6}}}$ and $b = \frac{\tau}{\sqrt{\sigma_{6,6}}}$ where $\{u_1, \ldots, u_6\} \in \mathcal{B}_{\sigma_{6,2},q+1}, \tau \in \mathrm{GF}(q)^*$.

Proof. *When $u \in U_{q+1}$, one has*

$$f(u) = \mathrm{Tr}_{q^2/q}(au^2 + bu^3) = \frac{1}{u^3}(bu^6 + au^5 + a^q u + b^q), \tag{11}$$

Thus $|\mathrm{zero}(f)| \leq 6$.

Assume that $|\mathrm{zero}(f)| = 6$. *From (11), there exists* $\{u_1, \ldots, u_6\} \in \binom{U_{q+1}}{6}$ *such that* $f(u) = \frac{b\prod_{i=1}^{6}(u-u_i)}{u^3}$. *By Vieta's formula,* $b\sigma_{6,1} = -a$, $\sigma_{6,2} = 0$, $\sigma_{6,3} = 0$, $\sigma_{6,4} = 0$, $b\sigma_{6,5} = -a^q$ *and* $b\sigma_{6,6} = b^q$. *One obtains* $b = \frac{\tau}{\sqrt{\sigma_{6,6}}}$ *from* $b\sigma_{6,6} = b^q$, *where* $\tau \in \mathrm{GF}(q)^*$. *Thus* $a = -\sigma_{6,1}\frac{\tau}{\sqrt{\sigma_{6,6}}}$.

Conversely, assume that $a = -\sigma_{6,1}\frac{\tau}{\sqrt{\sigma_{6,6}}}$ *and* $b = \frac{\tau}{\sqrt{\sigma_{6,6}}}$, *where* $\{u_1, \ldots, u_6\} \in \mathcal{B}_{\sigma_{6,2},q+1}$ *and* $\tau \in \mathrm{GF}(q)^*$. *Then* $f(u) = \frac{b\prod_{i=1}^{6}(u-u_i)}{u^3}$. *Consequently zero*$(f) = \{u_1, \ldots, u_6\}$ *and* $|\mathrm{zero}(f)| = 6$. *This completes the proof.*

Theorem 2. *Let* $q = 5^m$ *with* $m \geq 2$ *being a positive integer. Then the trace code* $\mathrm{Tr}_{q^2/q}(\mathcal{C}_{\{2,3\}})$ *has parameters* $[q+1, 4, q-5]_q$.

Proof. *Recall that (6) says that*

$$(\mathrm{Tr}_{q^2/q}(\mathcal{C}_{\{2,3\}}))^{\perp} = \mathcal{C}_{\{2,3\}}{}^{\perp}|_{\mathrm{GF}(q)}.$$

Thus $\mathrm{Tr}_{q^2/q}(\mathcal{C}_{\{2,3\}})$ *has dimensioin 4 by Theorem 1. Then we have* $d = (q+1) - 6 = q - 5$ *by Lemma 6. This completes the proof.*

We checked results with Magma [4] in the following examples.

Example 1. Let $q = 5^2$. Then $\mathrm{Tr}_{q^2/q}(\mathcal{C}_{\{2,3\}})$ has parameters $[26, 4, 20]_{25}$ and the code $\mathcal{C}_{\{2,3\}}^{\perp}|_{\mathrm{GF}(q)}$ has parameters $[26, 22, 4]_{25}$.

Example 2. Let $q = 5^3$. Then $\mathrm{Tr}_{q^2/q}(\mathcal{C}_{\{2,3\}})$ has parameters $[126, 4, 120]_{75}$.

Theorem 3. *Let* $q = 5^m$ *with* $m \geq 2$. *Let* k *be an integer with* $1 \leq k \leq q+1$ *and* $A_k\left(\mathrm{Tr}_{q^2/q}(\mathcal{C}_{\{2,3\}})\right) > 0$. *Then* $\mathcal{B}_k\left(\mathrm{Tr}_{q^2/q}(\mathcal{C}_{\{2,3\}})\right)$ *is invariant under the action of* $\mathrm{Stab}_{U_{q+1}}$. *In particular, the incidence structure* $\left(U_{q+1}, \mathcal{B}_k\left(\mathrm{Tr}_{q^2/q}(\mathcal{C}_{\{2,3\}})\right)\right)$ *is a 3-design when* $k > 3$.

Proof. *We only need to show that if* $\mathbf{c} \in \mathrm{Tr}_{q^2/q}(\mathcal{C}_{\{2,3\}})$ *and* π *is a linear fractional transformation listed in Lemma 1, then there exists a codeword* $\mathbf{c}' \in \mathrm{Tr}_{q^2/q}(\mathcal{C}_{\{2,3\}})$ *such that* $\mathrm{Supp}(\pi(\mathbf{c})) = \mathrm{Supp}(\mathbf{c}')$. *Denote by* $\mathbf{c}(a_2, a_3)$ *the codeword* $\left(\mathrm{Tr}_{q^2/q}(a_2u^2 + a_3u^3)\right)_{u \in U_{q+1}}$ *of* $\mathrm{Tr}_{q^2/q}(\mathcal{C}_{\{2,3\}})$, *where* $a_2, a_3 \in \mathrm{GF}(q^2)$. *We investigate the following three cases for* π.

If π *is the transformation given by* $u \mapsto u_0 u$, *where* $u_0 \in U_{q+1}$, *then it is clear that* $\pi(\mathbf{c}(a_2, a_3)) = \mathbf{c}(a_2u_0^2, a_3u_0^3)$. *Thus* $\mathrm{Supp}(\pi(\mathbf{c}(a_2, a_3))) = \mathrm{Supp}\left(\mathbf{c}(a_2u_0^2, a_3u_0^3)\right)$.

If π *is the transformation given by* $u \mapsto u^{-1}$, *then it is obvious that* $\pi(\mathbf{c}(a_2, a_3)) = \mathbf{c}(a_2, a_3)$. *Thus* $\mathrm{Supp}(\pi(\mathbf{c}(a_2, a_3))) = \mathrm{Supp}(\mathbf{c}(a_2, a_3))$.

Let π *be the transformation given by* $u \mapsto \frac{u+c^q}{cu+1}$ *where* $c \in \mathrm{GF}(q^2)^* \setminus U_{q+1}$. *Write* $f(u) = \mathrm{Tr}_{q^2/q}(a_2u^2 + a_3u^3)$ *and* $A = cu + 1$. *Then* $u + c^q = uA^q$. *A standard computation gives*

$$\begin{aligned}
&f\left(\frac{u+c^q}{cu+1}\right)\\
&=\mathrm{Tr}_{q^2/q}\left(a_2\left(\frac{u+c^q}{cu+1}\right)^2+a_3\left(\frac{u+c^q}{cu+1}\right)^3\right)\\
&=\mathrm{Tr}_{q^2/q}\left(\frac{a_2(u+c^q)^2(cu+1)+a_3(u+c^q)^3}{(cu+1)^3}\right)\\
&=\mathrm{Tr}_{q^2/q}\left(\frac{a_2u^2A^{2q}A+a_3u^3A^{3q}}{A^3}\right)\\
&=\frac{a_2A^{2q}Au^2+a_3A^{3q}u^3}{A^3}+\frac{a_2^qA^2A^qu^{2q}+a_3^qA^3u^{3q}}{A^{3q}}\\
&=\frac{a_2A^{5q}Au^2+a_3A^{6q}u^3+a_2^qA^5A^qu^{2q}+a_3^qA^6u^{3q}}{A^3A^{3q}}\\
&=\frac{a_2A^{5q}Au^2+a_3A^{6q}u^3+(a_2A^{5q}Au^2+a_3A^{6q}u^3)^q}{A^3A^{3q}}\\
&=\frac{1}{A^3A^{3q}}\mathrm{Tr}_{q^2/q}(a_2A^{5q}Au^2+a_3A^{6q}u^3).
\end{aligned} \tag{12}$$

Expanding $a_2A^{5q}Au^2$ yields

$$\begin{aligned}
&a_2A^{5q}Au^2\\
&=a_2(cu+1)^{5q}(cu+1)u^2\\
&=a_2(c^{5q}u^{5q}+1)(cu+1)u^2\\
&=a_2u^2(c^{5q+1}u^{5q+1}+c^{5q}u^{5q}+cu+1)\\
&=a_2(c^{5q+1}u^{5q+3}+c^{5q}u^{5q+2}+cu^3+u^2)\\
&=a_2(u^2+cu^3+c^{5q+1}u^{-2}+c^{5q}u^{-3}).
\end{aligned} \tag{13}$$

Expanding $a_3A^{6q}u^3$ yields

$$\begin{aligned}
&a_3A^{6q}u^3\\
&=a_3(c^{6q}u^{6q}+1)u^3\\
&=a_3(c^{6q}u^{6q+3}+u^3)\\
&=a_3(u^3+c^{6q}u^{-3}).
\end{aligned} \tag{14}$$

Combining (13) and (14) gives

$$\begin{aligned}
&\mathrm{Tr}_{q^2/q}(a_2A^{5q}Au^2+a_3A^{6q}u^3)\\
&=\mathrm{Tr}_{q^2/q}(a_2(u^2+cu^3+c^{5q+1}u^{-2}+c^{5q}u^{-3})+a_3(u^3+c^{6q}u^{-3}))\\
&=\mathrm{Tr}_{q^2/q}((a_2+a_2^qc^{5+q})u^2+(a_2c+a_2^qc^5+a_3+a_3^qc^6)u^3).
\end{aligned} \tag{15}$$

Plugging (15) into (12) yields

$$f\left(\frac{u+c^q}{cu+1}\right)=\frac{1}{A^3A^{3q}}\mathrm{Tr}_{q^2/q}(a_2'u^2+a_3'u^3),$$

where $a_2' = a_2 + a_2^q c^{5+q}$ *and* $a_3' = a_2 c + a_2^q c^5 + a_3 + a_3^q c^6$. *This clearly forces* $\mathrm{Supp}(\pi(\mathbf{c}(a_2, a_3))) = \mathrm{Supp}(\mathbf{c}(a_2', a_3'))$. *The desired conclusion then follows.*

The proof of Theorem 2 gives more, namely

$$\begin{aligned} &\mathrm{Tr}_{q^2/q}\left(a_2\left(\frac{u+c^q}{cu+1}\right)^2 + a_3\left(\frac{u+c^q}{cu+1}\right)^3\right) \\ &= \frac{1}{(cu+1)^3(cu+1)^{3q}}\mathrm{Tr}_{q^2/q}(a_2'u^2 + a_3'u^3), \end{aligned} \tag{16}$$

where $a_2, a_3 \in \mathrm{GF}(q^2)$, $c \in \mathrm{GF}(q^2) \setminus U_{q+1}$, $a_2' = a_2 + a_2^q c^{5+q}$ and $a_3' = a_2 c + a_2^q c^5 + a_3 + a_3^q c^6$.

The following theorem shows the invariance of the set of the supports of all the codewords of any fixed weight in $\mathcal{C}_{\{2,3\}}{}^{\perp}|_{\mathrm{GF}(q)}$ under the action of $\mathrm{PGL}(2, q)$.

Theorem 4. *Let* $q = 5^m$ *with* $m \geq 2$. *Let* k *be any integer with* $1 \leq k \leq q+1$ *and* $A_k\left(\mathcal{C}_{\{2,3\}}{}^{\perp}|_{\mathrm{GF}(q)}\right) > 0$. *Then* $\mathcal{B}_k\left(\mathcal{C}_{\{2,3\}}{}^{\perp}|_{\mathrm{GF}(q)}\right)$ *is invariant under the action of* $\mathrm{Stab}_{U_{q+1}}$. *In particular, the incidence structure* $\left(U_{q+1}, \mathcal{B}_k(\mathcal{C}_{\{2,3\}}{}^{\perp}|_{\mathrm{GF}(q)})\right)$ *is a 3-design when* $k > 3$.

Proof. *Recall that by (6) we have*

$$\left(\mathrm{Tr}_{q^2/q}(\mathcal{C}_{\{2,3\}})\right)^{\perp} = \mathcal{C}_{\{2,3\}}{}^{\perp}|_{\mathrm{GF}_{(q)}}.$$

Let $\mathbf{w}$ *be any codeword of* $\left(\mathrm{Tr}_{q^2/q}(\mathcal{C}_{\{2,3\}})\right)^{\perp} = \mathcal{C}_{\{2,3\}}{}^{\perp}|_{\mathrm{GF}_{(q)}}$ *and* π *be any linear fractional translation listed in Lemma 1. It is easily seen that if* π *is a transformation given by* $u \mapsto u_0 u$ *or* $u \mapsto 1/u$, *where*$u_0 \in U_{q+1}$, *then*

$$\pi(\mathbf{w}) \in \mathcal{C}_{\{2,3\}}{}^{\perp}|_{\mathrm{GF}(q)}. \tag{17}$$

Assume π *is a translation given by* $u \mapsto \frac{u+c^q}{cu+1}$ *where* $c \in \mathrm{GF}(q^2)^* \setminus U_{q+1}$. *It is obvious that* $\pi(\mathbf{w}) \in \left(\pi\ \left(\mathrm{Tr}_{q^2/q}(\mathcal{C}_{\{2,3\}})\right)\right)^{\perp}$. *From (16) we conclude that*

$$\pi\left(\mathrm{Tr}_{q^2/q}(\mathcal{C}_{\{2,3\}})\right) = \left(\frac{1}{(cu+1)^{3q+3}}\right)_{u\in U_{q+1}} \cdot \mathrm{Tr}_{q^2/q}(\mathcal{C}_{\{2,3\}}),$$

By (1) we have that

$$\left(\pi\left(\mathrm{Tr}_{q^2/q}(\mathcal{C}_{\{2,3\}})\right)\right)^{\perp} = \left((cu+1)^{3q+3}\right)_{u\in U_{q+1}} \cdot \left(\mathrm{Tr}_{q^2/q}(\mathcal{C}_{\{2,3\}})\right)^{\perp},$$

Consequently,

$$\pi(\mathbf{w}) \in \left((cu+1)^{3q+3}\right)_{u\in U_{q+1}} \cdot \left(\mathrm{Tr}_{q^2/q}(\mathcal{C}_{\{2,3\}})\right)^{\perp}. \tag{18}$$

Combining (17) *and (18) with Lemma 1 we can assert that the set of all the supports of* $\mathcal{C}_{\{2,3\}}{}^{\perp}|_{\mathrm{GF}(q)}$ *stays invariant under* $\mathrm{Stab}_{U_{q+1}}$. *This completes the proof.*

4 Summary and Concluding Remarks

In this paper, we investigated two infinite families of cyclic codes of length $5^m + 1$ over GF(5^m) and completely determined their parameters. A code from the first family has parameters $[q + 1, 4, q - 5]_q$, where $q = 5^m$ and $m \geq 2$ is an integer. A code from the second family has parameters $[q + 1, q - 3, 4]_q$, where $q = 5^m$ and $m \geq 2$ is an integer. 3-designs can be obtained from sets of the supports of all codewords of any fixed weight being invariant under PGL($2, q$).

References

1. Ding, C., Tang, C., Tonchev, V.D.: The projective general linear group $\mathrm{PGL}(2, 2^{\mathrm{m}})$ and linear codes of length $2^m + 1$. Des. Codes Cryptogr. **89**(7), 1713–1734 (2021)
2. Beth, T., Jungnickel, D., Lenz, H.: Design Theory. Cambridge University Press, Cambridge (1999)
3. Tang, C., Ding, C.: An infinite family of linear codes supporting 4-designs. IEEE Trans. Inform. Theory. **67**(1), 244–254 (2021)
4. Bosma, W., Cannon, J.: Handbook of Magma Functions. School of Mathematics and Statistics, University of Sydney, Sydney (1999)
5. Ding, C.: Designs from Linear Codes. World Scientific, Singapore (2018)
6. Ding, C., Tang, C.: Infinite families of near MDS codes holding t-designs. IEEE Trans. Inform. Theory. **66**(9), 5419–5428 (2020)
7. Du, X., Wang, R., Fan, C.: Infinite families of 2-designs from a class of cyclic codes. J. Combin. Des. **28**(3), 157–170 (2020)
8. Huffman, W.C., Pless, V.: Fundamentals of Error-Correcting Codes. Cambridge University Press, Cambridge (2003)
9. Giorgetti, M., Previtali, A.: Galois invariance, trace codes and subfield subcodes. Finite Fields Appl. **16**(2), 96–99 (2010)
10. Xiang, C., Tang, C., Liu, Q.: An infinite family of antiprimitive cyclic codes supporting Steiner systems $S(3, 8, 7^m + 1)$. Des. Codes Cryptogr. **90**, 1319–1333 (2022)

Private Information Retrieval Schemes Using Cyclic Codes

Şeyma Bodur(✉), Edgar Martínez-Moro, and Diego Ruano

IMUVa-Mathematics Research Institute, Universidad de Valladolid, Valladolid, Spain
{seyma.bodur,edgar.martinez,diego.ruano}@uva.es

Abstract. A Private Information Retrieval (PIR) scheme allows users to retrieve data from a database without disclosing to the server information about the identity of the data retrieved. A coded storage in a distributed storage system with colluding servers is considered in this work, namely the approach in [6] which considers a storage and retrieval code with a transitive group and provides binary PIR schemes with the highest possible rate. Reed-Muller codes were considered in [6]. In this work we consider cyclic codes and we show that binary PIR schemes using cyclic codes provide a larger constellation of PIR parameters and they may outperform the ones coming from Reed-Muller codes in some cases.

Keywords: Private information retrieval · Cyclic codes · Reed-Muller codes

1 Introduction

Many protocols protect the user and server from third parties while accessing the data. Nevertheless, no security measure protects the user from the server. As a result of this demand, Private Information Retrieval (PIR) protocols emerged [5]. They allow users to retrieve data from a database without disclosing to the server information about the identity of the data retrieved. We consider in this work that data is stored in a Distributed Storage System (DSS), since, if data is stored in a single database, one can only guarantee information-theoretic privacy by downloading the full database, which has a high communication cost.

Shah et al. [14] have shown that privacy is guaranteed when one bit more than the requested file size is downloaded, but it requires many servers. In case of non-response or a fail from some servers, the PIR scheme should allow servers to communicate with each other. Hence, it is natural to assume that the servers

This work was supported in part by Grant PGC2018-096446-B-C21 funded by MCIN/AEI/10.13039/501100011033 and by "ERDF A way of making Europe", by Grant RYC-2016-20208 funded by MCIN/AEI/10.13039/501100011033 and by "ESF Investing in your future", and by Grant CONTPR-2019-385 funded by Universidad de Valladolid and Banco Santander.

S. Mesnager and Z. Zhou (Eds.): WAIFI 2022, LNCS 13638, pp. 194–207, 2023.
https://doi.org/10.1007/978-3-031-22944-2_12

may collude, that is, they may inform each other of their input from the user. A scheme that addresses the situation where any t servers may collude is called a t-private information retrieval scheme and it was considered in [7,17]. This approach, that we consider in this work, uses Coding Theory and the security and performance depend on the parameters of linear codes and their star products (also called Shur products).

The PIR maximum possible rate was examined without collusion in [15] and with collusion in [16]. The PIR capacity obtained without colluding servers and using a Maximum Distance Separable (MDS) code was given in [1]. In [7], a PIR scheme with colluding servers for Generalized Reed Solomon (GRS) codes is given. Their PIR scheme rate is based on the minimum distance of a star product of the storage code and the retrieval code.

The use of a GRS, or an MDS code, requires working over a big base field. In order to address this issue, since binary base fields are desirable for practical implementations, [6] provided a PIR scheme that is based on binary Reed-Muller (RM) codes. They observed that the scheme reaches the highest possible rate if the codes used to define the PIR scheme have a transitive automorphism group, which is the case of RM codes.

In this work we propose to use cyclic codes to construct PIR schemes in the same fashion as [6]. Cyclic codes have also a transitive automorphism group and they can be defined over a binary (or small) finite field as well. Moreover, the star product of two cyclic codes is a cyclic code and its parameters can be computed [4]. Namely, the star product of two cyclic codes is given by the sum of their generating sets and we can compute its dimension and estimate its minimum distance considering cyclotomic cosets.

The main contributions of this work are given in Sect. 5. Our aim is to optimize the number of databases that can collude without disclosing to the server information about the identity of the data retrieved. In order to show the goodness of cyclic codes for PIR schemes, we first provide pairs of cyclic codes C and D, the storage code and retrieval code, such that the parameters of C, D, $D^\perp$, $C \star D$ and $(C \star D)^\perp$ are -at the same time- optimal or the best known. As we will recall in Sect. 2, their parameters determine the performance of the PIR scheme defined by C and D. Since a punctured RM code is a cyclic code, we may obtain PIR schemes using punctured RM codes by using cyclic codes. Moreover, we show that by using cyclic codes we obtain a larger constellation of possible parameters of binary PIR schemes. The construction of PIR schemes and the computations of their parameters follow from a detailed analysis of cyclotomic cosets. Then we focus on the privacy and on the rate of a PIR scheme since the upload cost in a PIR scheme can be neglected [1]. More concretely, in case that the storage code C has dimension 2, we obtain binary PIR schemes that greatly outperform the ones obtained using RM codes, more concretely they protect against a more significant number of colluding servers. Finally, we compare our schemes with shortened RM codes and we show that in this case the PIR schemes using cyclic codes outperform them as well, namely they offer more privacy for a fixed rate.

2 General Private Information Scheme

This section reviews some basic definitions of linear codes and briefly recalls PIR schemes (see [6,7,17] for further details). We denote by $\mathbb{F}_q$ the finite field with q elements. A linear code C is a linear subspace of $\mathbb{F}_q^n$. We denote its parameters by $[n, k, d]$.

Definition 1. *Given two linear codes C and D of length n over $\mathbb{F}_q$, we define their star product (or Shur product) $C \star D$ as the linear code in $\mathbb{F}_q^n$ spanned by the set $\{c \star d \mid c \in C, d \in D\}$, where $\star$ denotes the component-wise product $c \star d = (c_1 d_1, \ldots, c_n d_n)$.*

2.1 PIR Schemes

A PIR scheme consists of three stages; Data Storage, File Request and Response Process. In the Data Storage Process, files are uploaded to a DSS. In the File Request, users decide the file they want to retrieve, called the desired file, and according to it, they select queries that are sent to the servers. In the final Response Process, servers 'operate' the files with the queries generating a matrix of responses that are sent back to the user. It should be noted that the servers do not have any information about the file requested by the user.

Data Storage Process. We have r-files, each file has ρ-rows, k-columns, and the elements of the files are in $\mathbb{F}_q$. Since the number of files is r, the total file can be understood as $r\rho \times k$ matrix denoted by A, and each file is denoted by $\boldsymbol{a}^i$, where $i \in \{1, \ldots, r\}$.

The files are stored in a DSS. In order to upload the files into the servers, files are encoded by a k-dimensional storage code $C \subseteq \mathbb{F}_q^n$ with parameters $[n, k, d]$. Concerning encoding, we multiply the matrix A, which covers all files, by G_C the generator matrix of the linear code C and we obtain the matrix $Y := A \cdot G_C$. Since A is a $\rho r \times k$ matrix, Y has ρr rows and n columns.

Request and Response Process. Let's assume that the user wants to retrieve the file $\boldsymbol{a}^i$. Then the user chooses a random query Q^i and sends this query to the servers. Each server computes the inner product of Q_j^i and Y_j, where j is the server index, i.e., j^{th}-server computes $\langle Q_j^i, Y_j \rangle$. Then, servers send back the response vectors to the user.

The file is divided into parts, and a part is obtained in each round. With the final round, all parts of the file are completed. All the parts of the file from several servers are gathered to get the whole file. Therefore, the PIR rate is defined as the ratio of the information obtained during the process to the downloaded information.

If t-colluding servers communicate with each other and they cannot access any information about the desired file, it is said that the PIR scheme is resistant to t-colluding servers. The following theorem is the key for finding the number of colluding servers and the system's PIR rate.

Theorem 1 ([6]). *If the automorphism groups of C and $C \star D$ are transitive on the set $\{1, \dots, n\}$, then there exists a PIR scheme with rate $\frac{dim(C \star D)^{\perp}}{n}$ that resists a $(d_{D^{\perp}} - 1)$-collusion attack. That is, the privacy is $t = d_{D^{\perp}} - 1$.*

We will compare Reed-Muller codes, considered in [6], and cyclic codes in Sect. 5. For this reason, the next section gives a brief exposition of RM codes.

3 Reed-Muller Codes

The binary r^{th} order Reed-Muller Code, denoted by $RM(r, m)$, is defined to be

$$RM(r, m) = \{ev(f) : f \in \mathbb{F}_2[x_1, ..., x_m], \, deg(f) \leq r\}, \tag{1}$$

where $ev(f)$ is the evaluation of f at all points in $\mathbb{F}_2^m$.

Remark 1. $RM(r, m)$ is a linear code of length $n = 2^m$, dimension $k = \sum_{i=0}^{r} \binom{m}{i}$, and minimum distance 2^{m-r} [12]. One has that $RM(r_1, m) \star RM(r_2, m) = RM(r_1 + r_2, m)$, where $r_1 + r_2 \leq m$.

Since if we shorten or puncture RM codes at the position evaluating at $\mathbf{0}$ we obtain cyclic codes [18], the following two definitions will be helpful in Sect. 5.

Definition 2. *The shortened code at the position evaluating at $\mathbf{0}$ of a binary Reed-Muller code is denoted by the linear $[2^m - 1, k - 1, 2^{m-r}]$ code $C_{\bullet}$. The punctured code at the position evaluating at $\mathbf{0}$ of a binary Reed-Muller code is denoted by the linear $[2^m - 1, k, 2^{m-r} - 1]$ code $C^{\bullet}$.*

4 Cyclic Codes

In this section, we will be concerned with basic cyclic code definitions and compute the star product of two cyclic codes.

Definition 3. *A $[n, k]$ linear code C is said to be cyclic if every cyclic shift of a codeword $c = (c_0, c_1, \dots, c_{n-1}) \in C$ is also codeword in C, that is $c = (c_{n-1}, c_0, \dots, c_{n-2}) \in C$.*

Theorem 2 ([11]). *A linear code C is a cyclic code if and only if C is isomorphic, as a $\mathbb{F}_q$-linear space, to an ideal in the ring $R_n = \mathbb{F}_q[x]/\langle x^n - 1\rangle$.*

In order to have a semisimple algebra, i.e., non-repeated roots for the polynomial $x^n - 1$ and a 1–1 correspondence between it factors and its roots, from now on we will require that $\gcd(n, q) = 1$.

Definition 4. *Let C be a cyclic code in R_n. We call $g(x)$ the generator polynomial of C if there exists a unique monic polynomial $g(x)$ such that $C = \langle g(x)\rangle$. Clearly, $g(x)$ is a divisor of $x^n - 1$ in $\mathbb{F}_q[x]$.*

Definition 5. *A set $J \subseteq \{0, \dots, n-1\}$ is said to be the defining set of $C = \langle g(x) \rangle$ if $J = \{ j \in \mathbb{Z}/n\mathbb{Z} \mid g(\alpha^j) = 0 \}$ and a set I is called the generating set of $C = \langle g(x) \rangle$ if $I = \{ j \in \mathbb{Z}/n\mathbb{Z} \mid g(\alpha^j) \neq 0 \}$, where α is a primitive element of $\mathbb{F}_q$.*

Remark 2. Let $g(x)$ be a generator polynomial of cyclic code C then $g(x) = \prod_{j \in J}(x - \alpha^j)$ and $g(x) = \frac{x^n - 1}{\prod_{i \in I}(x - \alpha^i)}$. Furthermore, one has that $dim(C) = n - |J| = |I|$.

Remark 3. Assume that J is a defining set of the code C, then the generator polynomial of $C^\perp$ is $h(x) = \prod_{i \in -I}(x - \alpha^i)$, where $-I$ is the set of additive inverses in $\mathbb{Z}/n\mathbb{Z}$ of the elements in I.

Definition 6. *The cyclotomic coset containing s, denoted by U_s, is defined to be the set $\{s, sq, \dots, sq^i\}$ (mod n) where i is the smallest integer such that $q^i \equiv 1$ (mod n).*

We have the following result about the star product of cyclic codes.

Theorem 3. *Let I_1 and I_2 be the generating sets of the cyclic codes C and D, respectively. The star product of $C \star D$ is generated by*

$$g_{C \star D} = \frac{x^n - 1}{\prod_{j \in I_1 + I_2}(x - \alpha^j)}, \tag{2}$$

where + denotes the Minkowski sum on sets, that is,

$$I_1 + I_2 := \{i_1 + i_2 \mid i_1 \in I_1,\ i_2 \in I_2\}.$$

Proof. We will follow a similar way to [4, Theorem III.3], which proves this result for $C \star C$. It is well known from [2] that a cyclic code can be defined as follows, consider $\mathbb{K}$ the extension field of $\mathbb{F}_q$ such that $x^n - 1$ splits in linear factors in $\mathbb{K}[x]$. For a set $M \subseteq \{1, \dots, n-1\}$ let $\mathcal{B}(M)$ be the $\mathbb{K}$-vector space

$$\mathcal{B}(M) = \left\{ (f(\alpha^0), f(\alpha^1), \dots, f(\alpha^{n-1})) \mid f = \sum_{i \in M} f_i x^i \in \mathbb{K}[x] \right\}.$$

For a cyclic code C with defining set I, as a byproduct of Delsarte's theorem, one has that C is equal to the subfield subcode $\mathcal{B}(-I)|_{\mathbb{F}_q^n} = B(-I) \cap \mathbb{F}_q^n$ (see [4, Lemma 5]). Now note that the vector space obtained by the extension of scalars of C, denoted by $\mathbb{K} \otimes C$, is a $\mathbb{K}$-cyclic code with the same dimension as $\mathcal{B}(-I)$ (given by $|I|$) and, henceforth $(\mathbb{K} \otimes C) = \mathcal{B}(-I)$. Note that the extension by scalars commutes with the star product (see [13, Lemma 2.23]) thus it is clear that $C \star D = (\mathbb{K} \otimes (C \star D))|_{\mathbb{F}_q^n} = (\mathbb{K} \otimes C \star \mathbb{K} \otimes D)|_{\mathbb{F}_q^n} = (\mathcal{B}(-(I_1)) \star \mathcal{B}(-(I_2)))|_{\mathbb{F}_q^n} = \mathcal{B}(-(I_1 + I_2))|_{\mathbb{F}_q^n}$. □

Proposition 1 (BCH Bound). *Let J be a defining set of a cyclic code C with minimum distance d. If J contains $\delta - 1$ consecutive elements $\{i, \dots, i + \delta - 2\} \subseteq J$, where $i, \delta \in \mathbb{Z}/n\mathbb{Z}$, then $d \geq \delta$.*

The following example illustrates the previous statements on cyclic codes.

Example 1. Set $q = 2$, $n = 31$. Let I_C be a generating set and let J_C be a defining set of the code C. The first cyclotomic cosets (modulo 31) are:

$$U_0 = \{0\},\ U_1 = \{1, 2, 4, 8, 16\},\ U_3 = \{3, 6, 12, 17, 24\}.$$

Consider C the cyclic code with defining set $J_C = U_0 \cup U_1 \cup U_3$. One has that J_C contains $\{0, 1, 2, 3, 4\}$, thus the BCH bound of C is equal to 6. The dimension of C is equal to $k = |I_C| = 31 - 11 = 20$. Therefore, the parameters of this code are $[31, 20, \geq 6]$.

5 PIR Schemes from Cyclic Codes

In this section, we focus on cyclic codes towards obtaining PIR schemes over small fields and compute the code parameters with cyclotomic cosets. First, we will analyze the codes obtained from computer search, their cyclotomic cosets, PIR rates, and the number of colluding servers.

The formulation of the amount of colluding servers and the PIR rate is given in Theorem 1. This theorem is valid for PIR schemes arising from cyclic codes since the automorphism group of a cyclic code is also transitive [10]. Table 1 gives some cyclic codes, the rate of the corresponding PIR scheme arising from them, and the maximum number of servers that may collude, that is, the privacy parameter t.

Table 1. Computer search experiments

C	D	$D^{\perp}$	$C * D$	$(C * D)^{\perp}$	Privacy	Rate
$[127, 8, 63]$	$[127, 29, 43]$	$[127, 98, 10]$	$[127, 113, 5]$	$[127, 14, 56]$	9	14/127
$[127, 8, 63]$	$[127, 42, 32]$	$[127, 85, 13]$	$[127, 112, 6]$	$[127, 15, 55]$	12	15/127
$[127, 15, 55]$	$[127, 15, 55]$	$[127, 112, 6]$	$[127, 106, 7]$	$[127, 21, 48]$	5	21/127
$[127, 15, 55]$	$[127, 21, 48]$	$[127, 106, 7]$	$[127, 112, 6]$	$[127, 15, 55]$	6	15/127
$[127, 21, 48]$	$[127, 21, 48]$	$[127, 106, 7]$	$[127, 112, 6]$	$[127, 15, 55]$	6	15/127

For instance, the first row in Table 1 considers C as a storage code with parameters $[127, 8, 63]$ and D as a retrieval code with parameters $[127, 29, 43]$. Applying Theorem 1, we can conclude that this scheme is secure against 9-colluding servers since $d(D^{\perp}) = 10$ and that the PIR's rate is $\frac{dim(C \star D)^{\perp}}{n} = \frac{14}{127}$.

We have obtained the codes in Table 1 by computer search, their generating set can be found in Table 2. For instance, Consider the codes in the first row, the generating set of the code C consists of the union of the cyclotomic cosets U_1 and U_{31}, and the one of D consists of U_0, U_5, U_{23}, U_{27}, U_{31}. As mentioned before, the generating set of star products of cyclic codes are given by the Minkowski sum of their generating sets. Hence, the generating set of $C \star D$ consists of all cyclotomic cosets except U_{13} and U_{47}.

From now on, for sake of brevity, we will denote as $U_{\{s_1,\dots,s_t\}}$ the union of the t cyclotomic cosets given by $U_{\{s_1,\dots,s_t\}} = \bigcup_{j=1}^{t} U_{s_i}$.

Table 2. Cyclotomic cosets used for codes in Table 1.

C	D
$U_{\{0,31\}}$	$U_{\{0,5,23,27,31\}}$
$U_{\{0,11\}}$	$U_{\{1,3,11,23,43,55\}}$
$U_{\{0,5,43\}}$	$U_{\{0,23,43\}}$
$U_{\{0,23,63\}}$	$U_{\{19,31,55\}}$
$U_{\{1,10,29\}}$	$U_{\{7,31,55\}}$

Table 3 classifies the codes in Table 1 according to the best-known linear codes in the database [8], which gives lower and upper bounds on the parameters of linear codes. As it is shown in the following table, their parameters are the best known or optimal.

Table 3. Classification of codes in Table 1

C	D	$D^{\perp}$	$C \star D$	$(C \star D)^{\perp}$
Optimal	*Best – known*	*Best – known*	*Optimal*	*Optimal*
Optimal	*Best – known*	*Best – known*	*Optimal*	*Best – known*
Best – known	*Best – known*	*Optimal*	*Best – known*	*Best – known*
Best – known	*Best – known*	*Best – known*	*Optimal*	*Best – known*
Best – known	*Best – known*	*Best – known*	*Optimal*	*Best – known*

5.1 Comparison with Punctured and Shortened RM Codes

We will show now why cyclic codes may provide better performance than RM codes. Even though a RM code C is not cyclic, $C_\bullet$ and $C^\bullet$ are cyclic codes [18]. Therefore, we compare the PIR rate and privacy given by a cyclic code with the corresponding punctured and shortened RM codes.

First, let us focus on the comparison with punctured RM codes. For length 127 and 255, we fixed as storage code a $[127, 8, 63]$, $[255, 9, 172]$ cyclic code and collected the star product of some codes in Table 4 and Table 6, respectively. We remark that in Table 4 the BCH bound of the retrieval codes (D) equal to their minimum distance (Table 5).

Table 4. Comparison with punctured RM codes (Shadow rows)

C	D	$D^{\perp}$	$C * D$	$(C * D)^{\perp}$	Privacy	Rate
[127, 8, 63]	[127, 8, 63]	[127, 119, 4]	[127, 29, 31]	[127, 98, 7]	3	98/127
[127,8,63]	**[127,22,47]**	**[127,105,8]**	**[127,64,15]**	**[127,63,16]**	**7**	**63/127**
[127, 8, 63]	[127, 29, 31]	[127, 98, 8]	[127, 64, 15]	[127, 63, 16]	7	63/127
[127,8,63]	**[127,50,27]**	**[127,77,16]**	**[127,99,7]**	**[127,28,32]**	**15**	**28/127**
[127,8,63]	**[127,57,23]**	**[127,70,16]**	**[127,99,7]**	**[127,28,32]**	**15**	**28/127**
[127, 8, 63]	[127, 64, 15]	[127, 63, 16]	[127, 99, 7]	[127, 28, 32]	15	28/127
[127,8,63]	**[127,85,13]**	**[127,42,32]**	**[127,120,3]**	**[127,7,64]**	**31**	**7/127**
[127,8,63]	**[127,92,11]**	**[127,35,32]**	**[127,120,3]**	**[127,7,64]**	**31**	**7/127**
[127, 8, 63]	[127, 99, 7]	[127, 28, 32]	[127, 120, 3]	[127, 7, 64]	31	7/127

Table 5. Cyclotomic cosets used for codes in Table 4.

C	D
$U_{\{0,1\}}$	$U_{\{0,1\}}$
	$U_{\{0,1,5,9\}}$
	$U_{\{0,1,5,9,3\}}$
	$U_{\{0,1,5,9,3,11,19,21\}}$
	$U_{\{0,1,5,9,3,11,19,21,7\}}$
	$U_{\{0,1,5,9,3,11,19,21,7,13\}}$
	$U_{\{0,1,5,9,3,11,19,21,7,13,23,27,43\}}$
	$U_{\{0,1,5,9,3,11,19,21,7,13,23,27,43,29\}}$
	$U_{\{0,1,5,9,3,11,19,21,7,13,23,27,29,43,15\}}$

Unbold rows in Table 4 and Table 6 display the parameters of those codes obtained by the star product of two cyclic codes, equivalent to the punctured RM codes. **Bold** rows are obtained by the star product of a cyclic code and the fixed code C. Consequently, when the rate and the storage codes are fixed, cyclic codes provide the same parameters as punctured RM ones except D with parameters [255, 77, 31] which gives a better rate than RM codes. However, for a fixed-length n, the dimension of RM codes overgrow, thus for a fixed $C \star D$, there are not many values that the dimensions of the code C and D may take. Hence, the first advantage of using cyclic codes in the PIR scheme is to easily provide a larger constellation of parameters.

As an illustration of this fact, in the fourth and fifth rows of Table 4, the dimension of D can be 50 or 57 other than 64, or the dimension of D can be 85 or 92, different than 99. Thus, we have different options for the same rate and privacy. The following remark will show the method we used for obtaining the codes in Table 4 and Table 6.

Table 6. Comparison with punctured RM codes (Shadow rows)

C	D	$D^{\perp}$	$C * D$	$(C * D)^{\perp}$	Privacy	Rate
$[255, 9, 127]$	$[255, 9, 127]$	$[255, 246, 4]$	$[255, 37, 63]$	$[255, 218, 8]$	3	$\frac{218}{255}$
[255,9,127]	$\mathbf{[255, 25, \geq 63]}$	$\mathbf{[255, 230, \geq 8]}$	$\mathbf{[255, 93]}$	$\mathbf{[255, 162]}$	$\mathbf{\geq 7}$	$\mathbf{\frac{162}{255}}$
$\mathbf{[255, 9, 127]}$	$\mathbf{[255, 33, \geq 63]}$	$\mathbf{[255, 222, \geq 8]}$	$\mathbf{[255, 93]}$	$\mathbf{[255, 162]}$	$\mathbf{\geq 7}$	$\mathbf{\frac{162}{255}}$
$[255, 9, 127]$	$[255, 37, 63]$	$[255, 218, 8]$	$[255, 93, 31]$	$[255, 162, 16]$	7	$\frac{162}{255}$
$\mathbf{[255, 9, 127]}$	$\mathbf{[255, 77, \geq 31]}$	$\mathbf{[255, 178, \geq 16]}$	$\mathbf{[255, 161]}$	$\mathbf{[255, 94]}$	$\mathbf{\geq 15}$	$\mathbf{\frac{94}{255}}$
$\mathbf{[255, 9, 127]}$	$\mathbf{[255, 85, \geq 31]}$	$\mathbf{[255, 170, \geq 16]}$	$\mathbf{[255, 163]}$	$\mathbf{[255, 92]}$	$\mathbf{\geq 15}$	$\mathbf{\frac{92}{255}}$
$[255, 9, 127]$	$[255, 93, 31]$	$[255, 162, 16]$	$[255, 163, 15]$	$[255, 92, 32]$	15	$\frac{92}{255}$
$\mathbf{[255, 9, 127]}$	$\mathbf{[255, 133, \geq 15]}$	$\mathbf{[255, 122, \geq 32]}$	$\mathbf{[255, 219]}$	$\mathbf{[255, 36]}$	$\mathbf{\geq 31}$	$\mathbf{\frac{36}{255}}$
$\mathbf{[255, 9, 127]}$	$\mathbf{[255, 141, \geq 15]}$	$\mathbf{[255, 114, \geq 32]}$	$\mathbf{[255, 219]}$	$\mathbf{[255, 36]}$	$\mathbf{\geq 31}$	$\mathbf{\frac{36}{255}}$
$\mathbf{[255, 9, 127]}$	$\mathbf{[255, 149, \geq 15]}$	$\mathbf{[255, 106, \geq 32]}$	$\mathbf{[255, 219]}$	$\mathbf{[255, 36]}$	$\mathbf{\geq 31}$	$\mathbf{\frac{36}{255}}$
$\mathbf{[255, 9, 127]}$	$\mathbf{[255, 153, \geq 15]}$	$\mathbf{[255, 102, \geq 32]}$	$\mathbf{[255, 219]}$	$\mathbf{[255, 36]}$	$\mathbf{\geq 31}$	$\mathbf{\frac{36}{255}}$
$\mathbf{[255, 9, 127]}$	$\mathbf{[255, 161, \geq 15]}$	$\mathbf{[255, 94, \geq 32]}$	$\mathbf{[255, 219]}$	$\mathbf{[255, 36]}$	$\mathbf{\geq 31}$	$\mathbf{\frac{36}{255}}$
$[255, 9, 127]$	$[255, 163, 15]$	$[255, 92, 32]$	$[255, 219, 7]$	$[255, 36, 64]$	31	$\frac{36}{255}$
$\mathbf{[255, 9, 127]}$	$\mathbf{[255, 211, \geq 7]}$	$\mathbf{[255, 44, \geq 64]}$	$\mathbf{[255, 247]}$	$\mathbf{[255, 8]}$	$\mathbf{\geq 63}$	$\mathbf{\frac{8}{255}}$
$[255, 9, 127]$	$[255, 219, 7]$	$[255, 36, 64]$	$[255, 247, 3]$	$[255, 8, 128]$	63	$\frac{8}{255}$

Remark 4. The r-th order punctured generalized RM code is the cyclic code length $n = q^m - 1$ with generator polynomial

$$g(x) := \prod_{i \in I}(x - \alpha^i), \text{ where } I = \{i : w_q(i) \leq (q-1)c\}, \tag{3}$$

for some $c \in \mathbb{Z}^+$ and $w_q(i)$ is the number of non-zeros in the q-ary expression of i. Now using Equation (3), we have created the unbolded row in Table 4 and Table 6. Namely, if we add or remove some cyclotomic classes to the punctured RM code's generating sets, we can get another cyclic code, which provides the same rate and privacy. While making these additions and removals of cosets, we use Remark 3 to decide the heuristics of which cyclotomic cosets we select. Note that the minimum distance of the code $D^{\perp}$ provides the privacy of the scheme, so we wish $d(D^{\perp}) - 1$ being as big as possible. For this purpose, we set $-I$ to be a large set of consecutive elements.
For instance, the third row in Table 4, the generating set of C is comprised of U_0 and U_1, and D is comprised of U_0, U_1, U_3, U_5, U_9. The generating set of code D in the second row consists of U_0, U_1, U_5, U_9 by removing U_3. We have removed U_3, because U_3 does not change the BCH bound of the code. Moreover, in the fourth row in Table 6, the generating set of D is comprised of U_0, U_1, U_{17}, U_9, U_5, U_3. Removing U_{17} is not affecting the BCH bound of the code and, thus we obtain the parameter in the third row.

We also achieved a second advantage, more privacy, by reducing the dimension of the storage code C, which is not equivalent to punctured Reed-Muller

Table 7. Cyclotomic cosets used for codes in Table 6.

C	D
$U_{\{0,1\}}$	$U_{\{0,1\}}$
	$U_{\{0,1,3,5\}}$
	$U_{\{0,1,3,5,9\}}$
	$U_{\{0,1,3,5,9,17\}}$
	$U_{\{0,1,3,5,9,17,7,11,13,19,25\}}$
	$U_{\{0,1,3,5,9,17,7,11,13,19,21,25\}}$
	$U_{\{0,1,3,5,9,17,7,11,13,19,21,25,37\}}$
	$U_{\{0,1,3,5,9,17,7,11,13,19,21,25,37,15,23,27,29,39\}}$
	$U_{\{0,1,3,5,9,17,7,11,13,19,21,25,37,15,23,27,29,39,53\}}$
	$U_{\{0,1,3,5,9,17,7,11,13,19,21,25,37,15,23,27,29,39,45,53\}}$
	$U_{\{0,1,3,5,9,17,7,11,13,19,21,25,37,15,23,27,29,39,45,51,53\}}$
	$U_{\{0,1,3,5,9,17,7,11,13,19,21,25,37,15,23,27,29,39,43,45,51,53\}}$
	$U_{\{0,1,3,5,9,17,7,11,13,19,21,25,37,15,23,27,29,39,43,45,51,53,85\}}$
	$U_{\{0,1,3,5,9,17,7,11,13,19,21,25,37,15,23,27,29,39,43,45,51,53,85,31,47,55,59,61,87\}}$
	$U_{\{0,1,3,5,9,17,7,11,13,19,21,25,37,15,23,27,29,39,43,45,51,53,85,31,47,55,59,61,87,91\}}$

codes. We remark that the upload cost in the PIR scheme can be neglected [1], thus we focus on the value $d(D^{\perp}) - 1$, which provides privacy, and on $dim(C \star D)^{\perp}/n$, which gives the PIR rate. Therefore, we can reduce the dimension of the code C.

Example 2. Consider the punctured RM codes C_{RM} and D_{RM} with parameters $[63, 7, 31]$ and $[63, 42, 7]$, respectively. One has that the product code $C_{RM} \star D_{RM}$ has parameters $[63, 57, 3]$. The PIR scheme given using C_{RM} and D_{RM} protects against $d_{D^{\perp}} - 1 = 15$ collusions. Consider now the cyclic code C with parameters $[63, 2, 42]$, where the generating set of C is equal to U_{21}, and the cyclic code D with parameters $[63, 51, 3]$. One has that $C \star D = C_{RM} \star D_{RM}$. In this case $d_{D^{\perp}} - 1 = 19$. Therefore, our cyclic code proposal protects against a more significant number of colluding servers for the same rate.

Remark 5. Note that for length 63, there are two good cyclic codes in terms of the PIR parameters, one has dimension 2 (see Example 2) and the second one, D with parameters $[63, 40, 7]$, provides the same rate and privacy of a RM code. In the case of length 31, there are no binary cyclic codes that improve the rate or privacy of a RM code other than the ones equivalent to them. This is why we consider binary cyclic codes of length greater than ors equal to 127.

Table 8. Reducing the dimension of the storage

C	D	$D^{\perp}$	$(C \star D)^{\perp}$	Privacy	Rate
$[255, 2, 170]$	$[255, 192]$	$[255, 63, \mathbf{20{+}45}]$	$[255, \mathbf{19}]$	64	19/255
$[255, 2, 170]$	$[255, 195]$	$[255, 60, \mathbf{15{+}51}]$	$[255, \mathbf{8}]$	65	8/255
$[255, 2, 170]$	$[255, 198]$	$[255, 57, \mathbf{15{+}53}]$	$[255, \mathbf{8}]$	67	8/255
$[255, 2, 170]$	$[255, 200]$	$[255, 55, \mathbf{29{+}41}]$	$[255, \mathbf{11}]$	69	11/255
$[255, 2, 170]$	$[255, 201]$	$[255, 54, \mathbf{40{+}32}]$	$[255, \mathbf{9}]$	71	9/255
$[255, 2, 170]$	$[255, 202]$	$[255, 53, \mathbf{28{+}44}]$	$[255, \mathbf{8}]$	71	8/255
$[255, 2, 170]$	$[255, 204]$	$[255, 51, \mathbf{14{+}60}]$	$[255, \mathbf{11}]$	73	11/255

Table 9. Cyclotomic cosets used for codes in Table 8, where V is the set of all cyclotomic cosets for $q = 2$, modulo $n = 255$.

C	D
U_{85}	$V \setminus (U_{\{0,1,11,13,17,21,25,61,85,87\}})$
	$V \setminus (U_{\{1,13,25,27,29,31,45,119\}})$
	$V \setminus (U_{\{0,1,7,13,25,31,39,45\}})$
	$V \setminus (U_{\{0,1,13,17,25,29,31,63,85\}})$
	$V \setminus (U_{\{39,55,61,63,85,87,119,127\}})$
	$V \setminus (U_{\{0,1,9,13,25,31,111,119\}})$
	$V \setminus (U_{\{0,1,11,13,29,47,85,111\}})$

Table 8 contains more examples where, by using cyclic codes, the dimension of the storage code has been reduced. In this table, the minimum distance of $D^{\perp}$, which is related to the privacy, was first evaluated by the BCH bound and then its real value was computed using the powerful minimum distance algorithm in [9] (for instance, by 20 + 15 we mean that the BCH bound is equal to 20 and the real minimum distance is equal to 35). The table displays the privacy (number of colluding servers) and rate of the PIR scheme obtained using a code C with length 255 and dimension 2. Note that the PIR scheme obtained using the Punctured RM codes C_{RM} and $D_{RM}^{\perp}$ with parameters $[255, 9, 127]$ and $[255, 36, 64]$, respectively, protects against a maximum of 63 colluding servers. Moreover, the PIR rate of this scheme is equal to 8/255. In Table 8, the code pairs in all rows protect against more than 63 collusions. The cyclotomic cosets used for constructing the codes in Table 8 are given in Table 9.

In Table 10, shortened RM codes at the evaluation of $\mathbf{0}$ and cyclic codes with length 127 are analyzed. Again, we specify **bold** rows for star product of cyclic codes and unbold rows for star product of shortened RM codes. The storage code C with parameters $[127, 7, 64]$, equivalent to a shortened RM, is fixed. The only difference with respect to Table 4 is that we do not include the cyclotomic coset U_0 in the generating set of C.

Table 10. Comparison with shortened RM codes (Shadow rows)

C	D	$D^{\perp}$	$C * D$	$(C * D)^{\perp}$	Privacy	Rate
[127, 7, 64]	[127, 7, 64]	[127, 120, 3]	[127, 28, 32]	[127, 99, 7]	2	$\frac{99}{127}$
[127, 7, 64]	**[127, 22, 47]**	**[127, 105, 8]**	**[127, 63, 16]**	**[127, 64, 15]**	**7**	$\frac{\mathbf{64}}{\mathbf{127}}$
[127, 7, 64]	[127, 28, 32]	[127, 99, 7]	[127, 63, 16]	[127, 64, 15]	6	$\frac{64}{127}$
[127, 7, 64]	**[127, 50, 23]**	**[127, 77, 16]**	**[127, 98, 8]**	**[127, 29, 31]**	**15**	$\frac{\mathbf{29}}{\mathbf{127}}$
[127, 7, 64]	**[127, 57, 23]**	**[127, 70, 16]**	**[127, 98, 8]**	**[127, 29, 31]**	**15**	$\frac{\mathbf{29}}{\mathbf{127}}$
[127, 7, 64]	[127, 63, 16]	[127, 64, 15]	[127, 98, 8]	[127, 29, 31]	14	$\frac{29}{127}$
[127, 7, 64]	**[127, 85, 13]**	**[127, 42, 32]**	**[127, 119, 4]**	**[127, 8, 63]**	**31**	$\frac{\mathbf{8}}{\mathbf{127}}$
[127, 7, 64]	**[127, 92, 11]**	**[127, 35, 32]**	**[127, 119, 4]**	**[127, 8, 63]**	**31**	$\frac{\mathbf{8}}{\mathbf{127}}$
[127, 7, 64]	[127, 98, 8]	[127, 29, 31]	[127, 119, 4]	[127, 8, 63]	30	$\frac{8}{127}$

Table 11. Cyclotomic cosets used for codes in Table 10.

C	D
$U_{\{1\}}$	$U_{\{1\}}$
	$U_{\{0,1,5,9\}}$
	$U_{\{1,5,9,3\}}$
	$U_{\{0,1,5,9,3,11,19,21\}}$
	$U_{\{0,1,5,9,3,11,19,21,7\}}$
	$U_{\{1,5,9,3,11,19,21,7,13\}}$
	$U_{\{0,1,5,9,3,11,19,21,7,13,23,27,43\}}$
	$U_{\{0,1,5,9,3,11,19,21,7,13,23,27,43,29\}}$
	$U_{\{1,5,9,3,11,19,21,7,13,23,27,29,43,15\}}$

One has that the PIR schemes using cyclic codes protect against one more colluding server than shortened RM codes, as it can be seen at Table 10. Moreover, in this way we may increase the constellation of possible parameters. For instance, for a case rate equal to 29/127, the PIR scheme coming from a cyclic code protects against 15-collusion, but the one from a shortened RM code protects against 14-collusion. The cyclotomic cosets used for constructing the codes in Table 10 are given in Table 11.

6 Conclusion

By using cyclic codes, we provide binary PIR schemes with colluding servers in the fashion of [6]. We provide a family of optimal binary PIR schemes. Our PIR schemes have the advantage, with respect to PIR schemes from MDS codes, that they can be defined over a binary field. Moreover, they provide a larger constellation of parameters than the binary PIR schemes using Reed-Muller codes and they even outperform them in some cases. Note also that we come up with a reduced cost in generating the query vectors since a smaller dimension of the retrieval code means that less randomness will be needed to be generated by the user. All the examples in the paper were generated using the computer algebra system Magma [3].

Acknowledgements. We would like to thank F. Hernando (Universitat Jaume I) for providing us the code of the algorithm in [9].

References

1. Banawan, K., Ulukus, S.: The capacity of private information retrieval from coded databases. IEEE Trans. Inform. Theory **64**(3), 1945–1956 (2018). https://doi.org/10.1109/TIT.2018.2791994
2. Bierbrauer, J.: The theory of cyclic codes and a generalization to additive codes. Des. Codes Cryptogr. **25**(2), 189–206 (2002). https://doi.org/10.1023/A:1013808515797
3. Bosma, W., Cannon, J., Playoust, C.: The Magma algebra system. I. The user language. J. Symbolic Comput. **24**(3–4), 235–265 (1997). https://doi.org/10.1006/jsco.1996.0125. (computational algebra and number theory (London, 1993))
4. Cascudo, I.: On squares of cyclic codes. IEEE Trans. Inform. Theory **65**(2), 1034–1047 (2019)
5. Chor, B., Goldreich, O., Kushilevitz, E., Sudan, M.: Private information retrieval. J. ACM **45**(6), 965–982 (1998)
6. Freij-Hollanti, R., Gnilke, O.W., Hollanti, C., Horlemann-Trautmann, A.L., Karpuk, D., Kubjas, I.: t-private information retrieval schemes using transitive codes. IEEE Trans. Inform. Theory **65**(4), 2107–2118 (2019). https://doi.org/10.1109/TIT.2018.2871050
7. Freij-Hollanti, R., Gnilke, O.W., Hollanti, C., Karpuk, D.A.: Private information retrieval from coded databases with colluding servers. SIAM J. Appl. Algebra Geom. **1**(1), 647–664 (2017)
8. Grassl, M.: Bounds on the minimum distance of linear codes and quantum codes (2007). http://www.codetables.de. Accessed 3 Nov 2021
9. Hernando, F., Igual, F.D., Quintana-Ortí, G.: Algorithm 994: fast implementations of the Brouwer-Zimmermann algorithm for the computation of the minimum distance of a random linear code. ACM Trans. Math. Software **45**(2), Art. 23, 28 (2019). https://doi.org/10.1145/3302389
10. Huffman, W.C., Pless, V.: Fundamentals of Error-Correcting Codes. Cambridge University Press (2003). https://doi.org/10.1017/CBO9780511807077
11. van Lint, J.: Introduction to Coding Theory. Graduate Texts in Mathematics. Springer, Berlin Heidelberg (1998). https://doi.org/10.1007/978-3-642-58575-3
12. MacWilliams, F.J., Sloane, N.J.A.: The theory of error-correcting codes. I. North-Holland Mathematical Library, vol. 16, North-Holland Publishing Co., Amsterdam-New York-Oxford (1977)
13. Randriambololona, H.: On products and powers of linear codes under componentwise multiplication. In: Algorithmic arithmetic, geometry, and coding theory, Contemporary Mathematics, vol. 637, pp. 3–78. American Mathematical Society, Providence, RI (2015). https://doi.org/10.1090/conm/637/12749

14. Shah, N.B., Rashmi, K.V., Ramchandran, K.: One extra bit of download ensures perfectly private information retrieval. In: 2014 IEEE International Symposium on Information Theory, pp. 856–860 (2014). https://doi.org/10.1109/ISIT.2014.6874954
15. Sun, H., Jafar, S.A.: The capacity of private information retrieval. IEEE Trans. Inform. Theory **63**(7), 4075–4088 (2017)
16. Sun, H., Jafar, S.A.: The capacity of robust private information retrieval with colluding databases. IEEE Trans. Inform. Theory **64**(4, part 1), 2361–2370 (2018)
17. Tajeddine, R., Gnilke, O.W., el Rouayheb, S.: Private information retrieval from MDS coded data in distributed storage systems. IEEE Trans. Inform. Theory **64**, 7081–7093 (2018)
18. Yardi, A.D., Pellikaan, R.: On shortened and punctured cyclic codes. ArXiv:abs/1705.09859 (2017)

Two Classes of Optimal Few-Weight Codes Over $\mathbb{F}_q + u\mathbb{F}_q$

Zhao Hu, Bing Chen, Nian Li(✉), and Xiangyong Zeng

Hubei Key Laboratory of Applied Mathematics, Faculty of Mathematics and Statistics, Hubei University, Wuhan 430062, China
zhao.hu@aliyun.com, {chenbing,nian.li,xzeng}@hubu.edu.cn

Abstract. In this paper, we construct two families of linear codes over the ring $\mathbb{F}_q + u\mathbb{F}_q$ by the defining set approach, where q is a prime power and $u^2 = 0$. We completely determine their Lee weight distributions, which shows that these codes have few Lee weights. Via the Gray map, we obtain a family of near Griesmer codes over $\mathbb{F}_q$, which is also distance-optimal, and a family of linear codes over $\mathbb{F}_q$, whose optimality is characterized with an explicit computable criterion using the Griesmer bound.

Keywords: Optimal linear code · Few-weight code · Lee weight distribution

1 Introduction

Let $\mathbb{F}_{q^m}$ be the finite field with q^m elements and $\mathbb{F}_{q^m}^* = \mathbb{F}_{q^m}\backslash\{0\}$, where q is a power of a prime p and m is a positive integer. An $[n, k, d]$ linear code $\mathcal{C}$ over $\mathbb{F}_q$ is a k-dimensional subspace of $\mathbb{F}_q^n$ with minimum Hamming distance d. Let A_i denote the number of codewords with Hamming weight i in a code $\mathcal{C}$ of length n. The weight enumerator of $\mathcal{C}$ is defined by $1 + A_1z + A_2z^2 + \cdots + A_nz^n$. The sequence $(1, A_1, A_2, \cdots, A_n)$ is called the weight distribution of $\mathcal{C}$. A code is said to be a t-weight code if the number of nonzero A_i in the sequence $(A_1, A_2, \cdots, A_n)$ is equal to t. Linear codes with few weights have applications in secret sharing schemes [1,3], authentication codes [5,7], association schemes [2], strongly regular graphs and some other fields.

An $[n, k, d]$ linear code $\mathcal{C}$ over $\mathbb{F}_q$ is said to be distance-optimal if no $[n, k, d+1]$ code exists (i.e., this code has the largest minimum distance for given length n and dimension k) and it is called almost distance-optimal if there exists an $[n, k, d + 1]$ distance-optimal code. An $[n, k, d]$ linear code $\mathcal{C}$ is called optimal (resp. almost optimal) if its parameters n, k and d (resp. $d+1$) meet a bound on

This work was supported by the Knowledge Innovation Program of Wuhan-Basic Research under Grant 2022010801010319, the Natural Science Foundation of Hubei Province of China under Grant 2021CFA079 and the National Natural Science Foundation of China under Grant 62072162. National Natural Science Foundation of China under Grant 12001176.

S. Mesnager and Z. Zhou (Eds.): WAIFI 2022, LNCS 13638, pp. 208–220, 2023.
https://doi.org/10.1007/978-3-031-22944-2_13

linear codes with equality [10]. The Griesmer bound [9,16] for an $[n, k, d]$ linear code $\mathcal{C}$ over $\mathbb{F}_q$ is given by

$$n \geq g(k, d) := \sum_{i=0}^{k-1} \lceil \frac{d}{q^i} \rceil,$$

where $\lceil \cdot \rceil$ denotes the ceiling function. An $[n, k, d]$ linear code $\mathcal{C}$ is called a Griesmer code (resp. near Griesmer code) if its parameters n (resp. $n-1$), k and d achieve the Griesmer bound.

In 2007, Ding and Niederreiter [6] introduced a nice and generic way to construct linear codes via trace functions. Let $D \subset \mathbb{F}_{q^m}$ and define

$$\mathcal{C}_D = \{c_a = (\mathrm{Tr}_q^{q^m}(ax))_{x \in D} : a \in \mathbb{F}_{q^m}\},$$

where $\mathrm{Tr}_q^{q^m}(\cdot)$ is the trace function from $\mathbb{F}_{q^m}$ to $\mathbb{F}_q$. Then $\mathcal{C}_D$ is a linear code of length $n := |D|$ over $\mathbb{F}_q$. The code $\mathcal{C}_D$ is called a trace code over $\mathbb{F}_q$ and the set D is called the defining set of $\mathcal{C}_D$. Let R be a finite commutative ring and R_m be an extension of R of degree m. A trace code over R with defining set $L \subset R_m$ is defined by

$$\mathcal{C}_L = \{c_a = (\mathrm{Tr}(ax))_{x \in L} : a \in R_m\} \tag{1}$$

where $\mathrm{Tr}(\cdot)$ is a linear function from R_m to R. Using the construction above, many attempts have been made in this direction to obtain good linear codes over rings, see [12–15,17] for example.

The finite rings of the form $R = \mathbb{F}_q + u\mathbb{F}_q$, $u^2 = 0$, have been used widely as alphabets in certain codes. Let $\mathcal{R} = \mathbb{F}_{q^m} + u\mathbb{F}_{q^m}$ be an extension of R. The following cases of trace codes have been studied in the previous works:

1) When $R = \mathbb{F}_2 + u\mathbb{F}_2$, $u^2 = 0$; $\mathcal{R} = \mathbb{F}_{2^m} + u\mathbb{F}_{2^m}$, $L = \mathbb{F}_{2^m}^* + u\mathbb{F}_{2^m}$, the code $\mathcal{C}_L$ is a 2-weight code with respect to the Lee weight (see [14]).
2) When $R = \mathbb{F}_p + u\mathbb{F}_p$, $u^2 = 0$; $\mathcal{R} = \mathbb{F}_{p^m} + u\mathbb{F}_{p^m}$, $L = \mathcal{Q} + u\mathbb{F}_{p^m}$, where p is an odd prime and $\mathcal{Q}$ is the set of all square elements of $\mathbb{F}_{p^m}^*$, the code $\mathcal{C}_L$ is a 2-weight or 3-weight code with respect to the Lee weight (see [15]).
3) When $R = \mathbb{F}_p + u\mathbb{F}_p$, $u^2 = u$; $\mathcal{R} = \mathbb{F}_{p^m} + u\mathbb{F}_{p^m}$, $L = \mathcal{Q} + u\mathbb{F}_{p^m}^*$ and $L = \mathbb{F}_{p^m}^* + u\mathbb{F}_{p^m}^*$, where p is an odd prime and $\mathcal{Q}$ is the set of all square elements of $\mathbb{F}_{p^m}^*$, the code $\mathcal{C}_L$ is a 2-weight or few-weight code with respect to the Lee weight (see [13]).
4) When $R = \mathbb{F}_q + u\mathbb{F}_q$, $u^2 = 0$; $\mathcal{R} = \mathbb{F}_{q^m} + u\mathbb{F}_{q^m}$, $L = C_0^{(e,r)} + u\mathbb{F}_{q^m}$, where e is a divisor of $q-1$ and $C_0^{(e,r)}$ is the cyclotomic class of order e, the code $\mathcal{C}_L$ is a 2-weight or few-weight code with respect to the Lee weight (see [12]).
5) When $R = \mathbb{F}_2 + u\mathbb{F}_2$, $u^2 = 0$; $\mathcal{R} = \mathbb{F}_2^m + u\mathbb{F}_2^m$, $L = \Delta_1 + u\mathbb{F}_2^m \backslash \Delta_2$ and $L = \mathbb{F}_2^m \backslash \Delta_1 + u\mathbb{F}_2^m \backslash \Delta_2$, where Δ_1 and Δ_2 are simplicial complexes generated by a single maximal element, the code $\mathcal{C}_L$ is a few-weight code with respect to the Lee weight (see [17]).

It should be noted that some optimal linear codes have been obtained from the above constructions.

In this paper, let $R = \mathbb{F}_q + u\mathbb{F}_q$ with $u^2 = 0$ and $\mathcal{R} = \mathbb{F}_{q^m} + u\mathbb{F}_{q^m}$. The objective of this paper is to investigate two families of linear codes $\mathcal{C}_L$ defined by (1) with the following defining sets respectively:

1) $L = L_1 = \mathbb{F}_{q^r} + u(\mathbb{F}_{q^m} \backslash \mathbb{F}_{q^s})$;
2) $L = L_2 = \mathbb{F}_{q^m} \backslash \mathbb{F}_{q^r} + u(\mathbb{F}_{q^m} \backslash \mathbb{F}_{q^s})$,

where $m > 1$, r and s are positive integers satisfying $r|m$ and $s|m$. Note that, to some extent, these two families of linear codes generalize the results of [17] (see Remark 3 for details). Through some detailed calculations on certain exponential sums, we determine the Lee weight distributions of these codes $\mathcal{C}_L$ completely, which shows that $\mathcal{C}_{L_1}$ is 3-weight and $\mathcal{C}_{L_2}$ is 6-weight. Moreover, under the Gray map ϕ, we show that the codes $\phi(\mathcal{C}_{L_1})$ over $\mathbb{F}_q$ are near Griesmer codes and also distance-optimal codes. For the codes $\phi(\mathcal{C}_{L_2})$ over $\mathbb{F}_q$, we characterize the optimality of $\phi(\mathcal{C}_{L_2})$ with an explicit computable criterion and consequently obtain many distance-optimal linear codes over $\mathbb{F}_q$.

2 Preliminaries

In this section, we introduce some basic notation, definitions and lemmas which are needed for the subsequent section. Let q be a power of a prime p and denote the canonical additive character of $\mathbb{F}_q$ by

$$\chi(x) = \zeta_p^{\mathrm{Tr}_p^q(x)},$$

where ζ_p is a primitive complex p-th root of unity and $\mathrm{Tr}_p^q(\cdot)$ is the trace function from $\mathbb{F}_q$ to $\mathbb{F}_p$.

Let $R = \mathbb{F}_q + u\mathbb{F}_q$ with $u^2 = 0$. A linear code $\mathcal{C}$ of length n over R is an R-submodule of R^n. For any $a + ub \in R$ where $a, b \in \mathbb{F}_q$, the Gray map ϕ from R to $\mathbb{F}_q^2$ is defined by

$$\phi : R \to \mathbb{F}_q^2,\ a + ub \mapsto (b, a + b).$$

Any vector $\mathbf{x} \in R^n$ can be written as $\mathbf{x} = \mathbf{a} + u\mathbf{b}$ where $\mathbf{a}, \mathbf{b} \in \mathbb{F}_q^n$. The map ϕ is a bijection, which can be extended naturally from R^n to $\mathbb{F}_q^{2n}$ as follows:

$$\phi : R^n \to \mathbb{F}_q^{2n},\ \mathbf{x} = \mathbf{a} + u\mathbf{b} \mapsto (\mathbf{b}, \mathbf{a} + \mathbf{b}).$$

The Hamming weight $wt(\mathbf{a})$ of a vector $\mathbf{a} \in \mathbb{F}_q^n$ is the number of nonzero coordinates in $\mathbf{a}$. The Lee weight $wt_L(\mathbf{a} + u\mathbf{b})$ of a vector $\mathbf{a} + u\mathbf{b} \in R^n$ is the Hamming weight of its Gray image $\phi(\mathbf{a} + u\mathbf{b})$ as follows:

$$wt_L(\mathbf{a} + u\mathbf{b}) = wt(\mathbf{b}) + wt(\mathbf{a} + \mathbf{b}). \tag{2}$$

The Lee distance of $\mathbf{x}, \mathbf{y} \in R^n$ is defined as $wt_L(\mathbf{x} - \mathbf{y})$. One can check that the Gray map is an isometry from (R^n, d_L) and $(\mathbb{F}_q^{2n}, d_H)$.

Let $\mathcal{R} = \mathbb{F}_{q^m} + u\mathbb{F}_{q^m}$ with $u^2 = 0$. Let F be the Frobenius operator over $\mathcal{R}$ defined by $F(a+ub) = a^q + ub^q$. The trace function $\mathrm{Tr}(\cdot)$ is defined by

$$\mathrm{Tr} = \sum_{i=0}^{m-1} F^i : \mathcal{R} \to R, a+ub \mapsto \sum_{i=0}^{m-1} F^i(a+ub) = \sum_{i=0}^{m-1} (a^{q^i} + ub^{q^i}).$$

By the definition above, it can be readily verified that

$$\mathrm{Tr}(a+ub) = \mathrm{Tr}_q^{q^m}(a) + u\mathrm{Tr}_q^{q^m}(b) \tag{3}$$

where $\mathrm{Tr}_q^{q^m}(\cdot)$ denotes the trace function from $\mathbb{F}_{q^m}$ to $\mathbb{F}_q$.

With the discussion above, we show the following lemma to compute the Lee weight of the trace code $\mathcal{C}_L$ defined by (1) for a general defining set L.

Lemma 1. *Let $L = D_1 + uD_2 = \{a+ub : a \in D_1, b \in D_2\}$ where $D_1, D_2 \subset \mathbb{F}_{q^m}$. Then $\mathcal{C}_L$ defined by (1) is a code of length $|L|$ over R, and for any $a+ub \in \mathcal{R}\backslash\{0\}$, the Lee weight of the codeword c_{a+ub} in $\mathcal{C}_L$ is $wt_L(c_{a+ub}) = 2|L| - \Omega$ where*

$$\Omega = \frac{1}{q} \sum_{v \in \mathbb{F}_q} \sum_{y \in D_2} \chi(v\mathrm{Tr}_q^{q^m}(ay)) \sum_{x \in D_1} (\chi(v\mathrm{Tr}_q^{q^m}(bx)) + \chi(v\mathrm{Tr}_q^{q^m}((a+b)x))).$$

Proof. Observe that the length of $\mathcal{C}_L$ is $|L|$. Let $a+ub \in \mathcal{R}$ and $x+uy \in L$ where $x \in D_1$ and $y \in D_2$. By (3) and (2), for $(a,b) \neq (0,0)$, the Lee weight $wt_L(c_{a+ub})$ of the codeword c_{a+ub} in $\mathcal{C}_L$ is equal to

$$\begin{aligned}
& wt_L((\mathrm{Tr}((a+ub)(x+uy)))_{x \in D_1, y \in D_2}) \\
=\, & wt_L((\mathrm{Tr}_q^{q^m}(ax) + u\mathrm{Tr}_q^{q^m}(ay+bx))_{x \in D_1, y \in D_2}) \\
=\, & wt((\mathrm{Tr}_q^{q^m}(ay+bx))_{x \in D_1, y \in D_2}) + wt((\mathrm{Tr}_q^{q^m}(ay+(a+b)x))_{x \in D_1, y \in D_2}).
\end{aligned}$$

Further, using the orthogonal property of nontrivial additive characters of $\mathbb{F}_q$, for $(a,b) \neq (0,0)$, the Lee weight $wt_L(c_{a+ub})$ can be expressed as

$$wt_L(c_{a+ub}) = 2|L| - \Omega$$

where

$$\begin{aligned}
\Omega =\, & \frac{1}{q} \sum_{x \in D_1} \sum_{y \in D_2} \sum_{v \in \mathbb{F}_q} (\chi(v\mathrm{Tr}_q^{q^m}(ay+bx)) + \chi(v\mathrm{Tr}_q^{q^m}(ay+(a+b)x))) \\
=\, & \frac{1}{q} \sum_{v \in \mathbb{F}_q} \sum_{y \in D_2} \chi(v\mathrm{Tr}_q^{q^m}(ay)) \sum_{x \in D_1} (\chi(v\mathrm{Tr}_q^{q^m}(bx)) + \chi(v\mathrm{Tr}_q^{q^m}((a+b)x))).
\end{aligned}$$

This completes the proof. □

To compute the Lee weight distributions of the codes constructed in this paper, the following lemma will be needed in the sequel.

Lemma 2. *Let $m > 1$, $r < m$ and $s < m$ be positive integers with $r|m$, $s|m$ and* $\gcd(r,s) = t$. *Then for $z \in \mathbb{F}_{q^r}$, we have*

$$|\{a \in \mathbb{F}_{q^m} : \mathrm{Tr}_{q^s}^{q^m}(a) = 0, \mathrm{Tr}_{q^r}^{q^m}(a) = z\}| = \begin{cases} q^{m-r-s+t}, & \text{if } \mathrm{Tr}_{q^t}^{q^r}(z) = 0; \\ 0, & \text{if } \mathrm{Tr}_{q^t}^{q^r}(z) \neq 0. \end{cases}$$

Proof. Using the orthogonal property of nontrivial additive characters of $\mathbb{F}_q$, we have

$$\begin{aligned} &|\{a \in \mathbb{F}_{q^m} : \mathrm{Tr}_{q^s}^{q^m}(a) = 0, \mathrm{Tr}_{q^r}^{q^m}(a) = z\}| \\ &= \frac{1}{q^{r+s}} \sum_{x \in \mathbb{F}_{q^m}} \sum_{w \in \mathbb{F}_{q^s}} \chi(\mathrm{Tr}_q^{q^s}(w \mathrm{Tr}_{q^s}^{q^m}(x))) \sum_{v \in \mathbb{F}_{q^r}} \chi(\mathrm{Tr}_q^{q^r}(v(\mathrm{Tr}_{q^r}^{q^m}(x) - z))) \\ &= \frac{1}{q^{r+s}} \sum_{w \in \mathbb{F}_{q^s}} \sum_{v \in \mathbb{F}_{q^r}} \chi(\mathrm{Tr}_q^{q^r}(-vz)) \sum_{x \in \mathbb{F}_{q^m}} \chi(\mathrm{Tr}_q^{q^m}((w+v)x)) \\ &= q^{m-r-s} \sum_{v \in \mathbb{F}_{q^t}} \chi(\mathrm{Tr}_q^{q^t}(v \mathrm{Tr}_{q^t}^{q^r}(z))) \\ &= \begin{cases} q^{m-r-s+t}, & \text{if } \mathrm{Tr}_{q^t}^{q^r}(z) = 0; \\ 0, & \text{if } \mathrm{Tr}_{q^t}^{q^r}(z) \neq 0. \end{cases} \end{aligned}$$

This completes the proof. □

3 Two Classes of Optimal Linear Codes

In this section, we study the Lee weight distributions of two classes of linear codes over R of the form (1). Under the Gray map, we investigate the optimality of the images of these linear codes over R and consequently we can obtain two classes of optimal linear codes over $\mathbb{F}_q$.

3.1 The First Class of Optimal Linear Codes

In this subsection, we investigate the linear codes $\mathcal{C}_{L_1}$ of the form (1) with the defining set

$$L_1 = \mathbb{F}_{q^r} + u(\mathbb{F}_{q^m} \backslash \mathbb{F}_{q^s}). \tag{4}$$

Theorem 1. *Let $m > 1$, $r < m$ and $s < m$ be positive integers satisfying $r|m$ and $s|m$, and* $\gcd(r,s) = t$. *Then $\mathcal{C}_{L_1}$ defined by* (1) *and* (4) *is a 3-weight code of length $q^r(q^m - q^s)$ and size q^{m+r}, and its Lee weight distribution is given by*

Weight w	Multiplicity A_w
0	1
$2(q-1)q^{m+r-1}$	$(q^{m-r-s+t}-1)$
$2(q-1)q^{r-1}(q^m-q^s)$	$q^{m+r}-q^{m-r-s+t}(2q^{r-t}-1)$
$(q-1)q^{r-1}(2q^m-q^s)$	$2q^{m-r-s+t}(q^{r-t}-1)$

Moreover, the code $\phi(\mathcal{C}_{L_1})$ is a near Griesmer code and it is distance-optimal.

Proof. Clearly, the length of $\mathcal{C}_{L_1}$ is $q^r(q^m-q^s)$. By Lemma 1, for $a+ub \in \mathcal{R}\backslash\{0\}$, the Lee weight of the codeword c_{a+ub} in $\mathcal{C}_{L_1}$ is

$$wt_L(c_{a+ub}) = 2q^r(q^m-q^s) - \Omega$$

where

$$\Omega = \frac{1}{q}\sum_{v\in\mathbb{F}_q}\sum_{y\in\mathbb{F}_{q^m}\backslash\mathbb{F}_{q^s}} \chi(v\mathrm{Tr}_q^{q^m}(ay)) \sum_{x\in\mathbb{F}_{q^r}} (\chi(v\mathrm{Tr}_q^{q^m}(bx)) + \chi(v\mathrm{Tr}_q^{q^m}((a+b)x))).$$

Note that

$$\sum_{y\in\mathbb{F}_{q^m}\backslash\mathbb{F}_{q^s}} \chi(v\mathrm{Tr}_q^{q^m}(ay)) = \sum_{y\in\mathbb{F}_{q^m}} \chi(v\mathrm{Tr}_q^{q^m}(ay)) - \sum_{y\in\mathbb{F}_{q^s}} \chi(v\mathrm{Tr}_q^{q^s}(\mathrm{Tr}_{q^s}^{q^m}(a)y)).$$

To determine the value of Ω, we consider the following three cases.

Case (1): $a = 0$. Then we have

$$\begin{aligned}\Omega &= \frac{2}{q}(q^m-q^s)\sum_{v\in\mathbb{F}_q}\sum_{x\in\mathbb{F}_{q^r}} \chi(v\mathrm{Tr}_q^{q^r}(\mathrm{Tr}_{q^r}^{q^m}(b)x)) \\ &= \begin{cases} 2q^r(q^m-q^s), & \text{if } \mathrm{Tr}_{q^r}^{q^m}(b)=0; \\ 2q^{r-1}(q^m-q^s), & \text{if } \mathrm{Tr}_{q^r}^{q^m}(b)\neq 0. \end{cases}\end{aligned}$$

Thus, for $a = 0$, one gets

$$wt_L(c_{a+ub}) = \begin{cases} 0, & \text{if } \mathrm{Tr}_{q^r}^{q^m}(b)=0; \\ 2(q-1)q^{r-1}(q^m-q^s), & \text{if } \mathrm{Tr}_{q^r}^{q^m}(b)\neq 0. \end{cases}$$

Case (2): $a \neq 0$ and $\mathrm{Tr}_{q^s}^{q^m}(a) = 0$. Then we have

$$\begin{aligned}\Omega =& 2q^{r-1}(q^m-q^s) - q^{s-1}\sum_{v\in\mathbb{F}_q^*}\sum_{x\in\mathbb{F}_{q^r}} (\chi(v\mathrm{Tr}_q^{q^m}(bx)) + \chi(v\mathrm{Tr}_q^{q^m}((a+b)x))) \\ =& \begin{cases} 2q^r(q^{m-1}-q^s), & \text{if } \mathrm{Tr}_{q^r}^{q^m}(b)=\mathrm{Tr}_{q^r}^{q^m}(a+b)=0; \\ 2q^{r-1}(q^m-q^s), & \text{if } \mathrm{Tr}_{q^r}^{q^m}(b)\neq 0 \text{ and } \mathrm{Tr}_{q^r}^{q^m}(a+b)\neq 0; \\ 2q^{m+r-1}-(q+1)q^{r+s-1}, & \text{otherwise.} \end{cases}\end{aligned}$$

Thus, for $a \neq 0$ and $\mathrm{Tr}_{q^s}^{q^m}(a) = 0$, $wt_L(c_{a+ub})$ is equal to

$$\begin{cases} 2(q-1)q^{m+r-1}, & \text{if } \mathrm{Tr}_{q^r}^{q^m}(b) = \mathrm{Tr}_{q^r}^{q^m}(a+b) = 0; \\ 2(q-1)q^{r-1}(q^m - q^s), & \text{if } \mathrm{Tr}_{q^r}^{q^m}(b) \neq 0,\ \mathrm{Tr}_{q^r}^{q^m}(a+b) \neq 0; \\ (q-1)q^{r-1}(2q^m - q^s), & \text{otherwise.} \end{cases} \quad (5)$$

Case (3): $\mathrm{Tr}_{q^s}^{q^m}(a) \neq 0$. Then we have $\Omega = 2q^{r-1}(q^m - q^s)$ which indicates

$$wt_L(c_{a+ub}) = 2(q-1)q^{r-1}(q^m - q^s).$$

With the discussion above, $wt_L(c_{a+ub}) = 0$ if and only if $a = 0$ and $\mathrm{Tr}_{q^r}^{q^m}(b) = 0$, which indicates $A_0 = q^{m-r}$. This shows that the size of $\mathcal{C}_{L_1}$ is q^{m+r}. Moreover, $\mathcal{C}_{L_1}$ has three possible nonzero weights as follows: $w_1 = 2(q-1)q^{m+r-1}$, $w_2 = 2(q-1)q^{r-1}(q^m - q^s)$ and $w_3 = (q-1)q^{r-1}(2q^m - q^s)$.

Using Lemma 2, it follows from (5) that

$$\begin{aligned} A_{w_1} &= |\{(a,b) \in \mathbb{F}_{q^m}^2 : a \neq 0, \mathrm{Tr}_{q^s}^{q^m}(a) = \mathrm{Tr}_{q^r}^{q^m}(b) = \mathrm{Tr}_{q^r}^{q^m}(a+b) = 0\}| \\ &= q^{m-r}|\{a \in \mathbb{F}_{q^m}^* : \mathrm{Tr}_{q^s}^{q^m}(a) = \mathrm{Tr}_{q^r}^{q^m}(a) = 0\}| \\ &= q^{m-r}(q^{m-s-r+t} - 1). \end{aligned}$$

Moreover, we have

$$\begin{aligned} N_1 :&= |\{(a,b) \in \mathbb{F}_{q^m}^2 : a \neq 0, \mathrm{Tr}_{q^s}^{q^m}(a) = 0, \mathrm{Tr}_{q^r}^{q^m}(b) = 0 \text{ and } \mathrm{Tr}_{q^r}^{q^m}(a+b) \neq 0\}| \\ &= |\{(a,b) \in \mathbb{F}_{q^m}^2 : a \neq 0, \mathrm{Tr}_{q^s}^{q^m}(a) = 0, \mathrm{Tr}_{q^r}^{q^m}(b) = 0\}| - A_{w_1} \\ &= q^{2m-2r-s+t}(q^{r-t} - 1) \end{aligned} \quad (6)$$

and by denoting $a + b = -c$, it gives

$$\begin{aligned} N_2 :&= |\{(a,b) \in \mathbb{F}_{q^m}^2 : a \neq 0, \mathrm{Tr}_{q^s}^{q^m}(a) = 0, \mathrm{Tr}_{q^r}^{q^m}(b) \neq 0 \text{ and } \mathrm{Tr}_{q^r}^{q^m}(a+b) = 0\}| \\ &= |\{(a,c) \in \mathbb{F}_{q^m}^2 : a \neq 0, \mathrm{Tr}_{q^s}^{q^m}(a) = 0, \mathrm{Tr}_{q^r}^{q^m}(a+c) \neq 0 \text{ and } \mathrm{Tr}_{q^r}^{q^m}(c) = 0\}| \\ &= N_1. \end{aligned} \quad (7)$$

Thus, one gets $A_{w_3} = 2q^{2m-2r-s+t}(q^{r-t} - 1)$ and consequently $A_{w_2} = q^{2m} - A_{w_0} - A_{w_1} - A_{w_3} = q^{2m} - q^{2m-s-2r+t}(2q^{r-t} - 1)$. Then the Lee weight distribution of $\mathcal{C}_{L_1}$ follows.

Employing the Gray map on the linear code $\mathcal{C}_{L_1}$, we can obtain a linear code $\phi(\mathcal{C}_{L_1})$ over $\mathbb{F}_q$. It can be easily verified that the code $\phi(\mathcal{C}_{L_1})$ has parameters $[2q^r(q^m - q^s), m + r, 2(q-1)q^{r-1}(q^m - q^s)]$. By the Griesmer bound, we have

$$\begin{aligned} g(m+r, d) &= \sum_{i=0}^{m+r-1} \left\lceil \frac{2(q-1)q^{r-1}(q^m - q^s)}{q^i} \right\rceil \\ &= \sum_{i=0}^{s+r-1} 2(q-1)(q^{m+r-i-1} - q^{s+r-i-1}) + \sum_{i=s+r}^{m+r-1} 2(q-1)q^{m+r-i-1} - 1 \\ &= 2q^{m+r} - 2q^{s+r} - 1 \end{aligned}$$

which indicates that $\phi(\mathcal{C}_{L_1})$ is a near Griesmer code. Similarly, we have

$$g(m+r, d+1) = \sum_{i=0}^{m+r-1} \lceil \frac{2(q-1)q^{r-1}(q^m - q^s) + 1}{q^i} \rceil$$
$$= \begin{cases} 2q^{m+r} - 2q^{s+r} + s + r, & \text{if } q = 2; \\ 2q^{m+r} - 2q^{s+r} + s + r - 1, & \text{if } q > 2, \end{cases}$$

which implies that $\phi(\mathcal{C}_{L_1})$ is distance-optimal since $2q^r(q^m - q^s) < g(m, d+1)$. This completes the proof. □

Remark 1. Let $\mathcal{S}$ be the simplex code of dimension $m + r$ over $\mathbb{F}_q$ in the nonprojective case and then the length of $\mathcal{S}$ is $n = q^{m+r} - 1$. Let T be the subset of $\{1, \ldots, n\}$ such that $\mathcal{S}$ punctured on $\{1, \cdots, n\} \backslash T$ is the simplex code of dimension $r + s$ in the nonprojective case. Denote the concatenation of two codes $\mathcal{S}$ punctured on T by $\mathcal{C}$. One can check that $\mathcal{C}$ has the same parameters as $\phi(\mathcal{C}_{L_1})$ in Theorem 1 and $\mathcal{C}$ is a two-weight code. Thus although $\phi(\mathcal{C}_{L_1})$ and $\mathcal{C}$ have the same parameters, they are inequivalent since $\phi(\mathcal{C}_{L_1})$ is a three-weight code.

Example 1. Let $q = 3$, $m = 4$, $r = 2$ and $s = 1$. Magma experiments show that $\phi(\mathcal{C}_{L_1})$ is a $[1404, 6, 936]$ linear code with the weight enumerator $1 + 684z^{936} + 36z^{954} + 8z^{972}$, which is consistent with our result in Theorem 1. This code is a near Griesmer code and it is distance-optimal due to the Griesmer bound.

3.2 The Second Class of Linear Codes

In this subsection, we investigate the linear codes $\mathcal{C}_{L_2}$ of the form (1) with the defining set

$$L_2 = \mathbb{F}_{q^m} \backslash \mathbb{F}_{q^r} + u(\mathbb{F}_{q^m} \backslash \mathbb{F}_{q^s}). \tag{8}$$

Theorem 2. *Let $m > 1$, $r < m$ and $s < m$ be positive integers satisfying $r|m$ and $s|m$, and $t = \gcd(r, s)$. Then $\mathcal{C}_{L_2}$ defined by (1) and (8) is a 6-weight code of length $(q^m - q^r)(q^m - q^s)$ and size q^{2m}, and its Lee weight distribution is given by*

Weight w	Multiplicity A_w
0	1
$2(q-1)q^{m-1}(q^m - q^s)$	$q^{m-r} - 1$
$2(q-1)(q^{m-1} - q^{r-1})(q^m - q^s)$	$q^{2m-2r-s+t}(q^{2r+s-t} - 2q^{r-t} + 1)$
$(q-1)q^{m-1}(2q^m - 2q^r - q^s)$	$2(q^{m-r-s+t} - 1)$
$(q-1)(q^{m-1} - q^{r-1})(2q^m - q^s)$	$2q^{m-r-s+t}(q^{r-t} - 1)$
$2(q-1)q^{m-1}(q^m - q^r - q^s)$	$(q^{m-r-s+t} - 1)(q^{m-r} - 2)$
$(q-1)(2q^m(q^{m-1} - q^{r-1} - q^{s-1}) + q^{r+s-1})$	$2q^{m-r-s+t}(q^{r-t} - 1)(q^{m-r} - 1)$

Moreover, for $q^{m-r} > 2$, *the code* $\phi(\mathcal{C}_{L_2})$ *is distance-optimal if* $m+\min\{r,s\}+\delta > 2q^{r+s}$, *where*

$$\delta = \begin{cases} 1, & \text{if } q = 2; \\ -1, & \text{if } r = s \text{ and } q > 4; \\ 0, & \text{otherwise.} \end{cases}$$

Proof. The length of $\mathcal{C}_{L_2}$ is $(q^m - q^r)(q^m - q^s)$. By Lemma 1, for $a+ub \in \mathcal{R}\backslash\{0\}$, the Lee weight of the codeword c_{a+ub} in $\mathcal{C}_{L_2}$ is

$$wt_L(c_{a+ub}) = 2(q^m - q^r)(q^m - q^s) - \Omega$$

where

$$\Omega = \frac{1}{q} \sum_{v\in\mathbb{F}_q} \sum_{y\in\mathbb{F}_{q^m}\backslash\mathbb{F}_{q^s}} \chi(v\mathrm{Tr}_q^{q^m}(ay)) \sum_{x\in\mathbb{F}_{q^m}\backslash\mathbb{F}_{q^r}} (\chi(v\mathrm{Tr}_q^{q^m}(bx)) + \chi(v\mathrm{Tr}_q^{q^m}((a+b)x))).$$

Note that

$$\Omega_1 := \sum_{y\in\mathbb{F}_{q^m}\backslash\mathbb{F}_{q^s}} \chi(v\mathrm{Tr}_q^{q^m}(ay)) = \sum_{y\in\mathbb{F}_{q^m}} \chi(v\mathrm{Tr}_q^{q^m}(ay)) - \sum_{y\in\mathbb{F}_{q^s}} \chi(v\mathrm{Tr}_q^{q^m}(ay))$$

and

$$\begin{aligned} \Omega_2 :=& \sum_{x\in\mathbb{F}_{q^m}\backslash\mathbb{F}_{q^r}} (\chi(v\mathrm{Tr}_q^{q^m}(bx)) + \chi(v\mathrm{Tr}_q^{q^m}((a+b)x))) \\ =& \sum_{x\in\mathbb{F}_{q^m}} (\chi(v\mathrm{Tr}_q^{q^m}(bx)) + \chi(v\mathrm{Tr}_q^{q^m}((a+b)x))) \\ &- \sum_{x\in\mathbb{F}_{q^r}} (\chi(v\mathrm{Tr}_q^{q^m}(bx)) + \chi(v\mathrm{Tr}_q^{q^m}((a+b)x))). \end{aligned}$$

One can check that $\Omega_1 = q^m - q^s$ and $\Omega_2 = 2(q^m - q^r)$ if $v = 0$. Thus, we only need compute Ω_1 and Ω_2 for the case $v \in \mathbb{F}_q^*$ in the following. To further determine the value of Ω, we consider the following three cases.

Case (1): $a = 0$. Then for $v \in \mathbb{F}_q^*$, we have $\Omega_1 = q^m - q^s$ and

$$\Omega_2 = \begin{cases} 2(q^m - q^r), & \text{if } b = 0; \\ -2q^r, & \text{if } b \neq 0 \text{ and } \mathrm{Tr}_{q^r}^{q^m}(b) = 0; \\ 0, & \text{if } \mathrm{Tr}_{q^r}^{q^m}(b) \neq 0. \end{cases}$$

Thus one gets

$$\Omega = \begin{cases} 2(q^m - q^r)(q^m - q^s), & \text{if } b = 0; \\ 2(q^{m-1} - q^r)(q^m - q^s), & \text{if } b \neq 0 \text{ and } \mathrm{Tr}_{q^r}^{q^m}(b) = 0; \\ 2(q^{m-1} - q^{r-1})(q^m - q^s), & \text{if } \mathrm{Tr}_{q^r}^{q^m}(b) \neq 0 \end{cases}$$

which leads to

$$wt_L(c_{a+ub}) = \begin{cases} 0, & \text{if } b = 0; \\ 2(q-1)q^{m-1}(q^m - q^s), & \text{if } b \neq 0,\ \mathrm{Tr}_{q^r}^{q^m}(b) = 0; \\ 2(q-1)(q^{m-1} - q^{r-1})(q^m - q^s), & \text{if } \mathrm{Tr}_{q^r}^{q^m}(b) \neq 0 \end{cases} \quad (9)$$

Case (2): $a \neq 0$ and $\mathrm{Tr}_{q^s}^{q^m}(a) = 0$. Then we have $\Omega_1 = -q^s$ if $v \in \mathbb{F}_q^*$. To compute Ω_2, we consider the following three cases:
(*i*) For $b = 0$ and $a + b \neq 0$, it gives

$$\Omega_2 = \begin{cases} q^m - 2q^r, & \text{if } \mathrm{Tr}_{q^r}^{q^m}(a+b) = 0; \\ q^m - q^r, & \text{if } \mathrm{Tr}_{q^r}^{q^m}(a+b) \neq 0, \end{cases}$$

which implies

$$wt_L(c_{a+ub}) = \begin{cases} (q-1)q^{m-1}(2q^m - 2q^r - q^s), & \text{if } \mathrm{Tr}_{q^r}^{q^m}(a+b) = 0; \\ (q-1)(q^{m-1} - q^{r-1})(2q^m - q^s), & \text{if } \mathrm{Tr}_{q^r}^{q^m}(a+b) \neq 0. \end{cases}$$

(*ii*) For $b \neq 0$ and $a + b = 0$, it leads to

$$\Omega_2 = \begin{cases} q^m - 2q^r, & \text{if } \mathrm{Tr}_{q^r}^{q^m}(b) = 0; \\ q^m - q^r, & \text{if } \mathrm{Tr}_{q^r}^{q^m}(b) \neq 0, \end{cases}$$

which implies

$$wt_L(c_{a+ub}) = \begin{cases} (q-1)q^{m-1}(2q^m - 2q^r - q^s), & \text{if } \mathrm{Tr}_{q^r}^{q^m}(b) = 0; \\ (q-1)(q^{m-1} - q^{r-1})(2q^m - q^s), & \text{if } \mathrm{Tr}_{q^r}^{q^m}(b) \neq 0. \end{cases}$$

(*iii*) For $b \neq 0$ and $a + b \neq 0$, one gets

$$\Omega_2 = \begin{cases} -2q^r, & \text{if } \mathrm{Tr}_{q^r}^{q^m}(b) = 0 \text{ and } \mathrm{Tr}_{q^r}^{q^m}(a+b) = 0; \\ 0, & \text{if } \mathrm{Tr}_{q^r}^{q^m}(b) \neq 0 \text{ and } \mathrm{Tr}_{q^r}^{q^m}(a+b) \neq 0; \\ -q^r, & \text{otherwise}, \end{cases}$$

which implies

$$wt_L(c_{a+ub}) = \begin{cases} w_5, & \text{if } \mathrm{Tr}_{q^r}^{q^m}(b) = 0 \text{ and } \mathrm{Tr}_{q^r}^{q^m}(a+b) = 0; \\ w_2, & \text{if } \mathrm{Tr}_{q^r}^{q^m}(b) \neq 0 \text{ and } \mathrm{Tr}_{q^r}^{q^m}(a+b) \neq 0; \\ w_6, & \text{otherwise} \end{cases}$$

where $w_2 = 2(q-1)(q^{m-1} - q^{r-1})(q^m - q^s)$, $w_5 = 2(q-1)q^{m-1}(q^m - q^r - q^s)$ and $w_6 = (q-1)(2q^{2m-1} - 2q^{m+r-1} - 2q^{m+s-1} + q^{r+s-1})$.

Case (3): $\mathrm{Tr}_{q^s}^{q^m}(a) \neq 0$. Then we have $\Omega_1 = 0$ for $v \in \mathbb{F}_q^*$ and consequently $\Omega = 2(q^{m-1} - q^{r-1})(q^m - q^s)$. Thus it leads to $wt_L(c_{a+ub}) = 2(q-1)(q^{m-1} - q^{r-1})(q^m - q^s)$.

With the discussion above, $wt_L(c_{a+ub}) = 0$ if and only if $a = b = 0$, which implies that the size of $\mathcal{C}_{L_2}$ is q^{2m}. Moreover, $\mathcal{C}_{L_2}$ has six possible nonzero weights as follows: $w_1 = 2(q-1)q^{m-1}(q^m - q^s)$, $w_2 = 2(q-1)(q^{m-1} - q^{r-1})(q^m - q^s)$, $w_3 = (q-1)q^{m-1}(2q^m - 2q^r - q^s)$, $w_4 = (q-1)(q^{m-1} - q^{r-1})(2q^m - q^s)$, $w_5 = 2(q-1)q^{m-1}(q^m - q^r - q^s)$ and $w_6 = (q-1)(2q^{2m-1} - 2q^{m+r-1} - 2q^{m+s-1} + q^{r+s-1})$. Next, we compute the Lee weight distribution of $\mathcal{C}_{L_2}$.

From (9), it gives $A_{w_1} = |\{b \in \mathbb{F}_{q^m}^* : \mathrm{Tr}_{q^r}^{q^m}(b) = 0\}| = q^{m-r} - 1$. From (*i*) and (*ii*) of Case (2), we have

$$\begin{aligned}
A_{w_3} &= |\{(a,b) \in \mathbb{F}_{q^m}^2 : a \neq 0, b = 0, \mathrm{Tr}_{q^s}^{q^m}(a) = \mathrm{Tr}_{q^r}^{q^m}(a) = 0\}| \\
&\quad + |\{(a,b) \in \mathbb{F}_{q^m}^2 : a \neq 0, b \neq 0, a + b = 0, \mathrm{Tr}_{q^s}^{q^m}(a) = \mathrm{Tr}_{q^r}^{q^m}(b) = 0\}| \\
&= 2|\{a \in \mathbb{F}_{q^m}^* :, \mathrm{Tr}_{q^s}^{q^m}(a) = \mathrm{Tr}_{q^r}^{q^m}(a) = 0\}| \\
&= 2(q^{m-r-s+t} - 1)
\end{aligned}$$

by Lemma 2 and

$$\begin{aligned}
A_{w_4} &= |\{(a,b) \in \mathbb{F}_{q^m}^2 : a \neq 0, b = 0, \mathrm{Tr}_{q^s}^{q^m}(a) = 0\}| \\
&\quad + |\{(a,b) \in \mathbb{F}_{q^m}^2 : ab \neq 0, a + b = 0, \mathrm{Tr}_{q^s}^{q^m}(a) = 0\}| - A_{w_3} \\
&= 2|\{a \in \mathbb{F}_{q^m}^* : \mathrm{Tr}_{q^s}^{q^m}(a) = 0\}| - A_{w_3} \\
&= 2q^{m-r-s+t}(q^{r-t} - 1).
\end{aligned}$$

From (*iii*) of Case (2), it follows that $A_{w_5} = (q^{m-r-s+t} - 1)(q^{m-r} - 2)$ since

$$\begin{aligned}
&|\{(a,b) \in \mathbb{F}_{q^m}^2 : ab \neq 0, a + b \neq 0, \mathrm{Tr}_{q^s}^{q^m}(a) = \mathrm{Tr}_{q^r}^{q^m}(a) = \mathrm{Tr}_{q^r}^{q^m}(b) = 0\}| \\
&= |\{(a,b) \in \mathbb{F}_{q^m}^2 : ab \neq 0, \mathrm{Tr}_{q^s}^{q^m}(a) = \mathrm{Tr}_{q^r}^{q^m}(a) = \mathrm{Tr}_{q^r}^{q^m}(b) = 0\}| \\
&\quad - |\{(a,b) \in \mathbb{F}_{q^m}^2 : ab \neq 0, a + b = 0, \mathrm{Tr}_{q^s}^{q^m}(a) = \mathrm{Tr}_{q^r}^{q^m}(a) = \mathrm{Tr}_{q^r}^{q^m}(b) = 0\}| \\
&= (q^{m-r-s+t} - 1)(q^{m-r} - 1) - |\{a \in \mathbb{F}_{q^m}^* : \mathrm{Tr}_{q^s}^{q^m}(a) = \mathrm{Tr}_{q^r}^{q^m}(a) = 0\}| \\
&= (q^{m-r-s+t} - 1)(q^{m-r} - 2)
\end{aligned}$$

where the last equality holds due to Lemma 2. Moreover, similar to the computation of (7) in Theorem 1, it gives $A_{w_6} = 2q^{m-r-s+t}(q^{r-t} - 1)(q^{m-r} - 1)$ since

$$\begin{aligned}
&|\{(a,b) \in \mathbb{F}_{q^m}^2 : ab \neq 0, a + b \neq 0, \mathrm{Tr}_{q^s}^{q^m}(a) = \mathrm{Tr}_{q^r}^{q^m}(b) = 0, \mathrm{Tr}_{q^r}^{q^m}(a + b) \neq 0\}| \\
&= |\{(a,b) \in \mathbb{F}_{q^m}^2 : ab \neq 0, \mathrm{Tr}_{q^s}^{q^m}(a) = 0, \mathrm{Tr}_{q^r}^{q^m}(a) \neq 0, \mathrm{Tr}_{q^r}^{q^m}(b) = 0\}| \\
&= (|\{a \in \mathbb{F}_{q^m}^* : \mathrm{Tr}_{q^s}^{q^m}(a) = 0\}| - |\{a \in \mathbb{F}_{q^m}^* : \mathrm{Tr}_{q^s}^{q^m}(a) = \mathrm{Tr}_{q^r}^{q^m}(a) = 0\}|) \\
&\quad |\{b \in \mathbb{F}_{q^m} : b \neq 0, \mathrm{Tr}_{q^r}^{q^m}(b) = 0\}| \\
&= (q^{m-s} - q^{m-r-s+t})(q^{m-r} - 1) = q^{m-r-s+t}(q^{r-t} - 1)(q^{m-r} - 1).
\end{aligned}$$

Thus, we have $A_{w_2} = q^{2m} - 1 - A_{w_1} - A_{w_3} - A_{w_4} - A_{w_5} - A_{w_6} = q^{2m-2r-s+t}(q^{2r+s-t} - 2q^{r-t} + 1)$. Then the Lee weight distribution of $\mathcal{C}_{L_2}$ follows.

It's easy to check that the code $\phi(\mathcal{C}_{L_2})$ over $\mathbb{F}_q$ has parameters $[2(q^m - q^r)(q^m - q^s), 2m]$. Note that $w_5 < w_1, w_2, w_3, w_4, w_6$. If $q^{m-r} > 2$, which implies

$A_{w_5} > 0$, then the minimum distance of $\phi(\mathcal{C}_{L_2})$ is $d = 2(q-1)q^{m-1}(q^m - q^r - q^s)$. By the Griesmer bound, we have

$$g(2m, d) = \begin{cases} 2(q^{2m} - q^{m+r} - q^{m+s}) - 1, & \text{if } r = s \text{ and } q \neq 3; \\ 2(q^{2m} - q^{m+r} - q^{m+s}), & \text{otherwise} \end{cases}$$

and $g(2m, d+1) = 2(q^{2m} - q^{m+r} - q^{m+s}) + m + \min\{r, s\} + \delta$. Then $\phi(\mathcal{C}_{L_2})$ is distance-optimal if $m + \min\{r, s\} + \delta > 2q^{r+s}$. This completes the proof. □

Remark 2. If $r = t$ and $q^{m-r} > 2$, the code $\mathcal{C}_{L_2}$ in Theorem 2 reduces to a 4-weight code.

Example 2. Let $q = 2$, $m = 7$, $r = 1$ and $s = 1$. Magma experiments show that $\phi(\mathcal{C}_{L_2})$ is a $[31752, 14, 15872]$ linear code with the weight enumerator $1 + 3906z^{15872} + 12288z^{15876} + 126z^{16000} + 63z^{16128}$, which is consistent with our result in Theorem 2. This code is distance-optimal due to the Griesmer bound.

Remark 3. Let $\mathbb{F}_{2^r}$ be a subfield of $\mathbb{F}_{2^m}$ and $\{\alpha_1, \dots, \alpha_m\}$ be a basis of $\mathbb{F}_{2^m}$ over $\mathbb{F}_2$ where $\{\alpha_1, \dots, \alpha_r\}$ is a basis of $\mathbb{F}_{2^r}$ over $\mathbb{F}_2$. Then $\mathbb{F}_{2^m}$ is isomorphic to $\mathbb{F}_2^m$ under the map

$$\psi : \mathbb{F}_2^m \to \mathbb{F}_{2^m}, (x_1, \dots, x_m) \mapsto \alpha_1 x_1 + \dots + \alpha_m x_m, \qquad \forall x_1, \dots, x_m \in \mathbb{F}_2.$$

Thus, by the definition of simplicial complexes (see [4] and [11]), $\mathbb{F}_{2^r}$ can be viewed as a simplicial complex of $\mathbb{F}_{2^m}$ generated by the maximal element $\{1, \dots, r\}$. Therefore, to some extent, our results in Theorem 1 and Theorem 2 generalize the results of [17] to a general $R = \mathbb{F}_q + u\mathbb{F}_q$, $u^2 = 0$.

4 Conclusions

In this paper, we constructed two families of linear codes $\mathcal{C}_{L_1}$ and $\mathcal{C}_{L_2}$ over $\mathbb{F}_q + u\mathbb{F}_q$, $u^2 = 0$ with defining sets associated with subfields. To some extent, these two families of linear codes generalize the codes in [17] from $\mathbb{F}_2 + u\mathbb{F}_2$, $u^2 = 0$ to a general $\mathbb{F}_q + u\mathbb{F}_q$, $u^2 = 0$. By computing certain exponential sums, we completely determined the Lee weight distributions of these two families of codes, which shows that they have few Lee weights. Moreover, under the Gray map ϕ, we showed that the linear codes $\phi(\mathcal{C}_{L_1})$ over $\mathbb{F}_q$ are near Griesmer codes and also distance-optimal codes, and we characterized the optimality of the linear codes $\phi(\mathcal{C}_{L_2})$ over $\mathbb{F}_q$ with an explicit computable criterion using the Griesmer bound, which allows many distance-optimal linear codes over $\mathbb{F}_q$ to be produced.

References

1. Anderson, R.J., Ding, C., Hellseth, T., Kløve, T.: How to build robust shared control systems. Des. Codes Cryptogr. **15**(2), 111–123 (1998)
2. Calderbank, A.R., Goethals, J.: Three-weight codes and association schemes. Philips J. Res. **39**(4–5), 143–152 (1984)

3. Carlet, C., Ding, C., Yuan, J.: Linear codes from perfect nonlinear mappings and their secret sharing schemes. IEEE Trans. Inf. Theory **51**(6), 2089–2102 (2005)
4. Chang, S., Hyun, J.Y.: Linear codes from simplicial complexes. Des. Codes Cryptogr. **86**, 2167–2181 (2018)
5. Ding, C., Helleseth, T., Kløve, T., Wang, X.: A generic construction of cartesian authentication codes. IEEE Trans. Inf. Theory **53**(6), 2229–2235 (2007)
6. Ding, C., Niederreiter, H.: Cyclotomic linear codes of order 3. IEEE Trans. Inf. Theory **53**(6), 2274–2277 (2007)
7. Ding, C., Wang, X.: A coding theory construction of new systematic authentication codes. Theor. Comput. Sci. **330**(1), 81–99 (2005)
8. Grassl M.: Bounds on the minimum distance of linear codes. Online available at http://www.codetables.de, Accessed on 2022-07-13
9. Griesmer, J.H.: A bound for error correcting codes. IBM J. Res. Dev. **4**, 532–542 (1960)
10. Huffman W., Pless V.: Fundamentals of error-correcting codes. Cambridge University Press (1997)
11. Hyun, J.Y., Lee, J., Lee, Y.: Infinite families of optimal linear codes constructed from simplicial complexes. IEEE Trans. Inf. Theory **66**(11), 6762–6773 (2020)
12. Liu, H., Maouche, Y.: Two or few-weight trace codes over $\mathbb{F}_q + u\mathbb{F}_q$. IEEE Trans. Inf. Theory **65**(5), 2696–2703 (2019)
13. Shi, M., Guan, Y., Solé, P.: Two new families of two-weight codes. IEEE Trans. Inf. Theory **63**(10), 6240–6246 (2017)
14. Shi, M., Liu, Y., Solé, P.: Optimal two-weight codes from trace codes over $\mathbb{F}_2+u\mathbb{F}_2$. IEEE Commun. Lett. **20**(12), 2346–2349 (2016)
15. Shi M., Wu R., Liu Y., Solé P.: Two and three weight codes over $\mathbb{F}_p+u\mathbb{F}_p$. Cryptogr. Commun. **9**, 637–646 (2017)
16. Solomon, G., Stiffler, J.J.: Algebraically punctured cyclic codes. Inform. and Control **8**, 170–179 (1965)
17. Wu, Y., Zhu, X., Yue, Q.: Optimal few-weight codes from simplicial complexes. IEEE Trans. Inf. Theory **66**(6), 3657–3663 (2020)

Explicit Non-malleable Codes from Bipartite Graphs

Shohei Satake[1], Yujie Gu[2(✉)], and Kouichi Sakurai[2]

[1] School of Interdisciplinary Mathematical Sciences, Meiji University, 4-21-1 Nakano, Nakano-ku, Tokyo 164-8525, Japan
shohei_satake@meiji.ac.jp

[2] Faculty of Information Science and Electrical Engineering, Kyushu University, 744 Motooka Nishi-ku, Fukuoka 819-0395, Japan
{gu,sakurai}@inf.kyushu-u.ac.jp

Abstract. Non-malleable codes are introduced to protect the communication against adversarial tampering of data, as a relaxation of the error-correcting codes and error-detecting codes. To explicitly construct non-malleable codes is a central and challenging problem which has drawn considerable attention and been extensively studied in the past few years. Recently, Rasmussen and Sahai built an interesting connection between non-malleable codes and (non-bipartite) expander graphs, which is the first explicit construction of non-malleable codes based on graph theory other than the typically exploited extractors. So far, there is no other graph-based construction for non-malleable codes yet. In this paper, we aim to explore more connections between non-malleable codes and graph theory. Specifically, we first extend the Rasmussen-Sahai construction to bipartite expander graphs. Accordingly, we establish several explicit constructions for non-malleable codes based on Lubotzky-Phillips-Sarnak Ramanujan graphs and generalized quadrangles, respectively. It is shown that the resulting codes can either work for a more flexible split-state model or have better code rate in comparison with the existing results.

Keywords: Non-malleable code · Biregular graph · Expander graph · Split-state model

1 Introduction

Non-malleable codes, introduced by Dziembowski, Pietrzak and Wichs [19,20], are resilient to adversarial tampering on *arbitrary* number of symbols which is beyond the scope of error-correcting and error-detecting codes. Consider the following "tampering experiment". A message $m \in \mathcal{M}$ is encoded via a (randomized) encoding function enc : $\mathcal{M} \to \mathcal{X}$, yielding a codeword $c = \text{enc}(m)$. However the codeword c is modified by an adversary using some tampering function $f \in \mathcal{F}$ with $f : \mathcal{X} \to \mathcal{X}$ to an erroneous word $\tilde{c} = f(c)$, and $\tilde{c}$ is decoded using a deterministic function dec, resulting $\tilde{m} = \text{dec}(\tilde{c})$. In terms of the practical application, the reliability $\tilde{m} = m$ is desired. An error-correcting

S. Mesnager and Z. Zhou (Eds.): WAIFI 2022, LNCS 13638, pp. 221–236, 2023.
https://doi.org/10.1007/978-3-031-22944-2_14

code with minimum distance d can guarantee the reliable communication with respect to the family $\mathcal{F}$ which satisfies that for $f \in \mathcal{F}$ the Hamming distance between $\tilde{c} = f(c)$ and c is at most $\lfloor (d-1)/2 \rfloor$. However it is impossible to achieve the reliability using error-correcting codes if the tampering family $\mathcal{F}$ is large. In order to deal with this, Dziembowski *et al.* [19] proposed the non-malleable codes (with respect to $\mathcal{F}$), which ensure that either the tampered codeword is correctly decoded, i.e., $\tilde{m} = m$, or the decoded message $\tilde{m}$ is completely unrelated to the original message m. As remarked in [19] and [20], the concept of non-malleable codes is in a spirit of non-malleability proposed by Dolev, Dwork and Naor [16] in cryptographic primitives. Informally, the non-malleability in the context of encryption requires that given the ciphertext it is impossible to generate a different ciphertext so that the respective plaintexts are related [16].

It is known that no non-malleable code exists if the tampering family $\mathcal{F}$ is the entire space of functions. Thus the study on non-malleable codes has focused on the specific families $\mathcal{F}$. One typical tampering family is with the *split-state model*, which has also been investigated in the context of leakage cryptography [12,18]. Roughly speaking, this model assumes that the encoded memory/state of the system is partitioned into two parts and adversaries can arbitrarily tamper the data stored in each part independently. More precisely, each message is encoded into a word $c = (L, R) \in \mathcal{L} \times \mathcal{R}$ and adversaries try to tamper it using some functions $g : \mathcal{L} \to \mathcal{L}$ and $h : \mathcal{R} \to \mathcal{R}$ which change c to $\tilde{c} = (g(L), h(R)) \in \mathcal{L} \times \mathcal{R}$. Moreover, if $|\mathcal{L}| = |\mathcal{R}|$, we call it *equally-sized* split-state model.

To explicitly construct non-malleable codes is a fundamental and challenging problem. In the literature, explicit non-malleable codes for the split-state model have been derived based on two-source extractors and additive combinatorics, see [1–7,10,11,17,26,27] for example. Notably, Dziembowski, Kazana and Obremski [17] pointed out: "This brings a natural question if we could show some relationship between the extractors and the non-malleable codes in the split-state model. Unfortunately, there is no obvious way of formalizing the conjecture that non-malleable codes need to be based on extractors". Recently, Rasmussen and Sahai [32] discovered that (non-bipartite) expander graphs could provide non-malleable codes for the split-state model, which in some sense answers Dziembowski-Kazana-Obremski's question in [17]. Inspired by [32], we are interested with exploring more graph-theoretic constructions for split-state non-malleable codes. More precisely, we shall study the following problem.

Problem 1. Based on graph theory, provide explicit constructions of non-malleable codes for the split-state model.

Indeed, Rasmussen and Sahai [32] provided an elegant answer to Problem 1. However we noticed that the construction in [32] cannot be directly transferred to the general split-state model. Inspired by this, we initially extend the construction in [32] to *bipartite graphs*. Specifically, in this paper, we first establish a coding scheme based on bipartite graphs. Then we prove that when the underlying bipartite graph is an (r, s)*-biregular graph* with the second largest eigenvalue μ, our coding scheme provides $O(\frac{\mu^{3/2}}{\sqrt{rs}})$-non-malleable codes for the split-state

Table 1. Explicit graph-based ε-non-malleable codes in this paper and [32]

Ref.	$\|\mathcal{L}\|$	$\|\mathcal{R}\|$	Code rate	Comments
[32, Sec. C]	q^3	q^3	$\frac{1}{24\log_2(1/\varepsilon)+O(1)}$	$q = p^2$, p is a prime
Cor. 2	$\Theta(p^{5/2}\log(p))$	$\Theta(p^{5/2}\log(p))$	$\frac{1}{20\log_2(1/\varepsilon)+O(\log\log(1/\varepsilon))}$	p is an odd prime

Table 2. Code rates of non-malleable codes based on bipartite graphs with n vertices, where q is a prime power

Ref.	#vertices n	$\|\mathcal{L}\|$	$\|\mathcal{R}\|$	Code rate	Minimum rate for n ($\|\mathcal{L}\| = \|\mathcal{R}\| = n/2$)
Theorem 5	$2q^2(q+1)$	$(q+2)q^2$	q^3	$\frac{1}{\log_2(q^6+2q^5)}$	$\frac{1}{\log_2(q^6+2q^5+q^4)}$
Theorem 6	$(q^3+q^2+2)(q^5+1)$	$(q^2+1)(q^5+1)$	$(q^3+1)(q^5+1)$	$\frac{1}{\log_2(q^{15}+O(q^{14}))}$	$\frac{1}{\log_2(q^{16}+O(q^{15}))}$

model which is not necessarily to be equally-sized (see Theorem 3). This can be seen as an extension of the coding scheme in [32] in the sense that we could deduce the codes for equally-sized split-state model in [32] as special cases (see Remark 1). Based on this, we provide several more solutions to Problem 1 by means of Lubotzky-Phillips-Sarnak Ramanujan graphs and generalized quadrangles (see Tables 1 and 2). In particular, the resulting non-malleable codes can either work for more flexible *non-equally-sized* split-state model (see Theorems 5, 6) or have *better* code rate (see Theorem 4, Corollary 2) in comparison with the non-malleable codes in [32, Section C]. In particular, for a given size of graphs, codes for non-equally-sized split-state model in general realize larger code rate than the rate of codes for equally-sized split-state model (see Table 2 and Sect. 4).

The remainder of this paper is organized as follows. Section 2 briefly reviews non-malleable codes and basics in graph theory. Section 3 provides the coding scheme based on bipartite graphs and discusses its non-malleability. Section 4 analyzes the code rate of the established non-malleable codes. Section 5 presents several explicit constructions for non-malleable codes. Section 6 concludes this paper.

2 Preliminaries

In this section we recall the notion of non-malleable codes and some useful basics in graph theory. Throughout this paper, let $x \leftarrow \mathcal{X}$ denote that the random variable x sampled uniformly from a set $\mathcal{X}$. Let $\perp$ denote a special symbol.

For positive-valued functions f and g over $\mathbb{N}$, we say $f = O(g)$ as $n \to \infty$ if there exists a constant $C > 0$ that $f(n) < Cg(n)$ holds for any sufficiently large n. Similarly $f = \Omega(g)$ as $n \to \infty$ if there exists a constant $C > 0$ that $f(n) \geq Cg(n)$ holds for any sufficiently large n. In particular $f = \Theta(g)$ as $n \to \infty$ if $\Omega(g) = f = O(g)$ holds. Also $f = o(g)$ as $n \to \infty$ if $\lim_{n\to\infty} f(n)/g(n) = 0$.

2.1 Non-malleable Codes

Let $\mathcal{M}$ be a set of messages and $\mathcal{X}$ a set of codewords. A *coding scheme* is a pair of functions (enc, dec), where enc : $\mathcal{M} \to \mathcal{X}$ is a randomized encoding function, and *dec* : $\mathcal{X} \to \mathcal{M} \cup \{\perp\}$ is a deterministic decoding function. Assume that for all $m \in \mathcal{M}$,

$$\Pr[\mathrm{dec}(\mathrm{enc}(m)) = m] = 1,$$

where the probability is taken over the randomness of enc.

Let A, B be two random variables over the same set $\mathcal{X}$. Then the *statistical distance* between A and B is defined as

$$\Delta(A, B) := \frac{1}{2} \sum_{x \in \mathcal{X}} \Big| \Pr[A = x] - \Pr[B = x] \Big|.$$

Definition 1 (Split-state non-malleable codes). *In the* split-state model, *assume* $\mathcal{X} = \mathcal{L} \times \mathcal{R}$ *is the product set of sets* $\mathcal{L}$ *and* $\mathcal{R}$. *Let* $\mathcal{F}$ *be a set of functions from* $\mathcal{L} \times \mathcal{R}$ *to itself, where each* $f \in \mathcal{F}$ *can be represented as* $f(L, R) = (g(L), h(R))$ *for all* $(L, R) \in \mathcal{L} \times \mathcal{R}$ *with some* $g : \mathcal{L} \to \mathcal{L}$ *and* $h : \mathcal{R} \to \mathcal{R}$. *Then a coding scheme* (enc, dec) *such that* enc : $\mathcal{M} \to \mathcal{L} \times \mathcal{R}$ *and* dec : $\mathcal{L} \times \mathcal{R} \to \mathcal{M} \cup \{\perp\}$ *is called an* ε*-non-malleable code with respect to* $\mathcal{F}$ *if for every* $f \in \mathcal{F}$, *there exists a probability distribution* D_f *on* $\mathcal{M} \cup \{\mathrm{same}^*, \perp\}$ *such that for every* $m \in \mathcal{M}$, *we have* $\Delta(A_f^m, B_f^m) \leq \varepsilon$, *where for* $m \in \mathcal{M}$ *and* $f \in \mathcal{F}$, *let* A_f^m *and* B_f^m *be events defined as follows.*

$$A_f^m := \left\{ \begin{array}{c} (L, R) \leftarrow \mathrm{enc}(m); \\ \textit{Output } dec(g(L), h(R)) \end{array} \right\},$$

$$B_f^m := \left\{ \begin{array}{c} \tilde{m} \leftarrow D_f; \\ \textit{If } \tilde{m} = \mathrm{same}^* \textit{ output } m \textit{ else output } \tilde{m} \end{array} \right\}.$$

The symbol "same*" means that the decoded message is equal to the original message (tampering corrected). Hereafter, as in [17] and [32], the symbol "$\perp$" from Definition 1 will be dropped since it usually denotes the situation when the decoding function detects tampering and outputs an error message, which is not dealt in this paper. As mentioned in [17], this would be not so problematic for practical applications.

Definition 2 (Code rate). *For a coding scheme* $\mathcal{C}$ *with the set of messages* $\mathcal{M}$ *and the set of codewords* $\mathcal{X}$, *the* code rate $R(\mathcal{C})$ *is defined as*

$$R(\mathcal{C}) := \frac{\log_2 |\mathcal{M}|}{\log_2 |\mathcal{X}|}.$$

In particular if $\mathcal{M} = \{0, 1\}^\kappa$ *and* $\mathcal{X} = \{0, 1\}^n$ *then* $R(\mathcal{C})$ *is the ratio of the message length* κ *and codeword length* n.

This paper focuses on *single-bit* non-malleable codes, i.e., $\mathcal{M} = \{0,1\}$. It is shown in [17] that single-bit non-malleable codes can be formulated as in the following Theorem 1 as well.

Theorem 1 [17]. *Let* (enc, dec) *be a coding scheme with* $\mathrm{enc} : \{0,1\} \to \mathcal{X}$ *and* $\mathrm{dec} : \mathcal{X} \to \{0,1\}$. *Let* $\mathcal{F}$ *be a set of functions from* $\mathcal{X}$ *to itself. Then* (enc, dec) *is an* ε*-non-malleable code with respect to* $\mathcal{F}$ *if and only if it holds for every* $f \in \mathcal{F}$ *that*

$$\frac{1}{2} \sum_{b \in \{0,1\}} \Pr\Big[\mathrm{dec}\big(f(\mathrm{enc}(b))\big) = 1 - b\Big] \leq \frac{1}{2} + \varepsilon$$

where the probability is over the uniform choice of b and the randomness of enc.

2.2 Expander Graphs

Throughout this paper, we assume that all graphs are undirected and simple, i.e., without multiple edges and loops. Let $G = (V, E)$ denote a graph G with vertex set V and edge set E. Let $G = (V_1, V_2, E)$ be a bipartite graph with a partition (V_1, V_2) of vertex set and edge set $E \subset \{\{v_1, v_2\} : v_1 \in V_1, v_2 \in V_2\}$. For convenience, we identify $G = (V_1, V_2, E)$ with an orientation $\overrightarrow{G} = (V_1, V_2, \overrightarrow{E})$ where

$$\overrightarrow{E} = \{(v_1, v_2) : \{v_1, v_2\} \in E\} \subset V_1 \times V_2.$$

We call $\overrightarrow{G}$ the *associated orientation of* G.

We say a vertex has *degree* d if it connects exactly d edges. A graph G is called a *d-regular graph* if every vertex has degree d. A bipartite graph $G = (V_1, V_2, E)$ is called an *(r, s)-biregular graph* if every vertex of V_1 and V_2 has degree r and s, respectively. Clearly, for an (r, s)-biregular graph $G = (V_1, V_2, E)$ and its associate orientation $\overrightarrow{G} = (V_1, V_2, \overrightarrow{E})$, the following equation holds.

$$|E| = |\overrightarrow{E}| = r|V_1| = s|V_2|. \tag{1}$$

Let $G = (V, E)$ be a graph with n vertices. Then the *adjacency matrix* of G, denoted by $A(G)$, is a $|V| \times |V|$ binary matrix such that the (u, w)-entry is 1 if and only if $\{u, w\} \in E$. Clearly, $A(G)$ is a symmetric matrix and thus has exactly n real eigenvalues with multiplicity, denoted by $\lambda_1 \geq \lambda_2 \geq \cdots \geq \lambda_n$.

Lemma 1 (e.g. [9]**).** *Let G be a graph with n vertices.*

1. *If G is d-regular, then* $\lambda_1 = d$ *and* $\lambda_n \geq -d$, *where* $\lambda_n = -d$ *if and only if G is bipartite.*
2. *If G is* (r, s)*-biregular, then* $\lambda_1 = \sqrt{rs}$ *and* $\lambda_n = -\sqrt{rs}$.

By Lemma 1, the largest eigenvalue of a (bi-)regular graph is always determined. However, the second largest eigenvalue typically has rich properties. For a d-regular graph G, denote $\lambda(G) := \max_{2 \leq i \leq n} |\lambda_i|$. For an (r, s)-biregular graph G, denote

$$\mu(G) := \max_{2 \leq i \leq n-1} |\lambda_i|.$$

An (r, s)-biregular graph G is a *μ-spectral expander* if $\mu(G) \leq \mu$. It has the following nice expansion property.

Proposition 1 [37]. *Let $G = (V_1, V_2, E)$ be an (r, s)-biregular graph which is a μ-spectral expander. For a subset $S \subset V_1$, define the neighbour of S as $N(S) := \{u \in V_2 : u$ is adjacent to some vertex in $S\}$, and let $\rho(S) := \frac{|S|}{|V_1|}$. Then for every subset $S \subset V_1$,*

$$\frac{|N(S)|}{|S|} \geq \frac{r^2}{\rho(S)(rs - \mu^2) + \mu^2}.$$

By Proposition 1, it is readily seen that if G is a μ-spectral expander with small μ, then G has a good expansion property and thus we are interested in how $\mu(G)$ can be small.

Lemma 2 [24]. *Suppose that G is a sufficiently large graph. Then the followings hold.*

(1) If G is d-regular, then $\lambda(G) = \Omega(\sqrt{d})$.
(2) If G is (r, s)-biregular, then $\mu(G) = \Omega(\sqrt{r + s})$.

3 Codes from Bipartite Graphs

In this section we provide a bipartite graph based coding scheme and show that it produces non-malleable codes.

3.1 A Coding Scheme

First we propose a coding scheme based on bipartite graphs.

Construction 2. *Let $G = (V_1, V_2, E)$ be a bipartite graph and $\overrightarrow{G} = (V_1, V_2, \overrightarrow{E})$ the associated orientation of G. Then the* associated graph code $\mathcal{C}_G := (\mathrm{enc}_G, \mathrm{dec}_G)$ *consists of the functions*

$$\mathrm{enc}_G : \{0,1\} \to V_1 \times V_2, \quad \mathrm{dec}_G : V_1 \times V_2 \to \{0,1\}$$

such that

$$\mathrm{enc}_G(b) := \begin{cases} (u, w) \leftarrow (V_1 \times V_2) \setminus \overrightarrow{E} & \text{if } b = 0; \\ (u, w) \leftarrow \overrightarrow{E} & \text{if } b = 1, \end{cases}$$

$$\mathrm{dec}_G(v_1, v_2) := \begin{cases} 0 & \text{if } (v_1, v_2) \notin \overrightarrow{E}; \\ 1 & \text{if } (v_1, v_2) \in \overrightarrow{E}. \end{cases}$$

Remark 1. Rasmussen and Sahai [32] designed a coding scheme based on a graph $G = (V, E)$ so that the space of codewords is $V \times V$, but it works only for equally-sized split-state model with $|\mathcal{L}| = |\mathcal{R}| = |V|$. On the other hand, our code can be applied to a more flexible split-state model, i.e. $|\mathcal{L}| = |V_1|$ may not be necessarily equal to $|\mathcal{R}| = |V_2|$. An advantage of such model is discussed in Sect. 4 afterwards.

3.2 Non-malleability

We show that the coding scheme in Construction 2 based on biregular spectral expanders can produce non-malleable codes for the split-state model.

Theorem 3. *Let $G = (V_1, V_2, E)$ be an (r, s)-biregular graph with n vertices which is a μ-spectral expander. Suppose that $r = r(n), s = s(n)$ with $r, s \to \infty$ as $n \to \infty$ (hence $\mu = \mu(n) \to \infty$ by Lemma 2). Assume that $|E| = \Omega\Big(\frac{(rs)^2 \log(rs)}{\mu}\Big)$ $(n \to \infty)$. Let $\mathcal{F}$ be the set of all functions $f = (g, h)$ with $g : V_1 \to V_1$ and $h : V_2 \to V_2$, where $f(v_1, v_2) := (g(v_1), h(v_2))$ for any $(v_1, v_2) \in V_1 \times V_2$. Then the code $\mathcal{C}_G$ is an $O\Big(\frac{\mu^{\frac{3}{2}}}{\sqrt{rs}}\Big)$-non-malleable code with respect to $\mathcal{F}$ as $n \to \infty$.*

The proof of Theorem 3 is referred to Sect. 3.3.

Remark 2. Suppose that G is an (r, s)-biregular graph with $s \geq r$, $s = o(r^2)$ and $\mu(G) = O(\sqrt{s})$. Then Theorem 3 guarantees that the code $\mathcal{C}_G$ is an ε-non-malleable code with $\varepsilon = O(s^{1/4}/r^{1/2})$, where $s^{1/4}/r^{1/2} = o(1)$ by the assumption on r and s. On the other hand, according to Lemma 2, the quantity $\varepsilon = O(s^{1/4}/r^{1/2})$ in Theorem 3 is best possible up to a constant (i.e. the order of the magnitude of ε cannot be reduced in general).

The following corollary follows from Theorem 3 and (1).

Corollary 1. *Let $G = (V_1, V_2, E)$ be a bipartite d-regular graph with $|V_1| = |V_2| = n/2$ which is a μ-spectral expander. Suppose that $n = \Omega\Big(\frac{\log(d) \cdot d^3}{\mu}\Big)$ and $\mathcal{F}$ is as in Theorem 3. Then the code $\mathcal{C}_G$ is an $O\Big(\frac{\mu^{3/2}}{d}\Big)$-non-malleable code with respect to $\mathcal{F}$.*

Remark 3. Corollary 1 actually includes the explicit construction of non-malleable codes by Rasmussen and Sahai [32, Section C]. Indeed, for a finite abelian group X and a subset S of X, the Cayley graph $Cay(X, S)$ is an $|S|$-regular graph with vertex set X in which two vertices x and y are adjacent if and only if $xy^{-1} \in S$. Note that from $Cay(X, S)$, a bipartite $|S|$-regular graph can be easily obtained as follows. Take two disjoint copies X_1 and X_2 of X and construct a bipartite graph so that $x_1 \in X_1$ and $x_2 \in X_2$ are adjacent if and only if $x_1 x_2^{-1} \in S$; such a bipartite regular graph is called a bi-Cayley graph. For a prime p let $\mathbb{F}_p$ denote the p-element field and $q = p^2$. Rasmussen and Sahai [32] constructed an $O(q^{-1/4})$-non-malleable code from a non-bipartite graph $Cay(\mathbb{F}_p^6, S)$ with some $S \subset \mathbb{F}_p^6$ such that $|S| = q$. In terms of Corollary 1, the corresponding bi-Cayley graph provides the same non-malleable code as in [32, Section C].

3.3 Proof of Theorem 3

This subsection aims to prove Theorem 3. We adopt the following notations. Let X, Y be two sets and $f : X \to Y$ be a function. For each $y \in Y$, denote $f^{-1}(y) := \{x \in X : f(x) = y\}$. For a subset $S \subset Y$, denote $f^{-1}(S) := \cup_{s \in S} f^{-1}(s)$. Also let $G = (V_1, V_2, E)$ be an (r, s)-biregular graph with n vertices, $\mu(G) = \mu$ and $\overrightarrow{G} = (V_1, V_2, \overrightarrow{E})$ be the associated orientation of G. Then for any pair of subsets $S \subset V_1$ and $T \subset V_2$, let

$$\begin{aligned} E(S,T) &:= \left|\{(s,t) \in \overrightarrow{E} : s \in S, t \in T\}\right|, \\ D(S,T) &:= \frac{\sqrt{rs}}{\sqrt{|V_1||V_2|}} \cdot |S||T| - E(S,T). \end{aligned} \tag{2}$$

To prove Theorem 3, we shall employ Theorem 1 and the following lemmas.

Lemma 3. *Let $G = (V_1, V_2, E)$ be an (r, s)-biregular graph and $\overrightarrow{G} = (V_1, V_2, \overrightarrow{E})$ the associated orientation of G. For given functions $g : V_1 \to V_1$ and $h : V_2 \to V_2$, define $f : V_1 \times V_2 \to V_1 \times V_2$ such that $f(v_1, v_2) := (g(v_1), h(v_2))$ for any $(v_1, v_2) \in V_1 \times V_2$. Let*

$$T := \frac{1}{2} \sum_{b \in \{0,1\}} \Pr\Big[\mathrm{dec}\big(f(\mathrm{enc}(b))\big) = 1 - b\Big].$$

Then we have

$$T = \frac{1}{2} + \delta \cdot \sum_{(v,w) \in \overrightarrow{E}} D\big(g^{-1}(v), h^{-1}(w)\big)$$

where

$$\delta := \frac{|V_2|}{2r(|V_2| - r)|V_1|} = \frac{|V_1|}{2s(|V_1| - s)|V_2|}.$$

Due to the space limitation, we omit the proof of Lemma 3 here, which can be found in [34, Lemma 21].

Let $f = (g, h) : V_1 \times V_2 \to V_1 \times V_2$ be a given tampering function from $\mathcal{F}$. Recall that for each pair of $1 \le i \ne j \le 2$ and each vertex $v \in V_i$, $N(v) = \{u \in V_j : u, v \text{ are adjacent in } G\}$. Define the following partitions of V_1 and V_2.

$$G^1 := \left\{v \in V_1 : |g^{-1}(v)| > \frac{|V_1|}{rs}\right\}, \quad G^2 := \left\{v \in V_1 : |g^{-1}(v)| \le \frac{|V_1|}{rs}\right\},$$
$$H^1 := \left\{w \in V_2 : |h^{-1}(w)| > \frac{|V_2|}{rs}\right\}, \quad H^2 := \left\{w \in V_2 : |h^{-1}(w)| \le \frac{|V_2|}{rs}\right\}.$$

For $1 \le i, j \le 2$, let

$$R_{i,j} := \delta \cdot \sum_{(v,w) \in \overrightarrow{E} \cap (G^i \times H^j)} D\Big(g^{-1}(v), h^{-1}(w)\Big).$$

It follows from Lemma 3 that

$$T = \frac{1}{2} + \sum_{1 \le i,j \le 2} R_{i,j}. \tag{3}$$

In order to prove Theorem 3, we also need the following Lemmas 4, 5 and 6.

Lemma 4 $(i = 2)$. $R_{2,1} + R_{2,2} = O(\frac{1}{r})$.

Lemma 5 $(i = 1,\ j = 2)$. $R_{1,2} = O(\frac{1}{s})$.

The proofs of Lemma 4 and Lemma 5 can be found in [34, Lemmas 22-23].

Lemma 6 $(i = j = 1)$. $R_{1,1} = O\left(\frac{\mu^{\frac{3}{2}}}{\sqrt{rs}}\right)$.

The proof sketch of Lemma 6 is referred to Sect. 3.4.

Proof of Theorem 3. By Theorem 1 and (3), Theorem 3 immediately follows from Lemmas 4, 5 and 6. □

3.4 Proof Sketch of Lemma 6

In this subsection we sketch the proof of Lemma 6, in which the following lemma is required.

Lemma 7 (Expander mixing lemma, [14,21,22]**).** *Let $G = (V_1, V_2, E)$ be an (r, s)-biregular graph with $\mu(G) = \mu$. Then for any pair of subsets $S \subset V_1$ and $T \subset V_2$, we have*

$$|D(S,T)| \le \mu\sqrt{|S||T|}. \tag{4}$$

Remark 4. The non-malleable codes from [32] used the following fact. Let $G = (V, E)$ be a d-regular (possibly non-bipartite) graph with $\lambda(G) = \lambda$. Then for any pair of subsets $S, T \subset V$,

$$\left|\frac{d}{n}|S||T| - e(S,T)\right| \le \lambda\sqrt{|S||T|}. \tag{5}$$

Here $e(S,T)$ denotes the number of edges between S and T. However, if G is a bipartite graph, the estimation (5) cannot be used to prove the non-malleability for the coding scheme in [32] and the coding scheme in this paper (see Construction 2), since in this case $\lambda(G) = d$ (see Lemma 1), which only implies $O(\sqrt{d})$-non-malleable codes. However we could see from Theorem 3 that using Lemma 7 can produce $o(1)$-non-malleable codes.

The intuitive idea for proving Lemma 6 is to count the number of edges between G^1 and H^1 by using Lemma 7. To that end, we divide G^1 and H^1 into

$$G^1(k) := \left\{v \in G_1 : \frac{|V_1|}{2^{k-1}} \ge |g^{-1}(v)| \ge \frac{|V_1|}{2^k}\right\}\ (1 \le k \le \lceil\log(rs)\rceil),$$

$$H^1(l) := \left\{w \in H_1 : \frac{|V_2|}{2^{l-1}} \ge |h^{-1}(w)| \ge \frac{|V_2|}{2^l}\right\}\ (1 \le l \le \lceil\log(rs)\rceil).$$

For each pair of $1 \le k, l \le \lceil \log_2(rs) \rceil$, let

$$S_{k,l} := \delta \cdot \sum_{(v,w)\in \overrightarrow{E}\cap\left(G^1(k)\times H^1(l)\right)} D\left(g^{-1}(v), h^{-1}(w)\right).$$

Notice that $R_{1,1} = \sum_{1\le k,l\le\lceil\log_2(rs)\rceil} S_{k,l}$. Hence it suffices to prove

$$\sum_{1\le k,l\le\lceil\log_2(rs)\rceil} S_{k,l} = O\left(\frac{\mu^{\frac{3}{2}}}{\sqrt{rs}}\right). \tag{6}$$

To that end, we divide the sum in (6) into two parts, namely, to show

(Case 1) $$\sum_{1\le k\le l\le\lceil\log_2(rs)\rceil} S_{k,l} = O\left(\frac{\mu^{\frac{3}{2}}}{\sqrt{rs}}\right)$$

(Case 2) $$\sum_{1\le l< k\le\lceil\log_2(rs)\rceil} S_{k,l} = O\left(\frac{\mu^{\frac{3}{2}}}{\sqrt{rs}}\right).$$

Indeed, to derive the bound for (Case 1), we need to evaluate $S_{k,l}$ for $1 \le k \le l \le \lceil \log_2(rs) \rceil$. First observe that

$$\delta^{-1} S_{k,l} = \sum_{v\in G^1(k)} D\left(g^{-1}(v), \bigcup_{w\in N(v)\cap H^1(l)} h^{-1}(w)\right).$$

By Lemma 7 we have

$$\begin{aligned}
\delta^{-1} S_{k,l} &\le \sum_{v\in G^1(k)} \mu\sqrt{|g^{-1}(v)| \cdot \sum_{w\in N(v)\cap H^1(l)} |h^{-1}(w)|} \\
&\le \mu\sqrt{\frac{|V_1|}{2^{k-1}}\cdot\frac{|V_2|}{2^{l-1}}} \sum_{v\in G^1(k)} \sqrt{|N(v)\cap H^1(l)|} \\
&\le 2\mu\cdot 2^{-\frac{l+k}{2}} \cdot \sqrt{|V_1||V_2|}\cdot\sqrt{|G^1(k)|}\cdot\sqrt{E\left(G^1(k), H^1(l)\right)}.
\end{aligned}$$

Next by applying Lemma 7 for subsets $G^1(k)$ and $H^1(l)$ we have

$$\begin{aligned}
\delta^{-1} S_{k,l} &\le 2\mu\cdot 2^{-\frac{l+k}{2}} \cdot \sqrt{|V_1||V_2|}\cdot\sqrt{|G^1(k)|} \\
&\quad \cdot \sqrt{\frac{\sqrt{rs}}{\sqrt{|V_1||V_2|}}\cdot|G^1(k)||H^1(l)| + \mu\sqrt{|G^1(k)||H^1(l)|}}.
\end{aligned}$$

Then by means of Jensen's inequality and relation (1), we obtain

$$\begin{aligned}
S_{k,l} &\le O\left(\frac{\mu}{\sqrt{|E|}}\right)\cdot 2^{-\frac{l+k}{2}}\cdot|G^1(k)|\cdot\sqrt{|H^1(l)|} \\
&\quad + O\left(\frac{\mu^{\frac{3}{2}}}{\sqrt{rs}}\right)\cdot 2^{-\frac{l+k}{2}}\cdot\left(|G^1(k)|^3|H^1(l)|\right)^{\frac{1}{4}}.
\end{aligned} \tag{7}$$

Accordingly, the bound in (Case 1) can be obtained by summing up $S_{k,l}$ together with the estimation in (7) as well as the condition of $|E|$ in Theorem 3, and the detailed calculation is referred to [34, Lemma 24]. In addition, the bound in (Case 2) can be established by similar arguments as (Case 1). The complete proof can be found in [34, Lemma 24].

4 Non-equally-Sized Split-State Model

In this section we discuss the code rate of non-malleable codes for *non-equally-sized* split-state model. Recall from Theorem 3 that the non-malleable code $\mathcal{C}_G$ is established for a given integer n and a bipartite graph $G = (V_1, V_2, E)$ with n vertices. The robustness ε of the non-malleable code $\mathcal{C}_G$ relies on the parameters $|V_1|$, $|V_2|$, r, s and μ, which are functions of n. In terms of the code rate, we also have the following interesting observation.

Lemma 8. *Let n be a positive integer. Let $G = (V_1, V_2, E)$ be a bipartite graph with n vertices. Then*

$$R(\mathcal{C}_G) = \frac{1}{\log_2(|V_1||V_2|)}. \tag{8}$$

Moreover assuming $|V_1| \geq |V_2|$ we have

$$R(\mathcal{C}_G) \geq \frac{1}{\log_2(\lceil n/2 \rceil \cdot \lfloor n/2 \rfloor)}, \tag{9}$$

where the equality holds if and only if $|V_1| = \lceil n/2 \rceil$ and $|V_2| = \lfloor n/2 \rfloor$.

Proof. The Eq. (8) immediately follows from the construction of $\mathcal{C}_G$ (Construction 2). Also (9) can be proved by finding the maximum value of $\log_2(|V_1||V_2|)$ under the conditions that $|V_1| + |V_2| = n$ and $1 \leq |V_1|, |V_2| \leq n - 1$. Notice that $|V_2| = n - |V_1|$, then $|V_1||V_2| = |V_1|(n - |V_1|)$ is maximized if and only if $|V_1| = \lceil n/2 \rceil$ and $|V_2| = \lfloor n/2 \rfloor$. Since the function $\log_2(\cdot)$ is monotonically increasing, the inequality (9) follows. □

According to Lemma 8, for any bipartite graph $G = (V_1, V_2, E)$ with n vertices, it is readily seen that the smallest code rate occurs when $|V_1| = |V_2|$ (i.e. equally-sized). In other words, the non-malleable codes derived from Construction 2 can have *better* code rate for *non*-equally-sized split-state model in comparison with equally-sized split-state model in general. Furthermore it is worth noting that the larger the ratio of $|V_1|$ and $|V_2|$ is, the better the code rate is. In addition to the above analysis on code rate, one could also derive code rate according to the robustness parameter ε of the established code in Construction 2 (see Sect. 5.1, for example).

5 Explicit Constructions

In this section, we present explicit non-malleable codes based on specific biregular spectral expanders.

5.1 Via Lubotzky-Phillips-Sarnak Ramanujan Graphs

In this subsection, we construct non-malleable codes based on suitably chosen graphs from known families of bipartite regular graphs. The resulting codes can have better code rate in comparison with the codes in [32] (see also Remark 3).

To show our construction based on Corollary 1, we need the following claim.

Claim 1. *For a given large prime p, there exist explicit (bipartite) $(p+1)$-regular graphs G with $\Theta(p^{5/2}\log(p))$ vertices and $\mu(G) \leq 2\sqrt{p}$.*

Our construction of graphs is based on the following *Ramanujan graphs* due to Lubotzky, Phillips and Sarnak [29], and Margulis [30] (see also [13, Theorem 4.2.2]). Let p, r be two distinct odd primes such that $r > 2\sqrt{p}$ and p is a *quadratic non-residue* modulo r. Then one can explicitly construct a bipartite $(p+1)$-regular graph $X^{p,r}$ with $r(r^2-1)$ vertices and $\mu(X^{p,r}) \leq 2\sqrt{p}$ for every $r > 2\sqrt{p}$. Indeed the graph $X^{p,r}$ is constructed as a Cayley graph $Cay(\mathrm{PGL}_2(\mathbb{F}_r), S)$ with some explicit generating set $S \subset \mathrm{PGL}_2(\mathbb{F}_r)$ of size $p+1$, where $\mathrm{PGL}_2(\mathbb{F}_r)$ denotes the projective general linear group of rank 2 over the r-element field $\mathbb{F}_r$. The details of the construction can be found in [13]. Note that for each prime p, one could check whether two given vertices are adjacent in $X^{p,r}$ in $O(\log(r))$-time (e.g. [31]), and hence the graph can be constructed in poly(r)-time.

Proof of Claim 1. To prove Claim 1, it suffices to take the graph $X^{p,r}$ with $r = \Theta(p^{5/6}\log^{1/3}(p))$. Indeed for each sufficiently large prime p, by Bertrand's postulate, there exists a prime $r = \Theta(p^{5/6}\log^{1/3}(p)) > 2\sqrt{p}$, which can be found in poly(p)-time. If p is a quadratic non-residue modulo r, then $X^{p,r}$ is a bipartite $(p+1)$-regular graph with $\Theta(p^{5/2}\log(p))$ vertices and $\mu(X^{p,r}) \leq 2\sqrt{p}$. □

Thus we obtain the following theorem.

Theorem 4. *For any sufficiently large prime p, suppose that $r = \Theta(p^{5/2}\log(p))$ is a prime such that p is a quadratic non-residue modulo r. Then the code C_G with $G = X^{p,r}$ is an $O(p^{-1/4})$-non-malleable code for the split-state model with $|\mathcal{L}| = |\mathcal{R}| = \Theta(p^{5/2}\log(p))$.*

Note that Theorem 4 cannot deal with the case when $r = \Theta(p^{5/2}\log(p))$ is a prime such that p is a *quadratic residue* modulo r. However, in this case, one can instead explicitly construct a *non-bipartite* $(p+1)$-regular graph $Y^{p,r}$ with $r(r^2-1)/2$ vertices and $\lambda(Y^{p,r}) \leq 2\sqrt{p}$ (see [13,29,30]). By [32, Theorem 7], the graph $Y^{p,r}$ with $r = \Theta(p^{5/2}\log(p))$ provides an $O(p^{-1/4})$-non-malleable code for the split-state model with $|\mathcal{L}| = |\mathcal{R}| = \Theta(p^{5/2}\log(p))$. Moreover, for each pair of primes p and r, one could check whether two given vertices are adjacent or not in $X^{p,r}$ and $Y^{p,r}$ in $O(\log(p))$-time. By these facts and Theorem 4, we immediately obtain the following corollary.

Corollary 2. *For any sufficiently large prime p, there exists an explicit $(p+1)$-regular graph G with $\Theta(p^{5/2}\log(p))$ vertices which provides an $O(p^{-1/4})$-non-malleable code for the split-state model with $|\mathcal{L}| = |\mathcal{R}| = \Theta(p^{5/2}\log(p))$. In* *particular, for every* $0 < \varepsilon < 1$*, there exists an explicit* ε*-non-malleable code with code rate*

$$\frac{1}{20\log_2(1/\varepsilon) + O(\log\log(1/\varepsilon))}.$$

Moreover, both of encoding and decoding can be done in $O(\log(1/\varepsilon))$*-time.*

The last statement of Corollary 2 directly follows from the discussion in [32, Section 1.3]. Note that the explicit codes derived in [32] (see also Remark 3) have code rate $1/(24\log_2(1/\varepsilon) + O(1))$ while encoding and decoding time is $O(\log(1/\varepsilon))$. In other words, the resulting codes here can have better code rate in comparison with the codes in [32], while encoding and decoding time are the same (up to constant).

5.2 Via Generalized Quadrangles

In this subsection, we provide split-state non-malleable codes based on generalized quadrangles. The code rates of these codes can also be found in Table 2 of Sect. 1.

A *generalized quadrangle* of order (α, β) is an $(\alpha+1, \beta+1)$-biregular graph $GQ(\alpha, \beta) = (V_1, V_2, E)$ such that

- for all $x, y \in V_1 \cup V_2$, there exists a path of length ≤ 4 connecting x and y;
- for all $x, y \in V_1 \cup V_2$, if the length of the shortest path connecting x and y is $h < 4$, then there exists only one path of length h connecting x and y;
- for every $x \in V_1 \cup V_2$, there exists $y \in V_1 \cup V_2$ such that there exists a path of length 4 connecting x and y.

More details of generalized quadrangles can be found in [33,38]. Now based on generalized quadrangles, we can derive two families of non-malleable codes for the non-equally-sized scenario.

Theorem 5. *For any prime power* q*, the code* $\mathcal{C}_G$ *with* $G = GQ(q-1, q+1)$ *is an* $O(q^{-1/4})$*-non-malleable code for the split-state model with* $|\mathcal{L}| = (q+2)q^2$ *and* $|\mathcal{R}| = q^3$*.*

We need the following lemma to prove Theorem 5.

Lemma 9 [33,37,38]. *For the graph* $GQ(\alpha, \beta)$*, we have*

- $|V_1| = (\alpha+1)(\alpha\beta+1)$,
- $|V_2| = (\beta+1)(\alpha\beta+1)$,
- $\mu(GQ(\alpha, \beta)) = \sqrt{\alpha+\beta}$.

Proof of Theorem 5. To obtain the theorem, we apply an explicit construction of $GQ(q-1, q+1)$ for every prime power q ([8, Sections 4 and 5]). According to (1) and Lemma 9, we have $|E| = r|V_1| = \Theta(q^4)$ and $\frac{(rs)^2\log(rs)}{\mu} = \Theta(q^{7/2}\log q)$. Thus by Theorem 3, the code $\mathcal{C}_G$ with $G = GQ(q-1, q+1)$ gives the desired code. □

The following theorem can deal with more unbalanced non-equally-sized scenario.

Theorem 6. *For any prime power q, the code $\mathcal{C}_G$ with $G = GQ(q^2, q^3)$ is an $O(q^{-1/4})$-non-malleable code for the split-state model with $|\mathcal{L}| = (q^2+1)(q^5+1)$ and $|\mathcal{R}| = (q^3+1)(q^5+1)$.*

Proof. For every prime power q, there is an explicit construction of $GQ(q^2, q^3)$ (e.g. [33, Chapter 3]). By (1) and Lemma 9, we have $|E| = r|V_1| = \Theta(q^{10})$ and $\frac{(rs)^2 \log(rs)}{\mu} = \Theta(q^{17/2} \log q)$. Thus according to Theorem 3, $(\mathrm{enc}_G, \mathrm{dec}_G)$ with $G = GQ(q^2, q^3)$ gives the desired code. □

6 Concluding Remarks

In this paper, we proposed a coding scheme based on bipartite graphs and showed that the non-malleability can be satisfied if the underlying bipartite graph is a biregular μ-spectral expander with sufficiently small μ. Based on it, we provided explicit non-malleable codes via several types of biregular spectral expanders such as Ramanujan graphs and generalized quadrangles. The established non-malleable codes can either work for a more flexible split-state model or have better code rate in comparison with the existing results.

In addition, our results show that some related error-correcting codes have potential applications to constructing non-malleable codes for the split-state model. For example, it is well-known in coding theory and combinatorics that a low-density parity-check (LDPC) code has an associated bipartite graph called *Tanner graph*. Precisely, the Tanner graph of an LDPC code with parity-check matrix $H = (h_{ij})$ is a bipartite graph such that the vertex set is the index set of rows and columns of H, and two vertices i and j are adjacent if and only if $h_{ij} \neq 0$, see [36]. It is shown that the algebraic or combinatorial constructions of LDPC codes often provide Tanner graphs with small second largest eigenvalue, see [15,23,25,28,35] for example. According to the bipartite graph based coding scheme proposed in this paper, a connection between LDPC codes and non-malleable codes can be accordingly established. Particularly, the constructions in Theorems 5 and 6 are based on several typical bipartite graphs realized as Tanner graphs of LDPC codes.

In terms of practical applications, it is desirable to construct split-state non-malleable codes for k-bit messages with $k \geq 2$. As far as we know, there is no known graph-theoretic constructions of split-state non-malleable codes for $k \geq 2$. It would be of interest to generalize the graph-based codes in this paper and [32] for k-bit messages in the split-state model (see also [39, Section 2.1.3]).

Acknowledgements. The authors are grateful to Mr. Peter Rasmussen and Prof. Amit Sahai for their helpful comments to an earlier version of this paper. S. Satake has been supported by JSPS Grant-in-Aid for JSPS Fellows (Grant No. 20J00469) and JST ACT-X (Grant No. JPMJAX2109). Y. Gu has been supported by JSPS Grant-in-Aid for Early-Career Scientists (Grant No. 21K13830).

References

1. Aggarwal, D., Agrawal, S., Gupta, D., Maji, H.K., Pandey, O., Prabhakaran, M.: Optimal computational split-state non-malleable codes. In: Proceedings of Thirteenth IACR Theory of Cryptography Conference (TCC 2016-A), pp. 393–417 (2016)
2. Aggarwal, D., Briët, J.: Revisiting the Sanders-Bogolyubov-Ruzsa theorem in $\mathbb{F}_p^n$ and its application to non-malleable codes. In: Proceedings of 2016 IEEE International Symposium on Information Theory (ISIT), pp. 1322–1326 (2016)
3. Aggarwal, D., Dodis, Y., Lovett, S.: Non-malleable codes from additive combinatorics. SIAM J. Comput. **47**(2), 524–546 (2018)
4. Aggarwal, D., Dodis, Y., Kazana, T., Obremski, M.: Non-malleable reductions and applications. In: Proceedings of 47th Annual Symposium on the Theory of Computing (STOC 2015), pp. 459–468 (2015)
5. Aggarwal, D., Obremski, M.: Inception makes non-malleable codes shorter as well! Cryptology ePrint Archive, Report 2019/399 (2019)
6. Aggarwal, D. Obremski, M.: A constant rate non-malleable code in the split-state model. In: Proceedings of IEEE 61st Annual Symposium on Foundations of Computer Science (FOCS 2020), pp. 1285–1294 (2020)
7. Aggarwal, D., Obremski, M., Ribeiro, J.L., Simkin, M., Siniscalchi, L.: Computational and information-theoretic two-source (non-malleable) extractors. Cryptology ePrint Archive, Report 2020/259 (2020)
8. Ahrens, R.W., Szekeres, G.: On a combinatorial generalization of 27 lines associated with a cubic surface. J. Aust. Math. Soc. **10**(3–4), 485–492 (1969)
9. Brouwer, A.E., Haemers, W.H.: Spectra of Graphs. Springer, New York (2012). https://doi.org/10.1007/978-1-4614-1939-6
10. Chattopadhyay, E., Goyal, V., Li, X.: Non-malleable extractors and codes, with their many tampered extensions. In: Proceedings of 48th Annual Symposium on the Theory of Computing (STOC 2016), pp. 285–298 (2016)
11. Chattopadhyay, E., Zuckerman, D.: Non-malleable codes against constant split-state tampering. In: 55th Annual Symposium on Foundations of Computer Science (FOCS 2014), pp. 306–315 (2014)
12. Davì, F., Dziembowski, S., Venturi, D.: Leakage-resilient storage. In: Garay, J.A., De Prisco, R. (eds.) SCN 2010. LNCS, vol. 6280, pp. 121–137. Springer, Heidelberg (2010). https://doi.org/10.1007/978-3-642-15317-4_9
13. Davidoff, G., Sarnak, P., Valette, A.: Elementary Number Theory, Group Theory, and Ramanujan Graphs. Cambridge University Press, Cambridge (2003)
14. De Winter, S., Schillewaert, J., Verstraete, J.: Large incidence-free sets in geometries. Electron. J. Comb. **19**(4), #P24 (2012)
15. Diao, Q., Li, J., Lin, S., Blake, I.F.: New classes of partial geometries and their associated LDPC codes. IEEE Trans. Inf. Theory **62**(6), 2947–2965 (2016)
16. Dolev, D., Dwork, C., Naor, M.: Non-malleable cryptography. SIAM J. Comput. **30**(2), 391–437 (2000)
17. Dziembowski, S., Kazana, T., Obremski, M.: Non-malleable codes from two-source extractors. In: Proceedings of 33rd Annual Cryptology Conference (CRYPTO 2013), pp. 239–257 (2013)
18. Dziembowski, S., Pietrzak, K.: Leakage-resilient cryptography. In: Proceedings of 49th Annual IEEE Symposium on Foundations of Computer Science (FOCS 2008), pp. 293–302 (2008)

19. Dziembowski, S., Pietrzak, K., Wichs, D.: Non-malleable codes. In: Proceedings of Innovations in Computer Science (ICS 2010), pp. 434–452 (2010)
20. Dziembowski, S., Pietrzak, K., Wichs, D.: Non-malleable codes. J. ACM **65**(4), 20:1–20:32 (2018)
21. Haemers, W.: Eigenvalue techniques in design and graph theory. Ph.D. thesis, Eindhoven University of Technology (1979)
22. Haemers, W.: Interlacing eigenvalues and graphs. Linear Algebra Appl. **226**(228), 593–616 (1995)
23. Høholdt, T., Janwa, H.: Eigenvalues and expansion of bipartite graphs. Des. Codes Cryptogr. **65**(3), 259–273 (2012)
24. Li, W.-C.W., Solé, P.: Spectra of regular graphs and hypergraphs and orthogonal polynomials. Eur. J. Comb. **17**(5), 461–477 (1996)
25. Li, W.-C.W., Lu, M., Wang, C.: Recent developments in low-density parity-check codes. In: Chee, Y.M., Li, C., Ling, S., Wang, H., Xing, C. (eds.) IWCC 2009. LNCS, vol. 5557, pp. 107–123. Springer, Heidelberg (2009). https://doi.org/10.1007/978-3-642-01877-0_11
26. Li, X.: Improved non-malleable extractors, non-malleable codes and independent source extractors. In: Proceedings of 49th Annual ACM Symposium on the Theory of Computing (STOC 2017), pp. 1144–1156 (2017)
27. Li, X.: Non-malleable extractors and non-malleable codes: partially optimal constructions. Cryptology ePrint Archive, Report 2018/353 (2018)
28. Liu, Z., Pados, D.A.: LDPC codes from generalized polygons. IEEE Trans. Inform. Theory **51**(11), 3890–3898 (2005)
29. Lubotzky, A., Phillips, R., Sarnak, P.: Ramanujan graphs. Combinatorica **8**(3), 261–277 (1988)
30. Margulis, G.A.: Explicit group-theoretic constructions of combinatorial schemes and their applications in the construction of expanders and concentrators. Probl. Inform. Transm. **24**(1), 39–46 (1988)
31. Mohanty, S., O'Donnell, R., Paredes, P.: Explicit near-Ramanujan graphs of every degree. In: Proceedings of 52nd Annual ACM Symposium on Theory of Computing (STOC 2020), pp. 510–523 (2020)
32. Rasmussen, P.M.R., Sahai, A.: Expander graphs are non-malleable codes. In: Proceedings of Information-Theoretic Cryptography (ITC 2020), pp. 6:1–6:10 (2020)
33. Payne, S.E., Thas, J.A.: Finite Generalized Quadrangles. Pitman (Advanced Publishing Program), Boston (1984)
34. Satake, S. Gu, Y., Sakurai, K.: Graph-based construction for non-malleable codes, Cryptology ePrint Archive: Report 2021/164 (2021)
35. Sin, P., Sorci, J., Xiang, Q.: Linear representations of finite geometries and associated LDPC codes. J. Comb. Theory Ser. A. **173**(1), 105238 (2020)
36. Tanner, R.M.: A recursive approach to low complexity codes. IEEE Trans. Inform. Theory **27**(5), 533–547 (1981)
37. Tanner, R.M.: Explicit concentrators from generalized N-gons. SIAM J. Algebraic Discrete Methods **5**(3), 287–293 (1984)
38. van Maldeghem, H.: Generalized Polygons. MBirkhäuser Verlag, Basel (1998)
39. Wang, M.: On the efficiency of cryptographic constructions. Ph.D thesis, Purdue University (2021)

Cryptography

Algebraic Relation of Three MinRank Algebraic Modelings

Hao Guo[1] and Jintai Ding[2,3](✉)

[1] Tsinghua University, Beijing, China
guoh22@mails.tsinghua.edu.cn

[2] Ding Lab, Yanqi Lake Beijing Institute of Mathematical Sciences and Applications, Beijing, China
jintai.ding@gmail.com

[3] Yau Mathematical Sciences Center, Tsinghua University, Beijing, China

Abstract. We give algebraic relations among equations of three algebraic modelings for MinRank problem: support minors modeling, Kipnis–Shamir modeling and minors modeling.

Keywords: MinRank problem · Quadratic equation · Algebraic modeling

1 Introduction

In 2020, Bardet et al. introduced the support minors modeling [1] for solving MinRank problem, whose applications include the novel attacks on GeMSS [20] and Rainbow [2]. The powerful attacks make us wonder whether some new algebraic structures exist in support minors modeling and help it reduce the complexity of MinRank. In this paper we explore the algebraic relation between this modeling and other two modelings, namely minors modeling [11] and Kipnis–Shamir modeling [17].

The MinRank problem asks for a nonzero linear combination of given matrices with low rank. It has been used to attack some NIST-PQC candidates, for example ROLLO, RQC, GeMSS and Rainbow. In rank-metric-based code (for example ROLLO and RQC [1]) and rank syndrome problem [14] it is natural to consider MinRank problem since metric is defined by matrix rank. In multivariate cryptography, traditional ways to design a cryptography system and make trapdoors include two ways: using BigField structure [8,18,19] and using properties of BigField to build trapdoors; using UOV structure [7,16] and assigning vinegar variables to build trapdoors. Some of these trapdoors include special restrictions which can be detected by matrix rank, for example in HFE [18] the degree restriction of univariate polynomial and in Rainbow [7] the multi-layer oil and vinegar variable structure, therefore MinRank problem can be used to attack these schemes. On the other hand, since Buss et al. proved that MinRank problem is generally NP-hard [4], there exists some zero-knowledge scheme based on MinRank, for example [6].

S. Mesnager and Z. Zhou (Eds.): WAIFI 2022, LNCS 13638, pp. 239–249, 2023.
https://doi.org/10.1007/978-3-031-22944-2_15

There are many ways to solve the MinRank problem, including minors modeling, Kipnis–Shamir modeling and linear algebra search [15]. Besides these basic ideas, Wang et al. [22] also considered the hybrid method that combines Kipnis–Shamir modeling and minors modeling. Moreover, previous works also concern the complexity of MinRank. Faugere et al. focused on the case of underdetermined and well-determined cases [11–13] and proved that minors modeling is better than Kipnis–Shamir modeling by a little. For the over-determined case, Verbel et al. considered the case of the so-called 'superdetermined' case for Kipnis–Shamir modeling [21] which uses Jacobian of the matrix to induce equations.

Most of these modelings and analyses above fall into the step of calculating Gröbner basis [3] for the ideal corresponding to equations, which is the conceptual generalization of Gaussian Elimination and Euclid's greatest common factor. Efficient algorithm for solving Gröbner basis are F_4 [9] and its variant F_5 [10]. Meanwhile, support minors modeling does not require Gröbner basis computation and turns to XL-like methods [5,23] which has its full power when the number of equations is more than that of variables.

In this paper we focus on the quadratic equations given by Kipnis–Shamir modeling and support minors modeling. We found that by substituting c_T variables in equations of the support minors modeling with determinant-like polynomials in $y_{i,j}$ variables from equations of the Kipnis–Shamir modeling, all former equations become linear combination of latter equations with coefficients in the polynomial ring of $y_{i,j}$ variables. As a byproduct, we offer a constructive proof of the fact that the equations of the minors modeling come from linear combination of that of Kipnis–Shamir modeling with coefficients in the polynomial ring of $y_{i,j}$'s and linear variables x_k's.

2 Notation

We list some useful notations for the following statements and proofs:

- $\mathcal{I}$ (calligraphic font) stands for some index set with $r+1$ elements corresponding to either rows or columns. The row (column) number always starts from 1.
- $\{i_1 < \cdots < i_l \le r < i_{l+1} < \cdots < i_{r+1}\}$ stands for $\{i_1, \ldots, i_{r+1}\}$, with orders in the set specified as $i_1 < \cdots < i_l \le r < i_{l+1} < \cdots < i_{r+1}$.
- For matrix A, $A_{\mathcal{I},\mathcal{J}}$ stands for submatrix of A with rows $\mathcal{I}$ and columns $\mathcal{J}$.
- For a $m \times n$ matrix A, $\mathcal{I} \subset \{1, \ldots, m\}$ and $\mathcal{J} \subset \{1, \ldots, n\}$, if $|\mathcal{I}| = |\mathcal{J}|$, $|A|_{\mathcal{I},\mathcal{J}}$ stands for the minor of A with rows $\mathcal{I}$ and $\mathcal{J}$.
- For a $m \times n$ matrix A and $|\mathcal{J}| = m$, $A_{*,\mathcal{J}}$ is acronym for $A_{\{1,\ldots,m\},\mathcal{J}}$, and $|A|_{*,\mathcal{J}}$ is acronym for $|A|_{\{1,\ldots,m\},\mathcal{J}}$.
- T (letter 'T') stands for some index set subset of $\{1, \ldots, n\}$ with r elements.
- c_T represents $|C|_{*,T}$, where C is the coefficient matrix in support minors modeling.
- For matrix A, A^t stands for transpose of A.

3 Preliminaries

MinRank Problem. We give the statement of the MinRank problem:

Definition 1 (MinRank problem). *Fix a field $\mathbb{K}$. We denote $\mathbb{K}^{m\times n}$ as the vector space of matrices with m rows and n columns and entries in $\mathbb{K}$.*

Given matrices $M_1,\ldots,M_l \in \mathbb{K}^{m\times n}$ and a target rank r, the MinRank problem asks for elements $x_1,\ldots,x_l \in \mathbb{K}$ that are not all zero, such that the linear combination $M = \sum_{k=1}^{l} x_k M_k$ has rank no more than r.

Notice that sometimes the solution is restricted to some subfield $L \subset \mathbb{K}$ (for example in some BigField schemes). However, in this paper we only consider the case that solution takes value in $\mathbb{K}$.

Algebraic Modelings for Solving MinRank Problem. Below we describe three algebraic modelings for MinRank problem.

Minors Modeling. The matrix M has rank $\leq r$ iff all its $r+1$ minors are zero. Minors modeling simply uses these minor conditions as equations. There are $\binom{m}{r+1}\binom{n}{r+1}$ minors in matrix M, and they are all $r+1$ degree polynomials in the variables $x_1,\ldots,x_l$, since each entry of M is a linear form of these variables.

If we denote $M = (a_{i,j})$, then each $a_{i,j}$ can be written as

$$a_{i,j} = \sum_{k=1}^{l} a_{i,j}^{(k)} x_k \tag{1}$$

where $a_{i,j}^{(k)}$ is the (i,j)-th element of M_k. Each $(r+1)$-minor of M is a homogeneous polynomial of degree $r+1$ in the $a_{i,j}$'s, so when substituting $a_{i,j}$ with x_k's, we get a homogeneous polynomial of degree $r+1$ in the x_k's. If we make these polynomials equal to zero we get the corresponding equations of minors modeling.

Kipnis–Shamir Modeling. We recall the following rank–nullity theorem from linear algebra:

Lemma 1. *For a linear map $A\colon \mathbb{K}^n \to \mathbb{K}^m$, we have*

$$\dim A(\mathbb{K}^n) + \dim\ker(A) = n$$

Since the matrix M has rank $\leq r$, the dimension of the kernel of M is no less than $n-r$, hence it must contain a $(n-r)$-dim subspace. So there exists a full-rank matrix $Y \in \mathbb{K}^{n\times(n-r)}$ such that $MY = 0$. Notice further that for any invertible matrix $R \in \mathrm{GL}_{n-r}(\mathbb{K})$, we have $M(YR) = (MY)R = 0$, and YR also has full rank, so we can restrict some entries of Y and still expect a solution. Therefore, we solve the following matrix equation

$$M\begin{bmatrix} -Y' \\ I_{n-r} \end{bmatrix} = 0 \tag{2}$$

where I_{n-r} is the $(n-r) \times (n-r)$ identity matrix, and

$$Y' = \begin{bmatrix} y_{1,1} & \cdots & y_{1,n-r} \\ \vdots & & \vdots \\ y_{r,1} & \cdots & y_{r,n-r} \end{bmatrix} \tag{3}$$

is a $r \times (n-r)$ matrix. If (2) has a solution, then the rank of M must be less than r.

From (2) we can get $m(n-r)$ equations, each of the form

$$f_{i,j} = a_{i,r+j} - \sum_{k=1}^{r} a_{i,k} y_{k,j} = 0 \tag{4}$$

for $i = 1, \ldots, m$, $j = 1, \ldots, n-r$. If we plug in (1), we get quadratic equations with no square terms and the equations are linear in x_k's. Total number of variables is $p + r(n-r)$.

Support Minors Modeling. We recall the following rank decomposition theorem from linear algebra:

Lemma 2. *A $m \times n$ matrix M has rank $\leq r$ iff there exists a $m \times r$ matrix S and a $r \times n$ matrix C, such that $M = SC$.*

Since the rank of M is no more than r, we can find some matrices S and C such that $\sum_{k=1}^{p} x_k M_k = M = SC$. While we cannot make both S and C full rank (otherwise we know M is of rank r), we can assure that C has full rank by expanding the row space of M into a r-dim vector space and solve for entries of S. Since C has full rank, we know that each row $\mathbf{r}_i$ is in the row space of C, so the augmented matrix

$$C_i = \begin{bmatrix} \mathbf{r}_i \\ C \end{bmatrix}$$

has rank r. Therefore the maximal minors of C_i should be zero. If we denote c_T for the maximal minors of C with columns T, then using Laplace expansion of determinant, each maximal minor of C_i is a bilinear form in a_{ij} and c_T. By evaluating these maximal minors to be zero, we get $m\binom{n}{r+1}$ quadratic equations

$$|C_i|_{*,\mathcal{J}} = 0 \tag{5}$$

for $i = 1, \ldots, m$ and all subset $\mathcal{J} \subset \{1, \ldots, n\}$ with $r+1$ elements. If we plug in (1), we get equations bilinear in x_k's and c_T's. Total number of variables is $p + \binom{n}{r}$.

4 Main Results and Proofs

4.1 Relation Between Kipnis–Shamir Modeling and Support Minors Modeling

We will adopt the following matrix

$$C' = [I_r \; Y']$$

where Y' is the $r \times (n-r)$ matrix defined by (3). The core idea of this subsection is to make substitution $\phi\colon c_T \mapsto |C'|_{*,T}$ in support minors modeling. This is the same as replacing the coefficient matrix C from support minors modeling with C'.

The reason we consider matrix C' comes from cryptographical situations. In practical use of MinRank problem, the target rank r is often the smallest rank that $\sum_{k=1}^{p} x_k M_k$ can attain besides zero. In this case the rank decomposition $M = SC$ tells us that row space of C is the same as that of the M. Therefore from (2) we also get

$$C \begin{bmatrix} -Y' \\ I_{n-r} \end{bmatrix} = 0 \tag{2'}$$

We claim that

Lemma 3. *The reduced row echelon form of C is C'.*

Proof. Denote C'' to be the reduced row echelon form of C. Since C is full row rank, all rows in C'' have pivot elements. Since C'' is in reduced row echelon form, the r-th row of C'' must begin with $r-1$ zeros. Therefore it suffices to show that the (r,r)-th element of C'' is 1 instead of 0.

Assume instead that the r-th row of C'' begin with r zeros, then this row has the shape of

$$\begin{bmatrix} 0 \cdots 0 \; z_1 \cdots z_{n-r} \end{bmatrix}$$

for some $z_1, \ldots, z_{n-r} \in \mathbb{K}$. Using (2'), we get that

$$0 = \begin{bmatrix} 0 \cdots 0 \end{bmatrix} (-Y') + \begin{bmatrix} z_1 \cdots z_{n-r} \end{bmatrix} I_{n-r} = \begin{bmatrix} z_1 \cdots z_{n-r} \end{bmatrix}$$

So the r-th row of C'' is a zero row, which contradicts the fact that C has full rank. Therefore the (r,r)-th element of C'' is 1, and we get C'' has the shape of C'.

Since C' is the reduced row echelon form of C in cryptographical situations, it suffices to replace C with C' and use this to relate the support minors modeling and Kipnis–Shamir modeling. Notice that in general the row space of M is only contained in that of C, therefore (2') cannot be derived from (2).

Denote

$$C_i' = \begin{bmatrix} \mathbf{r}_i \\ C' \end{bmatrix} \tag{6}$$

the augmented matrix C_i with block C replaced by C'.

Some properties of ϕ are listed below:

Lemma 4. $\phi(c_{\{1,\ldots,r\}}) = 1$.

Lemma 5. $\phi(c_{\{1,\ldots,r\}\setminus\{i\}\cup\{r+j\}}) = (-1)^{r-i} y_{i,j}$.

Proof. Direct calculation. We have

$$|C'|_{*,\{1,\ldots,r\}\setminus\{i\}\cup\{r+j\}} = \begin{vmatrix} I_{i-1} & 0_{(i-1)\times(r-i)} & *_{(i-1)\times 1} \\ 0_{1\times(i-1)} & 0_{1\times(r-i)} & y_{ij} \\ 0_{(r-i)\times(i-1)} & I_{r-i} & *_{(r-i)\times 1} \end{vmatrix} = (-1)^{r-i} y_{i,j}$$

Maximal Minors of C_i'. To calculate maximal minors of C_i' and relate this with f_{ij} from Kipnis–Shamir modeling (see (4)), we consider the following matrix

$$L_i = \begin{bmatrix} 1 & -a_{i,1} \cdots -a_{i,r} \\ 0_{r\times 1} & I_r \end{bmatrix} \tag{7}$$

L_i is invertible matrix and has determinant 1. Also, when calculating maximal minors of L_iC_i', we have

$$|L_iC_i'|_{*,\mathcal{J}} = (\det L_i)|C_i'|_{*,\mathcal{J}} = |C_i'|_{*,\mathcal{J}} \tag{8}$$

since the determinant function is multiplicative. Therefore it suffices to consider the matrix L_iC_i'. Denote

$$L_iC_i' = \begin{bmatrix} Q_1 & Q_2 \\ Q_3 & Q_4 \end{bmatrix}$$

where Q_1 is $1 \times r$ matrix, Q_4 is $r \times (n-r)$ matrix, and the shape of Q_2 and Q_3 follows from the block matrix rules. We have $Q_3 = I_rI_r = I_r$, $Q_4 = I_rY' = Y'$. Also,

$$Q_1 = \begin{bmatrix} a_{i,1} \cdots a_{i,r} \end{bmatrix} + \begin{bmatrix} -a_{i,1} \cdots -a_{i,r} \end{bmatrix} I_r = 0$$

$$Q_2 = \begin{bmatrix} a_{i,r+1} \cdots a_{i,n} \end{bmatrix} + \begin{bmatrix} -a_{i,1} \cdots -a_{i,r} \end{bmatrix} \begin{bmatrix} y_{1,1} & \cdots & y_{1,n-r} \\ \vdots & & \vdots \\ y_{r,1} & \cdots & y_{r,n-r} \end{bmatrix} = \begin{bmatrix} f_{i,1} \cdots f_{i,n-r} \end{bmatrix}$$

So

$$L_iC_i' = \begin{bmatrix} 0_{1\times r} & f_{i,1} \cdots f_{i,n-r} \\ I_r & Y' \end{bmatrix} \tag{9}$$

From (9) and (8) we know that $|C_i'|_{*,\{1,\ldots,r\}\cup\{r+j\}} = (-1)^r f_{i,j}$. Therefore after applying substitution ϕ, equations of Kipnis–Shamir modeling can be viewed as a subset of equations of support minors modeling (up to a constant of -1). In general, we have

Proposition 1. *Suppose $\mathcal{J} = \{j_1 < \cdots < j_l \leq r < j_{l+1} < \cdots < j_{r+1}\}$, then*

$$|C_i'|_{*,\mathcal{J}} = \sum_{k=l+1}^{r+1} (-1)^{k-1} f_{i,j_k-r} |C'|_{*,\mathcal{J}\setminus\{j_k\}} \tag{10}$$

Proof. Simply use Laplace expansion.

Notice that $|C'|_{*,\mathcal{J}\setminus\{j_k\}}$ is maximal minor of C', which in turn is polynomial in $y_{i,j}$'s. So we know that $|C_i'|_{*,\mathcal{J}}$ is a linear combination of $f_{i,j}$'s with coefficients in $\mathbb{K}[y_{1,1},\ldots,y_{r,n-r}]$.

4.2 Solution Space of c_T's from Support Minors Modeling

We know that c_T is denoted to be the maximal minor of C with columns T. However when c_T becomes variables of equations, it becomes not so clear if any solution of equations from support minors modeling still has the meaning that corresponding M is of rank $\leq r$. It is intuitive that if M has rank $\leq r$, then we can expand the row space of M into a r dimensional space, and get a matrix A of r rows and n columns, whose maximal minors is a nonzero solution of (5) since A has full rank. Also, if M has rank $< r$, then different ways of expanding the row space of M will possibly give linear independent solutions for (5). In particular, we are interested in the following questions:

1. For some specific choice of x_k's such that M has rank $> r$, is the solution space of c_T's the zero space?
2. For some specific choice of x_k's such that M has rank r, is the solution space of c_T's dimension 1?
3. For some specific choice of x_k's such that M has rank $< r$, what can we say about the solution space of c_T's?

Nonetheless, we give the following proposition:

Proposition 2. *Suppose for some specific choice of x_k's, the rank of M is r'. Then the solution space of c_T has dimension $\binom{n-r'}{n-r}$. In particular, when $r' > r$ the only solution for c_T's is zero solution.*

Proof. We know that the equations (5) come from augmenting matrix C with a row of M and calculating the $r+1$ minors. In general, we can also augment C with b rows of M to get a b-augmented matrix

$$\begin{bmatrix} \mathbf{r}_{i_1} \\ \vdots \\ \mathbf{r}_{i_b} \\ C \end{bmatrix}$$

where $1 \leq i_1, \ldots, i_b \leq m$, and calculate its $r+b$ minors. Since all rows of M are in the row space of C, all these $r+b$ minors are zero as long as $r+b \leq n$. Using Laplace expansion along the first row we get a linear combination of $r+b-1$ minors of $(b-1)$-augmented matrix. Therefore the equations (5) are not linearly independent.

Since we know that M has rank r', it suffices to use these r' independent rows to generate augmented matrices. There are $\binom{r'+b-1}{b}$ different ways to b-augment the matrix C. Using some knowledge of syzygy, it is easy to deduce the number of independent equations as

$$\sum_{b=1}^{n-r}(-1)^{b-1}\binom{r'+b-1}{b}\binom{n}{r+b} \tag{11}$$

Lemma 6. *We have the following combinatorial identity:*

$$\sum_{b=0}^{n-r}(-1)^{b}\binom{r'+b-1}{b}\binom{n}{r+b}=\binom{n-r'}{n-r} \tag{12}$$

Proof. Denote

$$G(r',r,n)=\sum_{b=0}^{n-r}(-1)^{b}\binom{r'+b-1}{b}\binom{n}{r+b} \tag{13}$$

Since $\binom{n}{r+b}=\binom{n-1}{r+b}+\binom{n-1}{r-1+b}$, we have $G(r',r,n)=G(r',r,n-1)+G(r',r-1,n-1)$. Also $G(r',r,r)=(-1)^0\binom{r'-1}{0}\binom{r}{r}=1$. So it suffices to prove that $G(r',r',n)=1$.

Denote $F(r',n)=G(r',r',n)$. Notice that

$$\binom{n}{r'+b}=\binom{n-1}{r'+b}+\binom{n-1}{r'-1+b}$$
$$\binom{r'+b-1}{b}=\binom{r'-1+b-1}{b}+\binom{r'-1+b-1}{b-1}$$

Therefore $F(r',n)=F_1+F_2+F_3$, where

$$F_1=\sum_{b=0}^{n-1-r'}(-1)^{b}\binom{r'+b-1}{b}\binom{n-1}{r'+b}=F(r',n-1)$$
$$F_2=\sum_{b=0}^{n-r'}(-1)^{b}\binom{r'-1+b-1}{b}\binom{n-1}{r'-1+b}=F(r'-1,n-1)$$
$$F_3=\sum_{b=1}^{n-r'}(-1)^{b}\binom{r'+b-1-1}{b-1}\binom{n-1}{r'+b-1}=-F(r',n-1)$$

So $F(r',n)=F(r'-1,n-1)$, hence

$$F(r',n)=F(1,n-r'+1)=\sum_{b=0}^{n-r'}(-1)^{b}\binom{n-r'+1}{b+1}=1$$

i.e. $G(r',r',n)=1$. Therefore $G(r',r,n)=\binom{n-r'}{r-r'}=\binom{n-r'}{n-r}$.

Therefore

$$\sum_{b=1}^{n-r}(-1)^{b-1}\binom{r'+b-1}{b}\binom{n}{r+b}=\binom{n}{r}-\binom{n-r'}{n-r}$$

Since we have $\binom{n}{r}$ variables c_T, the solution space dimension is $\binom{n-r'}{n-r}$. Therefore when $r' > r$ the binomial coefficient takes the value 0. This ends the proof of Proposition 2.

Notice that when $r' < r$, we know that the solution space of c_T's has dimension more than 1. Therefore if we do not make the target rank r' optimal, then original equations from support minors modeling have more than 1 dimension of solutions, which means XL-like algorithms cannot make out a solution as [1] said.

4.3 Relation Between Kipnis–Shamir Modeling and Minors Modeling

We will adopt the following matrix:

$$M' = \begin{bmatrix} a_{1,1} & \cdots & a_{1,r} & f_{1,1} & \cdots & f_{1,n-r} \\ \vdots & & \vdots & \vdots & & \vdots \\ a_{m,1} & \cdots & a_{m,r} & f_{r,1} & \cdots & f_{r,n-r} \end{bmatrix} \tag{14}$$

Since only r columns of M' are of the form $a_{i,j}$, if we calculate $r+1$ minors of M', at least one column is of the form $f_{i,j}$, so all $r+1$ minors lie in the ideal of $\mathbb{K}[\{x_k\},\{y_{i,j}\}]$ generated by $f_{i,j}$'s. Notice that from (4), M' and M are related by the matrix equation

$$M = M'R \tag{15}$$

where

$$R = \begin{bmatrix} I_r & Y' \\ 0 & I_{n-r} \end{bmatrix}. \tag{16}$$

Using Cauchy–Binet formula, we can calculate $r+1$ minors of M:

$$|M|_{\mathcal{I},\mathcal{J}} = \sum_{\mathcal{K}} |M'|_{\mathcal{I},\mathcal{K}} |R|_{\mathcal{K},\mathcal{J}} \tag{17}$$

where $\mathcal{K}$ takes value of each $r+1$ subset of $\{1,\ldots,n\}$. Since all $|M'|_{\mathcal{I},\mathcal{K}}$'s lie in the ideal generated by $f_{i,j}$'s, so does $|M|_{\mathcal{I},\mathcal{J}}$.

5 Conclusion and Discussion

We discussed the quadratic equations from Kipnis–Shamir modeling and support minors modeling, and give the proof that they can be derived from each other. We also give proof that from equations of Kipnis–Shamir modeling we can get the minors equations. Heuristically, the equations derived from support minors modeling can be viewed as an application of bilinear XL on those from Kipnis–Shamir modeling with bi-degree (b, r). This helps us make sure that support minors modeling contains no new algebraic structures from Kipnis–Shamir modeling. However, these calculations above are from the viewpoint of commutative algebra (symbolic calculation) and cannot explain why supports minors modeling has major improvement from other modelings. We believe that the efficiency of support minors modeling comes from the way it solves equations since it contains no additional Gröbner basis calculation.

Acknowledgements. This work has been supported by the National Key R&D Program of China (No. 2021YFB3100100).

References

1. Bardet, M., et al.: Improvements of algebraic attacks for solving the rank decoding and MinRank problems. In: Moriai, S., Wang, H. (eds.) ASIACRYPT 2020. LNCS, vol. 12491, pp. 507–536. Springer, Cham (2020). https://doi.org/10.1007/978-3-030-64837-4_17
2. Beullens, W.: Improved cryptanalysis of UOV and rainbow. In: Canteaut, A., Standaert, F.-X. (eds.) EUROCRYPT 2021. LNCS, vol. 12696, pp. 348–373. Springer, Cham (2021). https://doi.org/10.1007/978-3-030-77870-5_13
3. Buchberger, B.: A theoretical basis for the reduction of polynomials to canonical forms. ACM SIGSAM Bull. **10**(3), 19–29 (1976)
4. Buss, J.F., Frandsen, G.S., Shallit, J.O.: The computational complexity of some problems of linear algebra. J. Comput. Syst. Sci. **58**(3), 572–596 (1999)
5. Courtois, N., Klimov, A., Patarin, J., Shamir, A.: Efficient algorithms for solving overdefined systems of multivariate polynomial equations. In: Preneel, B. (ed.) EUROCRYPT 2000. LNCS, vol. 1807, pp. 392–407. Springer, Heidelberg (2000). https://doi.org/10.1007/3-540-45539-6_27
6. Courtois, N.T.: Efficient zero-knowledge authentication based on a linear algebra problem MinRank. In: Boyd, C. (ed.) ASIACRYPT 2001. LNCS, vol. 2248, pp. 402–421. Springer, Heidelberg (2001). https://doi.org/10.1007/3-540-45682-1_24
7. Ding, J., Schmidt, D.: Rainbow, a new multivariable polynomial signature scheme. In: Ioannidis, J., Keromytis, A., Yung, M. (eds.) ACNS 2005. LNCS, vol. 3531, pp. 164–175. Springer, Heidelberg (2005). https://doi.org/10.1007/11496137_12
8. Ding, J., Yang, B.-Y.: Multivariates polynomials for hashing. In: Pei, D., Yung, M., Lin, D., Wu, C. (eds.) Inscrypt 2007. LNCS, vol. 4990, pp. 358–371. Springer, Heidelberg (2008). https://doi.org/10.1007/978-3-540-79499-8_28
9. Faugere, J.C.: A new efficient algorithm for computing gröbner bases (f4). J. Pure Appl. Algebra **139**(1–3), 61–88 (1999)
10. Faugere, J.C.: A new efficient algorithm for computing gröbner bases without reduction to zero (f 5). In: Proceedings of the 2002 International Symposium on Symbolic and Algebraic Computation, pp. 75–83 (2002)
11. Faugere, J.C., El Din, M.S., Spaenlehauer, P.J.: Computing loci of rank defects of linear matrices using gröbner bases and applications to cryptology. In: Proceedings of the 2010 International Symposium on Symbolic and Algebraic Computation, pp. 257–264 (2010)
12. Faugere, J.C., El Din, M.S., Spaenlehauer, P.J.: On the complexity of the generalized minrank problem. J. Symb. Comput. **55**, 30–58 (2013)

13. Faugère, J.-C., Levy-dit-Vehel, F., Perret, L.: Cryptanalysis of MinRank. In: Wagner, D. (ed.) CRYPTO 2008. LNCS, vol. 5157, pp. 280–296. Springer, Heidelberg (2008). https://doi.org/10.1007/978-3-540-85174-5_16
14. Gaborit, P., Ruatta, O., Schrek, J.: On the complexity of the rank syndrome decoding problem. IEEE Trans. Inf. Theory **62**(2), 1006–1019 (2015)
15. Goubin, L., Courtois, N.T.: Cryptanalysis of the TTM cryptosystem. In: Okamoto, T. (ed.) ASIACRYPT 2000. LNCS, vol. 1976, pp. 44–57. Springer, Heidelberg (2000). https://doi.org/10.1007/3-540-44448-3_4
16. Kipnis, A., Patarin, J., Goubin, L.: Unbalanced oil and vinegar signature schemes. In: Stern, J. (ed.) EUROCRYPT 1999. LNCS, vol. 1592, pp. 206–222. Springer, Heidelberg (1999). https://doi.org/10.1007/3-540-48910-X_15
17. Kipnis, A., Shamir, A.: Cryptanalysis of the HFE Public Key Cryptosystem by Relinearization. In: Wiener, M. (ed.) CRYPTO 1999. LNCS, vol. 1666, pp. 19–30. Springer, Heidelberg (1999). https://doi.org/10.1007/3-540-48405-1_2
18. Patarin, J.: Hidden Fields Equations (HFE) and Isomorphisms of Polynomials (IP): two new families of asymmetric algorithms. In: Maurer, U. (ed.) EUROCRYPT 1996. LNCS, vol. 1070, pp. 33–48. Springer, Heidelberg (1996). https://doi.org/10.1007/3-540-68339-9_4
19. Szepieniec, A., Ding, J., Preneel, B.: Extension field cancellation: a new central trapdoor for multivariate quadratic systems. In: Takagi, T. (ed.) PQCrypto 2016. LNCS, vol. 9606, pp. 182–196. Springer, Cham (2016). https://doi.org/10.1007/978-3-319-29360-8_12
20. Tao, C., Petzoldt, A., Ding, J.: Efficient key recovery for All HFE signature variants. In: Malkin, T., Peikert, C. (eds.) CRYPTO 2021. LNCS, vol. 12825, pp. 70–93. Springer, Cham (2021). https://doi.org/10.1007/978-3-030-84242-0_4
21. Verbel, J., Baena, J., Cabarcas, D., Perlner, R., Smith-Tone, D.: On the complexity of "uperdetermined" minrank instances. In: International Conference on Post-Quantum Cryptography, pp. 167–186. Springer (2019)
22. Wang, Y., Ikematsu, Y., Nakamura, S., Takagi, T.: Revisiting the Minrank problem on multivariate cryptography. In: You, I. (ed.) WISA 2020. LNCS, vol. 12583, pp. 291–307. Springer, Cham (2020). https://doi.org/10.1007/978-3-030-65299-9_22
23. Yang, B.-Y., Chen, J.-M., Courtois, N.T.: On asymptotic security estimates in XL and Gröbner bases-related algebraic cryptanalysis. In: Lopez, J., Qing, S., Okamoto, E. (eds.) ICICS 2004. LNCS, vol. 3269, pp. 401–413. Springer, Heidelberg (2004). https://doi.org/10.1007/978-3-540-30191-2_31

Decomposition of Dillon's APN Permutation with Efficient Hardware Implementation

José L. Imaña[2](✉), Lilya Budaghyan[1], and Nikolay Kaleyski[1]

[1] Selmer Center, Department of Informatics, University of Bergen, 5020 Bergen, Norway
{Lilya.Budaghyan,Nikolay.Kaleyski}@uib.no

[2] Department of Computer Architecture and Automation, Faculty of Physics, Complutense University, 28040 Madrid, Spain
jluimana@ucm.es

Abstract. Modern block ciphers incorporate a vectorial Boolean function (*S-box*) as their only nonlinear component. Almost Perfect Nonlinear (APN) functions exhibit optimal resistance to differential cryptanalysis and thus present ideal security properties as *S-boxes*. These optimal cryptographic properties have the side effect of making the function harder to represent and implement. As the number of variables of the function grows, lookup-table representations become less feasible, and so from a practical point of view, it is crucial to develop a good understanding of how cryptographically strong functions can be represented in hardware.

This paper focuses on one of the most important APN functions, namely Dillon's permutation in dimension 6. This is the only known APN permutation in an even number of variables. It is thus an ideal candidate for studying the efficiency of different representations since it combines at least two very important cryptographic properties, and since the number of variables is not large enough to make its computational investigation intractable. In this paper, we give a new description of Dillon's permutation as a composition of two functions and compare it with its classic univariate polynomial representation. We give hardware architectures for both representations, and we report on the results obtained from their FPGA implementations. From the experimental results, the implementation of the new decomposed Dillon's permutation presents reductions in the number of 2-input XOR gates of up to 27.3% and in the Area × Delay metrics of up to 27.4% with respect to the implementation of the corresponding univariate representation. Therefore, the new decomposed Dillon's permutation representation is more efficient than the univariate polynomial one when reconfigurable devices are used for the hardware implementation. This indicates that by representing APN functions as a compo-

The work of J.L. Imaña was supported by Ministerio de Economía y Competitividad (MINECO) under Grant RTI2018-093684-B-I00 and by the Comunidad de Madrid under Grant S2018/TCS-4423. The work of L. Budaghyan was supported by Trond Mohn Foundation under Grant "Construction of Optimal Boolean Functions" and by the Research Council of Norway under Grant 314395.

S. Mesnager and Z. Zhou (Eds.): WAIFI 2022, LNCS 13638, pp. 250–268, 2023.
https://doi.org/10.1007/978-3-031-22944-2_16

sition of simpler functions, significant reductions in the complexity of the implementation can be achieved.

Keywords: Almost perfect nonlinear (APN) · Boolean functions · Block cipher · *S-box* · finite field · hardware implementation

1 Introduction

Given two positive integers n and m, vectorial Boolean functions, or (n, m)-functions, i.e. functions from the vector space $\mathbb{F}_2^n$ to $\mathbb{F}_2^m$, are used as the only non-linear component of virtually all modern block ciphers. In this way, the cipher's security directly depends on the properties of the underlying functions, so it is necessary to use functions that possess specific cryptographic properties that indicate resistance to various types of cryptanalytic attacks. One of the most powerful cryptographic attacks against block ciphers known today is differential cryptanalysis [1]. The differential uniformity of a function measures its resistance against this kind of attack: the lower the differential uniformity, the better the resistance. Typically, we consider (n, m)-functions with $n = m$; in this case, the functions with the lowest possible differential uniformity are called almost perfect nonlinear (APN) functions and have been a subject of intense study in recent years.

Certain cipher designs, such as substitution-permutation networks (SPN), require vectorial Boolean functions that are bijective. Unfortunately, such functions seem very hard to find. Indeed, for a long time, it was believed that APN permutations over $\mathbb{F}_2^n$ with even n do not exist; this was disproved only in 2010 when Dillon et al. constructed an APN permutation of $\mathbb{F}_2^6$ [15]. Until today, this is the only known APN permutation on an even number of variables; the problem of finding other such APN permutations is known as the "big APN problem" and is one of the most important open questions in the cryptographic Boolean functions community.

Since Dillon's permutation combines two significant cryptographic properties, namely APN-ness and bijectivity, it is an ideal candidate for investigating the behaviour of various constructions and approaches to cryptographically optimal functions. This is also aided by the fact that functions with $n = 6$ variables are tractable with the help of modern computers. Nonetheless, it is expected that (n, n)-functions for much higher values of n will have to be used in order to provide good security in the future, and investigating the properties of good functions in relatively small dimensions is an excellent way to accumulate structural knowledge about cryptographically strong functions that can then be applied to higher dimensions.

An important practical consideration is how to represent and implement an (n, n)-function in software and hardware. When the dimension n is small enough, this can easily be done as a look-up table, i.e. by explicitly storing all values of the function in memory. However, the size of the look-up table increases exponentially with n, and especially in the case of a resource-constrained environment, such a representation may not be feasible or practical. This raises the important question of how to optimally represent such a function and how much difference the choice of a representation can make.

In this paper, we investigate the implementation efficiency of representing Dillon's permutation as the composition of two simpler functions. We provide a hardware implementation of both this new representation and the "classical" univariate representation of the permutation and compare the two according to several metrics. We conclude that the compositional representation significantly reduces the complexity of the implementation, and thus similar representations should be considered when implementing any function that cannot be directly represented as a look-up table.

The paper is organized as follows. Section 2 introduces the fundamental concepts used throughout the paper. Dillon's permutation, the new decomposed and univariate polynomial representations and their corresponding analysis are given in Sect. 3. Section 4 introduces the hardware architecture of the decomposed Dillon's permutation and presents the $\mathbb{F}_{2^6}$ multipliers used in the paper. The hardware architecture of the corresponding univariate representation of Dillon's permutation is given in Sect. 5. Theoretical complexities of the relevant hardware architectures are given in Sect. 6. Section 7 gives FPGA implementation results and discussion. Finally, Sect. 8 concludes the paper.

2 Notation and Preliminaries

Let $\mathbb{F}_2 = \{0, 1\}$ be the finite field with two elements and let $\mathbb{F}_{2^n}$ be the *extension field* with 2^n elements. Let also $p(y) = \sum_{i=0}^{m} p_i y^i$ be a monic irreducible polynomial of degree m over $\mathbb{F}_2$, where $p_i \in \mathbb{F}_2$ for $i = 0, 1, \ldots, m$. Any element x of the binary extension field $\mathbb{F}_{2^m}$ can be represented in the *standard basis* $\{1, \zeta, \ldots, \zeta^{m-1}\}$, where ζ is a root of $p(y)$, as

$$x = \sum_{i=0}^{m-1} x_i \zeta^i = (1, \zeta, \ldots, \zeta^{m-1}) \cdot (x_0, \ldots, x_{m-1})^T \tag{1}$$

with $x_i \in \mathbb{F}_2$. In this case, $(x_0, \ldots, x_{m-1})$ are the coordinates of x with respect to the standard basis.

Vectorial Boolean functions, or (n, m)-functions, are mappings between the vector spaces $\mathbb{F}_2^n$ and $\mathbb{F}_2^m$, where n and m are positive integers. These functions have a fundamental importance in cryptography because one or more (n, m)-functions are virtually always included in modern block ciphers as the only nonlinear components. Vectorial Boolean functions can also be considered as mappings between the binary extension fields $\mathbb{F}_{2^n}$ and $\mathbb{F}_{2^m}$, due to the identification of the vector space $\mathbb{F}_2^m$ with the extension field $\mathbb{F}_{2^m}$. If $n = m$, then any (m, m)-function can be uniquely expressed as a polynomial

$$g(x) = \sum_{i=0}^{2^m-1} c_i x^i \tag{2}$$

with $c_i \in \mathbb{F}_{2^m}$. This is called the *univariate representation* of g. In this polynomial representation, the *algebraic degree* of g is defined as the largest *Hamming weight* (number of 1's in the binary representation) of any exponent i with $c_i \neq 0$. Functions of algebraic degree no greater than 1 are called *affine*, while functions of algebraic degree 2 and 3 are called *quadratic* and *cubic*, respectively. An affine function g satisfying $g(0) = 0$ is called a *linear* function.

The *derivative* of an (m, m)-function g in direction $a \in \mathbb{F}_{2^m}$ is $D_a g(x) = g(x+a)+g(x)$. The number of solutions x to the equation $D_a g(x) = b$ is denoted by $\Delta_g(a, b)$, and the *differential uniformity* of g (denoted by Δ_g) is the largest value of $\Delta_g(a, b)$ among all $a \neq 0$ and all b. Differential uniformity can be used to determine the security of a block cipher against differential cryptanalysis [1]: the lower the differential uniformity, the better the security. An *almost perfect nonlinear (APN)* function is an (m, m)-function g with the lowest possible value $\Delta_g = 2$. That is the reason why APN functions have a fundamental importance in the construction of secure block ciphers [12,13]. To be used in a *Substitution Permutation Network (SPN)*, an APN function must be a permutation.

3 On Dillon's Permutation

The existence of APN (m, m)-permutations for even m is one of the most important open questions in the study of Boolean cryptographic functions. For a long time, it was believed that there are no APN permutations on an even number of variables. However, in 2010, Dillon presented an APN permutation over $\mathbb{F}_{2^6}$ [9]. To date, this is the only known APN permutation for even dimension, and it is unknown whether APN permutations exist over $\mathbb{F}_{2^m}$ for other even values of m.

Dillon's permutation has been used in designing the lightweight authenticated encryption algorithm FIDES [10], so the study of its mathematical properties and the efficiency of its hardware implementations are of fundamental interest [14] per se. In addition, being a cryptographically optimal permutation in a tractably small number of variables, it is a perfect candidate for studying the structure and behaviour of cryptographically strong functions. In the rest of the paper, we consider how Dillon's permutation can be advantageously implemented in hardware by utilizing its decomposition into two simpler functions.

3.1 New Decomposition of Dillon's APN Permutation

Dillon's APN permutation $g(x)$, with $x \in \mathbb{F}_{2^6}$, can be presented as a composition of f_1 and f_2^{-1}, i.e., $g = f_1 \circ f_2^{-1}$, where f_1 and f_2 are quadratic. Let $f(x)$ be given by

$$f(x) = \zeta x^3 + \zeta^5 x^{10} + \zeta^4 x^{24}, \tag{3}$$

where ζ is *Magma*'s [16] default primitive element, i.e. a root of the primitive pentanomial $p(y) = y^6 + y^4 + y^3 + y + 1$ over $\mathbb{F}_2$. If f_1 and f_2 are given as

$$f_1(x) = x + \zeta^7 x^8 + \zeta^4 f(x) + \zeta^{32} f(x)^8, \tag{4}$$
$$f_2(x) = \zeta^3 x + \zeta^{31} x^8 + f(x) + f(x)^8, \tag{5}$$

then the composition $g = f_1 \circ f_2^{-1}$ represents Dillon's APN permutation.

From Eq. (5), the univariate polynomial representation of f_2^{-1} is given by

$$f_2^{-1}(x) = \zeta x^{56} + \zeta x^{49} + \zeta^{22} x^{48} + \zeta x^{42} + \zeta^{23} x^{41} + \zeta^{36} x^{40} + \zeta x^{35} + \zeta^{15} x^{34} + \zeta^{29} x^{33} + \zeta^{36} x^{28} + \zeta^{36} x^{21} + \zeta^{22} x^{20} + \zeta^{36} x^{14} + \zeta^{23} x^{13} + \zeta^{8} x^{12} + \zeta^{58} x^{8} + \zeta^{36} x^{7} + \zeta^{15} x^{6} + \zeta x^{5} + \zeta^{58} x. \quad (6)$$

The univariate polynomial representation of Dillon's APN permutation $g(x)$, with $x \in \mathbb{F}_{2^6}$, generated by the *Magma* computational algebra software, is given by

$$g(x) = \zeta^{18} x^{57} + \zeta^{22} x^{56} + \zeta^{18} x^{50} + \zeta^{22} x^{49} + \zeta^{7} x^{48} + \zeta^{18} x^{43} + \zeta^{22} x^{42} + \zeta^{44} x^{41} + \zeta^{57} x^{40} + \zeta^{18} x^{36} + \zeta^{22} x^{35} + \zeta^{22} x^{34} + \zeta^{50} x^{33} + \zeta^{24} x^{32} + \zeta^{18} x^{29} + \zeta^{57} x^{28} + \zeta^{25} x^{25} + \zeta^{18} x^{24} + \zeta^{18} x^{22} + \zeta^{57} x^{21} + \zeta^{7} x^{20} + \zeta^{18} x^{18} + \zeta^{18} x^{17} + \zeta^{18} x^{15} + \zeta^{57} x^{14} + \zeta^{44} x^{13} + \zeta^{29} x^{12} + \zeta^{11} x^{11} + \zeta^{18} x^{10} + \zeta^{24} x^{8} + \zeta^{57} x^{7} + \zeta^{22} x^{6} + \zeta^{22} x^{5} + \zeta^{3} x^{4} + \zeta^{18} x^{3} + \zeta^{13} x. \quad (7)$$

A look-up table of this permutation is given in Table 1, where v and w are given in hexadecimal. For example, if the input x is given in hexadecimal as $x = vw = 2B$ then $g(2B) = 3F$.

Table 1. Dillon permutation $g(x)$ in hexadecimal.

		w															
		0	1	2	3	4	5	6	7	8	9	A	B	C	D	E	F
v	0	00	18	22	39	01	0B	04	3D	32	19	12	1D	02	1E	35	2A
	1	0D	10	0E	03	2C	05	27	2E	20	0C	38	34	23	09	36	1B
	2	3C	06	28	07	0F	16	13	2F	3B	21	17	3F	31	3A	33	2D
	3	24	1F	0A	37	15	2B	1C	14	30	3E	29	11	08	26	25	1A

3.2 Analysis of Decomposed and Univariate Polynomial Representations of Dillon's Permutation

Dillon's APN permutation $g(x)$, with $x \in \mathbb{F}_{2^6}$, can be expressed as $g = f_1 \circ f_2^{-1}$, i.e. $g(x) = f_1(f_2^{-1}(x))$, where f_1 and f_2 are given in terms of the function $f(x)$. It can be observed that f, f_1 and f_2 given in Eqs. (3), (4) and (5), respectively, are quadratic functions, while the univariate polynomial representation of f_2^{-1} given in Eq. (6) has an algebraic degree of 3. The function $f(x)$ only includes the addition of three different powers of x (multiplied by corresponding powers of ζ), while that of f_1 and f_2 includes the addition of four terms including two powers of x and the addition of $f(x)$ and $f(x)^8$ (some of them multiplied by corresponding powers of ζ). Furthermore, the univariate representation of f_2^{-1} includes 20 different powers of x that are multiplied by 8 different powers of ζ.

We can observe that the original univariate polynomial representation of Dillon's permutation given in (7) has algebraic degree 4 and includes 36 different powers of x that are multiplied by 12 different powers of ζ. This fact implies (as for the univariate representation of f_2^{-1}) that some terms ζ^i are common in several products $\zeta^i x^j$ for different values of j, so simplified expressions of the form $\zeta^i(x^{j_1} + x^{j_2} + \ldots) = \zeta^i v_i$ can be obtained, thereby reducing the number of arithmetic operations that need to be performed when evaluating the function.

These differences among the decomposed and univariate representation of Dillon's permutation influence the hardware implementation complexity, as illustrated in the sequel.

4 Hardware Architecture of the Decomposition of Dillon's Permutation

Following the analysis in Subsect. 3.2, the hardware architecture for the decomposition of Dillon's permutation is shown in Fig. 1, where the composition $g = f_1 \circ f_2^{-1}$ is represented by the two hardware modules f_2^{-1} and f_1 in such a way that the 6-bit input x (represented in the binary extension field $\mathbb{F}_{2^6}$ generated by the primitive pentanomial $p(y) = y^6 + y^4 + y^3 + y + 1$) is fed to the f_2^{-1} module. The output of this first module $y = f_2^{-1}(x)$ is supplied to the second module f_1 in order to compute the output value of Dillon's permutation $g(x) = f_1(y) = f_1(f_2^{-1}(x))$. As we deal with elements in the finite field $\mathbb{F}_{2^6}$, XOR and AND gates will be used as the basic components for the hardware implementation.

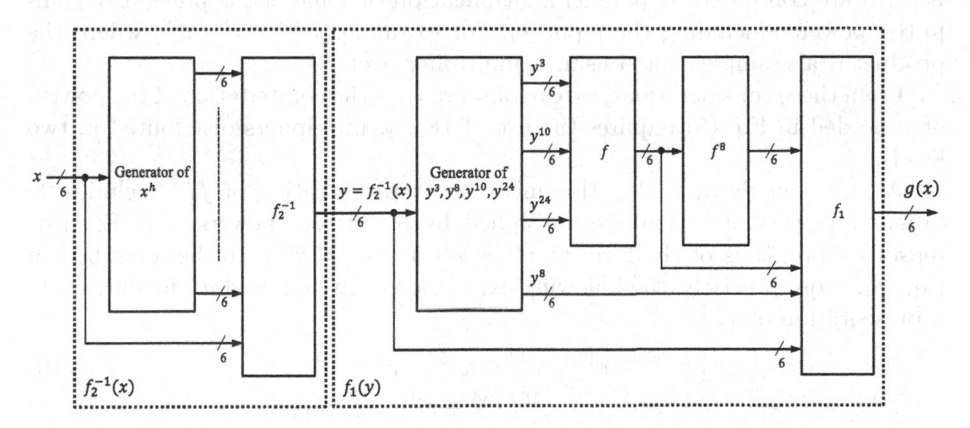

Fig. 1. Hardware architecture of decomposed Dillon's permutation $g(x) = f_1(f_2^{-1}(x))$.

4.1 Computation of f_2^{-1}

Following Eq. (6), the inverse function f_2^{-1} is given by the addition (XOR) of terms $\zeta^i x^j$ where the powers of x are computed by the method of square and multiply. In order to do this, the successive squares of x (in this case, x^2, x^4, x^8, x^{16} and x^{32}) modulo the primitive polynomial $p(y)$ must first be determined. These powers are easily computed in $\mathbb{F}_{2^6}$ due to the fact that x^{2^i}, for $i = 1, 2, \ldots, 5$, is

$$x^{2^i} = x_5\zeta^{5\cdot 2^i} + x_4\zeta^{4\cdot 2^i} + x_3\zeta^{3\cdot 2^i} + x_2\zeta^{2\cdot 2^i} + x_1\zeta^{1\cdot 2^i} + x_0\zeta^{0\cdot 2^i} \tag{8}$$

and the powers of ζ are reduced using $p(y)$.

Example 1. In order to compute $x^8 = x_5\zeta^{40} + x_4\zeta^{32} + x_3\zeta^{24} + x_2\zeta^{16} + x_1\zeta^8 + x_0$, the powers ζ^{40}, ζ^{32}, ζ^{24}, ζ^{16} and ζ^8 must be reduced modulo $p(y) = y^6 + y^4 + y^3 + y + 1$. Since ζ is a primitive element, we have that $\zeta^6 = \zeta^4 + \zeta^3 + \zeta + 1$, so $\zeta^{40} = \zeta^4 + \zeta^2 + 1$, $\zeta^{32} = \zeta^5 + \zeta^4 + \zeta$, $\zeta^{24} = \zeta^5 + \zeta^3 + \zeta + 1$, $\zeta^{16} = \zeta^4 + \zeta + 1$ and $\zeta^8 = \zeta^5 + \zeta^4 + \zeta^2 + \zeta + 1$, and their substitution into the expression for x^8 gives

$$\begin{aligned} x^8 = (x_4 + x_3 + x_1)\zeta^5 + (x_5 + x_4 + x_2 + x_1)\zeta^4 + (x_3)\zeta^3 + (x_5 + x_1)\zeta^2 + \\ (x_4 + x_3 + x_2 + x_1)\zeta + (x_5 + x_3 + x_2 + x_1 + x_0). \end{aligned} \tag{9}$$

Therefore the coordinates of the successive powers of x are given as the XORs of the coordinates of x.

Once we have determined the powers x^{2^i}, for $i = 1, 2, \ldots, 5$, the powers x^5, x^6, x^{12}, x^{20}, x^{33}, x^{34}, x^{40} and x^{48} that we need in Eq. (6) can be computed by parallel multiplication of some of the terms x, x^2, x^4, x^8, x^{16} and x^{32}. For example, $x^{40} = x^8 \cdot x^{32}$, where this product is implemented with a multiplier over $\mathbb{F}_{2^6}$ using the primitive polynomial $p(y) = y^6 + y^4 + y^3 + y + 1$. Multipliers selected for the implementation are described in Subsect. 4.3.

The remaining powers x^7, x^{13}, x^{14}, x^{21}, x^{28}, x^{35}, x^{41}, x^{42}, x^{49} and x^{56} from Eq. (6) are computed by parallel multiplication of some of the previously computed powers (including the input x). For example, $x^{56} = x^8 \cdot x^{48}$, where the product is also implemented using a multiplier over $\mathbb{F}_{2^6}$.

From these considerations, we can observe that the computation of the powers of x needed in Eq. (6) requires the use of 18 $\mathbb{F}_{2^6}$ multipliers distributed in two levels.

As given in Subsect. 3.2, the univariate representation of f_2^{-1} includes 20 different powers of x that are multiplied by 8 different powers of ζ. For this reason, expressions of the form $\zeta^i(x^{j_1} + x^{j_2} + \cdots) = \zeta^i v_i$ can be identified in Eq. (6). More precisely, the following expressions can be found in the univariate representation of f_2^{-1}:

$$\zeta(x^{56} + x^{49} + x^{42} + x^{35} + x^5) = \zeta v_1, \tag{10}$$

$$\zeta^{15}(x^{34} + x^6) = \zeta^{15} v_{15}, \tag{11}$$

$$\zeta^{22}(x^{48} + x^{20}) = \zeta^{22} v_{22}, \tag{12}$$

$$\zeta^{23}(x^{41} + x^{13}) = \zeta^{23}v_{23}, \quad (13)$$

$$\zeta^{36}(x^{40} + x^{28} + x^{21} + x^{14} + x^{7}) = \zeta^{36}v_{36}, \quad (14)$$

$$\zeta^{58}(x^{8} + x) = \zeta^{58}v_{58}. \quad (15)$$

The use of these expressions allows us to reduce the complexity of the hardware implementation, where the sum of such powers of x is simply performed as the bitwise XOR of the corresponding coordinates. Therefore, the univariate representation of f_2^{-1} given in Eq. (6) can be rewritten as:

$$f_2^{-1}(x) = \zeta v_1 + \zeta^8 x^{12} + \zeta^{15}v_{15} + \zeta^{22}v_{22} + \zeta^{23}v_{23} + \zeta^{29}x^{33} + \zeta^{36}v_{36} + \zeta^{58}v_{58}. \quad (16)$$

The new representation of f_2^{-1} given in Eq. (16) requires the sum of terms $\zeta^i x^j$ and $\zeta^i v_i$, where $v_i = (x^{j_1} + x^{j_2} + \cdots)$ and $x^j, v_i \in \mathbb{F}_{2^6}$. In general, the term $\zeta^i B$, with $B \in \mathbb{F}_{2^6}$, will be

$$\zeta^i B = \zeta^i(b_5\zeta^5 + b_4\zeta^4 + b_3\zeta^3 + b_2\zeta^2 + b_1\zeta + b_0) = b_5\zeta^{5+i} + b_4\zeta^{4+i} + b_3\zeta^{3+i} + b_2\zeta^{2+i} + b_1\zeta^{1+i} + b_0\zeta^i, \quad (17)$$

where the powers of ζ are reduced using the primitive pentanomial $p(y)$.

Example 2. The term $\zeta^8 B$ is given as $\zeta^8 B = (b_5\zeta^{13} + b_4\zeta^{12} + b_3\zeta^{11} + b_2\zeta^{10} + b_1\zeta^9 + b_0\zeta^8)$, where ζ^{13}, ζ^{12}, ζ^{11}, ζ^{10}, ζ^9 and ζ^8 must be reduced modulo $p(y) = y^6+y^4+y^3+y+1$. Since ζ is a primitive element, we have that $\zeta^6 = \zeta^4+\zeta^3+\zeta+1$, so $\zeta^{13} = \zeta^3 + 1$, $\zeta^{12} = \zeta^5 + \zeta^3 + 1$, $\zeta^{11} = \zeta^5 + \zeta^4 + \zeta^3 + 1$, $\zeta^{10} = \zeta^5 + \zeta^4 + 1$, $\zeta^9 = \zeta^5 + \zeta^4 + \zeta^2 + 1$ and $\zeta^8 = \zeta^5 + \zeta^4 + \zeta^2 + \zeta + 1$, and substituting this into the expression for $\zeta^8 B$ gives

$$\zeta^8 B = (b_4 + b_3 + b_2 + b_1 + b_0)\zeta^5 + (b_3 + b_2 + b_1 + b_0)\zeta^4 + (b_5 + b_4 + b_3)\zeta^3 + (b_1 + b_0)\zeta^2 + (b_0)\zeta + (b_5 + b_4 + b_3 + b_2 + b_1 + b_0). \quad (18)$$

Therefore, the coordinates of $\zeta^i B$ are given as the XORs of the coordinates of B depending on the reduction modulo the primitive polynomial $p(y)$.

The final sum of the terms appearing in the new univariate representation of f_2^{-1} given in Eq. (16) is simply performed as the bitwise XOR of the corresponding coordinates.

4.2 Computation of f_1

As shown in Eq. (4), the function f_1 is given as the sum of four terms including the input x, x^8, $f(x)$ and $f(x)^8$, where x^8, $f(x)$ and $f(x)^8$ are multiplied by ζ^7, ζ^4 and ζ^{32}, respectively. Furthermore, the function $f(x)$ given in Eq. (3) only includes the sum of x, x^{10} and x^{24} which are multiplied by ζ, ζ^5 and ζ^4, respectively.

As shown in Fig. 1 (where $y = f_2^{-1}(x)$ is the input to f_1), the hardware architecture for f_1 requires the parallel computation of y^3, y^{10} and y^{24} that we need for computing $f(y)$. These values are given as products of powers of y, i.e., $y^3 = y \cdot y^2$, $y^{10} = y^2 \cdot y^8$, $y^{24} = y^8 \cdot y^{16}$, so the powers y^2, y^8 and y^{16} must first be computed in a similar way as previosuly described in Subsect. 4.1 (see Example 1). Furthermore, the above products are implemented with $\mathbb{F}_{2^6}$ multipliers using the primitive polynomial $p(y) = y^6+y^4+y^3+y+1$. Multipliers selected for the implementation are described in Subsect. 4.3.

Using y^3, y^{10} and y^{24}, evaluating the function $f(y)$ requires their multiplication by ζ, ζ^5 and ζ^4, respectively, which can also be done in a way similar to what was described in Subsect. 4.1 (see Example 2). The sum of terms in Eq. (3) completes the computation of $f(y)$.

The function f_1 also requires the eighth power of $f(y)$. This power can be computed using the same expressions given in Eq. (9) (Example 1).

Once we have computed y^8, $f(y)$ and $f(y)^8$, these values must be multiplied by ζ^7, ζ^4 and ζ^{32}, respectively. This can be done in a similar way to Subsect. 4.1 using Eq. (17).

Example 3. Let $f(y)^8 = f_5^8\zeta^5 + f_4^8\zeta^4 + f_3^8\zeta^3 + f_2^8\zeta^2 + f_1^8\zeta + f_0^8$, where $f_i^8 \in \mathbb{F}_2$ $(i = 0, 1, \ldots, 5)$ are the coordinates of $f(y)^8$ with respect to the standard basis. Then the term $\zeta^{32}f(y)^8$ is given as $\zeta^{32}f(y)^8 = (f_5^8\zeta^{37} + f_4^8\zeta^{36} + f_3^8\zeta^{35} + f_2^8\zeta^{34} + f_1^8\zeta^{33} + f_0^8\zeta^{32})$, where ζ^{37}, ζ^{36}, ζ^{35}, ζ^{34}, ζ^{33} and ζ^{32} must be reduced modulo $p(y) = y^6 + y^4 + y^3 + y + 1$. Since ζ is a primitive element, we have that $\zeta^6 = \zeta^4 + \zeta^3 + \zeta + 1$, so $\zeta^{37} = \zeta^4 + \zeta^3 + \zeta^2 + \zeta + 1$, $\zeta^{36} = \zeta^5 + \zeta$, $\zeta^{35} = \zeta^4 + 1$, $\zeta^{34} = \zeta^5 + \zeta^2 + 1$, $\zeta^{33} = \zeta^5 + \zeta^4 + \zeta^3 + \zeta^2 + \zeta + 1$ and $\zeta^{32} = \zeta^5 + \zeta^4 + \zeta$, and substituting this into the expression for $\zeta^{32}f(y)^8$ gives

$$\begin{aligned}\zeta^{32}f(y)^8 = (f_4^8 + f_2^8 + f_1^8 + f_0^8)\zeta^5 + (f_5^8 + f_3^8 + f_1^8 + f_0^8)\zeta^4 + \\ (f_5^8 + f_1^8)\zeta^3 + (f_5^8 + f_2^8 + f_1^8)\zeta^2 + (f_5^8 + f_4^8 + f_1^8 + f_0^8)\zeta + \\ (f_5^8 + f_3^8 + f_2^8 + f_1^8).\end{aligned} \tag{19}$$

Therefore the coordinates of $\zeta^{32}f(y)^8$ are also given as the XORs of the coordinates of $f(y)^8$ depending on the reduction modulo the primitive polynomial $p(y)$.

Finally, the sum of terms appearing in the expression of f_1 given in Eq. (4) is simply performed as the bitwise XOR of the corresponding coordinates and also gives the output of the decomposed Dillon's permutation $g(x) = f_1(f_2^{-1}(x))$.

4.3 Multipliers Over $\mathbb{F}_{2^6}$

Multipliers over $\mathbb{F}_{2^6}$ for the primitive pentanomial $p(y) = y^6 + y^4 + y^3 + y + 1$ are needed for the computations first of x^5, x^6, x^{12}, x^{20}, x^{33}, x^{34}, x^{40}, x^{48} and then of x^7, x^{13}, x^{14}, x^{21}, x^{28}, x^{35}, x^{41}, x^{42}, x^{49}, x^{56} in Subsect. 4.1 for $f_2^{-1}(x)$. Finite field multipliers over $\mathbb{F}_{2^6}$ are also needed in Subsect. 4.2 for $f_1(y)$ for the computation of y^3, y^{10} and y^{24}.

We have used the method given in [18] for the construction of multipliers over $\mathbb{F}_{2^m}$ in the standard basis. In [18], the product $C = A \cdot B$, with $A, B \in \mathbb{F}_{2^6}$, is computed through a product matrix $\mathbf{K}$ (that depends on the primitive polynomial $p(y)$ and on the coordinate b_i of the operand B) in such a way that the product of the matrix $\mathbf{K}$ and the coordinates vector of the operand A gives the coordinates of the product C as sum-of-products of $a_i b_j$, with a_i and b_i being the coordinates of A and B, respectively. Furthermore, the matrix $\mathbf{K}$ can be decomposed into a sum of matrices depending on the primitive polynomial $p(y)$ used for the binary extension field $\mathbb{F}_{2^m}$. In order to compute the coordinates of the product C, the functions $\mathbf{S_i}$ $(1 \leq i \leq m)$ and $\mathbf{T_i}$ $(0 \leq i \leq m-2)$ were defined in [18] by the sum of terms $x_k = (a_k b_k)$ and $z_i^j = (a_i b_j + a_j b_i)$. These functions (obtained in [18] from the product of $\mathbf{K}$ and the coordinate vector of A) were given by $\mathbf{S_i} = x_p + \sum_{h=0}^{p-1} z_h^{i-h-1}$ and $\mathbf{T_i} = x_q + \sum_{j=1}^{r-(i+1)} z_{i+j}^{m-j}$, where $p = \lfloor i/2 \rfloor$, $q = (\lceil m/2 \rceil + \lfloor i/2 \rfloor)$, the term $x_p = a_p b_p$ only appears for i *odd* and x_q only appears for m and i *even* or for m and i *odd*. In this case, $r = q$. Otherwise, i.e., for m *even* and i *odd* or for m *odd* and i *even*, the term x_q does not appear and $r = (\lceil m/2 \rceil + \lceil i/2 \rceil)$. Using the above expressions for $\mathbb{F}_{2^6}$, the terms $\mathbf{S_i}$ and $\mathbf{T_i}$ are: $\mathbf{S_1} = x_0 = a_0 b_0$, $\mathbf{S_2} = z_0^1 = (a_0 b_1 + a_1 b_0)$, $\mathbf{S_3} = x_1 + z_0^2 = a_1 b_1 + (a_0 b_2 + a_2 b_0)$, $\mathbf{S_4} = z_0^3 + z_1^2 = (a_0 b_3 + a_3 b_0) + (a_1 b_2 + a_2 b_1)$, $\mathbf{S_5} = x_2 + z_0^4 + z_1^3 = a_2 b_2 + (a_0 b_4 + a_4 b_0) + (a_1 b_3 + a_3 b_1)$, $\mathbf{S_6} = z_0^5 + z_1^4 + z_2^3 = (a_0 b_5 + a_5 b_0) + (a_1 b_4 + a_4 b_1) + (a_2 b_3 + a_3 b_2)$, and $\mathbf{T_0} = x_3 + z_1^5 + z_2^4 = a_3 b_3 + (a_1 b_5 + a_5 b_1) + (a_2 b_4 + a_4 b_2)$, $\mathbf{T_1} = z_2^5 + z_3^4 = (a_2 b_5 + a_5 b_2) + (a_3 b_4 + a_4 b_3)$, $\mathbf{T_2} = x_4 + z_3^5 = a_4 b_4 + (a_3 b_5 + a_5 b_3)$, $\mathbf{T_3} = z_4^5 = (a_4 b_5 + a_5 b_4)$, $\mathbf{T_4} = x_5 = a_5 b_5$. The coordinates of the product $C = A \cdot B$ can then be computed as the addition of some of these terms.

This method was used in [18–20] for the computation of the finite field multiplication over $\mathbb{F}_{2^m}$ in the standard basis for irreducible *trinomials*, *type-I pentanomials* and *type-II pentanomials*, respectively, where explicit expressions for the coordinates of the products were given in terms of the $\mathbf{S_i}$ and $\mathbf{T_i}$ functions. Irreducible *type-I pentanomials* were defined in [21] as $p(y) = y^m + y^{n+1} + y^n + y + 1$, where $2 \leq n \leq \lfloor \frac{m}{2} \rfloor - 1$. It can be observed that the primitive pentanomial used in Dillon's APN permutation $p(y) = y^6 + y^4 + y^3 + y + 1$ follows the structure of type-I pentanomials, but in this case $n = 3$ and therefore it does not meet the valid value for the parameter n ($n = 2$ for type-I pentanomial in $\mathbb{F}_{2^6}$). Therefore, the expressions given in [19] can not be used in this case and the coordinates of the product must be obtained directly from the decomposition of the product matrix $\mathbf{K}$ given in [18] for the specific values $m = 6$, $n = 3$.

For an irreducible polynomial $p(y) = y^m + y^{k_t} + y^{k_t - 1} + \cdots + y^{k_1} + 1$, the matrix $\mathbf{K}$ can be decomposed into a sum of matrices whose number depends on the values of $k_t, k_t - 1, \ldots, k_1$ corresponding to the nonzero coefficients $p_{k_t}, p_{k_t - 1}, \ldots, p_{k_1}$, respectively, of $p(y)$. The decomposition of $\mathbf{K}$ into a sum of $m \times m$ matrices was given in [18] by

$$\mathbf{K} = \mathbf{K_0} + \sum_{j=1}^{t} \left(\sum_{i=1}^{\tau_j} \mathbf{K_i^{k_j}} + \sum_{p=1, p \neq j}^{t} \sum_{i=1}^{\tau_j - 1} \mathbf{K_i^{k_p, k_j}} \right) \tag{20}$$

where $\tau_i = \lceil \frac{m-1}{\Delta_i} \rceil$ ($\Delta_i = m - k_i$). The product of the decomposed matrix $\mathbf{K}$ given in Eq. (20) and the coordinate vector of A gives a set of functions $\mathbf{S_i}$ and $\mathbf{T_i}$ whose addition provides the coordinates of the product C.

For the primitive pentanomial $p(y) = y^6 + y^4 + y^3 + y + 1$, the left-hand side of Table 2 shows the specific expressions obtained in this case for the product $C = A \cdot B$ over $\mathbb{F}_{2^6}$ given in terms of the $\mathbf{S_i}$ and $\mathbf{T_i}$ functions, where the c_i ($i = 0, 1, \ldots, 5$) coordinates of the product C are given by the sum of terms in their corresponding rows, and where the columns show the $\mathbf{S_i}$ or $\mathbf{T_i}$ functions given by different matrices. We can observe on the left-hand side of Table 2 that several terms $\mathbf{T_i}$ can be cancelled, leading to the reduced expressions shown on the right-hand side of Table 2:
$c_0 = \mathbf{S_1}+\mathbf{T_0}+\mathbf{T_2}+\mathbf{T_3}+\mathbf{T_4}$, $c_1 = \mathbf{S_2}+\mathbf{T_0}+\mathbf{T_1}+\mathbf{T_2}$, $c_2 = \mathbf{S_3}+\mathbf{T_1}+\mathbf{T_2}+\mathbf{T_3}$, $c_3 = \mathbf{S_4}+\mathbf{T_0}$, $c_4 = \mathbf{S_5}+\mathbf{T_0}+\mathbf{T_1}+\mathbf{T_2}+\mathbf{T_3}+\mathbf{T_4}$ and $c_5 = \mathbf{S_6}+\mathbf{T_1}+\mathbf{T_2}+\mathbf{T_3}+\mathbf{T_4}$. Furthermore, it can be observed that there exist subexpressions (sums of $\mathbf{T_i}$ terms) that can be shared among different coordinates of the product. This *subexpressions sharing* can be used to reduce the area complexity of the hardware implementations. Table 2 shows the subexpressions that can be shared in this case with shadowed cells. If we define as intermediate variables $\mathbf{v_0} = \mathbf{T_1} + \mathbf{T_2}$, $\mathbf{v_1} = \mathbf{T_3} + \mathbf{T_4}$ and $\mathbf{v_2} = \mathbf{v_0} + \mathbf{v_1}$, then the final expressions of the product coordinates using subexpressions sharing are

$$\begin{aligned} c_0 &= \mathbf{S_1} + \mathbf{T_0} + \mathbf{T_2} + \mathbf{v_1} \\ c_1 &= \mathbf{S_2} + \mathbf{T_0} + \mathbf{v_0} \\ c_2 &= \mathbf{S_3} + \mathbf{v_0} + \mathbf{T_3} \\ c_3 &= \mathbf{S_4} + \mathbf{T_0} \\ c_4 &= \mathbf{S_5} + \mathbf{T_0} + \mathbf{v_2} \\ c_5 &= \mathbf{S_6} + \mathbf{v_2}. \end{aligned} \tag{21}$$

Table 2. Coordinates c_i of the product for the pentanomial $p(y) = y^6 + y^4 + y^3 + y + 1$.

c_0	$\mathbf{S_1}$	$\mathbf{T_0}$	$\mathbf{T_2}$	$\mathbf{T_4}$			$\mathbf{T_3}$							
c_1	$\mathbf{S_2}$	$\mathbf{T_1}$	$\cancel{\mathbf{T_3}}$				$\cancel{\mathbf{T_4}}$				$\mathbf{T_0}$	$\cancel{\mathbf{T_3}}$	$\mathbf{T_2}$	$\cancel{\mathbf{T_4}}$
c_2	$\mathbf{S_3}$	$\mathbf{T_2}$	$\cancel{\mathbf{T_4}}$								$\mathbf{T_1}$	$\cancel{\mathbf{T_4}}$	$\mathbf{T_3}$	
c_3	$\mathbf{S_4}$	$\cancel{\mathbf{T_3}}$					$\mathbf{T_0}$	$\cancel{\mathbf{T_3}}$	$\cancel{\mathbf{T_2}}$	$\cancel{\mathbf{T_4}}$	$\cancel{\mathbf{T_2}}$		$\cancel{\mathbf{T_4}}$	
c_4	$\mathbf{S_5}$	$\cancel{\mathbf{T_4}}$	$\mathbf{T_0}$	$\mathbf{T_2}$	$\cancel{\mathbf{T_4}}$	$\mathbf{T_3}$	$\mathbf{T_1}$	$\mathbf{T_4}$	$\cancel{\mathbf{T_3}}$		$\cancel{\mathbf{T_3}}$			
c_5	$\mathbf{S_6}$		$\mathbf{T_1}$	$\mathbf{T_3}$		$\mathbf{T_4}$	$\mathbf{T_2}$		$\cancel{\mathbf{T_4}}$		$\cancel{\mathbf{T_4}}$			

=

$\mathbf{S_1}$	$\mathbf{T_0}$	$\mathbf{T_2}$	$\mathbf{T_3}$	$\mathbf{T_4}$	
$\mathbf{S_2}$	$\mathbf{T_0}$	$\mathbf{T_1}$	$\mathbf{T_2}$		
$\mathbf{S_3}$	$\mathbf{T_1}$	$\mathbf{T_2}$	$\mathbf{T_3}$		
$\mathbf{S_4}$	$\mathbf{T_0}$				
$\mathbf{S_5}$	$\mathbf{T_0}$	$\mathbf{T_1}$	$\mathbf{T_2}$	$\mathbf{T_3}$	$\mathbf{T_4}$
$\mathbf{S_6}$	$\mathbf{T_1}$	$\mathbf{T_2}$	$\mathbf{T_3}$	$\mathbf{T_4}$	

5 Hardware Architecture of the Univariate Representation of Dillon's Permutation

Following the analysis in Subsect. 3.2, the hardware architecture for Dillon's permutation deduced by the univariate polynomial representation given in Eq. (7) is shown in Fig. 2, where the different modules needed for the computation are included. The 6-bit input x is represented in the finite field $\mathbb{F}_{2^6}$ generated by the primitive pentanomial $p(y) = y^6 + y^4 + y^3 + y + 1$.

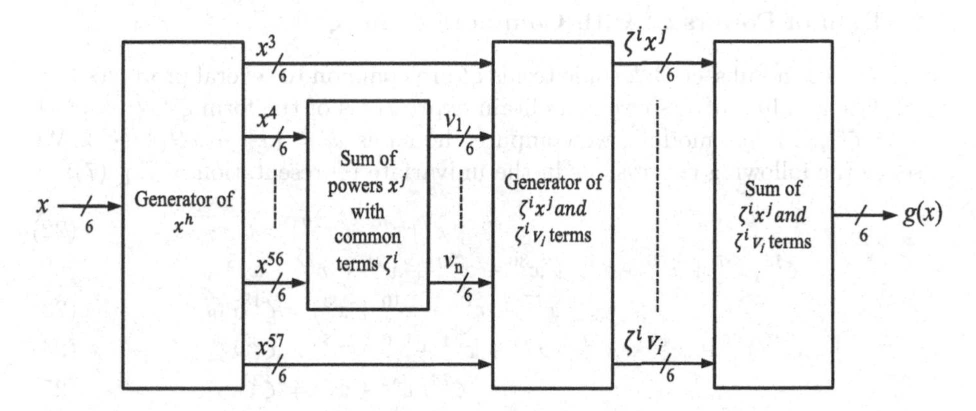

Fig. 2. Hardware architecture of univariate polynomial representation of Dillon's permutation.

5.1 Generator of x^h

This module computes the powers of x using the method of square and multiply in a similar way as shown in Subsect. 4. In order to do this, the successive squares x^2, x^4, x^8, x^{16} and x^{32} modulo the primitive polynomial $p(y)$ must first be determined. These powers are easily computed in $\mathbb{F}_{2^6}$ due to the fact that x^{2^i}, for $i = 1, 2, \ldots, 5$, is given as in Eq. (8) and the powers of ζ are reduced using $p(y)$ (see Example 1). Therefore the coordinates of the successive powers of x are given as the XORs of the coordinates of x.

Once we have determined the powers x^{2^i}, for $i = 1, 2, \ldots, 5$, the powers x^3, x^5, x^6, x^{10}, x^{12}, x^{17}, x^{18}, x^{20}, x^{24}, x^{34}, x^{36} and x^{40} needed in Eq. (7) can be computed by means of the parallel multiplication of some of the terms x, x^2, x^4, x^8, x^{16} and x^{32}. For example, $x^{24} = x^8 \cdot x^{16}$, where this product is implemented with the $\mathbb{F}_{2^6}$ multiplier for the primitive pentanomial $p(y) = y^6 + y^4 + y^3 + y + 1$ described in Subsect. 4.3 and given in Eq. (21).

The remaining powers x^7, x^{11}, x^{13}, x^{14}, x^{15}, x^{21}, x^{22}, x^{25}, x^{28}, x^{29}, x^{33}, x^{35}, x^{41}, x^{42}, x^{43}, x^{48}, x^{49}, x^{50}, x^{56} and x^{57} that appear in Eq. (7) are determined by the parallel multiplication of some of the powers previously computed (also including the input x). For example, $x^{57} = x^{17} \cdot x^{40}$, where the product is also implemented using a multiplier over $\mathbb{F}_{2^6}$ for the given pentanomial $p(y)$.

From the above considerations, it can be observed that the computation of the powers of x needed in Eq. (7) requires the use of 32 $\mathbb{F}_{2^6}$ multipliers distributed in two levels.

5.2 Sum of Powers x^j with Common Terms ζ^i

As discussed in Subsect. 3.2, some terms ζ^i are common to several products $\zeta^i x^j$ for different values of j, so we can obtain expressions of the form $\zeta^i(x^{j_1} + x^{j_2} + \cdots) = \zeta^i v_i$. In this module, we compute the sums $v_i = (x^{j_1} + x^{j_2} + \cdots)$. We observe the following expressions in the univariate representation of Eq. (7):

$$\zeta^7(x^{48} + x^{20}) = \zeta^7 v_7, \tag{22}$$

$$\zeta^{18}(x^{57} + x^{50} + x^{43} + x^{36} + x^{29} + x^{24} + x^{22} + x^{18} + x^{17} + x^{15} + x^{10} + x^3) = \zeta^{18} v_{18}, \tag{23}$$

$$\zeta^{22}(x^{56} + x^{49} + x^{42} + x^{35} + x^{34} + x^6 + x^5) = \zeta^{22} v_{22}, \tag{24}$$

$$\zeta^{24}(x^{32} + x^8) = \zeta^{24} v_{24}, \tag{25}$$

$$\zeta^{44}(x^{41} + x^{13}) = \zeta^{44} v_{44}, \tag{26}$$

$$\zeta^{57}(x^{40} + x^{28} + x^{21} + x^{14} + x^7) = \zeta^{57} v_{57}. \tag{27}$$

The use of these expressions allows the reduction of the complexity of the hardware implementation. The sum of such powers x^n is simply performed as the bitwise XOR of the corresponding coordinates. Therefore, the univariate representation of Eq. (7) can be rewritten as:

$$g(x) = \zeta^3 x^4 + \zeta^7 v_7 + \zeta^{11} x^{11} + \zeta^{13} x + \zeta^{18} v_{18} + \zeta^{22} v_{22} + \zeta^{24} v_{24} + \zeta^{25} x^{25} + \zeta^{29} x^{12} + \zeta^{44} v_{44} + \zeta^{50} x^{33} + \zeta^{57} v_{57}. \tag{28}$$

5.3 Generator of the Terms $\zeta^i x^j$ and $\zeta^i v_i$

The new representation of Dillon's univariate polynomial given in Eq. (28) requires the sum of terms $\zeta^i x^j$ and $\zeta^i v_i$, where $v_i = (x^{j_1} + x^{j_2} + \cdots)$ and $x^j, v_i \in \mathbb{F}_{2^6}$. The general expression of the term $\zeta^i B$, with $B \in \mathbb{F}_{2^6}$, is given in Eq. (17), where the powers of ζ are reduced using the primitive polynomial $p(y)$.

Example 4. The term $\zeta^{24}B$ is given as $\zeta^{24}B = (b_5\zeta^{29} + b_4\zeta^{28} + b_3\zeta^{27} + b_2\zeta^{26} + b_1\zeta^{25} + b_0\zeta^{24})$, where ζ^{29}, ζ^{28}, ζ^{27}, ζ^{26}, ζ^{25} and ζ^{24} must be reduced modulo $p(y) = y^6 + y^4 + y^3 + y + 1$. Since ζ is a primitive element, we have that $\zeta^6 = \zeta^4 + \zeta^3 + \zeta + 1$, so $\zeta^{29} = \zeta^5 + \zeta^4 + \zeta^3 + \zeta^2 + 1$, $\zeta^{28} = \zeta^5 + \zeta^4 + \zeta + 1$, $\zeta^{27} = \zeta^5 + \zeta^4 + \zeta^2$,

$\zeta^{26} = \zeta^4 + \zeta^3 + \zeta$, $\zeta^{25} = \zeta^3 + \zeta^2 + 1$ and $\zeta^{24} = \zeta^5 + \zeta^3 + \zeta + 1$, and substituting this into the expression for $\zeta^{24}B$ gives Eq. (29):

$$\begin{aligned}\zeta^{24}B = (b_5 + b_4 + b_3 + b_0)\zeta^5 + (b_5 + b_4 + b_3 + b_2)\zeta^4 + \\ (b_5 + b_2 + b_1 + b_0)\zeta^3 + (b_5 + b_1 + b_3)\zeta^2 + \\ (b_4 + b_2 + b_0)\zeta + (b_5 + b_4 + b_1 + b_0).\end{aligned} \tag{29}$$

Therefore the coordinates of $\zeta^i B$ are given as the XORs of the coordinates of B depending on $p(y)$.

5.4 Sum of the Terms $\zeta^i x^j$ and $\zeta^i v_i$

The final sum of terms $\zeta^i x^j$ and $\zeta^i v_i$ appearing in the univariate polynomial representation of Dillon's permutation given in Eq. (28) is simply performed as the bitwise XOR of the corresponding coordinates.

6 Theoretical Complexities of Dillon's APN Hardware Architectures

Area and time theoretical complexities of the different architectures for computing Dillon's permutation can be obtained from the complexities of the modules previously described. Area complexity corresponds to the number of 2-input AND and 2-input XOR gates. The number of XOR gates for each architecture is given by the hardware synthesis tool used for the FPGA implementations given in Sect. 7. For this reason, we only give in this section the number of 2-input AND gates. Time complexity is determined by the maximum number of 2-input AND and 2-input XOR gates that a signal must traverse from input to output, and is given in terms of the values T_{AND} and T_{XOR} that describe the gate delay of a 2-input AND and 2-input XOR gate, respectively.

6.1 Complexity of $\mathbb{F}_{2^6}$ Multipliers

From the description of the $\mathbb{F}_{2^m}$ multiplier given in Subsect. 4.3, we can observe that the number of 2-input AND gates is given by m^2, so in this case the number of AND gates needed is 36 for each multiplier.

Using the $\mathbf{S_i}$ and $\mathbf{T_i}$ expressions given in Subsect. 4.3, we can observe that the number of XOR levels of the terms $\mathbf{S_1}$, $\mathbf{S_2}$, $\mathbf{S_3}$, $\mathbf{S_4}$, $\mathbf{S_5}$, $\mathbf{S_6}$, $\mathbf{T_0}$, $\mathbf{T_1}$, $\mathbf{T_2}$, $\mathbf{T_3}$ and $\mathbf{T_4}$ is 0, 1, 2, 2, 3, 3, 3, 2, 2, 1 and 0, respectively. Furthermore, the $\mathbf{S_i}$ and $\mathbf{T_i}$ terms have one level of AND gates. Therefore, using the coordinate expressions given in Equation (21), we can show that the $\mathbb{F}_{2^6}$ multiplier has a maximum theoretical delay of $T_{AND} + 5T_{XOR}$.

These complexities will be used in the sequel for the computation of the theoretical complexities of the decomposed and univariate representations of Dillon's permutation.

6.2 Decomposed Dillon's Permutation

As shown in Sect. 4 and in Fig. 1, the composition $g = f_1 \circ f_2^{-1}$ is represented by the two hardware modules f_2^{-1} and f_1, so we have to determine the complexities for these two modules.

The module for the computation of f_2^{-1} given in Subsect. 4.1 involves the generation of the powers x^h that requires the use of 18 $\mathbb{F}_{2^6}$ multipliers distributed in two levels, so the total number of AND gates is $18{\times}36 = 648$. Concerning the time delay, the generator of x^h involves the generation of the powers x^{2^i} (with a maximum delay of $3T_{XOR}$) and the two levels of $\mathbb{F}_{2^6}$ multipliers for the generation of the remaining powers of x^h (therefore with a maximum delay of $2T_{AND} + 10T_{XOR}$). The sum of powers x^j with common terms ζ^i presents a maximum delay of $3T_{XOR}$ corresponding to the addition of 5 terms for the variables v_1 and v_{36} as given in Eqs. (10) and (14), respectively. The generation of the $\zeta^i x^j$ and $\zeta^i v_i$ terms given in Subsect. 4.1 needs a maximum of $3T_{XOR}$. The final sum of the eight terms appearing in the univariate representation of f_2^{-1} given in Eq. (16) also requires a maximum of $3T_{XOR}$. Therefore, the maximum delay of the module for the computation of f_2^{-1} is $2T_{AND} + 22T_{XOR}$.

The module for the computation of f_1 given in Subsect. 4.2 involves the generation of the powers y^h that requires the use of 3 $\mathbb{F}_{2^6}$ multipliers in one level, so the total number of AND gates is $3{\times}36 = 108$. With respect to the time delay, the generator of y^h involves the generation of the powers y^{2^i} (with a maximum delay of $3T_{XOR}$) and one level of $\mathbb{F}_{2^6}$ multipliers for the generation of y^3, y^{10} and y^{24} (with a maximum delay of $T_{AND} + 5T_{XOR}$). The generation of the $\zeta^i x^j$ terms needed for the computation of $f(y)$ requires a maximum of $3T_{XOR}$ and the sum of the three terms of $f(y)$ given in Eq. (3) needs $2T_{XOR}$. The generation of $f(y)^8$ requires $3T_{XOR}$ and the computation of $\zeta^{32} f(y)^8$ needs a delay of $2T_{XOR}$ for the sum of a maximum of 4 terms, as given in Eq. (19). The final sum of the four terms appearing in the expression of f_1 given in Eq. (4) also requires a maximum of $2T_{XOR}$. Therefore, the maximum delay of the module for the computation of f_1 is $T_{AND} + 20T_{XOR}$.

Finally, the combination of the theoretical complexities above for f_2^{-1} and f_1 lets us determine that the total number of 2-input AND gates needed for the hardware realization of the decomposition of Dillon's permutation is 756 and that the maximum delay is $3T_{AND} + 42T_{XOR}$ as shown in Table 3.

6.3 Univariate Representation of Dillon's Permutation

From the descriptions of the different modules of the univariate architecture given in Sect. 5 and Fig. 2, we can observe that the module for generating x^h is the only one involving $\mathbb{F}_{2^6}$ multiplications, so the number of 2-input AND gates is given by this module. As shown in Subsect. 5.1, this module requires the use of 32 $\mathbb{F}_{2^6}$ multipliers distributed in two levels, so the total number of AND gates is $32{\times}36 = 1152$. Table 3 shows the theoretical number of 2-input AND gates needed for the univariate expression.

With respect to time delay, the x^h module involves the generation of the powers x^{2^i} (with a maximum delay of $3T_{XOR}$) and the two levels of $\mathbb{F}_{2^6}$ multipliers for the generation of the remaining powers of x^h (therefore with a maximum delay of $2T_{AND} + 10T_{XOR}$). The next module performing the sum of powers x^j with common terms ζ^i given in Subsect. 5.2 presents a maximum delay of $4T_{XOR}$ corresponding to the addition of 12 terms as given in Eq. (23). The generator of the $\zeta^i x^j$ and $\zeta^i v_i$ terms given in Subsect. 5.3 needs a maximum of $3T_{XOR}$. Finally, the last module performing the sum of $\zeta^i x^j$ and $\zeta^i v_i$ terms given in Subsect. 5.4 has a maximum delay of $4T_{XOR}$ corresponding to the addition of 12 terms as given in Eq. (28). Therefore, the maximum theoretical delay of the univariate polynomial representation of Dillon's permutation is $2T_{AND} + 24T_{XOR}$ as shown in Table 3.

6.4 Comparison of Theoretical Complexities

As shown in Table 3, the decomposed architecture of Dillon's APN permutation has a reduction of 34.4% in the number of AND gates with respect to the univariate polynomial representation. However, it also presents a higher delay, with increases of 50% and 75% in T_{AND} and T_{XOR}, respectively, with respect to the univariate representation.

It must be noted that the theoretical complexity is not an exact predictor of area consumption and time delay when reconfigurable hardware (FPGA) is used for the implementation. For this reason, the following section presents experimental results when both architectures are implemented in commercial Xilinx FPGA devices.

Table 3. Theoretical complexities.

	#AND	Delay
Composition of f_1 and f_2^{-1}	756	$3T_{AND} + 42T_{XOR}$
Univariate polynomial	1152	$2T_{AND} + 24T_{XOR}$

7 Hardware Implementations

In order to compare the two architectures of Dillon's permutation (decomposed and univariate) given previously, we have produced FPGA implementations of the hardware architectures given in Sects. 4 and 5. The architectures have been described in VHDL, synthesized and implemented on Xilinx FPGA Artix-7 XC7A12T-3-CPG238 using VIVADO 2021.2. Experimental post place and route results are given in Table 4. It must be noted that both implementations fit in 6 LUTs (Lookup Tables). For this reason, and in order to show the differences between both architectures, Table 4 includes the number of 2-input XOR gates given by the Xilinx tool. The synthesizer supplies the total number of inferred

XORs, including XOR gates with multiple inputs and several bits each. In order to have a fair comparison of the different implementations, these XORs must be converted to equivalent 2-input XOR gates.

In Table 4, #XOR represents the number of 2-input XOR gates supplied by the synthesizer, *Delay* represents the delay (in nanoseconds) needed by each architecture to compute the output of Dillon's APN permutation, and *Area* × *Delay* represents the product of the number of 2-input XOR gates and the delay (less is better). The columns % represent the percentages (with respect to the highest value) corresponding to #XOR, *Delay* and *Area* × *Delay* metrics. From Table 4, we can observe that the implementation of the decomposed Dillon's permutation architecture presents the best results in the number of 2-input XORs and in *Area* × *Delay*, with reductions of up to 27.3% and 27.4%, respectively, in comparison to the univariate polynomial implementation. With respect to the experimental time results, both architectures exhibit almost the same delay.

Therefore, implementation results given in Table 4 show the efficiency of the new decomposed architecture for Dillon's APN permutation compared with its univariate polynomial representation when reconfigurable devices are used for the hardware implementation.

Table 4. Experimental complexities (using subexpressions sharing for the multipliers).

	#XOR	%	Delay (ns)	%	#XOR × Delay	%
Composition of f_1 and f_2^{-1}	1167	72.7	5.645	99.9	6587.72	72.6
Univariate polynomial	1605	100.0	5.651	100.0	9069.86	100.0

8 Conclusion

We have considered the recent representation of Dillon's permutation as the composition of a quadratic function and the inverse of a quadratic function. We have produced and implemented hardware implementations for this decomposition representation and the original univariate representation of Dillon's permutation. We have computed the Area and delay metric for both representations and have implemented the resulting architectures in FPGA. We have observed that using the decomposed representation can reduce the number of 2-input XOR gates by up to 27.3%, while the Area × Delay metrics can be reduced by up to 27.4% with respect to the univariate representation.

We have thus demonstrated that the choice of representation can significantly affect the implementation complexity. The choice of representation needs to be carefully considered when realizing vectorial Boolean functions in practice, especially in cases where the number of variables is high, or the computational environment has limited resources.

References

1. Biham, E., Shamir, A.: Differential cryptanalysis of DES-like cryptosystems. J. Cryptology **4**(1), 3–72 (1991)
2. Matsui, M.: Linear cryptanalysis method for DES cipher. In: Helleseth, T. (ed.) EUROCRYPT 1993. LNCS, vol. 765, pp. 386–397. Springer, Heidelberg (1994). https://doi.org/10.1007/3-540-48285-7_33
3. Nyberg, K.: Differentially uniform mappings for cryptography. In: Helleseth, T. (ed.) EUROCRYPT 1993. LNCS, vol. 765, pp. 55–64. Springer, Heidelberg (1994). https://doi.org/10.1007/3-540-48285-7_6
4. Chabaud, F., Vaudenay, S.: Links between differential and linear cryptanalysis. In: De Santis, A. (ed.) EUROCRYPT 1994. LNCS, vol. 950, pp. 356–365. Springer, Heidelberg (1995). https://doi.org/10.1007/BFb0053450
5. Advanced Encryption Standard, FIPS 197 (2001)
6. Nyberg, K., Knudsen, L.R.: Provable security against differential cryptanalysis. In: Brickell, E.F. (ed.) CRYPTO 1992. LNCS, vol. 740, pp. 566–574. Springer, Heidelberg (1993). https://doi.org/10.1007/3-540-48071-4_41
7. Hou, X.-D.: Affinity of Permutations on $\mathbb{F}_2^n$. Discret. Appl. Math. **154**, 313–325 (2006)
8. Calderini, M., Sala, M., Villa, I.: A note on APN permutations in even dimension. Finite Fields Appl. **46**, 1–16 (2017)
9. Dillon, J.F.: APN polynomials: an update. In: International Conference on Finite Fields and Applications Fq9 (2009)
10. Bilgin, B., Bogdanov, A., Knežević, M., Mendel, F., Wang, Q.: FIDES: lightweight authenticated cipher with side-channel resistance for constrained hardware. In: Bertoni, G., Coron, J.-S. (eds.) CHES 2013. LNCS, vol. 8086, pp. 142–158. Springer, Heidelberg (2013). https://doi.org/10.1007/978-3-642-40349-1_9
11. Carlet, C.: Boolean functions for cryptography and error correcting codes. In: Boolean Models and Methods in Mathematics, Computer Science, and Engineering, Ch. 8, pp. 257–397. Cambridge Univ. Press (2010)
12. Budaghyan, L., Carlet, C.: Classes of quadratic APN trinomials and hexanomials and related structures. IEEE Trans. Inf. Theory **54**(5), 2354–2357 (2008)
13. Budaghyan, L., Helleseth, T., Kaleyski, N.: A new family of APN Quadrinomials. IEEE Trans. Inf. Theory **66**(11), 7081–7087 (2020)
14. Calderini, M.: On the EA-classes of known APN functions in small dimensions. Cryptogr. Commun. **12**(5), 821–840 (2020). https://doi.org/10.1007/s12095-020-00427-1
15. Browning, K., Dillon, J., McQuistan, M., Wolfe, A.: An APN permutation in dimension six. In: Finite Fields: Theory and Applications FQ9 (Contemporary Mathematics), vol. 518, pp. 33–42 (2010)
16. Magma Computational Algebra System. Computational Algebra Group, University of Sydney. http://magma.maths.usyd.edu.au/magma/
17. Perrin, L., Udovenko, A., Biryukov, A.: Cryptanalysis of a theorem: decomposing the only known solution to the big APN problem. In: Robshaw, M., Katz, J. (eds.) CRYPTO 2016. LNCS, vol. 9815, pp. 93–122. Springer, Heidelberg (2016). https://doi.org/10.1007/978-3-662-53008-5_4
18. Imaña, J.L., Sánchez, J.M., Tirado, F.: Bit-parallel finite field multipliers for irreducible trinomials. IEEE Trans. on Comput. **55**(5), 520–533 (2006)
19. Imaña, J.L., Hermida, R., Tirado, F.: Low complexity bit-parallel multipliers based on a class of irreducible pentanomials. IEEE Trans. VLSI Syst. **14**(12), 1388–1393 (2006)

20. Imaña, J.L.: Efficient polynomial basis multipliers for type ii irreducible pentanomials. IEEE Trans. Circuits Syst. II: Express Briefs **59**(11), 795–799 (2012)
21. Rodríguez-Henríquez, F., Koç, Ç.K.: Parallel Multipliers Based on Special Irreducible Pentanomials. IEEE Trans. Comput. **52**(12), 1535–1542 (2003)

New Versions of Miller-loop Secured Against Side-Channel Attacks

Nadia El Mrabet[1], Loubna Ghammam[2(✉)], Nicolas Meloni[3], and Emmanuel Fouotsa[1,2,3,4]

[1] Department SAS, Mines Saint-Etienne, CEA-LETI, Centre CMP, 13541 Gardanne, France
nadia.el-mrabet@emse.fr, fouotsa.emmanuel@uniba.cm

[2] ITK Engineering GmbH, Im Speyerer Tal 6, 76761 Rülzheim, Germany
loubna.ghammam@itk-engineering.de

[3] Université de Toulon, Avenue de l'Université, BP20132, 83957 LA GARDE CEDEX, France
nicola.meloni@univ-tln.fr

[4] Higher Teacher Training College, Bambili The University of Bamenda, P.o.Box 39, Bambili, Bamenda, Cameroon

Abstract. In this paper, we propose two new versions of Miller algorithm in order to secure pairing computations against existing side-channel attacks (SCA). We have chosen to use the co-Z arithmetic on elliptic curves from which we derive two methods for pairing computations: one based on Euclidean addition chains and one based on Zeckendorf representation. We show that our propositions are resistant to existing side-channel attacks against pairing-based cryptography. We consider differential power analysis and fault attacks. The complexities of our solutions are compared with state-of-the-art one. We demonstrate that our new proposed versions are more efficient by **17%**.

Keywords: Miller algorithm · Euclidean addition chains · Fibonacci number · Optimal ate pairing · Side-channel attacks

1 Introduction

Pairing-based cryptography (PBC) provides several protocols such as short signature protocols and hierarchical encryption [1,2], making it a promising tool for the Internet of things (IoT) or cloud computing. In the first case, pairings would be implemented on constrained devices [3] such as microcontrollers and would be subject to invasive and non-invasive Side-Channel Attacks (SCA) [4–9]. Efficient and secure pairing computation has thus been an active research area [8,10,11].

The most efficient pairing computation algorithms are usually based on the Tate model [12] and rely on the Miller algorithm in order to compute the rational function $f_{s,Q}$ such that $\mathrm{Div}(f_{s,Q}) = s(Q) - ([s]Q) - (s-1)(\mathcal{O})$, where P and Q are two points of an elliptic curve defined over a finite field $\mathbb{F}_q$ and $\mathcal{O}$ represents

S. Mesnager and Z. Zhou (Eds.): WAIFI 2022, LNCS 13638, pp. 269–287, 2023.
https://doi.org/10.1007/978-3-031-22944-2_17

the identity element for the group of rational points of the elliptic curve. This function is computed thanks to the *double-and-add* loop called Miller loop [13] and followed by a final exponentiation.

Such algorithms are known to be natural targets to SCA [9,14–18]. Our contribution is to provide a new approach to Miller loop to replace the usual *double-and-add* loop by a loop based on addition chains. Such approaches, e.g., based on the *double-and-add* algorithm, have been proven to be an interesting alternative to standard scalar multiplication methods when security is in balance with speed, especially when combining to co-Z arithmetic [19]. Concretely, we provide two versions of the Miller algorithm, one based on Euclidean addition chain and the other version based on the Zeckendorf representation of the parameter controlling the number of iterations. We show that both versions are resistant to Side-Chanel Attacks. Though the computational cost of these versions of the Miller algorithm is higher than that of the classic Miller algorithm, they are the most efficient among other countermeasures like masking against SCA.

The paper is organized as follows. Sections 2 and 3 are devoted to the necessary mathematical background on pairing computations, co-Z arithmetic, and scalar multiplication using Euclidean additions chains (EAC). In Sect. 4, we present the new versions of the Miller algorithm, their complexities, and we compare them with the standard version. Section 6 is devoted to the state of the art on SCA and in Sect. 7 we analyze the security of our algorithms against some of those attacks. Finally, in Sect. 8, we provide detailed comparisons between our counter-measure, and other counter-measures used in the literature.

Notations
In this paper we denote by:

- M_e a multiplication in the field $\mathbb{F}_{p^e}$.
- S_e a squaring in the field $\mathbb{F}_{p^e}$.

A multiplication, a squaring and an inversion in $\mathbb{F}_p$ are respectively denoted by M, S and I.

2 Background on Pairing Computations

Let E be an elliptic curve defined over a prime field $\mathbb{F}_p$, with p a large prime number. Let r be a large prime divisor of $\#E(\mathbb{F}_p)$. In practice, (E, p, r) are provided using parametric families [20]. Let k be the smallest integer such that r divides $p^k - 1$, k is called the embedding degree of E relative to r. Let $G_1 = E(\mathbb{F}_p)[r]$ be the r-torsion subgroup of $E(\mathbb{F}_p)$. Let $G_2 = E'(\mathbb{F}_{p^{k/d}})[r] \cap$ Ker$(\pi_p - [p])$ where E' is the twist of E (if it exists) of degree d, π_p represents the Frobenius map over E and $[p]$ is the scalar multiplication by p over E. The subgroup of $\mathbb{F}^{\star}_{p^k}$ consisting of r-th roots of unity is denoted by $G_3 = \mu_r$. Let s be an integer constructed for the optimal Ate pairing [21] and depending on r and p. Then the (optimal) Ate pairing [21] is given by

$$T_r : G_1 \times G_2 \to G_3; (P, Q) \mapsto f_{s,Q}(P)^{(p^k-1)/r}.$$

The pairing computation is divided into two main steps [8, Chap. 3]. First, one has to compute $f_{s,Q}(P)$ using an iterative algorithm denoted as the Miller algorithm and described in Algorithm 1 below. As any algorithm based on the double-and-add algorithm, the overall cost directly depends on the length and Hamming weight of integer s.

Algorithm 1: Miller algorithm $(P, Q) \to f_{s,Q}(P)$

Require: $P \in G_1, Q \in G_2$, $s = (s_{n-1}, \ldots, \ldots s_0)$: binary representation of s
Ensure: $f_{s,Q}(P) \in G_3(\subset \mathbb{F}^*_{p^k})$
1: $f_1 \leftarrow 1$
2: $T \leftarrow Q$
3: **for** $i = n - 2$ **down to** 0 **do**
4: $\quad T \leftarrow [2]T$
5: $\quad f_1 \leftarrow f_1^2 \cdot l_1(P)$ where l_1 is the tangent to the point T.
6: $\quad$ **if** $r_i = 1$ **then**
7: $\quad\quad T \leftarrow T + Q$
8: $\quad\quad f_1 \leftarrow f_1 \cdot l_2(P)$ where l_2 is the line passing through T and P.
9: $\quad$ **end if**
10: **end for**
11: **return** f_1

The second step of computing the Tate pairing and its variants is the final exponentiation and consists of raising the final result of the main loop, $f_{s,Q}(P)$, to the power of $\frac{p^k-1}{r}$. The computation of this part can be simplified thanks to the k-th cyclotomic polynomial [3].

3 Co-Z Arithmetic on Elliptic Curves

Given an elliptic curve E over a finite field $\mathbb{F}_q$ there are many ways to compute the addition of two points. To avoid the computation of inversions in $\mathbb{F}_q$ when considering affine coordinates, it is standard to represent the points of E in Jacobian coordinates. A point (x, y) given in affine coordinates is represented in Jacobian coordinates by a triplet $(X : Y : Z)$ such that $x = X/Z^2$ and $y = Y/Z^3$, the point at infinity being then represented by $(0 : 1 : 0)$. A typical point addition costs 11 multiplications (M) and 5 squarings (S) in $\mathbb{F}_q$. Considerable literature now exists on the various point addition formulae, coordinate systems and curve shapes that can be used in a different context [22].

In the particular case of two points sharing the same Z-coordinate, the addition can be performed using only 5M+2S [19]. This operation is usually referred to as co-Z addition or ZADD. Finding two points sharing the same Z-coordinate is very unlikely but, when combined with the right scheme, one can perform a whole scalar multiplication using the co-Z addition. In this work, we considered two of those schemes: one based on Euclidean addition chains (EAC) and another based on the Zeckendorf representation of integers [19].

3.1 Euclidean Addition Chains

A Euclidean addition chain (EAC) computing an integer k is a sequence $n = (n_1, n_2, \ldots, n_w)$ such that $n_1 = 1$, $n_2 = 2$, $n_3 = 3$, $n_w = k$ and $\forall\ 3 \leqslant i \leqslant w-1$, if $n_i = n_{i-1} + n_j$, for some $j < i-1$, then we have $n_{i+1} = n_i + n_{i-1}$ (big step) or $n_{i+1} = n_i + n_j$ (small step). Such an addition chain, computing an integer greater than 3, can easily be represented by a binary chain $c_4, c_5, \ldots, c_w$ representing the succession of big steps (noted 0) and small steps (noted 1) starting from $(1, 2, 3)$. For instance, the EAC $(1, 2, 3, 4, 7, 10, 13, 23)$ can be represented as $(1, 0, 1, 1, 0)$. Another example is the Fibonacci sequence $(1, 2, 3, 5, 8, 13, 21, \ldots)$ which is only made of big steps.

Example 31. *The addition chain* (1, 2, 3, 4, 7, 11, 15, 19, 34, 53) *is an Euclidean addition chain computing* 53. *For example, in step 4, we have computed* $4 = 3 + 1$. *Then for step 5, we must add 3 or 1 to 4, that means that from step 4, we can compute* $7 = 4 + 3$ *or* $5 = 4 + 1$. *In our example, we have chosen to compute,* $7 = 4+3$ *(which is a big step). For step 6, we have computed* $11 = 7 + 4$ *(big step), etc.*

Finding such chains is simple: choose an integer g co-prime with k and then apply the subtractive form of Euclid's algorithm to those numbers.

Example 32. *We have presented* 53 *as an Euclidean addition chain. Let* $g = 34$, $\gcd(34, 53) = 1$. *Let apply now the subtractive form of Euclid's algorithm:*

$$53 - 34 = 19$$
$$34 - 19 = 15$$
$$19 - 15 = 4$$
$$15 - 4 = 11$$
$$11 - 4 = 7$$
$$7 - 4 = 3$$
$$4 - 3 = 1$$
$$3 - 1 = 2$$
$$2 - 1 = 1$$
$$1 - 1 = 0$$

The Euclidean addition chain computing 53 *is* (1, 2, 3, 4, 7, 11, 15, 19, 34, 53). *We find this chain by reading the first number of each line of the subtractive form of Euclid's algorithm.*

It has been shown that those chains can be an efficient alternative to the classical binary chain used in most scalar multiplication algorithms when side-channel attack resistance is an issue [19,23]. Algorithm 2 [19] describes how to perform a scalar multiplication using an EAC.

The main drawback of this approach is that, although it is easy to find an EAC computing a given integer k, finding a short one seems to be a challenging problem [24] making it difficult to use EAC for scalar multiplication directly. However, in the case of PBC, the Miller algorithm uses a fixed scalar which means that it is worth investigating finding an efficient EAC computing the parameter s. However, this method will become less efficient when s grows in size as finding a short EAC will become harder.

Algorithm 2. [19] Calculation of $[k]P$ where k is presented as an Euclidean Additions Chain.

Require: $P \in E, c = (c_4, \ldots, \ldots c_w)$ an EAC computing k
Ensure: $[k]P \in E$
  $(U_1, U_2) \leftarrow ([2]P, P)$
  for $i = 4$ **to** w **do**
    if $c_i = 0$ **then**
      $(U_1, U_2) \leftarrow (U_1 + U_2, U_1)$
    else
      $(U_1, U_2) \leftarrow (U_1 + U_2, U_2)$
    end if
  end for
  $U_1 \leftarrow U_1 + U_2$
  return U_1

3.2 Zeckendorf Representation

Let k be an integer and $(F_i)_{i \geq 0}$ the Fibonacci sequence, a classical result states that k can be uniquely written in the form $k = \sum_{i=2}^{l} d_i F_i$, with $d_i \in \{0, 1\}$ and $d_i d_{i+1} = 0$ [25]. An integer k written in this form is said to be in Zeckendorf representation and will be denoted as $k = (d_{l-1}, \ldots, d_2)_Z$. Such a representation is easy to compute as it can be obtained using a greedy algorithm [19].

As mentioned in [19,25], the disadvantage of the Zeckendorf representation is that it requires 44% more digits than the binary method requires. However, contrary to EAC, it is easy to bound its length [25]. Given the Zeckendorf representation of k, one can perform a scalar multiplication using the following Algorithm 3 [19].

Algorithm 3. [19] Computing $[k]P$ using Zeckendorf representation

Require: $P \in E, k = (d_n, \ldots, \ldots d_2)_Z$ Zeckendorf representation of k with $d_n = 1$
Ensure: $[k]P \in E$
1: $(U_1, U_2) \leftarrow (P, P)$
2: **for** $i = n-1$ **to** 2 **do**
3:   **if** $d_i = 1$ **then**
4:     $(U_1, U_2) \leftarrow (U_1 + P, U_2)$
5:   **end if**
6:   $(U_1, U_2) \leftarrow (U_1 + U_2, U_1)$
7: **end for**
8: **return** U_1

4 Co-Z Approach to the Miller Algorithm

In this section, we will present two variants of the Miller algorithm. We should remember that the basic idea of Miller's algorithm is to consider the binary representation of the Miller loop length parameter s and apply the *double-and-add* algorithm. The main idea of our work is to consider different representations of s to take advantage of co-Z formulae adapted for pairing computations.

4.1 Miller-Euclide

The idea is to adapt the scalar multiplication scheme described in Algorithm 2 to Miller loop and take advantage of the co-Z coordinates to limit the computational cost. Our approach, referred to as Miller-Euclide, is shown in Algorithm 4.

Algorithm 4 : Miller-Euclide, The computation of $f_{s,Q}(P)$ using an Euclidean addition chain.

Require: $P \in G_1$, $Q \in G_2$, $c = (c_4, \ldots, \ldots c_w)$ the euclidean addition chain computing s,
Ensure: $f_{s,Q}(P)$
1: $(T_1, T_2) \leftarrow ([2]Q, Q)$
2: $(f_1, f_2) \leftarrow (l_{Q,Q}(P), 1)$
3: **for** $i = 4$ **to** w **do**
4:   **if** $c_i = 0$ **then**
5:     $(f_1, f_2) \leftarrow (f_1 \times f_2 \times \ell_{T_1,T_2}(P), f_1)$
6:     $(T_1, T_2) \leftarrow (T_1 + T_2, T_1)$
7:   **else**
8:     $(f_1, f_2) \leftarrow (f_1 \times f_2 \times \ell_{T_1,T_2}(P), f_2)$
9:     $(T_1, T_2) \leftarrow (T_1 + T_2, T_2)$
10:   **end if**
11: **end for**
12: $f_1 \leftarrow f_1 \times f_2 \times \ell_{T_1+T_2,T_1}(P)$
13: **return** f_1

First, let us notice that the initialization consists of one doubling and the evaluation of the tangent line passing through Q. It has been shown [19] that the doubling is compatible with the co-Z coordinates without additional cost so that the initialization can be performed in 2 multiplications and 5 squarings for the doubling. We add to this cost 3 multiplications and one squaring for the evaluation of the tangent. The details of this computation are presented in Algorithm 5.

Algorithm 5. Initialization step of Algorithm 4 lines 1 and 2

Require: $P = (x_P, y_P)$ and $Q = (X_Q, Y_Q, Z_Q)$

Ensure: $T_1 = [2]Q, T_2 = Q$ sharing the same z-coordinate Z', $f_1 = \ell_{Q,Q}(P)$, $X_P = x_P Z'^2$ and $Y_P = y_P Z'^3$

1: $A \leftarrow X_Q^2$ $(S_{k/d})$
2: $B \leftarrow Y_Q^2$ $(S_{k/d})$
3: $C \leftarrow B^2$ $(S_{k/d})$
4: $D \leftarrow 2((X_1 + B)^2 - A - C)$ $(S_{k/d})$
5: $E \leftarrow 3A$
6: $F \leftarrow E^2$ $(S_{k/d})$
7: $X_{2Q} \leftarrow F - 2D$
8: $Y_{2Q} \leftarrow E(D - X_{2Q}) - 8C$ $(M_{k/d})$
9: $Z_{2Q} \leftarrow 2Y_Q Z_Q$ $(M_{k/d})$
10: $X_P \leftarrow x_P Z_{2Q}^2$ $(S_{k/d} + M_{k/d})$
11: $Y_P \leftarrow y_P Z_{2Q}^3$ $(2M_{k/d})$
12: $f_1, f_2 \leftarrow Y_p + X_p - 3A^2 + 16C, 1$
13: $T2 = (X_2, Y_2) \leftarrow D, 8C$
14: $T1 = (X_1, Y_1) \leftarrow X_{2Q}, Y_{2Q}$

The main loop of the new Miller-Euclide algorithm is only made of addition steps, one for each bit of the EAC, so that the computational cost of our method is simply linked to the length of the EAC representing s.

In order to evaluate the cost of our method we must evaluate the cost of a full addition step. Let $T_1 = (X_1, Y_1, Z)$ and $T_2 = (X_2, Y_2, Z)$ be two points with the same z-coordinate. From [19] we have that the coordinates (X_3, Y_3, Z_3) of the point $T_3 = T_1 + T_2$ can be computed using the following equations:

$$A = (X_2 - X_1)^2, B = X_1 A, C = X_2 A, D = (Y_2 - Y_1)^2$$
$$X_3 = D - B - C,$$
$$Y_3 = (Y_2 - Y_1)(B - X_3) - Y_1(C - B),$$
$$Z_3 = Z(X_2 - X_1).$$

This computation requires 5 multiplications and 2 squarings over $\mathbb{F}_{p^{k/d}}$. On top of that the quantities $X_1 A = X_1(X_2 - X_1)^2$ and $Y_1(C - B) = Y_1(X_2 - X_1)^3$ computed during the addition can be seen as the x and y-coordinates of the point $(X_1(X_2 - X_1)^2, Y_1(X_2 - X_1)^3, Z(X_2 - X_1)) \sim (X_1, Y_1, Z)$. Thus it is possible to add P_1 and $P_1 + P_2$ with the same formulae during the next iteration of the main loop.

Let us now consider the value of $l_{T_1,T_2}(P)$ given by the equation of the line joining T_1 and T_2 evaluated on $P = (x_P, y_P)$. The standard equation of such line in affine coordinates is given by

$$y_p = \frac{y_1 - y_2}{x_1 - x_2} x_P + \frac{y_2 x_1 - y_1 x_2}{x_1 - x_2}, \tag{1}$$

To convert this equation to co-Z coordinate, we just need to apply the transformation $(x_i, y_i) \mapsto \left(\frac{X_i}{Z^2}, \frac{Y_i}{Z^3}\right)$ which gives us

$$\begin{aligned}
y_p &= \frac{Y_1 - Y_2}{Z(X_1 - X_2)} x_P + \frac{Y_2 X_1 - Y_1 X_2}{Z^3(X_1 - X_2)} \\
&\Leftrightarrow y_p(X_1 - X_2)Z^3 = (Y_1 - Y_2)Z^2 x_p + Y_2 X_1 - Y_1 X_2 \\
&\Leftrightarrow y_p(X_1 - X_2)Z^3 = (Y_1 - Y_2)(x_p Z^2 - X_1) - Y_1(X_1 - X_2) \\
&\Leftrightarrow y_p Z'^3 = (Y_1 - Y_2)(x_p Z'^2 - X_1') - Y_1'
\end{aligned}$$

where $X_1' = X_1(X_1 - X_2)^2$, $Y_1' = Y_1(X_1 - X_2)^3$ and $Z' = Z(X_1 - X_2)$. Now we remark that all those values have been computed during the point addition as well as the values of $(X_1 - X_2)^2$ and $(X_1 - X_2)^3$. If we suppose that the values of $y_p Z^3$ and $x_p Z^2$ were stored from the previous iteration of the main loop, the evaluation of the line joining the points T_1 and T_2 required exactly 3 multiplications over $\mathbb{F}_{p^{k/d}}$. Moreover, the value of $Z' = Z(X_1 - X_2)$ itself is not needed during the whole process, so that we can spare one multiplication per iteration, for a total cost of 7 multiplications and 2 squarings over $\mathbb{F}_{p^{k/d}}$.

Finally, we have to compute $f_1 \times f_2 \times \ell_{T_1,T_2}(P)$ which requires one sparse multiplication between elements of $\mathbb{F}_{p^{k/d}}$ and $\mathbb{F}_{p^k}$ and one full multiplication over $\mathbb{F}_{p^k}$.

Algorithm 6 sums up the computations required to perform one addition step with their respective costs.

Therefore, each addition step of Algorithm 4 requires $M_k + M_{k,k/d} + 7M_{k/d} + 2S_{k/d}$ operations. At the end the total cost of the algorithm is $5M_{k/d} + 6S_{k/d} + (n-2)(M_k + M_{k,k/d} + 7M_{k/d} + 2S_{k/d})$.

4.2 Miller-Fibonacci

Let us now adapt Algorithm 3 to the Miller loop to present our new Algorithm 7 which computes $f_{s,Q}(P)$ using the Zeckendorf representation of s.

First, let us note that the first computation will be a doubling instead of an addition, independently of the value of s_{n-1}, but for the sake of simplicity, we only use the additive notation. For this particular step, we use Algorithm 5.

Each step of this algorithm is similar to Miller-Euclide. If the current digit is a 0, we use the procedure described in Algorithm 6. If the current digit is a 1 then we have to perform an additional sparse multiplication and then the evaluation of the line passing through two points of the elliptic curve. To do

---

Algorithm 6. Computing one addition step of Algorithm 4

Require: T_1, T_2 with same z-coordinate Z, $X_p = x_p Z^2, Y_p = y_P Z^3$ and $f_1, f_2 \in \mathbb{F}_{p^k}$
Ensure: $Z' = Z(X_1 - X_2), T_3 = T_1 + T_2, T_1 = (X_3, Y_3, Z'), T_2 = (X_1 Z'^2 Y_1 Z'^3, Z'), X_p = x_p Z'^2, Y_p = y_P Z'^3, f_1 = f_1 \times f_2 \times \ell_{T_1,T_2}(P)$

1: $A \leftarrow (X_1 - X_2)^2$ $(S_{k/d})$
2: $B \leftarrow X_1 A$ $(M_{k/d})$
3: $C \leftarrow X_2 A$ $(M_{k/d})$
4: $D \leftarrow (Y_1 - Y_2)^2$ $(S_{k/d})$
5: $E \leftarrow C - B$
6: $F \leftarrow Y_1 E$
7: $X_3 \leftarrow D - B - C$
8: $Y_3 \leftarrow (Y_2 - Y_1)(B - X_3) - F$ $(2M_{k/d})$
9: $X_P \leftarrow X_P A$ $(M_{k/d})$
10: $Y_P \leftarrow Y_P E$ $(M_{k/d})$
11: $L \leftarrow Y_P - (Y_1 - Y_2)(X_P - X_3) - Y_3$ $(M_{k/d})$
12: $f_3 \leftarrow f_1 L$ $(M_{k,k/d})$
13: $f_3 \leftarrow f_1 f_2$ (M_k)
14: $f_1, f_2 \leftarrow f_3, f_1$
15: $X_2, Y_2 \leftarrow B, F$
16: $X_1, Y_1 \leftarrow X_3, Y_3$

Algorithm 7: Miller-Fibonacci: The computation of $f_{s,Q}(P)$ using Fibonacci sequences

---

Require: $P \in G_1$, $Q \in G_2$, $(d_n, \ldots, \ldots d_2)$ The Zeckendorf representation of s,
Ensure: $f_{u,Q}(P)$

1: $(T_1, T_2) \leftarrow (Q, Q)$
2: $(f_1, f_2) \leftarrow (1, 1)$
3: **for** $i = n - 1$ **down to 2 do**
4:     **if** $d_i = 1$ **then**
5:         $(f_1, f_2) \leftarrow (f_1 \times l_{T_1,T_2}(P), f_2)$
6:         $(T_1, T_2) \leftarrow (T_1 + Q, T_2)$
7:     **end if**
8:     $(f_1, f_2) \leftarrow (f_1 \times f_2 \times l_{T_1,T_2}(P), f_1)$
9:     $(T_1, T_2) \leftarrow (T_1 + T_2, T_1)$
10: **end for**
11: **return** f_1

---

so, we must first upgrade the coordinates of T_1 and Q so that they share the same z-coordinate as well as the values of X_p and Y_p (this is easily done in 7 multiplications and 2 squarings) and then use the formulae given in Algorithm 6. On top of that, we have to keep track of the z-coordinate of T_1 during the whole process which adds another multiplication. In the end the cost when the $d_i = 0$ is $M_k + M_{k,k/d} + 8M_{k/d} + 2S_{k/d}$ plus $7M_{k/d} + 2S_{k/d}$ if $d_i = 1$. The total cost of Algorithm 7 is then $5M_{k/d} + 6S_{k/d} + (n-2)(M_k + M_{k,k/d} + 7M_{k/d} + 2S_{k/d}) + hz(s)(7M_{k/d} + 2S_{k/d})$ where $hz(s)$ is the hamming weight of the Zeckendorf representation of s.

5 Comparison

In this section, we compare three versions of the Miller algorithm that allow us to compute the rational function $f_{s,Q}$ evaluated at P: Miller's algorithm, Miller-Euclide, and Miller-Fibonacci. For the sake of simplicity, we restraint our comparisons to the computation of the Optimal Ate pairing over BN curves [26] and BLS 12 curves [27].

After Barbulescu and Duquesne results presented in [28], it is recommended to use BLS12 for computing Optimal Ate pairing for the 128-bit security level instead of BN curves. However, BN curves are still considered in practice for several schemes [29,30]. That's why, in this paper, we consider the two curves in our computation.

Both of these curves have the embedding degree $k = 12$, therefore, the arithmetic in their extension tower has the same complexity (since we don't consider additions in our cost evaluation). The tower extension used for pairing computation on both curves is given by: The field $\mathbb{F}_{p^{12}}$ is built using the following extension tower.

- $\mathbb{F}_{p^2} = \mathbb{F}_p[\alpha]/(\alpha^2 + 1)$
- $\mathbb{F}_{p^6} = \mathbb{F}_{p^2}[\beta]/(\beta^3 - (\alpha + 1)))$
- $\mathbb{F}_{p^{12}} = \mathbb{F}_{p^6}[\gamma]/(\gamma^2 - \beta)$

In Table 1, we also present the cost of each operation needed for computing Miller algorithm using projective coordinates [31] and also the cost of the operations needed for our new versions of Miller's algorithm.

Table 1. Cost of necessary operations for Miller loop

Operation	Cost in $\mathbb{F}_p$
Multiplication in $\mathbb{F}_{p^{12}}$	54 M
Sparse Multiplication in $\mathbb{F}_{p^{12}}$	39 M
Squaring in $\mathbb{F}_{p^{12}}$	36 M
Multiplication in $\mathbb{F}_{p^2}$	3 M
Squaring in $\mathbb{F}_{p^2}$	2 M
Addition step for Miller classic (Projective coordinates)	80 M
Doubling step for Miller classic (Projective coordinates)	100 M
Initialization step for Miller-Euclide/Fibonacci	27 M
Addition step for Miller-Euclide	118 M
Addition step for Miller-Fibonacci(for $u_i = 0$)	121 M
Addition step for Miller-Fibonacci (for $u_i = 1$)	146 M

Note that these two curves, BN and BLS12, have a twist of degree $d = 6$ and the twist isomorphism is given by the following map.

$$\begin{aligned}\Psi : E'\left(\mathbb{F}_{p^2}\right) &\to E(\mathbb{F}_{p^{12}})\\ (x_{Q'}, y_{Q'}) &\mapsto (\gamma^2 x_{Q'}, \gamma^3 y_{Q'})\end{aligned}$$

5.1 BN Curves

A Barreto-Naehrig (BN) curve [26] is an elliptic curve $E : y^2 = x^3 + b$ defined over a finite prime field $\mathbb{F}_p$ such that

- $b \in \mathbb{F}_p$,
- $E(\mathbb{F}_p)$ has prime order $n = \#E(\mathbb{F}_p)$,
- $p = p(u) = 36u^4 + 36u^3 + 24u^2 + 6u + 1$, and
  $n = n(u) = 36u^4 + 36u^3 + 18u^2 + 6u + 1$ for some $u \in \mathbb{Z}$.

We choose the following elliptic curve which is efficient in practice [31]:

$$E(\mathbb{F}) : \; y^2 = x^3 + 2.$$

The twisted elliptic curve is defined by the equation:

$$E'(\mathbb{F}_{p^2}) : y^2 = x^3 + (1 - i).$$

Recall that the Optimal Ate pairing over BN curves is the following map:

$$\begin{aligned}&E(\mathbb{F}_p)[r] \times \Psi(E'(\mathbb{F}_{p^2}))[r] \longrightarrow \mathbb{F}^\star_{p^{12}}\\ &\qquad (P, Q) \longmapsto = \left(\left(f_{6u+2,Q}(P) l_{[6u+2]Q,\pi_p(Q)}(P) l_{[6u+2]Q,\pi_p^2(Q)}(P)\right)\right)^{\frac{p^{12}-1}{r}}\end{aligned}$$

Using the costs presented in Table 1, we obtain the comparison between the different versions of Miller's algorithm to compute $f_{6u+2,Q}(P)$, in terms of operations in $\mathbb{F}_p$. This comparison is presented in Table 2. We recall that the pairing parameter in our case is $s = 6u + 2$ with $u = -2^{114} + 2^{101} - 2^{14} - 1$. In this case, it is possible to find a EAC of length 168 (using the subtractive euclidean algorithm with parameter s with $s = 174601120802286487002979030408532519$ when we consider $u = 47085091321987640844053613709579372$). Moreover, the Zeckendorf representation of $s = 6u + 2 = Fibo(169) + Fibo(61) + Fibo(99)$ with a chosen parameter $u = 15533701296897238962071235758772159$, has a length of 168 and its hamming weight is 3. Thus, we have:

From this Table 2, we can notice that the classical Miller algorithm is more efficient than our two new versions. However, it is not secure when considering DPA attack. We show in the next section that our method protects against this attack.

Table 2. Cost of each Miller algorithm for BN curves

Method	Complexity in $\mathbb{F}_p$	
Miller classic Algorithm 1	12068 M	
Miller-Euclide Algorithm 4	19991 M	**+65,5%**
Miller-Fibonacci Algorithm 7	20683 M	**+71,3%**

5.2 BLS 12 Curves

In 2002, Barreto, Lynn and Scott presented in [27] a method to generate pairing-friendly elliptic curves over a prime field $\mathbb{F}_p$ with embedding degree $k = 12$. BLS12 are defined over $\mathbb{F}_p$ by the following equation:

$$E : y^2 = x^3 + b$$

and by a parameter $u \in \mathbb{Z}$ such that:

$$\begin{cases} p = (u-1)^2(u^4 - u^2 + 1)/3 + u \\ r = u^4 - u^2 + 1 \\ t = u + 1 \end{cases} \tag{2}$$

where t is the trace of the Frobenius map on the curve. The parameter u is chosen such that p and r are prime and have the sizes corresponding to the desired security level. For the 128-bit security level and as recommended in [28], p and r are of at least 461 and 308 and the proposed parameter $u = -2^{77} + 2^{50} + 2^{33}$. The choosed BLS12 elliptic curve is defined over $\mathbb{F}_{\text{,}}$ by $E(\mathbb{F}) : \ y^2 = x^3 + 4$ which admits a twist of degree $d = 6$ and given by $E'(\mathbb{F}_{p^2}) : y^2 = x^3 + 4(1 - i)$.

The Optimal Ate pairing on BLS12 curves is defined by the following map.

$$\begin{aligned} E(\mathbb{F}_p)[r] \times \Psi(E'(\mathbb{F}_{p^2}))[r] &\longrightarrow \mathbb{F}^{\star}_{p^{12}} \\ (P, Q) &\longmapsto = ((f_{u,Q}(P))^{\frac{p^{12}-1}{r}}. \end{aligned}$$

Using the costs presented in Table 1, we obtain the comparison between the different versions of the Miller algorithm to compute $f_{u,Q}(P)$, in terms of operations in $\mathbb{F}_p$. This comparison is presented in Table 3. We recall that the pairing parameter in our case is $u = -2^{77} + 2^{50} + 2^{30}$. In this case, it is possible to find a EAC of length 112 (using the subtractive euclidean algorithm with parameter $u' = 204035480723636378792533$ and $s' = 126100861998131272870737$). Note that u' is chosen such that p and r are primes. Moreover, the Zeckendorf representation of $u = 704925247670891258604\,97 = Fibo(3) + Fibo(7) + Fibo(24) + Fibo(111)$ has length 111 and its hamming weight is 4 (u is chosen such that we obtain an efficient Zeckendorf representation where p and r are primes. Thus, we have:

Table 3. Cost of each Miller algorithm for BLS12

Method	Complexity in $\mathbb{F}_p$	
Miller classic Algorithm 1	7438 M	
Miller-Euclide Algorithm 4	13147 M	**+76,8%**
Miller-Fibonacci Algorithm 7	13936 M	**+87,3%**

From this Table 3, we can notice that the classical Miller algorithm is more efficient than our two new versions. However, it is not secure when considering DPA attack. We show in the next section that our method protects against this attack.

6 SCA Against Miller's Algorithm

The SCA re the perturbation analysis that are invasive and observation analysis that are non-invasive. Non-invasive attacks against pairing, such as DPA, CPA, template attack, have been widely studied [4,6,9,14,15,17,32,33], so as the invasive attacks (basically fault attacks) [7,8]. In this section, we briefly recall the non-invasive attacks against pairing, then, we will illustrate the fact that our new algorithms seem to be resistant to existing attacks.

6.1 Non-invasive Attacks

Non-invasive attacks can use for instance the power consumption or electromagnetic emission of a device. Those attacks are also called observation analysis, they include Differential Power Analysis, Correlation Power Analysis, Template attacks and can be performed in vertical or horizontal mode. We will describe the resistance of our algorithms by using the DPA attack model to resume the non invasive attacks. Those attacks can allow an attacker to compute the intermediate values within cryptographic computations through statistical analysis of data collected from multiple cryptographic operations. In pairing-based cryptography, the secret is one of the inputs, either the point P either the point Q. The public data is the other point. The vulnerabilities of a pairing implemented with a classical Miller algorithm have been illustrated in several articles [4,17,32–38]. The scheme is as follow:

1. First choose the secret (P or Q).
2. Then find during the algorithm an operation that involves both the secret data and a data that can be manipulated freely during the attack. This operation is the target of the attack. In pairing-based cryptography, the freely chosen data are the coordinates of the public point.
3. For a vertical attack: the operation is executed at a specific time t_0, the statistical analysis is performed on a collection of traces for the same secret point and hundred to millions of public point.

4. For a horizontal attack: the same secret data is used more than once during a single algorithm execution, the statistical analysis is performed using one public entry at different times t_i.

6.2 Existing Countermeasure

To avoid SCA in pairing-based cryptography, several counter-measures have been proposed in the literature:

1. The counter-measure proposed by Öztürk et al. in [39]. The principle of this counter-measure is to avoid any perturbation against Miller's algorithm using resilient counters. The disadvantage of this counter-measure is that it is a proof of concept implementation and doesn't have a theoretical basis for measuring its security.
2. The second counter-measure is proposed by Gosh et al. in [40]. It consists of implementing a modified version of the Miller algorithm. Unfortunately, this method introduces an additional calculation which is expensive and which, moreover, does not really improve security for the Miller algorithm.
3. The use of the homogeneity property of projective coordinates (or Jacobian coordinates) of the point P or of the point Q, is a counter-measure against this attack on Miller's algorithm [34]. This counter-measure causes a modification of the result of the Miller algorithm: thus, we obtain a multiple of the final result of Miller algorithm. However, this is not a problem because this multiple being in the subfield of $\mathbb{F}_{p^k}$, will be eliminated by the final exponentiation. The advantage of this counter-measure is that it is not expensive. However, unfortunately, El Mrabet et al. have shown in [41] that this is not enough to protect Miller's algorithm.
4. Another efficient counter-measure is to choose two integers a et b such that $a \times b = 1 \mod r$ [34]. Then, the idea is to compute the pairings between the points $[a]P$ and $[b]Q$ instead of computing the pairing between P and Q. This operation is possible thanks to the bilinearity of pairings, we have:
$$e([a]P,\ [b]Q) = e(P,\ Q)^{ab}.$$
Also, thanks to the final exponentiation, the exponent, ab will be canceled. Finally, we obtain:
$$e(P,\ Q)^{ab\times(\frac{p^k-1}{r})} = e(P,\ Q)^{(\frac{p^k-1}{r})}.$$
5. Finally, we cite the most considered counter-measure: **masking** [34]. When the secret is the point P, this counter-measure consists of choosing a random point R of the elliptic curve E such that $e(P, R)$ is defined, then computing the pairing between P and Q passing through the point R. Using the property of bilinearity, we obtain:
$$e(P,\ Q) = e(P,\ Q + R) \times e(P,\ -R).$$
Therefore, this counter-measure consists of computing two pairings instead of one and then performing their product.

The last two counter-measures are the most considered in practice. In a theoretical study of the various existing counter-measures, El Mrabet et al. suggested in [7] to consider the counter-measure by masking to protect the Miller algorithm against SCA attacks. The counter-measure based on the bilinearity of pairings implies the computation of two scalar multiplications, one on $E(\mathbb{F}_p)$, and one on $E(\mathbb{F}_{p^k})$, both of them are scalar multiplication by an integer of order r. Make these two computations secure against side channel attack is a classical problem in elliptic curve cryptography. In [42], Feix et al. demonstrate that even a secure scalar multiplication can be sensitive to side channel attack. The use of that counter-measure in pairing-based cryptography would imply to add two other sensitive computations with their counter-measure [19,23,43]. On the other side the masking counter-measure, imply one addition over E and in the worst case two pairing computations shorter than the scalar multiplication. As a consequence, we consider the masking counter-measure in the sequel of the article.

7 Resistance of Our Methods Against Non-invasive Attacks

Since Miller-Fibonacci algorithm is based on the operation used in the Miller-Euclide so that proving that it is secure against a non invasive would show that the Miller-Euclide is also secured.

The Miller-Fibonacci algorithm is based on writing the pairing parameter s as a sequence of Fibonacci numbers. In our new algorithm, we have only additions steps. Therefore we only have to evaluate the line passing through the points T_1 and T_2 (or T_1 and Q) of respective coordinates (X_1, Y_1, Z) and (X_2, Y_2, Z) evaluated on the public point P.

Therefore, the only equation involving known data that the non invasive attack can use is the following one:

$$\begin{aligned} l_{T_1,T_2}(P) = y_P Z^3(X_1 - X_2)^3 - Y_1(X_1 - X_2)^3 \\ -(Y_1 - Y_2)\left(x_P Z^2(X_1 - X_2)^2 - X_1(X_1 - X_2)^2\right) \end{aligned}$$

The DPA attack model in pairing-based cryptography [4,34], is efficient when we have an operation that involved a known data and one of the temporary variables X_1, X_2, Y_1, Y_2 or Z. If the secret is the point Q, the known values are x_p and y_p. With a DPA we can find $Z^2(X_1 - X_2)^2$ using the multiplication $x_P Z^2(X_1 - X_2)^2$ and $Z^3(X_1 - X_2)^3$ using $y_P Z^3(X_1 - X_2)^3$. This gives us access to the value $Z(X_1 - X_2)$. Recall that our algorithm is described using the co-Z arithmetic, as a consequence the value of Z is the same for the points at each execution step of the Miller-Fibonacci algorithm. We analyze now which information can be found given two hypotheses on Z, either $Z_Q = 1$ or $Z_Q \neq 1$. If $Z_Q = 1$, i.e. $Z = 1$, then using a vertical DPA attack we can find the values $(X_1 - X_2)$, $(X_1 - X_2)^2$, and $(X_1 - X_2)^3$. Then using another DPA we can find either Y_1 using $Y_1(X_1 - X_2)^3$ or X_1 using $X_1(X_1 - X_2)^2$. With $Z = 1$ and

X_1 or Y_1 and the elliptic curve equation, we can find Y_1 or X_1. The attack would be successful. But if $Z_Q \neq 1$, we cannot perform the attack. Indeed, we made the hypothesis that Q is secret, then Z_Q is also secret and different from 1. We do not know either Z, neither X_1 and X_2 as they are derived from Q. We can make hypotheses on the value of Z and then on $(X_1 - X_2)$, but we cannot try all the possible values of Z in $\mathbb{F}_p$. As a consequence, as long as $Z_Q \neq 1$, our Miller-Fibonacci algorithm, based on presenting the parameter pairing s with the Zeckendorf representation, is natively protected against a DPA attack. Indeed, once we eliminate the leakage of information during the operations $y_P Z^3(X_1 - X_2)^3$ and $x_P Z^2(X_1 - X_2)^2$, the other registers depend only on the secret point and consequently are protected against any vertical DPA analysis. In fact, our algorithm is also secure against horizontal attacks. Indeed, in horizontal attacks, the hypothesis is that a same entry is used more than once during one single execution of the algorithm. Using the co-Z arithmetic, we use the same Z coordinates during one step of the Miller algorithm. It is not sufficient to perform an horizontal SCA.

The same analysis can be done for Miller-Euclide which is also protected against SCA as long as $Z \neq 1$.

8 Comparison

In the Sect. 7, we have proved that our new algorithms resist against DPA attacks. In this section, we compare their respective complexities with a classical Miller algorithm protected again DPA attack. We choose to compare these two counter-measures in the context of computing Optimal Ate pairing on BN and BLS 12 curves for the reasons explained in the Introduction.

Remark 81. *Our methods for computing the Miller loop using Euclidean addition chains or Fibonacci Sequences are available for computing all pairings and in any elliptic curve.*

We have to evaluate $f_{6u+2,Q}(P)$ using Miller algorithm where $6u + 2$ is an integer of 114 bits and Hamming weight 4.

We present the following table the cost of our counter-measure by comparing it with the counter-measure by masking.

Table 4. Comparison between the costs of the counter-measures

Algorithm	DPA secure	Cost BN	Cost BLS12
Miller	No	12068 M_p	7708 M_p
Miller+masking	Yes	24136 M_p	15416 M_p
Miller-Euclide	Yes	19991 M_p	13147 M_p
Miller-Fibonacci	Yes	20683 M_p	13936 M_p

Table 4 shows that our counter-measures are more efficient than the counter-measure proposed in the literature. The Miller-Euclide and the Miller-Fibonacci algorithm are, respectively, 16% and 17% faster than the Miller standard algorithm combined with the masking.

9 Conclusion

In this paper, we have proposed two new versions of Miller's algorithm: Miller-Euclide and Miller-Fibonacci. The complexity of these algorithms depends on the Miller loop parameter. Therefore, we showed how to choose a short representation for the parameter s which makes our algorithms as efficient in terms of complexity as possible.
We proved also in this paper that these two algorithms resist against Differential Power Analysis attacks.

We compare the complexity of our new algorithm Miller-Fibonacci with the most efficient counter-measure of the classical Miller algorithm. We proved that our proposal Miller-Fibonacci algorithm is more efficient than the counter-measure by masking about 10.7% in the worst case and in the best case about 20%. We only investigate the vertical power analysis, we leave as an open problem the analysis of our algorithms against horizontal power analysis. Indeed, this kind of analysis needs implementations and practical attacks.

References

1. National Institute of Standards and Technology. http://csrc.nist.gov/publications/PubsSPs.html
2. Shamir, A.: Identity-based cryptosystems and signature schemes. In: Blakley, G.R., Chaum, D. (eds.) Identity-based cryptosystems and signature schemes. LNCS, vol. 196, pp. 47–53. Springer, Heidelberg (1985). https://doi.org/10.1007/3-540-39568-7_5
3. Duquesne, S., Ghammam, L.: Memory-saving computation of the pairing final exponentiation on BN curves. Groups Complex. Cryptol. **8**(1), 75–90 (2016)
4. El Mrabet, N., Di Natale, G., Flottes, M.L.: A practical differential power analysis attack against the miller algorithm. In: PRIME 2009–5th Conference on Ph.D. Research in Microelectronics and Electronics, Circuits and Systems Magazine, IEEE Xplore (2009)
5. Blömer, J., Günther, P., Liske, G.: Improved side channel attacks on pairing based cryptography. In: Prouff, E. (ed.) COSADE 2013. LNCS, vol. 7864, pp. 154–168. Springer, Heidelberg (2013). https://doi.org/10.1007/978-3-642-40026-1_10
6. Unterluggauer, T., Wenger, E.: Practical attack on bilinear pairings to disclose the secrets of embedded devices. In: 2014 Ninth International Conference on Availability, Reliability and Security, pp. 69–77 (2014)
7. El Mrabet, N., Fournier, J.J.A., Goubin, L., Lashermes, R.: A survey of fault attacks in pairing based cryptography. Cryptogr. Commun. **7**(1), 185–205 (2015)
8. El Mrabet, N., Joye, M.: Guide to Pairing-Based Cryptography. CRC Cryptography and Network Security Series, Chapman & Hall/CRC Press, Boca Raton (2017)

9. Jauvart, D., El Mrabet, N., Fournier, J.J.A., Goubin, L.: Improving side-channel attacks against pairing-based cryptography. J. Cryptogr. Eng. **10**, 1–6 (2019)
10. Chatterjee, S., Hankerson, D., Menezes, A.: On the efficiency and security of pairing-based protocols in the type 1 and type 4 settings. In: Hasan, M.A., Helleseth, T. (eds.) WAIFI 2010. LNCS, vol. 6087, pp. 114–134. Springer, Heidelberg (2010). https://doi.org/10.1007/978-3-642-13797-6_9
11. Azarderakhsh, R., et al.: Fast software implementations of bilinear pairings. IEEE Trans. Depend. Secur. Comput. **14**(6), 605–619 (2017)
12. Frey, G., Müller, M., Rück, H.-G.: The TATE pairing and the discrete logarithm applied to elliptic curve cryptosystems. IEEE Trans. Inf. Theory **45**(5), 1717–1719 (1999)
13. Miller, V.S.: The Weill pairing, and its efficient calculation. J. Cryptol. **17**(4), 235–261 (2004)
14. Page, D., Vercauteren, F.: Fault and side channel attacks on pairing based cryptography. IEEE Trans. Comput. **55–9**, 1075–1080 (2006)
15. Whelan, C., Scott, M.: Side channel analysis of practical pairing implementations: which path is more secure? In: Nguyen, P.Q. (ed.) VIETCRYPT 2006. LNCS, vol. 4341, pp. 99–114. Springer, Heidelberg (2006). https://doi.org/10.1007/11958239_7
16. Unterluggauer, T., Wenger, E.: Practical attack on bilinear pairings to disclose the secrets of embedded devices. In: 2014 Ninth International Conference on Availability, Reliability and Security, pp. 69–77, September 2014
17. Jauvart, D., Fournier, J.J.A., El-Mrabet, N., Goubin, L.: Improving side-channel attacks against pairing-based cryptography. In: Cuppens, F., Cuppens, N., Lanet, J.-L., Legay, A. (eds.) CRiSIS 2016. LNCS, vol. 10158, pp. 199–213. Springer, Cham (2017). https://doi.org/10.1007/978-3-319-54876-0_16
18. Nadia El Mrabet, Louis Goubin, Sylvain Guilley, Jacques Fournier, Damien Jauvart, Martin Moreau, Pablo Rauzy, and Franck Rondepierre. Physical Attacks, chapter 12. CRC Press, 2016
19. Meloni, N.: New point addition formulae for ECC applications. In: Arithmetic of Finite Fields, First International Workshop, WAIFI 2007, Madrid, Spain, June 21–22, 2007, Proceedings, pp. 189–201 (2007)
20. Freeman, D., Scott, M., Teske, E.: A taxonomy of pairing-friendly elliptic curves. J. Cryptol. **23**(2), 224–280 (2010)
21. Vercauteren, F.: Optimal pairings. IEEE Trans. Inf. Theory **56**(1), 455–461 (2010)
22. Cohen, H., et al.: Handbook of Elliptic and Hyperelliptic Curve Cryptography, 2nd edn. Chapman & Hall/CRC, Boca Raton (2012)
23. Herbaut, F., Liardet, P.-Y., Méloni, N., Téglia, Y., Véron, P.: Random Euclidean addition chain generation and its application to point multiplication. In: Gong, G., Gupta, K.C. (eds.) INDOCRYPT 2010. LNCS, vol. 6498, pp. 238–261. Springer, Heidelberg (2010). https://doi.org/10.1007/978-3-642-17401-8_18
24. Knuth, D., Yao, A.: Analysis of the subtractive algorithm for greater common divisors, pp. 4720–4722, December 1975
25. Zeckendof, E.: Représentations des nombres naturels par une somme de nombre de Fibonacci ou de nombres de Lucas. Bulletin de la Société Royale des Sciences de Liège (1972)
26. Pereira Geovandro, C.C.F., Simplício Jr., M.A., Naehrig, M., Barreto,P.S.L.M.: A family of implementation-friendly BN elliptic curves. J. Syst. Softw. **84**(8), 1319–1326 (2011)
27. Barreto, P.S.L.M., Lynn, B., Scott, M.: Constructing elliptic curves with prescribed embedding degrees. In: Security in Communication Networks (2002)

28. Barbulescu, R., Duquesne, S.: Updating key size estimations for pairings. J. Cryptol. **32**, 1298–1336 (2019)
29. Black-box wallets: Fast anonymous two-way payments for constrained devices. IACR Cryptology ePrint Archive, p. 1199 (2019)
30. Canard, S., Diop, U., Kheir, N., Paindavoine, M., Sabt, M.: Blindids: market-compliant and privacy-friendly intrusion detection system over encrypted traffic. In: AsiaCCS 2017, Abu Dhabi, Emirats Arabes Unis, 2–6 avril, pp. 561–574 (2017)
31. Duquesne, S., El Mrabet, N., Haloui, S., Rondepierre, F.: Choosing and generating parameters for low level pairing implementation on BN curves. IACR Cryptol. ePrint Archive **2015**, 1212 (2015)
32. Ghosh, S., Roychowdhury, D.: Security of prime field pairing cryptoprocessor against differential power attack. In: Joye, M., Mukhopadhyay, D., Tunstall, M. (eds.) InfoSecHiComNet 2011. LNCS, vol. 7011, pp. 16–29. Springer, Heidelberg (2011). https://doi.org/10.1007/978-3-642-24586-2_4
33. Pan, W., Marnane, W.P.: A correlation power analysis attack against Tate pairing on FPGA. In: Koch, A., Krishnamurthy, R., McAllister, J., Woods, R., El-Ghazawi, T. (eds.) ARC 2011. LNCS, vol. 6578, pp. 340–349. Springer, Heidelberg (2011). https://doi.org/10.1007/978-3-642-19475-7_36
34. Page, D., Vercauteren, F.: Fault and Side-Channel Attacks on Pairing Based Cryptography (2004)
35. Whelan, C., Scott, M.: Side channel analysis of practical pairing implementations: which path is more secure? In: VIETCRYPT 2006, pp. 99–114 (2006)
36. Mrabet, N.: What about vulnerability to a fault attack of the Miller's algorithm during an identity based protocol? In: Park, J.H., Chen, H.-H., Atiquzzaman, M., Lee, C., Kim, T., Yeo, S.-S. (eds.) ISA 2009. LNCS, vol. 5576, pp. 122–134. Springer, Heidelberg (2009). https://doi.org/10.1007/978-3-642-02617-1_13
37. Unterluggauer, T., Wenger, E.: Practical attack on bilinear pairings to disclose the secrets of embedded devices. In: ARES, pp. 69–77 (2014)
38. El Mrabet, N., Fouotsa, E.: Failure of the point blinding countermeasure against fault attack in pairing-based cryptography. In: El Hajji, S., Nitaj, A., Carlet, C., Souidi, E.M. (eds.) C2SI 2015. LNCS, vol. 9084, pp. 259–273. Springer, Cham (2015). https://doi.org/10.1007/978-3-319-18681-8_21
39. Öztürk, E., Gaubatz, G., Sunar, B.: Tate pairing with strong fault resiliency. In: Fourth International Workshop on Fault Diagnosis and Tolerance in Cryptography, 2007, FDTC 2007, Vienna, Austria, 10 September 2007, pp. 103–111 (2007)
40. Ghosh, S., Mukhopadhyay, D., Chowdhury, D.R.: Fault attack, countermeasures on pairing based cryptography. I. J. Netw. Secur. **12**(1), 21–28 (2011)
41. Lashermes, R., Paindavoine, M., El Mrabet, N., Fournier, J.J.A., Goubin, L.: Practical validation of several fault attacks against the miller algorithm. In: 2014 Workshop on Fault Diagnosis and Tolerance in Cryptography, FDTC 2014, Busan, South Korea, 23 September 2014, pp. 115–122 (2014)
42. Feix, B., Roussellet, M., Venelli, A.: Side-channel analysis on blinded regular scalar multiplications. In: Meier, W., Mukhopadhyay, D. (eds.) INDOCRYPT 2014. LNCS, vol. 8885, pp. 3–20. Springer, Cham (2014). https://doi.org/10.1007/978-3-319-13039-2_1
43. Fournaris, A.P.: Fault and power analysis attack protection techniques for standardized public key cryptosystems. In: Sklavos, N., Chaves, R., Di Natale, G., Regazzoni, F. (eds.) Hardware Security and Trust, pp. 93–105. Springer, Cham (2017). https://doi.org/10.1007/978-3-319-44318-8_5

A Class of Power Mappings with Low Boomerang Uniformity

Haode Yan[1], Ziying Zhang[1](✉), and Zhengchun Zhou[2]

[1] School of Mathematics, Southwest Jiaotong University, Chengdu 610031, China
hdyan@swjtu.edu.cn,
zzy.swjtu.edu.cn@my.swjtu.edu.cn

[2] School of Information Science and Technology, Southwest Jiaotong University, Chengdu 610031, China
zzc@swjtu.edu.cn

Abstract. Let $f(x) = x^{\frac{q-3}{2}}$ be a power mapping over $\mathbb{F}_q$, where q is an odd prime power. The differential uniformity of f was determined by Helleseth and Sandberg [14] in 1997. In this paper, we study the boomerang uniformity of f via its differential properties. It is shown that f has low boomerang uniformity when $q \equiv 3 \pmod 4$.

Keywords: Power function · Differential uniformity · Boomerang uniformity

1 Introduction

Substitution boxes (S-boxes for short) play a crucial role in the field of symmetric block ciphers. Let $\mathbb{F}_q$ be the finite field with q elements. For a function f from $\mathbb{F}_q$ to itself, the main tools to handle f regarding the differential attack are the difference distribution table (DDT for short) introduced by Biham and Shamir [2] and the differential uniformity which was introduced by Nyberg [21] in 1994. For any $a, b \in \mathbb{F}_q$, the DDT entry at point (a, b), denoted by $\delta_f(a, b)$, is defined as

$$\delta_f(a,b) = \left|\{x \in \mathbb{F}_q : \ f(x+a) - f(x) = b\}\right|,$$

where $|S|$ denotes the cardinality of the set S. The differential uniformity of the function f, denoted by δ_f, is defined as

$$\delta_f = \max\{\delta_f(a,b) : \ a \in \mathbb{F}_q^*, \ b \in \mathbb{F}_q\},$$

where $\mathbb{F}_q^* = \mathbb{F}_q \setminus \{0\}$. When f is used as an S-box inside a cryptosystem, the smaller the value δ_f is, the better the contribution of f to the resistance against differential attack. When $\delta_f = 1$ (respectively, $\delta_f = 2$), the function f is called a perfect nonlinear (PN) function (respectively, an almost perfect nonlinear (APN) function). The recent results on cryptographic functions with low differential

S. Mesnager and Z. Zhou (Eds.): WAIFI 2022, LNCS 13638, pp. 288–297, 2023.
https://doi.org/10.1007/978-3-031-22944-2_18

uniformity can be found in [1,3,5,6,13,18,22,23,25,27,29,30,34] and their references. More precisely, the readers can refer to a recent monograph [8], Chapter 11, which is written by Carlet.

Another important cryptanalytical technique on block ciphers is the boomerang attack introduced by Wagner [28], which is a variant of differential cryptanalysis. In order to analyze the boomerang attack of block ciphers in a better way, analogous to the DDT concerning differential attack, Cid *et al.* [9] firstly proposed the boomerang connectivity table (BCT). Let f be a permutation from $\mathbb{F}_{2^n}$ to itself. For $a, b \in \mathbb{F}_{2^n}$, the BCT entry at point (a, b), denoted by $\beta_f(a, b)$, is defined as

$$\beta_f(a,b) = \left|\{x \in \mathbb{F}_{2^n} : \ f^{-1}(f(x+a)+b) + f^{-1}(f(x)+b) = a\}\right|.$$

Further, to quantify the resistance of a function against the boomerang attack, Boura and Canteaut [4] introduced the concept of boomerang uniformity, which is the maximum value in the BCT excluding the first row and the first column. That is, the boomerang uniformity of the permutation f, denoted by β_f, is given by

$$\beta_f = \max\left\{\beta_f(a,b) : \ a, b \in \mathbb{F}_q^*\right\}.$$

Similarly, the smaller the value β_f is, the better the contribution of f to the resistance against boomerang attack. Recently, Li *et al.* in [16] generalized the definition of $\beta_f(a, b)$ for any function f (not necessarily being a permutation) over $\mathbb{F}_q$. The BCT entry of f at point (a, b), denoted by $\beta_f(a, b)$, is the number of solutions $(x, y) \in \mathbb{F}_q \times \mathbb{F}_q$ of the following system of equations

$$\begin{cases} f(x) - f(y) & = b, \\ f(x+a) - f(y+a) & = b, \end{cases}$$

where $a, b \in \mathbb{F}_q^*$. The research on cryptographic functions with low boomerang uniformity has been a hot issue in recent years, see for example [4,9,11,16,17, 20,26,33]. More precisely, for recent progress of cryptographic functions with known boomerang uniformity, the readers can refer to the survey article [19], which is written by Mesnager, Mandal and Msahli.

Power functions with low differential uniformity serve as good candidates for the design of S-boxes not only because of their strong resistance to differential attacks but also for the usually low implementation cost in hardware. The differential properties of power functions can be studied more easily due to their particular algebraic structures. Hence, the study on the boomerang uniformity of power mappings attracts a lot of attention. More precisely, when f is a power function, i.e., $f(x) = x^d$ for an integer d, one easily sees that $\beta_f(a, b) = \beta_f(1, \frac{b}{a^d})$ for any $a, b \in \mathbb{F}_q^*$. The boomerang properties of f are completely determined by the values of $\beta_f(1, b)$ as b runs through $\mathbb{F}_q^*$. Equivalently, we need to consider the number of solutions $(x, y) \in \mathbb{F}_q \times \mathbb{F}_q$ of the following equation system

$$\begin{cases} x^d - y^d & = b \\ (x+1)^d - (y+1)^d & = b \end{cases}$$

for $b \in \mathbb{F}_q^*$. Although the power functions have good algebraic structures, there are only a few classes of power mappings with known boomerang uniformity in the literature. We list them in Table 1.

In this paper, we mainly study the boomerang uniformity of the power mapping $x^{\frac{q-3}{2}}$ over $\mathbb{F}_q$ via its differential properties, where $q \equiv 3 \pmod 4$ is an odd prime power. The rest of this paper is organized as follows. Section 2 first introduces some frequently-used notation, and then gives some lemmas which will be used later. Section 3 investigates the boomerang uniformity of $x^{\frac{q-3}{2}}$. Section 4 concludes this paper.

Table 1. Power functions with known boomerang uniformity over $\mathbb{F}_{p^n}$

p	d	Conditions	β_f	Reference
2	2^n-2	$n \equiv 2 \pmod 4$	4	[4]
2	2^n-2	$n \equiv 0 \pmod 4$	6	[4]
2	2^k+1	$e = \gcd(n,k)$, n/e is odd	2^e	[10,12]
2	2^k-1	$\gcd(n,k)=1$, δ_f is not a power of 2	$\delta_f \le \beta_f \le 2\delta_f - 2$	[33]
2	2^m-1	$n=2m$, m be odd (resp. even)	2 (resp. 4)	[11]
2	$2^{m+1}-1$	$n=2m$, $m \ge 2$	2^m+2	[32]
2	$2^{2k}+2^k+1$	$n=4k$, k is odd	≤ 24	[7]
3	$\frac{3^n+3}{2}$	n is odd	3	[15]
odd	p^n-2	any n	≤ 5	[15]
odd	p^k+1	$e=\gcd(n,k)$, n/e is odd	p^e	[24]
odd	p^k+1	$e=\gcd(n,k)$, n/e is even	$p^e(p^e-1)$	[24]
odd	p^m-1	$n=2m$, $p^m \not\equiv 2 \pmod 3$	2	[31]
odd	$\frac{(p^m-1)(p^m+3)}{2}$	$n=2m$, $p^m \not\equiv 2 \pmod 3$, $p^m \equiv 3 \pmod 4$	2	[31]
odd	$\frac{p^n-3}{2}$	$p^n \equiv 3 \pmod 4$, 5 is a nonsquare	≤ 4	This paper
odd	$\frac{p^n-3}{2}$	$p^n \equiv 3 \pmod 4$, 5 is a square	≤ 6	This paper

2 Preliminaries

In this section, we introduce some frequently-used notation in this paper and give some lemmas which will be used in the following.

- q is an odd prime power.
- $\mathbb{F}_q$ is the finite field with q elements.
- Let $f(x) = x^{\frac{q-3}{2}}$ be a power mapping over $\mathbb{F}_q$.
- $\Delta(x) = f(x+1) - f(x) = (x+1)^{\frac{q-3}{2}} - x^{\frac{q-3}{2}}$.

- For any $b \in \mathbb{F}_q$, let $\Delta^{-1}(b) = \{x : \Delta(x) = b\}$ and $\delta(b) = |\Delta^{-1}(b)|$.
- Let $\chi(\cdot)$ be the quadratic multiplicative character over $\mathbb{F}_q^*$, i.e., for any $x \in \mathbb{F}_q^*$,

$$\chi(x) = x^{\frac{q-1}{2}} = \begin{cases} 1, & \text{if } x \text{ is a square element,} \\ -1, & \text{if } x \text{ is a nonsquare element.} \end{cases}$$

- For $i, j \in \{1, -1\}$, we define

$$C_{i,j} = \{x \in \mathbb{F}_q \backslash \{0, -1\} : \chi(x) = i \text{ and } \chi(x+1) = j\}.$$

The differential uniformity of f was determined by Helleseth and Sandberg in [14]. We have the following theorem.

Theorem 1. *Let $q \equiv 3 \pmod 4$ be a prime power. For $q > 7$, the differential uniformity of $f(x) = x^{\frac{q-3}{2}}$ is given by*

$$\delta_f = \begin{cases} 1, & \text{if } q = 27, \\ 2, & \text{if } \chi(5) = -1, \\ 3, & \text{if } \chi(5) = 1. \end{cases}$$

Moreover, the following lemma was shown in the proof of Theorem 1 in [14].

Lemma 1. *Let $q \equiv 3 \pmod 4$ be a prime power. With the notation introduced as above, we have*

- *(i) $\Delta^{-1}(0) = \{-\frac{1}{2}\}$ and $\delta(0) = 1$.*
- *(ii) If $\chi(5) = -1$, then $\Delta^{-1}(1) = \{0\}$, $\Delta^{-1}(-1) = \{-1\}$ and $\delta(1) = \delta(-1) = 1$.*
- *(iii) If $\chi(5) = 1$, then $\delta(b) = 3$ if and only if $b = \pm 1$. Moreover, $\Delta^{-1}(1) = \{0, \frac{\sqrt{5}-1}{2}, \frac{\sqrt{5}+1}{2}\}$, $\Delta^{-1}(-1) = \{-1, \frac{-\sqrt{5}-1}{2}, \frac{-\sqrt{5}-3}{2}\}$ with $\chi(\frac{-1+\sqrt{5}}{2}) = -1$, $\Delta^{-1}(1) = \{0, \frac{-\sqrt{5}-1}{2}, \frac{1-\sqrt{5}}{2}\}$, $\Delta^{-1}(-1) = \{-1, \frac{\sqrt{5}-1}{2}, \frac{\sqrt{5}-3}{2}\}$ with $\chi(\frac{-1+\sqrt{5}}{2}) = 1$.*
- *(iv) For $b \neq \pm 1$, we have $\delta(b) \leq 2$. More precisely, if $\delta(b) = 2$, i.e., the equation $\Delta(x) = b$ has two distinct solutions, namely x_1 and x_2, then one of x_1 and x_2 is in $C_{1,1} \cup C_{-1,-1}$, and the other is in $C_{1,-1} \cup C_{-1,1}$.*

3 The Boomerang Uniformity of the Power Function $x^{\frac{q-3}{2}}$ Over $\mathbb{F}_q$

In this section, we investigate the boomerang uniformity of the power mapping f via its differential properties. We denote by β_f the boomerang uniformity of f. Our main result is shown as follows.

Theorem 2. *Let q be an odd prime power with $q \equiv 3 \pmod 4$. For $q \neq 7$ and $q \neq 27$, we have,*

$$\beta_f \leq \begin{cases} 4, & \text{if } \chi(5) = -1, \\ 6, & \text{if } \chi(5) = 1. \end{cases}$$

Proof. For any $b \in \mathbb{F}_q^*$, we consider the number of solutions $(x, y) \in \mathbb{F}_q \times \mathbb{F}_q$ of the following equation system

$$\begin{cases} x^{\frac{q-3}{2}} - y^{\frac{q-3}{2}} &= b, \\ (x+1)^{\frac{q-3}{2}} - (y+1)^{\frac{q-3}{2}} &= b. \end{cases} \tag{1}$$

If (x, y) is a solution of (1), we have $x \neq y$ since $b \neq 0$. Moreover, we have $\Delta(x) = \Delta(y)$ from (1).We assert that $\delta(\Delta(x)) = 2$ or 3 by Lemma 1 and $x \neq y$. We discuss in the following two cases.

Case 1. $\delta(\Delta(x)) = 2$. It is clear that $\Delta(x) \neq \pm 1$ by Lemma 1, then we have $x, y \neq 0, -1$. The equation system (1) becomes

$$\begin{cases} \chi(x)x^{-1} - \chi(y)y^{-1} &= b, \\ \chi(x+1)(x+1)^{-1} - \chi(y+1)(y+1)^{-1} &= b. \end{cases} \tag{2}$$

By Lemma 1 (iv), for each $x \in C_{i,j}$, $i, j \in \{1, -1\}$, there are two possible sets of y. We have the following 8 subcases, which we summarize in Table 2.

Table 2. Eight subcases from equation system (2)

	(x, y)	Equation system
I	$(x, y) \in C_{1,1} \times C_{1,-1}$	$\begin{cases} x^{-1} - y^{-1} = b, \\ (x+1)^{-1} + (y+1)^{-1} = b. \end{cases}$
II	$(x, y) \in C_{1,1} \times C_{-1,1}$	$\begin{cases} x^{-1} + y^{-1} = b, \\ (x+1)^{-1} - (y+1)^{-1} = b. \end{cases}$
III	$(x, y) \in C_{1,-1} \times C_{1,1}$	$\begin{cases} x^{-1} - y^{-1} = b, \\ -(x+1)^{-1} - (y+1)^{-1} = b. \end{cases}$
IV	$(x, y) \in C_{1,-1} \times C_{-1,-1}$	$\begin{cases} x^{-1} + y^{-1} = b, \\ -(x+1)^{-1} + (y+1)^{-1} = b. \end{cases}$
V	$(x, y) \in C_{-1,1} \times C_{1,1}$	$\begin{cases} -x^{-1} - y^{-1} = b, \\ (x+1)^{-1} - (y+1)^{-1} = b. \end{cases}$
VI	$(x, y) \in C_{-1,1} \times C_{-1,-1}$	$\begin{cases} -x^{-1} + y^{-1} = b, \\ (x+1)^{-1} + (y+1)^{-1} = b. \end{cases}$
VII	$(x, y) \in C_{-1,-1} \times C_{1,-1}$	$\begin{cases} -x^{-1} - y^{-1} = b, \\ -(x+1)^{-1} + (y+1)^{-1} = b. \end{cases}$
VIII	$(x, y) \in C_{-1,-1} \times C_{-1,1}$	$\begin{cases} -x^{-1} + y^{-1} = b, \\ -(x+1)^{-1} - (y+1)^{-1} = b. \end{cases}$

Subcase I. $(x, y) \in C_{1,1} \times C_{1,-1}$. After a simple calculation, we obtain two quadratic equations as follows.

$$\begin{cases} b(b-2)x^2 + (b^2 - 4b + 2)x - (b-2) &= 0, \quad (3) \\ b^2(y+1)^2 - (b^2+2)(y+1) + b &= 0. \quad (4) \end{cases}$$

It is easy to see that $b \neq 2$ and $b \neq 0$, otherwise, we have $x = 0$ or $y = -1$, a contradiction. If (3) has two solutions, namely x_1 and x_2, then $x_1x_2 = -\frac{1}{b}$. We mention that -1 is a nonsquare element. When b is a square element, then $\chi(x_1x_2) = -1$, and at most one of x_1 and x_2 satisfies $x \in C_{1,1}$. Similarly, if (4) has two solutions, namely y_1 and y_2, then $(y_1+1)(y_2+1) = \frac{1}{b}$. When b is a nonsquare element, then $\chi((y_1+1)(y_2+1)) = -1$, and at most one of y_1 and y_2 satisfies $y \in C_{1,-1}$. By a discussion as above, we conclude that for any $b \in \mathbb{F}_q^*$, this subcase contributes at most 1 solution.

Subcase II. $(x, y) \in C_{1,1} \times C_{-1,1}$. In this subcase, we obtain two quadratic equations as follows.

$$\begin{cases} b(b+2)(x+1)^2 - (b^2+4b+2)(x+1) + b + 2 &= 0, \quad (5) \\ b^2y^2 + (b^2+2)y - b &= 0. \quad (6) \end{cases}$$

It is easy to see that $b \neq -2$ and $b \neq 0$. If (5) (respectively, (6)) has two solutions, namely x_1 and x_2 (respectively, y_1 and y_2), then $(x_1+1)(x_2+1) = \frac{1}{b}$ (respectively, $y_1y_2 = -\frac{1}{b}$). By a similar proof as for subcase I, we conclude that for any $b \in \mathbb{F}_q^*$, this subcase contributes at most 1 solution.

Subcase III. $(x, y) \in C_{1,-1} \times C_{1,1}$. We obtain two quadratic equations as follows.

$$\begin{cases} b^2(x+1)^2 - (b^2+2)(x+1) - b &= 0, \\ b(b+2)y^2 + (b^2+4b+2)y + b + 2 &= 0. \end{cases}$$

Similar to the proof of subase I, this subcase contributes at most 1 solution.

Subcase IV. $(x, y) \in C_{1,-1} \times C_{-1,-1}$. Since $\frac{q-3}{2}$ is even, then (x, y) is a solution of (1) if and only if $(-x-1, -y-1)$ is a solution of (1). For any $y \in \mathbb{F}_q \setminus \{0, -1\}$, $y \in C_{1,1}$ if and only if $-y-1 \in C_{-1,-1}$, $y \in C_{1,-1}$ (respectively, $C_{-1,1}$) if and only if $-y-1 \in C_{1,-1}$ (respectively, $C_{-1,1}$). We conclude that the number of the solutions in this subcase is the same to that of subcase III.

Subcase V. $(x, y) \in C_{-1,1} \times C_{1,1}$. We obtain two quadratic equations as follows.

$$\begin{cases} b^2x^2 + (b^2+2)x + b &= 0, \quad (7) \\ b(b-2)(y+1)^2 - (b^2-4b+2)(y+1) - (b-2) &= 0. \quad (8) \end{cases}$$

Similar to the proof of subcase I, this subcase contributes at most 1 solution.

For subcases VI, VII and VIII, we assert that the numbers of solutions in subcases VI and V (respectively, subcases VII and I, subcases VIII and II) are the same, similar to subcases IV and III. We conclude that the equation system (2) has at most one solution in each subcase.

Next we show that the equation system (2) cannot have solutions in subcase I and subcase V simultaneously. Otherwise, let $(x_1, y_1) \in C_{1,1} \times C_{1,-1}$ be a solution of (2) in subcase I and $(u_1, v_1) \in C_{-1,1} \times C_{1,1}$ be a solution of (2) in subcase V. Then

$$\chi(x_1) = 1, \ \chi(x_1+1) = 1, \ \chi(y_1) = 1, \ \chi(y_1+1) = -1,$$

and

$$\chi(u_1) = -1,\ \chi(u_1+1) = 1,\ \chi(v_1) = 1,\ \chi(v_1+1) = 1.$$

When we discard the condition on the values of the quadratic character, there is the other solution (x_2, y_2) (respectively, (u_2, v_2)) of equations (3) and (4) (respectively, (7) and (8)). Considering quadratic equations (3) and (8), only one of their coefficients has a different sign. More precisely, we have

$$x_1 + x_2 = -\frac{b^2-4b+2}{b(b-2)} = -((v_1+1)+(v_2+1))$$

and

$$x_1x_2 = -\frac{1}{b} = (v_1+1)(v_2+1).$$

Then $x_1 = -(v_2+1)$ and $x_2 = -(v_1+1)$ since $x_1, v_1 \in C_{1,1}$. Consequently,

$$\chi(b) = \chi(\frac{1}{b}) = \chi(-(v_1+1)(v_2+1)) = \chi(v_1+1)\chi(x_1) = \chi(x_1) = 1.$$

Similarly, considering quadratic equations (4) and (7), we have

$$u_1 + u_2 = -\frac{b^2+2}{b^2} = -((y_1+1)+(y_2+1))$$

and

$$u_1u_2 = \frac{1}{b} = (y_1+1)(y_2+1).$$

Then $u_1 = -(y_2+1)$ and $u_2 = -(y_1+1)$ since $u_1 \in C_{-1,1}$ and $y_1 \in C_{1,-1}$. Hence

$$\chi(b) = \chi(\frac{1}{b}) = \chi((y_1+1)(y_2+1)) = \chi(-(y_1+1)u_1) = -\chi(y_1+1)\chi(u_1) = -1,$$

which is a contradiction. Therefore, for any $b \in \mathbb{F}_q^*$, subcase I and subcase V cannot give solutions simultaneously, so they contribute at most one solution altogether. Similarly, subcases II and III contribute at most one solution altogether. That is to say, for any $b \in \mathbb{F}_q^*$, there are at most four solutions of (1) in this case.

Case 2. $\delta(\Delta(x)) = 3$. By Lemma 1, we know that this case only occurs when $\chi(5) = 1$ and $\Delta(x) = \pm 1$. Note that $\frac{-1+\sqrt{5}}{2} \cdot \frac{-1-\sqrt{5}}{2} = -1$, without loss of generality, we assume that $\chi(\frac{-1+\sqrt{5}}{2}) = -1$. Then we can obtain $\Delta^{-1}(1) = \{0, \frac{\sqrt{5}-1}{2}, \frac{\sqrt{5}+1}{2}\}$ and $\Delta^{-1}(-1) = \{-1, -\frac{\sqrt{5}+1}{2}, -\frac{\sqrt{5}+3}{2}\}$ by Lemma 1 (iii).

We can list all possible pairs (x, y) with $\Delta(x) = \Delta(y) = \pm 1$. Plugging all pairs (x, y) into the first equation of the system (1), the corresponding b's are obtained. We have the following table (Table 3).

Table 3. The Solutions of $\Delta(x) = \Delta(y) = \pm 1$ and Corresponding b

(x, y) with $\Delta(x) = \Delta(y) = \pm 1$	The Corresponding b
$(0, \frac{\sqrt{5}-1}{2}), (-1, \frac{-\sqrt{5}-1}{2})$	$b = \frac{\sqrt{5}+1}{2}$
$(\frac{\sqrt{5}-1}{2}, 0), (\frac{-\sqrt{5}-1}{2}, -1)$	$b = -\frac{\sqrt{5}+1}{2}$
$(0, \frac{\sqrt{5}+1}{2}), (-1, \frac{-\sqrt{5}-3}{2})$	$b = \frac{\sqrt{5}-1}{2}$
$(\frac{\sqrt{5}+1}{2}, 0), (\frac{-\sqrt{5}-3}{2}, -1)$	$b = -\frac{\sqrt{5}-1}{2}$
$(\frac{\sqrt{5}-1}{2}, \frac{\sqrt{5}+1}{2}), (\frac{-\sqrt{5}-1}{2}, \frac{-\sqrt{5}-3}{2})$	$b = 1$
$(\frac{\sqrt{5}+1}{2}, \frac{\sqrt{5}-1}{2}), (\frac{-\sqrt{5}-3}{2}, \frac{-\sqrt{5}-1}{2})$	$b = -1$

It is obvious that, for each $b \in \{\pm 1, \pm \frac{\sqrt{5}+1}{2}, \pm \frac{\sqrt{5}-1}{2}\}$, the equation system (1) has two solutions in this case. For $b \in \mathbb{F}_q^* \backslash \{\pm 1, \pm \frac{\sqrt{5}+1}{2}, \pm \frac{\sqrt{5}-1}{2}\}$ the equation system (1) has no solution in this case. Note that Case 2 only occurs when $\chi(5) = 1$, the desired results follow.

Remark 1. By making a computer investigation, we have the boomerang uniformity of f is equal to 4 with $q = 3^5$. In addition, the boomerang uniformity of f is equal to 6 with $q = 131$. Therefore, we can conclude that our bound is tight.

4 Conclusion

In this paper, we mainly study the boomerang uniformity of the power function $x^{\frac{q-3}{2}}$ over $\mathbb{F}_q$ via their differential properties, where $q \equiv 3 \pmod 4$ is an odd prime power. It is shown that the power function has low boomerang uniformity. We mention that our approach may be used in determining the boomerang uniformity of other power mappings. It is worthy finding applications of power mappings with low boomerang uniformity in sequence designs, coding theory and combinatorial designs.

Acknowledgements. The authors would like to thank the anonymous reviewers for giving us invaluable comments and suggestions that greatly improved the quality of this paper. H. Yan's research was supported by the Natural Science Foundation of Sichuan (Grant No. 2022NSFSC1805) and the Fundamental Research Funds for the Central Universities of China (Grant No. 2682021ZTPY076).

References

1. Beth, T., Ding, C.: On almost perfect nonlinear permutations. In: Helleseth, T. (ed.) EUROCRYPT 1993. LNCS, vol. 765, pp. 65–76. Springer, Heidelberg (1994). https://doi.org/10.1007/3-540-48285-7_7

2. Biham, E., Shamir, A.: Differential cryptanalysis of des-like cryptosystems. J. Cryptol. **4**(1), 3–72 (1991)
3. Blondeau, C., Perrin, L.: More differentially 6-uniform power functions. Des. Codes Cryptogr. **73**(2), 487–505 (2014). https://doi.org/10.1007/s10623-014-9948-2
4. Boura, C., Canteaut, A.: On the boomerang uniformity of cryptographic Sboxes. IACR Transactions on Symmetric Cryptology, pp. 290–310 (2018)
5. Budaghyan, L., Carlet, C., Helleseth, T., Li, N., Sun, B.: On upper bounds for algebraic degrees of APN functions. IEEE Trans. Inf. Theory **64**(6), 4399–4411 (2017)
6. Budaghyan, L., Carlet, C., Leander, G.: Two classes of quadratic APN binomials inequivalent to power functions. IEEE Trans. Inf. Theory **54**(9), 4218–4229 (2008)
7. Calderini, M., Villa, I.: On the boomerang uniformity of some permutation polynomials. Cryptogr. Commun. **12**(6), 1161–1178 (2020). https://doi.org/10.1007/s12095-020-00439-x
8. Carlet, C.: Boolean Functions for Cryptography and Coding Theory. Cambridge University Press, Cambridge (2021)
9. Cid, C., Huang, T., Peyrin, T., Sasaki, Yu., Song, L.: Boomerang connectivity table: a new cryptanalysis tool. In: Nielsen, J.B., Rijmen, V. (eds.) EUROCRYPT 2018. LNCS, vol. 10821, pp. 683–714. Springer, Cham (2018). https://doi.org/10.1007/978-3-319-78375-8_22
10. Eddahmani, S., Mesnager, S.: Explicit values of the tables DDT, BCT, FBCT, and FBDT of the inverse, the gold, and the Bracken-Leander functions
11. Hasan, S.U., Pal, M., Stănică, P.: Boomerang uniformity of a class of power maps. Des. Codes Cryptogr. **89**(11), 2627–2636 (2021). https://doi.org/10.1007/s10623-021-00944-x
12. Hasan, S., Pal, M., Stănică, P.: The binary gold function and its c-boomerang connectivity table. Cryptogr. Commun., 1–24 (2022)
13. Helleseth, T., Rong, C., Sandberg, D.: New families of almost perfect nonlinear power mappings. IEEE Trans. Inf. Theory **45**(2), 475–485 (1999)
14. Helleseth, T., Sandberg, D.: Some power mappings with low differential uniformity. Appl. Algebra Eng. Commun. Comput. **8**(5), 363–370 (1997)
15. Jiang, S., Li, K., Li, Y., Qu, L.: Differential and boomerang spectrums of some power permutations. Cryptogr. Commun. **14**(2), 371–393 (2021). https://doi.org/10.1007/s12095-021-00530-x
16. Li, K., Qu, L., Sun, B., Li, C.: New results about the boomerang uniformity of permutation polynomials. IEEE Trans. Inf. Theory **65**(99), 7542–7553 (2019)
17. Li, N., Hu, Z., Xiong, M., Zeng, X.: 4-uniform BCT permutations from generalized butterfly structure. arXiv preprint arXiv:2001.00464 (2020)
18. Li, Y., Wang, M.: Constructing differentially 4-uniform permutations over $\mathbb{GF}(2^{2m})$ from quadratic APN permutations over $\mathbb{GF}(2^{2m+1})$. Des. Codes Cryptogr. **72**(2), 249–264 (2014)
19. Mesnager, S., Mandal, B., Msahli, M.: Survey on recent trends towards generalized differential and boomerang uniformities. Cryptogr. Commun., 1–45 (2021)
20. Mesnager, S., Tang, C., Xiong, M.: On the boomerang uniformity of quadratic permutations. Des. Codes Cryptogr. **88**(10), 2233–2246 (2020). https://doi.org/10.1007/s10623-020-00775-2
21. Nyberg, K.: Differentially uniform mappings for cryptography. In: Helleseth, T. (ed.) EUROCRYPT 1993. LNCS, vol. 765, pp. 55–64. Springer, Heidelberg (1994). https://doi.org/10.1007/3-540-48285-7_6
22. Qu, L., Tan, Y., Li, C., Gong, G.: More constructions of differentially 4-uniform permutations on $\mathbb{F}_{2^{2k}}$. Des. Codes Cryptogr. **78**(2), 391–408 (2016)

23. Qu, L., Tan, Y., Tan, C., Li, C.: Constructing differentially 4-uniform permutations over $\mathbb{F}_{2^{2k}}$ via the switching method. IEEE Trans. Inf. Theory **59**(7), 4675–4686 (2013)
24. Stănică, P.: Using double Weil sums in finding the c-boomerang connectivity table for monomial functions on finite fields. Applicable Algebra in Engineering, Communication and Computing, pp. 1–22 (2021)
25. Tang, D., Carlet, C., Tang, X.: Differentially 4-uniform bijections by permuting the inverse function. Des. Codes Cryptogr. **77**(1), 117–141 (2015)
26. Tu, Z., Li, N., Zeng, X., Zhou, J.: A class of quadrinomial permutations with boomerang uniformity four. IEEE Trans. Inf. Theory **66**(6), 3753–3765 (2020)
27. Tu, Z., Zeng, X.: Non-monomial permutations with differential uniformity six. J. Syst. Sci. Complex. **31**(4), 1078–1089 (2018)
28. Wagner, D.: The boomerang attack. In: Knudsen, L. (ed.) FSE 1999. LNCS, vol. 1636, pp. 156–170. Springer, Heidelberg (1999). https://doi.org/10.1007/3-540-48519-8_12
29. Xiong, M., Yan, H.: A note on the differential spectrum of a differentially 4-uniform power function. Finite Fields Their Appl. **48**, 117–125 (2017)
30. Xiong, M., Yan, H., Yuan, P.: On a conjecture of differentially 8-uniform power functions. Des. Codes Cryptogr. **86**(8), 1601–1621 (2018)
31. Yan, H., Li, Z., Song, Z., Feng, R.: Two classes of power mappings with boomerang uniformity 2. Adv. Math. Commun. (2022)
32. Yan, H., Zhang, Z., Li, Z.: Boomerang spectrum of a class of power functions. In: International Workshop on Signal Design and its Applications in Communications (IWSDA), pp. 1–4 (2022)
33. Zha, Z., Hu, L.: The boomerang uniformity of power permutations x^{2^k-1} over $\mathbb{F}_{2^n}$. In: 2019 Ninth International Workshop on Signal Design and its Applications in Communications (IWSDA), pp. 1–4. IEEE (2019)
34. Zha, Z., Hu, L., Sun, S.: Constructing new differentially 4-uniform permutations from the inverse function. Finite Fields Their Appl. **25**, 64–78 (2014)

New Classes of Bent Functions via the Switching Method

Xi Xie, Bing Chen, Nian Li(✉), and Xiangyong Zeng

Hubei Key Laboratory of Applied Mathematics, Faculty of Mathematics and Statistics, Hubei University, Wuhan 430062, China
xi.xie@aliyun.com, {chenbing,nian.li,xzeng}@hubu.edu.cn

Abstract. The switching method is a powerful method to construct bent functions. In this paper, using this method, we present two generic constructions of piecewise bent functions from known ones, which generalize some earlier works. Further, based on these two generic constructions, we obtain several infinite families of bent functions from quadratic bent functions and the Maiorana-MacFarland class of bent functions by calculating their duals. It is worth noting that our constructions can produce bent functions with the optimal algebraic degree.

Keywords: Bent function · Walsh transform · Switching method

1 Introduction

Boolean bent functions were first introduced by Rothaus in 1976 [10] as an interesting combinatorial object with maximum Hamming distance to the set of all affine functions. Over the last four decades, bent functions have attracted a lot of research interest due to their important applications in cryptography [1], sequences [8] and coding theory [2,3]. Later, Kumar, Scholtz and Welch in [4] generalized the notion of Boolean bent functions to the case of functions over an arbitrary finite field.

Let $\mathbb{F}_{p^n}$ denote the finite field with p^n elements, where p is a prime and n is a positive integer. Given a function $f(x)$ mapping from $\mathbb{F}_{p^n}$ to $\mathbb{F}_p$, the Walsh transform of $f(x)$ is defined by

$$\widehat{f}(b) = \sum\nolimits_{x\in\mathbb{F}_{p^n}} \omega^{f(x)-\mathrm{Tr}(bx)},\ b \in \mathbb{F}_{p^n},$$

where $\omega = e^{\frac{2\pi\sqrt{-1}}{p}}$ is a complex primitive p-th root of unity. According to [4], $f(x)$ is called a p-ary bent function if all its Walsh coefficients satisfy $|\widehat{f}(b)| = p^{n/2}$.

This work was supported by the Knowledge Innovation Program of Wuhan-Basic Research under Grant 2022010801010319, the Natural Science Foundation of Hubei Province of China under Grant 2021CFA079 and the National Natural Science Foundation of China under Grant 62072162.

S. Mesnager and Z. Zhou (Eds.): WAIFI 2022, LNCS 13638, pp. 298–309, 2023.
https://doi.org/10.1007/978-3-031-22944-2_19

A p-ary bent function $f(x)$ is called regular if $\widehat{f}(b) = p^{n/2}\omega^{\widetilde{f}(b)}$ holds for some function $\widetilde{f}(x)$ mapping $\mathbb{F}_{p^n}$ to $\mathbb{F}_p$, and it is called weakly regular if there exists a complex μ having unit magnitude such that $\widehat{f}(b) = \mu^{-1}p^{n/2}\omega^{\widetilde{f}(b)}$ for all $b \in \mathbb{F}_{p^n}$. The function $\widetilde{f}(x)$ is called the dual of $f(x)$ and it is also bent.

The switching method is a powerful method to construct bent functions. In 2016, Xu et al. [15] constructed a class of piecewise p-ary bent functions $f(x) = g(x) + c\mathrm{Tr}(x)^{p-1}$ from Gold functions $g(x)$ via the switching method, where p is an odd prime and $c \in \mathbb{F}_p^*$. In 2017, Xu et al. [17] constructed two classes of piecewise p-ary bent functions $f(x) = g(x) + \mathrm{Tr}(ux)\mathrm{Tr}(x)^{p-1}$ from p-ary Kasami functions and Sidelnikov functions $g(x)$ [6], where p is an odd prime and $u \in \mathbb{F}_{p^n}^*$. Recall that Tang et al. [11] constructed bent functions of the form $g(x) + F(\mathrm{Tr}(u_1x), \mathrm{Tr}(u_2x), \cdots, \mathrm{Tr}(u_\tau x))$, where $g(x) : \mathbb{F}_{p^n} \to \mathbb{F}_p$, $F(x_1, \cdots, x_\tau) \in \mathbb{F}_p[x_1, \cdots, x_\tau]$ and $u_i \in \mathbb{F}_{p^n}$ for $1 \leq i \leq \tau$. Motivated by Xu et al.'s and Tang et al.'s works, we investigate the bentness of piecewise bent functions of the forms

$$f(x) = g(x) + F(\mathrm{Tr}(u_1x), \mathrm{Tr}(u_2x), \cdots, \mathrm{Tr}(u_\tau x)) + c\big(\prod_{i=1}^{\kappa} \mathrm{Tr}(v_ix)\big)^{p-1} \quad (1)$$

and

$$f(x) = g(x) + F(\mathrm{Tr}(u_1x), \mathrm{Tr}(u_2x), \cdots, \mathrm{Tr}(u_\tau x))\big(\prod_{i=1}^{\kappa} \mathrm{Tr}(v_ix)\big)^{p-1}, \quad (2)$$

where $g(x) : \mathbb{F}_{p^n} \to \mathbb{F}_p$, $F(x_1, \cdots, x_\tau) \in \mathbb{F}_p[x_1, \cdots, x_\tau]$, $c \in \mathbb{F}_p^*$, $u_i \in \mathbb{F}_{p^n}$ for $1 \leq i \leq \tau$ and $v_j \in \mathbb{F}_{p^n}^*$ for $1 \leq j \leq \kappa$. In this paper, we first present generic constructions of bent functions with the forms (1) and (2) respectively. Notice that the works of Xu et al. [15,17] are two special cases of our work for $\kappa = 1$, $F(x_1) = 0$ in (1) and $\kappa = 1$, $F(x_1) = x_1$ in (2). In addition, by calculating the duals of quadratic bent functions and the Maiorana-MacFarland class of bent functions, several infinite families of bent functions are constructed by using these generic constructions.

The rest of this paper is organized as follows. Section 2 fixes some notation and introduces some preliminaries. Section 3 proposes two generic constructions of bent functions of the forms (1) and (2) respectively. Section 4 constructs several infinite families of bent functions using the two generic constructions given in Sect. 3. Finally, Sect. 5 concludes this paper.

2 Preliminaries

Throughout this paper, let $\mathbb{F}_{p^n}$ denote the finite field with p^n elements, where p is a prime and n is a positive integer. The trace function from $\mathbb{F}_{p^n}$ to its subfield $\mathbb{F}_{p^k}$ is defined by $\mathrm{Tr}_k^n(x) = \sum_{i=0}^{n/k-1} x^{p^{ik}}$. In particular, when $k = 1$, we use the notation $\mathrm{Tr}(x)$ instead of $\mathrm{Tr}_1^n(x)$. A function $F(x_1, \cdots, x_n) : \mathbb{F}_p^n \mapsto \mathbb{F}_p$ is often represented by its algebraic normal form

$$F(x_1, \cdots, x_n) = \sum_{e=(e_1,\cdots,e_n)\in\mathbb{F}_p^n} a(e)\big(\prod_{i=1}^{n} x_i^{e_i}\big), \ a(e) \in \mathbb{F}_p. \quad (3)$$

A polynomial in $\mathbb{F}_p[x_1, \cdots, x_n]$ with the form (3) is called a reduced polynomial. The algebraic degree of $F(x_1, \cdots, x_n)$, denoted by $\deg(F)$, is defined as $\deg(F) = \max_{e \in \mathbb{F}_p^n} \{\sum_{i=1}^n e_i : a(e) \neq 0\}$, where $e = (e_1, \cdots, e_n) \in \mathbb{F}_p^n$.

Lemma 1. ([5, Propositions 4.4 and 4.5]) *Let $f(x)$ be a bent function from $\mathbb{F}_{p^n}$ to $\mathbb{F}_p$. Then $\deg(f) \leq \frac{(p-1)n}{2} + 1$. Moreover, if $f(x)$ is weakly regular bent, then $\deg(f) \leq \frac{(p-1)n}{2}$.*

To simplify the proof of our main result in the sequel, we fix some notation firstly. Let $g(x)$ be a function from $\mathbb{F}_{p^n}$ to $\mathbb{F}_p$. For $r := (r_1, \cdots, r_\kappa) \in \mathbb{F}_p^\kappa$ and $v := (v_1, \cdots, v_\kappa) \in \mathbb{F}_{p^n}^* \times \cdots \times \mathbb{F}_{p^n}^*$, define

$$\mathbb{T}_{r,v} = \{x \in \mathbb{F}_{p^n} : \mathrm{Tr}(v_i x) = r_i,\ i = 1, \cdots, \kappa\}$$

and

$$\mathbb{S}_g(r, v, b) = \sum\nolimits_{x \in \mathbb{T}_{r,v}} \omega^{g(x) - \mathrm{Tr}(bx)}$$

for any $b \in \mathbb{F}_{p^n}$. It then can be verified that $\mathbb{F}_{p^n} = \cup_{r \in \mathbb{F}_p^\kappa} \mathbb{T}_{r,v}$. Moreover, we have

$$\begin{aligned}\widehat{g}\big(b - \sum\nolimits_{i=1}^{\kappa} v_i s_i\big) &= \sum\nolimits_{x \in \mathbb{F}_{p^n}} \omega^{g(x) - \mathrm{Tr}\left((b - \sum_{i=1}^{\kappa} v_i s_i)x\right)} \\ &= \sum_{r \in \mathbb{F}_p^\kappa} \sum_{x \in \mathbb{T}_{r,v}} \omega^{g(x) - \mathrm{Tr}(bx) + \sum_{i=1}^{\kappa} r_i s_i} \\ &= \sum_{r \in \mathbb{F}_p^\kappa} \mathbb{S}_g(r, v, b) \omega^{\sum_{i=1}^{\kappa} r_i s_i}.\end{aligned}$$

From the inverse Fourier transform we can derive

$$\mathbb{S}_g(r, v, b) = \frac{1}{p^\kappa} \sum_{s \in \mathbb{F}_p^\kappa} \omega^{-\sum_{i=1}^{\kappa} r_i s_i} \widehat{g}\big(b - \sum\nolimits_{i=1}^{\kappa} v_i s_i\big), \tag{4}$$

where $s := (s_1, \cdots, s_\kappa) \in \mathbb{F}_p^\kappa$.

To make the computation of the Walsh transform of the functions investigated in the sequel feasible, we consider a class of weakly regular bent functions $g(x) : \mathbb{F}_{p^n} \to \mathbb{F}_p$ whose dual satisfies

$$\widetilde{g}\big(x - \sum\nolimits_{i=1}^{\kappa} v_i s_i - \sum\nolimits_{i=1}^{\tau} u_i t_i\big) = \widetilde{g}\big(x - \sum\nolimits_{i=1}^{\tau} u_i t_i\big) + \sum\nolimits_{i=1}^{\kappa} \varphi_{v_i}(x) s_i, \tag{5}$$

where $\varphi_{v_i}(x)$ is a function from $\mathbb{F}_{p^n}$ to $\mathbb{F}_p$, $u_i \in \mathbb{F}_{p^n}$, $v_j \in \mathbb{F}_{p^n}^*$ and $t_i, s_j \in \mathbb{F}_p$ for each $1 \leq i \leq \tau$ and $1 \leq j \leq \kappa$. Without loss of generality, assume that $\widehat{g}(b) = \mu^{-1} p^{n/2} \omega^{\widetilde{g}(b)}$ for any $b \in \mathbb{F}_{p^n}$. Then

$$\begin{aligned}\widehat{g}\big(b - \sum\nolimits_{i=1}^{\tau} u_i t_i - \sum\nolimits_{i=1}^{\kappa} v_i s_i\big) &= \mu^{-1} p^{n/2} \omega^{\widetilde{g}\left(b - \sum_{i=1}^{\tau} u_i t_i\right) + \sum_{i=1}^{\kappa} \varphi_{v_i}(b) s_i} \\ &= \widehat{g}\big(b - \sum\nolimits_{i=1}^{\tau} u_i t_i\big) \omega^{\sum_{i=1}^{\kappa} \varphi_{v_i}(b) s_i}.\end{aligned} \tag{6}$$

Further we characterize the value of $\mathbb{S}_g(r, v, b)$ as below.

Lemma 2. *Let $g(x) : \mathbb{F}_{p^n} \to \mathbb{F}_p$ be a weakly regular bent function whose dual satisfies* (5)*. For any $b \in \mathbb{F}_{p^n}$, $\mathbb{S}_g(r, v, b) = \widehat{g}(b)$ if $r_i = \varphi_{v_i}(b)$ for any $1 \le i \le \kappa$ and otherwise, $\mathbb{S}_g(r, v, b) = 0$, where $r = (r_1, \cdots, r_\kappa) \in \mathbb{F}_p^\kappa$ and $v = (v_1, \cdots, v_\kappa) \in \mathbb{F}_{p^n}^* \times \cdots \times \mathbb{F}_{p^n}^*$.*

Proof. For any $b \in \mathbb{F}_{p^n}$, $r = (r_1, \cdots, r_\kappa) \in \mathbb{F}_p^\kappa$ and $v = (v_1, \cdots, v_\kappa) \in \mathbb{F}_{p^n}^* \times \cdots \times \mathbb{F}_{p^n}^*$, (4) yields

$$\mathbb{S}_g(r, v, b) = \frac{1}{p^\kappa} \sum_{s \in \mathbb{F}_p^\kappa} \omega^{-\sum_{i=1}^{\kappa} r_i s_i} \widehat{g}(b - \sum\nolimits_{i=1}^{\kappa} v_i s_i),$$

where $s = (s_1, \cdots, s_\kappa) \in \mathbb{F}_p^\kappa$. By using (6), one gives

$$\mathbb{S}_g(r, v, b) = \frac{1}{p^\kappa} \widehat{g}(b) \sum_{s \in \mathbb{F}_p^\kappa} \omega^{\sum_{i=1}^{\kappa} (\varphi_{v_i}(b) - r_i) s_i}.$$

Further the desired result follows from the fact that $\sum_{s_i \in \mathbb{F}_p} \omega^{(\varphi_{v_i}(b) - r_i) s_i} = p$ if $r_i = \varphi_{v_i}(b)$ and 0 otherwise for any $1 \le i \le \kappa$. This completes the proof.

Let $h : \mathbb{F}_{p^n} \to \mathbb{F}_p$ be defined as in

$$h(x) = g(x) + F(\mathrm{Tr}(u_1 x), \mathrm{Tr}(u_2 x), \cdots, \mathrm{Tr}(u_\tau x)), \tag{7}$$

where $g(x) : \mathbb{F}_{p^n} \to \mathbb{F}_p$, $F(x_1, \cdots, x_\tau)$ is an arbitrary reduced polynomial in $\mathbb{F}_p[x_1, \cdots, x_\tau]$, $u_i \in \mathbb{F}_{p^n}$ for $1 \le i \le \tau$. The Walsh transform of $h(x)$ can be given as follows.

Lemma 3. ([13, Theorem 1]) *Let $h(x)$ be defined as in* (7)*. Then for any $b \in \mathbb{F}_{p^n}$,*

$$\widehat{h}(b) = \frac{1}{p^\tau} \sum_{(t_1, \cdots, t_\tau) \in \mathbb{F}_p^\tau} \widehat{F}(t_1, \cdots, t_\tau) \widehat{g}(b - \sum\nolimits_{i=1}^{\tau} u_i t_i).$$

Next we characterize the value of $\mathbb{S}_h(r, v, b)$ for a class of weakly regular bent functions $g(x)$.

Lemma 4. *Let $h(x)$ be defined as in* (7) *and $g(x) : \mathbb{F}_{p^n} \to \mathbb{F}_p$ be a weakly regular bent function whose dual satisfies* (5)*. Then for any $b \in \mathbb{F}_{p^n}$, $\mathbb{S}_h(r, v, b) = \widehat{h}(b)$ if $r_i = \varphi_{v_i}(b)$ for any $1 \le i \le \kappa$ and otherwise, $\mathbb{S}_h(r, v, b) = 0$, where $r = (r_1, \cdots, r_\kappa) \in \mathbb{F}_p^\kappa$ and $v = (v_1, \cdots, v_\kappa) \in \mathbb{F}_{p^n}^* \times \cdots \times \mathbb{F}_{p^n}^*$.*

Proof. For any $b \in \mathbb{F}_{p^n}$, $r = (r_1, \cdots, r_\kappa) \in \mathbb{F}_p^\kappa$ and $v = (v_1, \cdots, v_\kappa) \in \mathbb{F}_{p^n}^* \times \cdots \times \mathbb{F}_{p^n}^*$, (4) gives

$$\mathbb{S}_h(r, v, b) = \frac{1}{p^\kappa} \sum_{s \in \mathbb{F}_p^\kappa} \omega^{-\sum_{i=1}^{\kappa} r_i s_i} \widehat{h}(b - \sum\nolimits_{i=1}^{\kappa} v_i s_i), \tag{8}$$

where $s=(s_1,\cdots,s_\kappa)\in\mathbb{F}_p^\kappa$. Note that Lemma 3 yields

$$\begin{aligned}\widehat{h}(b-\sum_{i=1}^{\kappa}v_is_i)&=\frac{1}{p^\tau}\sum_{t\in\mathbb{F}_p^\tau}\widehat{F}(t_1,\cdots,t_\tau)\widehat{g}(b-\sum_{i=1}^{\tau}u_it_i-\sum_{i=1}^{\kappa}v_is_i)\\&=\frac{1}{p^\tau}\omega^{\sum_{i=1}^{\kappa}\varphi_{v_i}(b)s_i}\sum_{t\in\mathbb{F}_p^\tau}\widehat{F}(t_1,\cdots,t_\tau)\widehat{g}(b-\sum_{i=1}^{\tau}u_it_i),\end{aligned}$$

where $t:=(t_1,\cdots,t_\tau)$. The last equality follows from (6). Then from Lemma 3, one obtains

$$\widehat{h}(b-\sum_{i=1}^{\kappa}v_is_i)=\omega^{\sum_{i=1}^{\kappa}\varphi_{v_i}(b)s_i}\widehat{h}(b). \tag{9}$$

Substituting (9) into (8) gives

$$\mathbb{S}_h(r,v,b)=\frac{1}{p^\kappa}\sum_{s\in\mathbb{F}_p^\kappa}\omega^{\sum_{i=1}^{\kappa}(\varphi_{v_i}(b)-r_i)s_i}\widehat{h}(b).$$

Further the desired result follows from the fact $\sum_{s_i\in\mathbb{F}_p}\omega^{(\varphi_{v_i}(b)-r_i)s_i}=p$ if $r_i=\varphi_{v_i}(b)$ and 0 otherwise for any $1\le i\le\kappa$. This completes the proof.

3 The Generic Constructions of Bent Functions

In this section, we will present two generic constructions of bent functions from known ones with certain properties.

3.1 The First New Class of Bent Functions

A class of non-quadratic p-ary bent functions $f(x)=\mathrm{Tr}(\lambda x^{p^k+1})+c\mathrm{Tr}(x)^{p-1}$ with $\deg(f)=p-1$ was presented in [15], where p is an odd prime and $n/\gcd(k,n)$ is odd. Based on Xu et al.'s work [15], we present the first construction.

Construction 1: Let $u_1,\cdots,u_\tau$ be $\tau\ge1$ elements in $\mathbb{F}_{p^n}$, $v_1,\cdots,v_\kappa$ be $\kappa\ge1$ elements in $\mathbb{F}_{p^n}^*$ and $c\in\mathbb{F}_p^*$. Let $g(x)$ be a weakly regular bent function over $\mathbb{F}_{p^n}$ whose dual satisfies (5) and $F(x_1,\cdots,x_\tau)$ be any reduced polynomial in $\mathbb{F}_p[x_1,\cdots,x_\tau]$. Generate the function $f(x):\mathbb{F}_{p^n}\to\mathbb{F}_p$ from g and F as in (1).

Then our first main result is stated as follows.

Theorem 1. *Let $f(x):\mathbb{F}_{p^n}\to\mathbb{F}_p$ be the function generated by Construction 1. Then $f(x)$ is a bent function if $h(x)$ given by (7) is weakly regular bent.*

Proof. For any $b\in\mathbb{F}_{p^n}$, the Walsh transform of $f(x)$ defined by (1) is

$$\widehat{f}(b)=\sum_{\prod_{i=1}^{\kappa}\mathrm{Tr}(v_ix)=0}\omega^{h(x)-\mathrm{Tr}(bx)}+\sum_{\prod_{i=1}^{\kappa}\mathrm{Tr}(v_ix)\ne0}\omega^{h(x)-\mathrm{Tr}(bx)+c}.$$

Note that

$$\{x\in\mathbb{F}_{p^n}:\prod_{i=1}^{\kappa}\mathrm{Tr}(v_ix)=0\}=\bigcup_{r\in\mathbb{F}_p^\kappa}\{x\in\mathbb{T}_{r,v}:\prod_{i=1}^{\kappa}r_i=0\},$$

where $r := (r_1, \cdots, r_\kappa)$. Then

$$\begin{aligned}\widehat{f}(b) &= \sum_{\substack{\prod_{i=1}^{\kappa} r_i = 0 \\ r_i \in \mathbb{F}_p}} \sum_{x \in \mathbb{T}_{r,v}} (1-\omega^c)\omega^{h(x)-\mathrm{Tr}(bx)} + \omega^c \widehat{h}(b) \\ &= \sum_{\substack{\prod_{i=1}^{\kappa} r_i = 0 \\ r_i \in \mathbb{F}_p}} (1-\omega^c)\mathbb{S}_h(r,v,b) + \omega^c \widehat{h}(b).\end{aligned}$$

From Lemma 4 we can deduce that

$$\widehat{f}(b) = (1-\omega^c)\widehat{h}(b) + \omega^c \widehat{h}(b) = \widehat{h}(b)$$

if $\prod_{i=1}^{\kappa} \varphi_{v_i}(b) = 0$ and $\widehat{f}(b) = \omega^c \widehat{h}(b)$ if $\prod_{i=1}^{\kappa} \varphi_{v_i}(b) \neq 0$. Hence the desired conclusion follows.

3.2 The Second New Class of Bent Functions

Based on the work in [15], Xu et al. [17] further constructed two classes of piecewise p-ary bent functions $f(x) = g(x) + \mathrm{Tr}(ux)\mathrm{Tr}(x)^{p-1}$ from the p-ary Kasami functions and Sidelnikov functions. Inspired by the idea coined in [15] and [17], we present the second construction.

Construction 2: Let $u_1, \cdots, u_\tau$ be $\tau \geq 1$ elements in $\mathbb{F}_{p^n}$, $v_1, \cdots, v_\kappa$ be $\kappa \geq 1$ elements in $\mathbb{F}_{p^n}^*$ and $c \in \mathbb{F}_{p}^*$. Let $g(x)$ be a weakly regular bent function over $\mathbb{F}_{p^n}$ whose dual satisfies (5) and $F(x_1, \cdots, x_\tau)$ be any reduced polynomial in $\mathbb{F}_p[x_1, \cdots, x_\tau]$. Generate the function $f(x) : \mathbb{F}_{p^n} \to \mathbb{F}_p$ from g and F as in (2). Note that $f(x)$ can be rewritten as

$$f(x) = \begin{cases} g(x), \text{ if } \prod_{i=1}^{\kappa} \mathrm{Tr}(v_i x) = 0, \\ h(x), \text{ if } \prod_{i=1}^{\kappa} \mathrm{Tr}(v_i x) \neq 0, \end{cases}$$

where $h(x)$ is defined as in (7).

Then our second main result is stated as follows.

Theorem 2. *Let $f(x) : \mathbb{F}_{p^n} \to \mathbb{F}_p$ be the function generated by Construction 2. Then $f(x)$ is a bent function if $h(x)$ given by (7) is bent.*

Proof. For any $b \in \mathbb{F}_{p^n}$, the Walsh transform of $f(x)$ defined by (2) is

$$\widehat{f}(b) = \sum_{\prod_{i=1}^{\kappa} \mathrm{Tr}(v_i x) = 0} \omega^{g(x)-\mathrm{Tr}(bx)} + \sum_{\prod_{i=1}^{\kappa} \mathrm{Tr}(v_i x) \neq 0} \omega^{h(x)-\mathrm{Tr}(bx)}.$$

Note that

$$\{x \in \mathbb{F}_{p^n} : \prod\nolimits_{i=1}^{\kappa} \mathrm{Tr}(v_i x) = 0\} = \bigcup\nolimits_{r \in \mathbb{F}_p^\kappa} \{x \in \mathbb{T}_{r,v} : \prod\nolimits_{i=1}^{\kappa} r_i = 0\},$$

where $r := (r_1, \cdots, r_\kappa)$. Then

$$\begin{aligned}\widehat{f}(b) &= \sum_{\substack{\prod_{i=1}^{\kappa} r_i = 0\\ r_i \in \mathbb{F}_p}} \sum_{x \in \mathbb{T}_{r,v}} (\omega^{g(x)-\mathrm{Tr}(bx)} - \omega^{h(x)-\mathrm{Tr}(bx)}) + \widehat{h}(b)\\ &= \sum_{\substack{\prod_{i=1}^{\kappa} r_i = 0\\ r_i \in \mathbb{F}_p}} (\mathbb{S}_g(r, v, b) - \mathbb{S}_h(r, v, b)) + \widehat{h}(b),\end{aligned}$$

where $v = (v_1, \cdots, v_\kappa)$. Together with Lemmas 2 and 4, we can conclude that

$$\widehat{f}(b) = \widehat{g}(b) + \widehat{h}(b) - \widehat{h}(b) = \widehat{g}(b)$$

if $\prod_{i=1}^{\kappa} \varphi_{v_i}(b) = 0$ and $\widehat{f}(b) = \widehat{h}(b)$ if $\prod_{i=1}^{\kappa} \varphi_{v_i}(b) \neq 0$. Hence the desired conclusion follows.

Remark 1. Here we show that our constructions of bent functions are different from those in the previous works. Observe that bent functions of the forms (1) and (2) given by this paper are of the form $g(x) + F(\mathrm{Tr}(\alpha_1 x), \cdots, \mathrm{Tr}(\alpha_\lambda x))$, where $g(x)$ is a weakly regular bent function from $\mathbb{F}_{p^n}$ to $\mathbb{F}_p$, F is a reduced polynomial given by (3) and $\alpha_i \in \mathbb{F}_{p^n}$ for $1 \leq i \leq \lambda$. Recently, some attempts have been made to construct bent functions of the above general form, see [7, 9, 11–16, 18]. It should be noted that all these known results in this direction depended on the dual of weakly regular bent function $g(x)$, that is,

$$\widetilde{g}(x - \sum\nolimits_{i=1}^{\lambda} \alpha_i t_i) = \widetilde{g}(x) + \sum\nolimits_{1 \leq i \leq j \leq \lambda} A_{ij} t_i t_j + \sum\nolimits_{i=1}^{\lambda} g_i(x) t_i, \tag{10}$$

where $A_{ij} \in \mathbb{F}_p$ for $1 \leq i \leq j \leq \lambda$, $g_i(x) : \mathbb{F}_{p^n} \to \mathbb{F}_p$ and $t_i \in \mathbb{F}_p$ for $1 \leq i \leq \lambda$. Denote the number of nonzero elements in $\{A_{ij} : 1 \leq i \leq j \leq \lambda\}$ by N. Notice that these known results of the above general form were given for $N \leq 3$ [7, 9, 11–16, 18] expect a class of bent functions with $\deg(F) = 2$ and $N > 3$ [13] when p is an odd prime. However, although our results also depend on the dual of $g(x)$, bent functions with $\deg(F) > 2$ and $N > 3$ can be constructed by this paper for odd p, see Examples 1–4 for details.

Remark 2. As pointed out in [11, Lemma 2.1], for $p = 2$, the algebraic degree of $F(\mathrm{Tr}(u_1 x), \cdots, \mathrm{Tr}(u_\tau x))$ is the same as that of $F(x_1, \cdots, x_\tau)$ for some u_i, and it can be generalized to any characteristic directly, where $u_i \in \mathbb{F}_{p^n}$ for $1 \leq i \leq \tau$. Hence there is no doubt that the generic constructions given in Theorems 1 and 2 can produce bent functions with the maximal algebraic degree given in Lemma 1.

4 Specific Constructions of Infinite Families of Bent Functions

In this section, by using constructions 1 and 2, we shall introduce specific constructions of bent functions from some known ones whose duals satisfy (5).

4.1 New Classes of Bent Functions from Quadratic Bent Functions

It is well-known that a homogeneous quadratic bent function is weakly regular and its dual is also a homogenous quadratic bent function [6]. Let $g(x)$ be a homogeneous quadratic bent function with the dual $\widetilde{g}(x) = \sum_{k=0}^{n-1} \mathrm{Tr}(a_k x^{p^k+1})$, $a_k \in \mathbb{F}_{p^n}^*$. Through some calculation, we can obtain that $\widetilde{g}(x - \sum_{i=1}^{\kappa} v_i s_i - \sum_{i=1}^{\tau} u_i t_i)$ is equal to

$$\widetilde{g}(x - \sum_{i=1}^{\tau} u_i t_i) + \sum_{i=1}^{\kappa} \varphi_{v_i}(x) s_i + \sum_{i=1}^{\kappa}\sum_{j=1}^{\tau} \varphi_{v_i}(u_j) s_i t_j + \sum_{1\le i\le j\le \kappa} G(v_i, v_j) s_i s_j$$

for all $x \in \mathbb{F}_{p^n}$, $u_i \in \mathbb{F}_{p^n}$, $v_j \in \mathbb{F}_{p^n}^*$ and $s_i, t_j \in \mathbb{F}_p$, $1 \le i \le \tau$, $1 \le j \le \kappa$, where $G(v_i, v_j) = \varphi_{v_i}(v_j)$ if $i < j$ and $G(v_i, v_i) = \sum_{k=0}^{n-1} \mathrm{Tr}(a_k v_i^{p^k+1})$ if $i = j$. Here $\varphi_{v_i}(x) := \sum_{k=0}^{n-1} \mathrm{Tr}(a_k(v_i x^{p^k} + v_i^{p^k} x))$. Obviously, when $\varphi_{v_i}(u_j) = 0$ for all $1 \le i \le \kappa, 1 \le j \le \tau$, $\varphi_{v_i}(v_j) = 0$ for all $1 \le i < j \le \kappa$ and $\sum_{k=0}^{n-1} \mathrm{Tr}(a_k v_i^{p^k+1}) = 0$ for all $1 \le i \le \kappa$, the dual of $g(x)$ satisfies (5).

The specific construction of bent functions from quadratic bent functions can be simply introduced by using the Sidelnikov function. From [6], we know that $g(x) = \mathrm{Tr}(ax^2)$ is weakly regular bent and its dual is $\widetilde{g}(x) = -\mathrm{Tr}(x^2/(4a))$, where $a \in \mathbb{F}_{p^n}^*$ and p is an odd prime. Then the following two theorems are directly obtained from Theorem 1 and Theorem 2 respectively.

Theorem 3. *Let p be an odd prime, $a \in \mathbb{F}_{p^n}^*$, $c \in \mathbb{F}_p^*$, $u_1, \cdots, u_\tau$ be $\tau \ge 1$ elements in $\mathbb{F}_{p^n}$ and $v_1, \cdots, v_\kappa$ be $\kappa \ge 1$ elements in $\mathbb{F}_{p^n}^*$ satisfying $\mathrm{Tr}(v_i v_j/(4a)) = 0$ for all $1 \le i \le j \le \kappa$ and $\mathrm{Tr}(u_i v_j/(4a)) = 0$ for all $1 \le i \le \tau$, $1 \le j \le \kappa$. Then*

$$f(x) = \mathrm{Tr}(ax^2) + F(\mathrm{Tr}(u_1 x), \mathrm{Tr}(u_2 x), \cdots, \mathrm{Tr}(u_\tau x)) + c\Big(\prod_{i=1}^{\kappa} \mathrm{Tr}(v_i x)\Big)^{p-1}$$

is a bent function if $h(x)$ given by (7) is weakly regular bent.

Example 1. Let $p = 3$, $n = 5$, $a = 1$, $c = 1$, $\tau = 4$, $\kappa = 2$, $F(x_1, x_2, x_3, x_4) = x_1x_2 + x_3x_4$ and ξ be a primitive element of $\mathbb{F}_{3^5}$. Take $u_1 = 1$, $u_2 = \xi^{212}$, $u_3 = 2$, $u_4 = \xi^{11}$, $v_1 = \xi^{341}$ and $v_2 = \xi^{705}$. Then $\mathrm{Tr}(u_1^2) = \mathrm{Tr}(u_3^2) = \mathrm{Tr}(u_4^2) = \mathrm{Tr}(u_1u_2) = -1$, $\mathrm{Tr}(u_2^2) = \mathrm{Tr}(u_1u_3) = \mathrm{Tr}(u_2u_3) = 1$, $\mathrm{Tr}(u_1u_4) = \mathrm{Tr}(u_2u_4) = \mathrm{Tr}(u_3u_4) = 0$, $\mathrm{Tr}(v_1^2) = \mathrm{Tr}(v_1v_2) = \mathrm{Tr}(v_2^2) = 0$ and $\mathrm{Tr}(u_iv_j) = 0$ for all $1 \le i \le 4$, $j = 1, 2$. It can be verified that $h(x) = \mathrm{Tr}(x^2) + \mathrm{Tr}(x)\mathrm{Tr}(\xi^{212}x) + \mathrm{Tr}(2x)\mathrm{Tr}(\xi^{11}x)$ is bent by the discussion of Case II of [13]. Theorem 3 now establishes that

$$f(x) = \mathrm{Tr}(x^2) + \mathrm{Tr}(x)\mathrm{Tr}(\xi^{212}x) + \mathrm{Tr}(2x)\mathrm{Tr}(\xi^{11}x) + (\mathrm{Tr}(\xi^{341}x)\mathrm{Tr}(\xi^{705}x))^2$$

is a bent function over $\mathbb{F}_{3^5}$. Moreover, it can be checked that $N = 7$ and $\deg(F) = 4$, where N is defined as in Remark 1.

Theorem 4. *Let p be an odd prime, $a \in \mathbb{F}_{p^n}^*$, $u_1, \cdots, u_\tau$ be $\tau \ge 1$ elements in $\mathbb{F}_{p^n}$ and $v_1, \cdots, v_\kappa$ be $\kappa \ge 1$ elements in $\mathbb{F}_{p^n}^*$ satisfying $\mathrm{Tr}(v_iv_j/(4a)) = 0$ for all $1 \le i \le j \le \kappa$ and $\mathrm{Tr}(u_iv_j/(4a)) = 0$ for all $1 \le i \le \tau$, $1 \le j \le \kappa$. Then*

$$f(x) = \mathrm{Tr}(ax^2) + F(\mathrm{Tr}(u_1 x), \mathrm{Tr}(u_2 x), \cdots, \mathrm{Tr}(u_\tau x))\Big(\prod_{i=1}^{\kappa} \mathrm{Tr}(v_i x)\Big)^{p-1}$$

is a bent function if $h(x)$ given by (7) is bent.

Example 2. Let $p = 3, n = 6, a = 1, \tau = 4, \kappa = 1, F(x_1, x_2, x_3, x_4) = x_1x_2+x_3x_4$ and ξ be a primitive element of $\mathbb{F}_{3^6}$. Take $u_1 = \xi^2$, $u_2 = \xi$, $u_3 = \xi^{26}$, $u_4 = \xi^{118}$ and $v_1 = \xi^7$. Then $\mathrm{Tr}(u_1^2) = \mathrm{Tr}(u_3^2) = 1$, $\mathrm{Tr}(u_2^2) = \mathrm{Tr}(u_4^2) = 2$, $\mathrm{Tr}(u_iu_j) = 0$ for others $1 \le i < j \le 4$, $\mathrm{Tr}(v_1^2) = 0$ and $\mathrm{Tr}(u_iv_1) = 0$ for all $1 \le i \le 4$. It can be verified that $h(x) = \mathrm{Tr}(x^2) + \mathrm{Tr}(\xi x)\mathrm{Tr}(\xi^2x) + \mathrm{Tr}(\xi^{26}x)\mathrm{Tr}(\xi^{118}x)$ is bent by [13, Theorem 5]. Theorem 4 now establishes that

$$f(x) = \mathrm{Tr}(x^2) + (\mathrm{Tr}(\xi x)\mathrm{Tr}(\xi^2x) + \mathrm{Tr}(\xi^{26}x)\mathrm{Tr}(\xi^{118}x))\mathrm{Tr}(\xi^7x)^2$$

is a bent function over $\mathbb{F}_{3^6}$. Moreover, it can be checked that $N = 4$ and $\deg(F) = 4$, where N is defined as in Remark 1.

4.2 New Classes of Bent Functions from the Maiorana-MacFarland Class of Bent Functions

Let $n = 2m$ be a positive integer. By identifying an element $x \in \mathbb{F}_{p^n}$ with a vector $(y, z) \in \mathbb{F}_{p^m} \times \mathbb{F}_{p^m}$, the Maiorana-MacFarland class of bent functions on $\mathbb{F}_{p^n}$ can be expressed as

$$g(x) = g(y, z) = \mathrm{Tr}_1^m(y\pi(z)) + h(z),\ y, z \in \mathbb{F}_{p^m}, \tag{11}$$

where $\pi : \mathbb{F}_{p^m} \to \mathbb{F}_{p^m}$ is a permutation and h is a function from $\mathbb{F}_{p^m}$ to $\mathbb{F}_p$. Such class of bent functions is regular and its dual [4] is equal to

$$\widetilde{g}(x) = \widetilde{g}(y, z) = \mathrm{Tr}_1^m\left(z\pi^{-1}(y)\right) + h\left(\pi^{-1}(y)\right),$$

where π^{-1} is the inverse permutation of π. There are several π and h such that the dual of $g(y, z)$ satisfies (5). In the following, we give the calculation of $\widetilde{g}(y, z)$ for the case while π is a linear permutation and $h(z) = 0$. In the same manner, it can be verified that $g(x)$ with some other π and h is suitable for Construction 1 and Construction 2. When π is a linear permutation and $h(z) = 0$, one has

$$\pi^{-1}\Big(y - \sum_{i=1}^{\kappa} v_{i,1}s_i - \sum_{i=1}^{\tau} u_{i,1}t_i\Big) = \pi^{-1}(y - \sum_{i=1}^{\tau} u_{i,1}t_i) - \sum_{i=1}^{\kappa} \pi^{-1}(v_{i,1})s_i,$$

where $u_i = (u_{i,1}, u_{i,2}) \in \mathbb{F}_{p^m} \times \mathbb{F}_{p^m}$, $v_j = (v_{j,1}, v_{j,2}) \in \mathbb{F}_{p^m}^* \times \mathbb{F}_{p^m}^*$ and $t_i, s_j \in \mathbb{F}_p$, $1 \le i \le \tau$, $1 \le j \le \kappa$. Then by a direct calculation, we can derive that

$$\widetilde{g}\Big(y - \sum_{i=1}^{\kappa} v_{i,1}s_i - \sum_{i=1}^{\tau} u_{i,1}t_i, z - \sum_{i=1}^{\kappa} v_{i,2}s_i - \sum_{i=1}^{\tau} u_{i,2}t_i\Big)$$

equals to

$$\widetilde{g}(y - \sum_{i=1}^{\tau} u_{i,1}t_i, z - \sum_{i=1}^{\tau} u_{i,2}t_i) + \sum_{i=1}^{\kappa} \varphi_{v_{i,1},v_{i,2}}(y, z)s_i + \mathcal{G}$$

with

$$\mathcal{G}=\sum_{i=1}^{\kappa}\sum_{j=1}^{\tau}\varphi_{v_{i,1},v_{i,2}}(u_{j,1},u_{j,2})s_it_j+\sum_{1\leq i\leq j\leq\kappa}G(v_{i,1},v_{i,2},v_{j,1},v_{j,2})s_is_j$$

where $G(v_{i,1},v_{i,2},v_{j,1},v_{j,2})=\varphi_{v_{i,1},v_{i,2}}(v_{j,1},v_{j,2})$ if $i<j$ and $G(v_{i,1},v_{i,2},v_{i,1},v_{i,2}) = \mathrm{Tr}_1^m(\pi^{-1}(v_{i,1})v_{i,2})$ if $i=j$. Here $\varphi_{v_{i,1},v_{i,2}}(y,z):=\mathrm{Tr}_1^m(\pi^{-1}(v_{i,1})z+v_{i,2}\pi^{-1}(y))$. If $\varphi_{v_{i,1},v_{i,2}}(u_{j,1},u_{j,2})=0$ for all $1\leq i\leq\kappa, 1\leq j\leq\tau$, $\varphi_{v_{i,1},v_{i,2}}(v_{j,1}, v_{j,2})=0$ for all $1\leq i<j\leq\kappa$ and $\mathrm{Tr}_1^m(\pi^{-1}(v_{i,1})v_{i,2})=0$ for all $1\leq i\leq\kappa$, the dual of $g(x)$ satisfies (5). Then the following two theorems are directly obtained from Theorem 1 and Theorem 2 respectively.

Theorem 5. *Let $c\in\mathbb{F}_p^*$, $u_1,\cdots,u_\tau$ be $\tau\geq 1$ elements in $\mathbb{F}_{p^{2m}}$ and $v_1,\cdots,v_\kappa$ be $\kappa\geq 1$ elements in $\mathbb{F}_{p^{2m}}^*$, where p is a prime. Write u_i as $(u_{i,1},u_{i,2})\in\mathbb{F}_{p^m}\times\mathbb{F}_{p^m}$ for each $1\leq i\leq\tau$ and v_j as $(v_{j,1},v_{j,2})\in\mathbb{F}_{p^m}\times\mathbb{F}_{p^m}$ for each $1\leq j\leq\kappa$. Let $g(y,z)=\mathrm{Tr}_1^m(y\pi(z))$, where π is a linear permutation over $\mathbb{F}_{p^m}$. If $\mathrm{Tr}_1^m(\pi^{-1}(v_{i,1})u_{j,2}+v_{i,2}\pi^{-1}(u_{j,1}))=0$ for all $1\leq i\leq\kappa, 1\leq j\leq\tau$, $\mathrm{Tr}_1^m(\pi^{-1}(v_{i,1})v_{j,2}+v_{i,2}\pi^{-1}(v_{j,1}))=0$ for all $1\leq i<j\leq\kappa$ and $\mathrm{Tr}_1^m(\pi^{-1}(v_{i,1})v_{i,2})=0$ for all $1\leq i\leq\kappa$. Then*

$$f(y,z)=h(y,z)+c\Big(\prod_{i=1}^{\kappa}\mathrm{Tr}_1^m(v_{i,1}y+v_{i,2}z)\Big)^{p-1}$$

is a bent function if

$$h(y,z)=\mathrm{Tr}_1^m(y\pi(z))+F(\mathrm{Tr}_1^m(u_{1,1}y+u_{1,2}z),\cdots,\mathrm{Tr}_1^m(u_{\tau,1}y+u_{\tau,2}z)) \quad (12)$$

is weakly regular bent.

Example 3. Let $p=5$, $m=3$, $\tau=4$, $\kappa=1$, $F(x_1,x_2,x_3,x_4)=x_1x_2+x_3x_4$, $c=1$ and ξ be a primitive element of $\mathbb{F}_{5^3}$. Take $\pi(z)=z^5$, $(u_{1,1},u_{1,2})=(\xi^{26},\xi^{50})$, $(u_{2,1},u_{2,2})=(\xi^{32},\xi^{51})$, $(u_{3,1},u_{3,2})=(\xi^{119},\xi^{8})$, $(u_{4,1},u_{4,2})=(\xi^{63},\xi^{5})$ and $(v_{1,1},v_{1,2})=(\xi^{36},\xi^{114})$. Then $\pi^{-1}(z)=z^{25}$, $\mathrm{Tr}_1^3((u_{1,1})^{25}u_{1,2})=\mathrm{Tr}_1^3((u_{3,1})^{25}u_{3,2})=1$, $\mathrm{Tr}_1^3((u_{2,1})^{25}u_{2,2})=\mathrm{Tr}_1^3((u_{4,1})^{25}u_{4,2})=2$, $\mathrm{Tr}_1^3((u_{i,1})^{25}u_{j,2}+u_{i,2}(u_{j,1})^{25})=0$ for $1\leq i<j\leq 4$, $\mathrm{Tr}_1^3((v_{1,1})^{25}u_{j,2}+v_{1,2}(u_{j,1})^{25})=0$ for all $1\leq j\leq 4$ and $\mathrm{Tr}_1^3((v_{1,1})^{25}v_{1,2})=0$. It can be verified that $h(y,z)=\mathrm{Tr}_1^3(yz^5)+F(\mathrm{Tr}_1^3(u_{1,1}y+u_{1,2}z),\cdots,\mathrm{Tr}_1^3(u_{4,1}y+u_{4,2}z))$ is bent by [13, Theorem 5]. Theorem 5 now establishes that

$$f(y,z)=\mathrm{Tr}_1^3(yz^5)+F(\mathrm{Tr}_1^m(u_{1,1}y+u_{1,2}z),\cdots,\mathrm{Tr}_1^m(u_{4,1}y+u_{4,2}z))+\mathrm{Tr}_1^m(\xi y+2z)^4$$

is a bent function over $\mathbb{F}_{5^6}$. Moreover, it can be checked that $N=4$ and $\deg(F)=4$, where N is defined as in Remark 1.

Theorem 6. *Let $u_i=(u_{i,1},u_{i,2})(1\leq i\leq\tau)$ be τ elements in $\mathbb{F}_{p^m}\times\mathbb{F}_{p^m}$ and $v_j=(v_{j,1},v_{j,2})(1\leq j\leq\kappa)$ be κ elements in $\mathbb{F}_{p^m}^*\times\mathbb{F}_{p^m}^*$, where p is a prime. Let $g(y,z)=\mathrm{Tr}_1^m(y\pi(z))$, where π is a linear permutation over $\mathbb{F}_{p^m}$. If $\mathrm{Tr}(\pi^{-1}(v_{i,1})u_{j,2}+v_{i,2}\pi^{-1}(u_{j,1}))=0$ for all $1\leq i\leq\kappa, 1\leq j\leq\tau$, $\mathrm{Tr}(\pi^{-1}(v_{i,1})v_{j,2}$*

$+v_{i,2}\pi^{-1}(v_{j,1})) = 0$ *for all* $1 \leq i < j \leq \kappa$ *and* $\mathrm{Tr}(\pi^{-1}(v_{i,1})v_{i,2}) = 0$ *for all* $1 \leq i \leq \kappa$. *Then*

$$\widetilde{g}(y,z) + F(\mathrm{Tr}(u_{1,1}y + u_{1,2}z), \cdots, \mathrm{Tr}(u_{\tau,1}y + u_{\tau,2}z))\big(\prod\nolimits_{i=1}^{\kappa} \mathrm{Tr}(v_{i,1}y + v_{i,2}z)\big)^{p-1}$$

is a bent function if $h(x)$ *given by* (12) *is bent.*

Example 4. Let $p = 3$, $m = 3$, $\tau = 4$, $\kappa = 1$, $F(x_1, x_2, x_3, x_4) = x_1x_2 + x_3x_4$ and ξ be a primitive element of $\mathbb{F}_{3^3}$. Take $\pi(z) = z^3$, $(u_{1,1}, u_{1,2}) = (\xi^6, \xi^{21})$, $(u_{2,1}, u_{2,2}) = (\xi^{25}, \xi^{15})$, $(u_{3,1}, u_{3,2}) = (\xi^{11}, \xi^{24})$, $(u_{4,1}, u_{4,2}) = (\xi^5, \xi^{12})$ and $(v_{1,1}, v_{1,2}) = (\xi, 2)$. Then $\pi^{-1}(z) = z^9$, $\mathrm{Tr}_1^3((u_{1,1})^9 u_{1,2}) = \mathrm{Tr}_1^3((u_{3,1})^9 u_{3,2}) = 1$, $\mathrm{Tr}_1^3((u_{2,1})^9 u_{2,2}) = \mathrm{Tr}_1^3((u_{4,1})^9 u_{4,2}) = 2$, $\mathrm{Tr}_1^3((u_{i,1})^9 u_{j,2} + u_{i,2}(u_{j,1})^9) = 0$ for $1 \leq i < j \leq 4$, $\mathrm{Tr}_1^3((v_{1,1})^9 u_{j,2} + v_{1,2}(u_{j,1})^9) = 0$ for all $1 \leq j \leq 4$ and $\mathrm{Tr}_1^3((v_{1,1})^9 v_{1,2}) = 0$. It can be verified that $h(y,z) = \mathrm{Tr}_1^3(yz^3) + F(\mathrm{Tr}_1^3(u_{1,1}y + u_{1,2}z), \cdots, \mathrm{Tr}_1^3(u_{4,1}y + u_{4,2}z))$ is bent by [13, Theorem 5]. Theorem 6 now establishes that

$$f(y,z) = \mathrm{Tr}_1^3(yz^3) + F(\mathrm{Tr}_1^m(u_{1,1}y + u_{1,2}z), \cdots, \mathrm{Tr}_1^m(u_{4,1}y + u_{4,2}z))\mathrm{Tr}_1^m(\xi y + 2z)^2$$

is a bent function over $\mathbb{F}_{3^6}$. Moreover, it can be checked that $N = 4$ and $\deg(F) = 4$, where N is defined as in Remark 1.

5 Conclusions

In this paper, we proposed two generic constructions of bent functions with the forms (1) and (2), which generalized some previous works [15,17]. Moreover, based on our constructions, several infinite families of bent functions can be obtained from quadratic bent functions and the Maiorana-MacFarland class of bent functions by calculating their duals. In addition, it was shown that bent functions with the maximal algebraic degree can be obtained from our constructions.

References

1. Carlet, C.: Boolean Functions for Cryptography and Coding Theory, Cambridge University Press, Cambridge (2021)
2. Cohen, G., Honkala, I., Litsyn, S., Lobstein, A.: Covering Codes. The North Holland, Amsterdam (1997)
3. Ding, C., Fan, C., Zhou, Z.: The dimension and minimum distance of two classes of primitive BCH codes. Finite Fields Appl. **45**, 237–263 (2017)
4. Kumar, P.V., Scholtz, R.A., Welch, L.R.: Generalized bent functions and their properties. J. Combin. Theory Ser. A **40**(1), 90–107 (1985)
5. Hou, X.: p-ary and p-ary versions of certain results about bent functions and resilient functions. Finite Fields Appl. **10**(4), 566–582 (2004)
6. Helleseth, T., Kholosha, A.: Monomial and quadratic bent functions over the finite fields of odd characteristic. IEEE Trans. Inf. Theory **52**(5), 2018–2032 (2006)

7. Mesnager, S.: Several new infinite families of bent functions and their duals. IEEE Trans. Inf. Theory **60**(7), 4397–4407 (2014)
8. Olsen, J., Scholtz, R., Welch, L.: Bent-function sequences. IEEE Trans. Inf. Theory **28**(6), 858–864 (1982)
9. Qi, Y., Tang, C., Zhou, Z., Fan, C.: Several infinite families of p-ary weakly regular bent functions. Adv. Math. Commun. **12**(2), 303–315 (2018)
10. Rothaus, O.S.: On bent functions. J. Combin. Theory Ser. A **20**(3), 300–305 (1976)
11. Tang, C., Zhou, Z., Qi, Y., Zhang, X., Fan, C., Helleseth, T.: Generic construction of bent functions and bent idempotents with any possible algebraic degrees. IEEE Trans. Inf. Theory **63**(10), 6149–6157 (2017)
12. Wang, L., Wu, B., Liu, Z., Lin, D.: Three new infinite families of bent functions. Sci. China Inf. Sci. **61**(3), 1–14 (2018)
13. Xie X., Li N., Zeng X., Tang X., Yao Y.: Several classes of bent functions over finite fields. arXiv: 2108.00612 (2021)
14. Xu, G., Cao, X., Xu, S.: Several new classes of Boolean functions with few Walsh transform values. Appl. Algebra Eng. Commun. Comput. **28**(2), 155–176 (2017)
15. Xu, G., Cao, X., Xu, S.: Constructing new APN functions and bent functions over finite fields of odd characteristic via the switching method. Cryptogr. Commun. **8**(1), 155–171 (2016)
16. Xu, G., Cao, X., Xu, S.: Several classes of quadratic ternary bent, near-bent and 2-plateaued functions. Int. J. Found. Comput. Sci. **28**(1), 1–18 (2017)
17. Xu, G., Cao, X., Xu, S.: Two classes of p-ary bent functions and linear codes with three or four weights. Cryptogr. Commun. **9**(1), 117–131 (2017)
18. Zheng, L., Peng, J., Kan, H., Li, Y.: Several new infinite families of bent functions via second order derivatives. Cryptogr. Commun. **12**(6), 1143–1160 (2020). https://doi.org/10.1007/s12095-020-00436-0

Sequences

Correlation Measure of Binary Sequence Families With Trace Representation

Ana I. Gómez[1], Domingo Gomez-Perez[2](✉), and Andrew Tirkel[3]

[1] Universidad Rey Juan Carlos, Móstoles, Spain
ana.gomez.perez@urjc.es
[2] Universidad de Cantabria, Santander, Spain
domingo.gomez@unican.es
[3] Scientific Technologies, Melbourne, Australia
atirkel@bigpond.net.au

Abstract. In this paper, we analyze the occurrence of peaks in the correlation measure of several families of binary sequences used in communications. This concept corresponds to the Low Probability of Intercept (LPI) properties, terminology used in the area of communications. For each family, we provide a low order for which the correlation measure exhibits a full peak.

1 Introduction

Wireless communication is one the most widespread way to transmit information from one device to another. Due to the rising incidence of cyberattacks to mobile devices and services, security of communications is an increasingly vital aspect.

Code Division Multiple Access (CDMA) has been replaced in 4G and 5G cellular protocols. However, it is still supported by old devices and used in locations where modern wireless infrastructure can not be deployed. In CDMA, each user is identified by a different pseudorandom binary sequence, called the *spreading code*. The spreading code modulates a carrier, spreads the spectrum of the waveform and hides the communication in the background noise. The aim of this article is providing a new viewpoint on some spreading codes in Direct Sequence Spread Spectrum (DSSS) modulation and analyzing the existence of peaks in the correlation measure of relative low order.

Although many authors have worked on the characterization of statistical properties of the spreading codes, there is still considerable uncertainty. The first results in this direction were given by Warner et al. [17,18], who studied worst case high-order correlations. They showed that classical sequence families used in CDMA such as Gold codes or small Kasami for certain periods are not secure against a blind triple correlation attack, because those are built from m-sequences. Further work [1] has shown that m-sequences and Gold codes can be

The first two authors are supported by Grant PID2019-110633GB-I00, funded by MCIN/AEI/10.13039/501100011033.

S. Mesnager and Z. Zhou (Eds.): WAIFI 2022, LNCS 13638, pp. 313–319, 2023.
https://doi.org/10.1007/978-3-031-22944-2_20

detected in signal intercepts by searching for a pattern in the peaks of the triple order correlation, in the noiseless case, and in correlations of higher order, for high Signal to Noise Ratio (SNR).

Aside from high-order correlations, other authors have tried to attack the communication secrecy using less sophisticated (and easier to compute) statistical properties. We list some related work and some drawbacks associated to them. Attacks using the period of the spreading code, as described by Gouda [8,9], and others [4,11,12], require the removal of the carrier and do not address the issues of additive noise, interference, multipath, etc. A recent work deals with multipath channels but not with the removal of the carrier [14]. Summing up, a major defect in those experiments is that they entail labor-intensive calculations regarding the spreading code alone, whereas those based on high-order correlation typically tolerate the presence of the carrier [10].

Boztaş et Parampalli [2,3] derive the location of triple correlation peaks using the Zech logarithm. These attacks work in the same way as the previously described ones, but include the terminology of low probability of intercept.

These authors show when Gold codes are immune to triple correlation attacks and obtain an upper bound for the triple correlation of the binary Legendre sequence, slightly larger than twice the square root of the sequence length.

Recently, Chen et al. [5] have studied bounds on the correlation measure of arbitrary order and the linear complexity, relating correlation measure with minimum distance of linear codes.

Our aim is to further explore this relation and to find better security estimates against statistical attacks for some families of binary sequences.

2 Mathematical Background

We review some standard results from finite field arithmetic. This work focuses only on finite fields of characteristic 2, which are denoted by $\mathbb{F}_{2^n}$. Throughout the rest of the paper, $\alpha \in \mathbb{F}_{2^n}$ denotes a primitive element and we write $\mathbb{F}_{2^d} \subset \mathbb{F}_{2^n}$ when d divides n. The trace function Tr_d^n (for simplicity, we write Tr when $d = 1$) is a linear map from $\mathbb{F}_{2^n}$ to a subfield $\mathbb{F}_{2^d} \subset \mathbb{F}_{2^n}$ given by

$$\mathrm{Tr}_d^n(x) = \sum_{j=1}^{n/d} x^{2^{d \cdot j}}.$$

It is known that the trace function is surjective and satisfies the following properties (see [15, Theorem 1.4.50]):

$$\mathrm{Tr}_d^n(x + y) = \mathrm{Tr}_d^n(x) + \mathrm{Tr}_d^n(y), \qquad \mathrm{Tr}_d^n(c \cdot x) = c \cdot \mathrm{Tr}_d^n(x) \quad \forall c \in \mathbb{F}_{2^d}.$$

We consider different families of binary sequences, due to their application in communication systems. Let us describe them briefly.

An m-sequence is a sequence of period $2^n - 1$ defined by a primitive element α as

$$s_i = \mathrm{Tr}(\alpha^i), \quad i = 0, \dots, 2^n - 2. \tag{1}$$

Gold codes [7] is a family of $2^n + 1$ sequences of period $2^n - 1$, containing two m-sequences defined by an integer κ (with $\gcd(\kappa, n) = 1$) and two primitive elements: α and $\alpha^{2^\kappa+1}$. The remaining family sequences are defined as

$$s_i^{(d)} = \mathrm{Tr}(\alpha^{i+d} + \alpha^{i(2^\kappa+1)}), \quad i = 0, \ldots, 2^n - 2, \tag{2}$$

where $0 \leq d \leq 2^n - 2$.

The small Kasami family consists of 2^m sequences with period $2^n - 1$, where $n = 2m$. It contains an m-sequence defined by a primitive element α. The remaining sequences are given by

$$s_i^{(d)} = \mathrm{Tr}(\alpha^i + \alpha^{(i+d)(2^m+1)}), \quad i = 0, \ldots, 2^n - 2, \tag{3}$$

where $0 \leq d \leq 2^m - 2$.

The large Kasami family is also defined for even $n = 2m$, but only for odd m. It contains sequences from the Gold family for $\kappa = m + 1$ and the following ones:

$$s_i^{(d)} = \mathrm{Tr}(\alpha^i + \alpha^{(i+d)(2^m+1)} + \alpha^{i(2^{m+1}+1)}), \quad i = 0, \ldots, 2^n - 2, \tag{4}$$

where $0 \leq d \leq 2^m - 2$.

The three term trace is a sequence of period $2^n - 1$, for $n = 2m + 1$, defined by

$$s_i = \mathrm{Tr}(\alpha^i + \alpha^{(2^m+1)i} + \alpha^{(2^m+2^{m-1}+1)i}), \quad i = 0, \ldots, 2^n - 2. \tag{5}$$

GWM sequences [7] are pseudonoise sequences and they are divided in these different types:

- GMW Type I: These sequences depend on the existence of $\mathbb{F}_{2^d} \subset \mathbb{F}_{2^n}$ and they are defined as

$$s_i = \mathrm{Tr}_1^d((\mathrm{Tr}_d^n(\alpha^i))^k), \quad i = 0, \ldots, 2^n - 2,$$

where $\gcd(k, 2^d - 1) = 1$.
- Cascaded GMW: This family of sequences also depends on the existence of $\mathbb{F}_{2^{d_1}} \subset \mathbb{F}_{2^{d_2}} \subset \ldots \mathbb{F}_{2^{d_r}} \subset \mathbb{F}_{2^n}$ and they are defined as

$$s_i = \mathrm{Tr}_1^{d_1}((\mathrm{Tr}_{d_1}^{d_2}((\ldots(\mathrm{Tr}_{d_r}^n(\alpha^i))^{k_r}\ldots)^{k_2}))^{k_1}), \quad i = 0, \ldots, 2^n - 2,$$

where $\gcd(k_j, 2^{d_j} - 1) = 1, \ j = 1 \ldots, r$.
- Generalized GMW: The last type of GMW uses a chain of field $\mathbb{F}_{2^d} \subset \mathbb{F}_{2^n}$ and a function $f : \mathbb{F}_{2^d} \mapsto \mathbb{F}_2$. They are defined as

$$s_i = f(\mathrm{Tr}_d^n(\alpha^i)), \quad i = 0, \ldots, 2^n - 2. \tag{6}$$

For a binary sequence $\mathcal{S}$ with period $2^n - 1$, the *periodic correlation measure* of order k is defined [16] as

$$\theta_k(\mathcal{S}) = \max_D \left| \sum_{n=0}^{2^n-2} (-1)^{s_{n+d_1}+s_{n+d_2}+\cdots+s_{n+d_k}} \right|,$$

where $D = (d_1, \cdots, d_k)$ and $0 \le d_1 < d_2 < \ldots < d_k < 2^n - 2$. A binary sequence $\mathcal{S}$ is said to have a *full peak* in the periodic correlation measure of order k is $\theta_k(\mathcal{S}) = 2^n - 1$.

We use standard terminology from coding theory, see [15] for a more complete introduction. Let C be a binary cyclic code of length $2^n - 1$. It is associated with an ideal in the ring $\mathbb{F}_2[x]/(x^{2^n-1}+1)$. This ideal is generated by a polynomial $g(x)$ that divides $x^{2^n-1}+1$. Cyclic codes are commonly described through a *defining set*. Namely, for integers $0 \le i_1 < \ldots < i_l < 2^n - 1$, the defining set $A = \{\alpha^{i_1}, \alpha^{i_2}, \ldots, \alpha^{i_l}\}$ determines

$$\left\{c(x) \in \mathbb{F}_2[x]/(x^{2^n-1}+1) \quad | \quad c(\xi) = 0 \quad \forall \xi \in A\right\},$$

which is an ideal and therefore a cyclic code.

3 Full Peaks in the Correlation Measure

It is known that m-sequences have the shift-and-add property, so they have a full peak in the correlation measure of order 3. For the rest of the described sequences, we find a (low) order for which there is a full peak, providing hence a bound for the lowest order of interest.

Theorem 1. *The following statements hold.*

- *Sequences in the small Kasami family have a full peak in the correlation measure of order* 3 *or* 4.
- *Sequences of a Gold code have a full peak in the correlation measure of order* 5.
- *Sequences in the large Kasami family have a full peak in the correlation measure of order* 7.
- *The three term trace has a full peak in the correlation measure of order* 7 *and no peak for an order less than* 5.

Proof. The proofs are based on the results by Van Lint and Wilson [13]. Any of the sequences that we are studying are defined as $s_i = \mathrm{Tr}(\sum_{j=1}^{t} \alpha^{ie_j})$, for some integers $e_0, \ldots, e_t$. We notice that the sequence has a full peak in the periodic correlation measure of order k for shifts $0 \le d_1 < \cdots < d_k < 2^n - 1$ if and only if, for every $i \ge 0$,

$$\sum_{m=1}^{k} \sum_{j=1}^{t} \alpha^{(i+d_m)e_j} = \sum_{j=1}^{t} \sum_{m=1}^{k} \alpha^{(i+d_m)e_j} = \sum_{j=1}^{t} \alpha^{i \cdot e_j} \sum_{m=1}^{k} \alpha^{d_m \cdot e_j} = 0.$$

This holds if and only if

$$\sum_{m=1}^{k} \alpha^{d_m \cdot e_j} = 0, \quad \forall\, j = 1, \ldots, t,$$

in other words, if and only if the minimum distance of the cyclic code defined by the set $\{\alpha^{e_1}, \ldots, \alpha^{e_t}\}$ is less than k. By Theorem 12 from [13], the minimum distance of a cyclic code given by $\{\alpha, \alpha^t\}$ is less than 5 and, if $\gcd(t, n) \neq 1$, less than 4. This proves the statement regarding the small Kasami family. For the Gold code, the minimum distance is exactly 5, according to Theorem 14. A similar proof than that of Theorem 16 (always from [13]) gives the minimum distance for the large Kasami family. In the case of the three term trace, the lower bound follows from the fact that the cyclic code for that sequence contains the cyclic code for the Gold case. □

For the GMW, the linear complexity of the sequence can be quite high and the bounds found by Chen et al. [5] are large and impractical for attacks. Unfortunately, the next theorem proves the existence of peaks in much more smaller orders, even for the minimum possible $k = 3$.

Theorem 2. *The following statements hold.*

- *GMW Type I and Cascaded GMW sequences have a full peak in the correlation measure of order* 3.
- *Suppose that* (s_i) *is a Generalized GWM sequence as defined as in Eq.* (6) *and the sequence* $S_i = f(\beta^i)$, $i = 0, \ldots, 2^d - 2$ *have a full peak in the correlation measure of order* k *where* $\beta = \alpha^{2^{n/d}-1}$. *Then* (s_i) *have a full peak in the correlation measure of order* k.

Proof. The second item implies the first because for GMW Type I sequences (S_i) is a proper decimation of a m-sequence, so it has a full peak in the correlation measure of order 3. For Cascaded GMW, (S_i) is again a Cascaded GMW where the chain of fields is smaller. Applying induction on the length of the chain and noticing that the base case correspond to GMW Type I sequences, we obtain that there is a peak in the correlation measure of order 3.

For the Generalized GMW sequences, suppose that sequence S_i has a full peak in the correlation measure of order k for shifts $0 \leq d_1 < \cdots < d_k < 2^d - 1$. This occurs if and only if, for every $i \geq 0$,

$$0 = \sum_{j=1}^{k} f(\beta^{i+d_j}).$$

By hypothesis, α is a primitive root in $\mathbb{F}_{2^n}$ and Tr_d^n is a surjective function so there exist $e_1, \ldots, e_k$ integers such that

$$\beta^{d_j} = \mathrm{Tr}_d^n(\alpha^{e_j}), \quad j = 1, \ldots, k.$$

Now, write i as it expansion in base $2^{n/d} - 1$, i.e $i_1(2^{n/d} - 1) + i_2$.

Suppose $\mathrm{Tr}_d^n(\alpha^{i_2}) \neq 0$ and define h as $\beta^h = \mathrm{Tr}_d^n(\alpha^{i_2})$. Notice that this h exists because β is a primitive root of $\mathbb{F}_{2^d}$. We have that,

$$\sum_{j=1}^{k} f(\mathrm{Tr}_d^n(\alpha^{i+e_j})) = \sum_{j=1}^{k} f(\mathrm{Tr}_d^n(\alpha^{(2^{n/d}-1)i_1+i_2+e_j})) = \sum_{j=1}^{k} f(\beta^{i_1+d_j+h}) = 0.$$

The case $\mathrm{Tr}_d^n(\alpha^{i_2}) = 0$ is dealt exactly the same, so this finishes the proof. □

4 Conclusions and Open Problems

In this paper we have improved all of the bounds found by Chen et al. [5] and summarized in Table 1 therein. Indeed, for Gold codes and small and large Kasami families, we find exact values for the order which guarantee a full peak in the correlation measure. For the three term trace sequence, we provide quite tight bounds. We have also shown that most GMW sequences have full peaks in the correlation measure of order 3, although they present high linear complexity. This happens because the existence of a multidimensional array structure behind them [6] and, due to this fact, it is possible to give exactly the shifts for which it occurs for GMW Type I and Cascaded GMW sequences depending on the field structure.

It would be interesting to study these attacks for other pseudonoise sequences, as sequences with value 1 in the correlation measure of order 2. Indeed, some preliminary computer experiments shows peaks in the correlation measure of order k for small k in hyperovals when the length of the sequence is divisible by a small prime. Another subject that it has not been explored is wether the position of the peaks and the number of them can help to recover the full sequence. Although it is not the case for GMW sequences for $k = 3$, it would be certainly interesting to study wether correlation peaks of order 5 reveals information about the structure of the sequence.

References

1. Adams, E.R.: Identification of pseudo-random sequences in DS/SS intercepts by higher-order statistics. Technical report. Cranfiel University, Royal Military College of Science (2004)
2. Boztaş, S., Parampalli, U.: Low probability of intercept properties of some binary sequence families with good correlation properties. In: 2012 IEEE International Symposium on Information Theory Proceedings, pp. 1226–1230. IEEE (2012)
3. Boztaş, S., Parampalli, U.: On the relative abundance of nonbinary sequences with perfect autocorrelations. In: 2011 IEEE International Symposium on Information Theory Proceedings, pp. 494–498. IEEE (2011)
4. Burel, G., Bouder, C.: Blind estimation of the pseudo-random sequence of a direct sequence spread spectrum signal. In: MILCOM 2000 Proceedings. 21st Century Military Communications. Architectures and Technologies for Information Superiority (Cat. No. 00CH37155), vol. 2, pp. 967–970. IEEE (2000)
5. Chen, Z., Gómez, A.I., Gómez-Pérez, D., Tirkel, A.: Correlation measure, linear complexity and maximum order complexity for families of binary sequences. Finite Fields Their Appl. **78**, 101977 (2022)
6. Gómez, A.I., Gómez-Pérez, D., Tirkel, A.: Generalised GMW sequences. In: 2021 IEEE International Symposium on Information Theory (ISIT), pp. 1806–1811 (2021)
7. Golomb, S.W., Gong, G.: Signal design for good correlation: for wireless communication, cryptography, and radar (2005)
8. Gouda, M.: High immunity triple channelized correlation receiver. In: 2009 First International Conference on Computational Intelligence, Communication Systems and Networks, pp. 409–413. IEEE (2009)

9. Gouda, M., El-Hennawy, A., Mohamed, A.E.: Detection of gold codes using higher-order statistics. In: 2011 First International Conference on Informatics and Computational Intelligence, pp. 361–364. IEEE (2011)
10. Houghton, A., Reeve, C.: Direction finding on spread-spectrum signals using the time-domain filtered cross spectral density. IEE Proc. Radar Sonar Navig. **144**(6), 315–320 (1997)
11. Jang, J.W., Kim, Y.S.: Low probability of intercept property of binary Sidelnikov sequences. In: 2015 International Conference on Information and Communication Technology Convergence (ICTC), pp. 733–735. IEEE (2015)
12. Jang, J.W., Lim, D.W.: Large low probability of intercept properties of the quaternary sequence with optimal correlation property constructed by Legendre sequences. Cryptogr. Commun. **8**(4), 593–604 (2016)
13. van Lint, J.H., Wilson, R.M.: On the minimum distance of cyclic codes. IEEE Trans. Inf. Theory **32**, 23–40 (1986)
14. Mirzadeh Sarcheshmeh, H., Khaleghi Bizaki, H., Alizadeh, S.: PN sequence blind estimation in multiuser DS-CDMA systems with multipath channels based on successive subspace scheme. Int. J. Commun. Syst. **31**(12), e3591 (2018)
15. Niederreiter, H., Winterhof, A.: Applied Number Theory. Springer, Cham (2015). https://doi.org/10.1007/978-3-319-22321-6
16. Pirsic, G.I., Winterhof, A.: On discrete Fourier transform, ambiguity, and Hamming-autocorrelation of pseudorandom sequences. Des. Codes Cryptogr. **73**(2), 319–328 (2014). https://doi.org/10.1007/s10623-013-9916-2
17. Warner, E., Mulgrew, B., Grant, P.: Triple correlation analysis of m-sequences. Electron. Lett. **29**(20), 1755–1756 (1993)
18. Warner, E., Mulgrew, B., Grant, P.: Triple correlation analysis of binary sequences for codeword detection. IEE Proc. Vis. Image Signal Process. **141**(5), 297–302 (1994)

Linear Complexity of Generalized Cyclotomic Sequences with Period $p^n q^m$

Vladimir Edemskiy[1] and Chenhuang Wu[2(✉)]

[1] Department of Applied Mathematics and Informatics, Yaroslav-the-Wise Novgorod State University, Veliky Novgorod 173003, Russia
vladimir.edemsky@novsu.ru

[2] Fujian Key Laboratory of Financial Information Processing, Putian University, Putian, Fujian 351100, China
ptuwch@163.com

Abstract. Linear complexity is a very important merit factor for measuring the unpredictability of pseudo-random sequences for applications. The higher the linear complexity, the better the unpredictability of a sequence. In this paper, we continue the investigation of generalized cyclotomic sequences constructed by new generalized cyclotomy presented by Zeng et al. In detail, we consider the new generalized cyclotomic sequence with period $p^n q^m$ where p, q are odd distinct primes and n, m are natural numbers. It is shown that these sequences have high linear complexity. Finally, we also give some examples to illustrate the correctness of our results.

Keywords: Binary sequences · Generalized cyclotomy · Linear complexity

1 Introduction

Linear complexity is a very important merit factor for measuring the unpredictability of pseudo-random sequences. The linear complexity of a sequence may be defined as the length of the shortest linear feedback shift register which generates the sequence [1]. According to Berlekamp- Massey algorithm, if the linear complexity of the sequence is l, then $2l$ consecutive terms of the sequence can be used to restore the whole sequence. Hence, a "high" linear complexity should be no less than one-half of the length (or minimum period) of the sequence [2]. For cryptographic applications, sequences with high linear complexity are required.

An important method of designing sequences with high linear complexity uses classical cyclotomic classes and generalized cyclotomic classes to construct sequences. Cyclotomy is related to difference sets, sequences, coding theory, and cryptography [3]. Classical cyclotomy was first considered in detail by Gauss. Later, Whiteman presented the generalized cyclotomy of order d with respect to the product of two distinct

V. Edemskiy was supported by Russian Science Foundation according to the research project No. 22-21-00516, https://rscf.ru/en/project/22-21-00516/. C. Wu was partially supported by the Projects of International Cooperation and Exchange NSFC-RFBR No. 61911530130, by the Natural Science Foundation of Fujian Province No. 2020J01905.

S. Mesnager and Z. Zhou (Eds.): WAIFI 2022, LNCS 13638, pp. 320–333, 2023.
https://doi.org/10.1007/978-3-031-22944-2_21

odd primes, which is not consistent with classical cyclotomy [4]. It was extended to odd integers in [5]. Further, a new generalized cyclotomy that includes classical cyclotomy as a special case was introduced by Ding and Helleseth [3]. Fan and Ge proposed a unified approach that determines both Whiteman's and Ding-Helleseth's generalized cyclotomy [6]. In the past decades, the linear complexity of binary and nonbinary Whiteman's and Ding-Helleseth's generalized cyclotomic sequences has been extensively studied [7–12] (see also references therein).

Zeng et al. in [13] presented a new approach and suggested a new generalized cyclotomy. Further, this new generalized cyclotomy was discussed in [14]. Based on the generalized cyclotomic classes from [13], Xiao et al.[15] presented a new family of cyclotomic binary sequences of period p^n and determined the linear complexity of the sequences for the case when $n = 2$ and $f = 2^r$. Later, these results were generalized in [16, 17]. The use of new generalized cyclotomic classes for constructing sequences with high linear complexity and even periods $2p^n, 2^m p^n$ was considered in [18, 19]. In this paper, we will study the linear complexity of new generalized cyclotomic sequences with period $p^n q^m$. These sequences are defined using new generalized cyclotomic classes from [13]. Thus, we continue the study of new generalized cyclotomic sequences started in [15–17].

The rest of the paper is organized as follows. The definition of sequences and the main result are introduced in Sect. 2. In Sect. 3 we discuss some subsidiary statements about the sequence polynomial and in Sect. 4 we prove our main result. We conclude the paper in Sect. 5.

2 Definitions of Sequences

First of all, we recall the definition of new generalized cyclotomic classes presented in [13] for $N = p^n q^m$, where p and q are odd distinct primes, $n > 0, m > 0$. Suppose e divides $p-1$ and $q-1$; then $p-1 = ef$ and $q-1 = eh$. It is well known that there exists primitive roots η and ξ modulo p^2 and q^2 respectively. In this case, η is the primitive root modulo p^k, $k = 1, 2, \ldots, n$ and ξ is the primitive root modulo q^l, $l = 1, 2, \ldots, m$ [20].

Let $v = p^k q^l, v \neq 1$, where $k = 0, 1, \ldots, n$; $l = 0, 1, \ldots, m$.

According to the Chinese Remainder Theorem, there exists g_v such that

$$g_v \equiv \eta^{fp^{k-1}} \pmod{p^k} \text{ when } k \geq 1 \text{ and } g_v \equiv \xi^{hq^{l-1}} \pmod{q^l} \text{ when } l \geq 1. \quad (1)$$

Also there exist ζ_p, ζ_q such that

$$\zeta_p \equiv \begin{cases} \eta \pmod{p^n}, \\ 1 \pmod{q^m}, \end{cases} \text{ and } \zeta_q \equiv \begin{cases} \xi \pmod{q^m}, \\ 1 \pmod{p^n}. \end{cases} \quad (2)$$

Throughout this paper, we let $\mathbb{Z}_v$ be the ring of integers modulo v for a positive integer v, and $\mathbb{Z}_v^*$ be the multiplicative group of $\mathbb{Z}_v$. According to [13] we know that $D^{(v)} = \{g_v^s \mid s = 0, \ldots, e-1\}$ is the subgroup of $\mathbb{Z}_v^*$.

Define

$$\Psi_v = \begin{cases} \mathbb{Z}_{fp^{k-1}} \times \mathbb{Z}_{(q-1)q^{l-1}}, & \text{if } k \geq 1, l \geq 1, \\ \mathbb{Z}_{fp^{k-1}} \times \{0\}, & \text{if } l = 0, \\ \{0\} \times \mathbb{Z}_{hq^{l-1}}, & \text{if } k = 0. \end{cases}$$

Let $I = (i_1, i_2) \in \Psi_v$ and $D_I^{(v)} = \zeta_p^{i_1} \zeta_q^{i_2} D^{(v)}$.

By [13] we have the following partitions

$$\mathbb{Z}_v^* \setminus \{0\} = \bigcup_{I \in \Psi_v} D_I^{(v)} \text{ and } \mathbb{Z}_N \setminus \{0\} = \bigcup_{v|N, v>1} \bigcup_{I \in \Psi_v} \frac{N}{v} D_I^{(v)}.$$

It is necessary to note that for $v = p^k$ (or $v = q^l$) we obtain a partition of $\mathbb{Z}_{p^k}^*$ as in [15] and in this case $\eta^t D^{(p^k)}$ is equal to $D_t^{(p^k)} = \{\eta^{t+fp^{k-1}i} \bmod p^k \mid i = 0, 1, \ldots, e-1\}$, $t = 0, 1, \ldots, fp^{k-1} - 1$. The properties of $D_t^{(p^k)}$ were studied in [15, 16].

Let f and h be even numbers and b, c be integers such that $0 \leq b < fp^{n-1}$, $0 \leq c < hq^{m-1}$. Then define

$$\Psi_v^{(1)} = \begin{cases} \{(i_1 + b, i_2) \in \Psi_v \mid 0 \leq i_1 < p^{k-1}f/2 - 1\}, & \text{if } k \geq 1, \\ \{(0, i_2 + c) \in \Psi_v \mid 0 \leq i_2 < q^{l-1}h/2 - 1\}, & \text{if } k = 0. \end{cases}$$

Let

$$C_1^{(v)} = \bigcup_{I \in \Psi_v^{(1)}} D_I^{(v)} \text{ and } C_0^{(v)} = \bigcup_{I \in \Psi_v \setminus \Psi_v^{(1)}} D_I^{(v)}.$$

Then we see that

$$|C_j^{(v)}| = \begin{cases} p^{k-1}(p-1)q^{l-1}(q-1)/2, & \text{if } k \geq 1, l \geq 1, \\ p^{k-1}(p-1)/2, & \text{if } k \geq 1, l = 0, \\ q^{l-1}(q-1)/2, & \text{if } k = 0, l \geq 1. \end{cases} \quad (3)$$

for $v = p^k q^l, j = 0, 1$.

Define

$$C_j = \bigcup_{v|N, v>1} \frac{N}{v} C_j^{(v)}, j = 0, 1$$

or, in more detail

$$C_j = \bigcup_{k=1}^{n} \bigcup_{l=1}^{m} p^{n-k} q^{m-l} C_j^{(p^k q^l)} \cup \bigcup_{k=1}^{n} p^{n-k} q^m C_j^{(p^k)} \cup \bigcup_{l=1}^{m} p^n q^{m-l} C_j^{(q^l)}. \quad (4)$$

By definition we get $\mathbb{Z}_N \setminus \{0\} = C_0 \cup C_1$ and $C_0 \cap C_1 = \emptyset$.

Then we can define a balanced binary sequence s^∞ with period N as follows:

$$s_i = \begin{cases} 1, & \text{if } i \bmod N \in C_1 \cup \{0\}, \\ 0, & \text{if } i \bmod N \in C_0. \end{cases} \quad (5)$$

A sequence is called balanced if the numbers of 1's and 0's in one minimum period differ by no more than one. Earlier, new generalized cyclotomic classes were used to construct sequences with period p^n. It is necessary to note that for $N = p^2$ this sequence is the same as in [15].

We conclude this section by recalling the notion of the linear complexity and one method of studying the linear complexity. For a N-periodic sequence $s^\infty = \{s_i\}_{i\geq 0}$ over the $\mathbb{F}_2$ (the finite field of two elements), we recall that the linear complexity over $\mathbb{F}_2$, denoted by $\mathrm{LC}(s^\infty)$, is the least order L such that $\{s_i\}$ satisfies

$$s_{i+L} = c_{L-1}s_{i+L-1} + \ldots + c_1 s_{i+1} + c_0 s_i \quad \text{for } i \geq 0,$$

where $c_0 \neq 0, c_1, \ldots, c_{L-1} \in \mathbb{F}_2$.

It is well known (see, for instance, [21]) that if $S(x) = s_0 + s_1 x + \cdots + s_{N-1}x^{N-1}$ then the linear complexity of s^∞ is given by

$$\mathrm{LC}(s^\infty) = N - \deg\left(\gcd\left(x^N - 1, S(x)\right)\right).$$

Thus, if α is a primitive root of order N of unity in the extension of the field $\mathbb{F}_2$, then in order to find the linear complexity of a sequence, it is sufficient to study the zeros of $S(x)$ in the set $\{\alpha^i \mid i = 0, 1, \ldots, N-1\}$.

In this paper, we will study the linear complexity of s^∞ defined by (5). The values $S(\alpha^i)$ we will consider in the following section.

2.1 Main Result

To begin with, we introduce some new notations. Let $\mathrm{ord}_p(2)$ be the order[1] of 2 modulo p and

$$l_k = \begin{cases} k, & \text{if } q^m \in D^{(p^k)} \text{ or } q^m \in \eta^{p^{k-1}f/2} D^{(p^k)}, \\ 0, & \text{otherwise} \end{cases}$$

for $k = 1, 2, \ldots, n$. Let $k_0 = \max\limits_{1 \leqslant k \leqslant n} l_k$.

A prime p is called a Wieferich prime if $2^{p-1} \equiv 1 \pmod{p^2}$. It is well known that there are only two Wieferich primes, 1093 and 3511, up to 6×10^{17}. Bellow, we will consider only non-Wieferich primes. Our main contribution is the following statement.

Theorem 1. *Let* $2^{p-1} \not\equiv 1 \pmod{p^2}$, $2^{q-1} \not\equiv 1 \pmod{q^2}$, $\gcd(p, q-1) = \gcd(p-1, q) = 1$ *and let* s^∞ *be a sequence defined by* (5). *Then*

(i) $\mathrm{LC}(s^\infty) = N - r_p \cdot \mathrm{ord}_p(2) - p^{k_0} r_q \cdot \mathrm{ord}_q(2) - \delta$ *for* $k_0 > 0$,
where r_p, r_q *are integers satisfying inequalities* $0 \leq r_p \leq \frac{p-1}{2\mathrm{ord}_p(2)}$, $0 \leq r_q \leq \frac{q-1}{2\mathrm{ord}_q(2)}$ *and*

$$\delta = \begin{cases} 1, & \text{if } (p^n q^m + 1)/2 \text{ is even}, \\ 0, & \text{if } (p^n q^m + 1)/2 \text{ is odd}. \end{cases}$$

[1] The order of 2 modulo p is the least positive integer T such that $2^T \equiv 1 \pmod{p}$.

(ii) $\mathrm{LC}(s^\infty) = N - r_N \cdot \mathrm{ord}_{pq}(2) - r_p \cdot \mathrm{ord}_p(2) - r_q \cdot \mathrm{ord}_q(2) - \delta$ *for* $k_0 = 0$,
where r_N *is an integer satisfying inequality* $0 \le r_N \le \frac{(p-1)(q-1)}{2\,\mathrm{ord}_{pq}(2)}$.

According to Theorem 1 the considered sequences have high linear complexity.

Remark 1. We will show that the value r_N also depends on n, m and is not defined only by p, q.

3 Subsidiary Lemmas

In this section we prove a few lemmas and discuss the properties of the generating polynomial of s^∞.

Lemma 1. *Let* $v = p^k q^l$, $k = 1, \ldots, n$; $l = 1, \ldots, m$. *Then*

(i) $C_1^{(v)} \bmod p^k = C_1^{(p^k)}$;
(ii) $C_1^{(v)} \bmod q^l = \mathbb{Z}_{q^l}^*$.

Proof. Suppose $i \in C_1^{(v)}$; then there exist u, t, s such that $i = \zeta_p^{u+b} \zeta_q^{t+c} g_v^s$ and $0 \le u < p^{k-1} f/2, 0 \le t < q^{l-1}(q-1)$, $0 \le s < e$. So, by (1), (2) and the definition of $C_1^{(v)}$ we get that $i \bmod p^k = \eta^{u+b} \eta^{sp^{k-1} f}$, i.e., $i \bmod p^k \in C_1^{(p^k)}$. Further, it is obvious that $i \pmod{q^l} \in \mathbb{Z}_{q^l}^*$. Moreover, it is clear that if $j \in C_1^{(v)}, j \ne i$ then $j \not\equiv i \pmod{p^k}$ or $j \not\equiv i \pmod{q^l}$. Since by (3) we have

$$|C_1^{(v)}| = \frac{p^{k-1}(p-1)}{2} \cdot q^{l-1}(q-1) = |C_1^{(p^k)}| \cdot |\mathbb{Z}_{q^l}^*|,$$

it follows that the conclusion of the lemma holds. □

Let $S(X) = \sum_{i=0}^{N-1} s_i X^i$ be the generating polynomial of s^∞. Define as in [15, 16] the subsidiary polynomials, i.e.,

$$T_b^{(p^k)}(X) = \sum_{i \in C_1^{(p^k)}} X^i,\ k = 1, 2, \ldots, n \text{ and } T_c^{(q^l)}(X) = \sum_{i \in C_1^{(q^l)}} X^i, l = 1, 2, \ldots, m.$$

Define

$$S_b^{(p^n)}(X) = \sum_{k=1}^{n} T_b^{(p^k)}(X^{p^{n-k}}) \text{ and } S_c^{(q^m)}(X) = \sum_{l=1}^{m} T_c^{(q^l)}(X^{q^{m-l}}).$$

As noted before, the sequence s^∞ defined by (5) for $N = p^2$ is the same as in [15]. In this case, $S_b^{(p^n)}(X) + 1$ is the polynomial of generalized cyclotomic sequence with period p^n considered in [16]. The properties of this polynomial are studied in [15, 16].

In the next lemma, we will recall the properties of this polynomial that are necessary in what follows.

Let α be a primitive N-th root of unity in the extension of $\mathbb{F}_2$. Since $\gcd(p^n, q^m) = 1$ then there exist integers x, y such that $xp^n + yq^m = 1$. Define $\beta = \alpha^{yq^m}$ and $\gamma = \alpha^{xp^n}$. Then $\alpha = \beta\gamma$, also β and γ are primitive p^n-th and q^m-th roots of unity, respectively. Denote $\beta_k = \beta^{p^{n-k}}$, $k = 1, 2 \ldots, n$ and $\gamma_l = \gamma^{q^{m-l}}$, $l = 1, 2 \ldots, m$.

Lemma 2 *[16]. For any $a \in \eta^t C^{(p^k)}$, we see that*

(i) $S_i^{(p^k)}(\beta_k^{p^d a}) = S_{i+t}^{(p^{k-d})}(\beta_{k-d}) + (p^d - 1)/2 \bmod 2$ *for* $0 \le d < k$.

(ii) $S_i^{(p^k)}(\beta_k^a) + S_{i+d_k/2}^{(p^k)}(\beta_k^a) = 1$, *where* $d_k = p^{k-1} f/2$.

(iii) Let p be a non-Wieferich prime. Then $S_i^{(p^k)}(\beta_k) \notin \{0,1\}$ *for* $k > 1$.

(iv) Let p be a non-Wieferich prime. Then $S_i^{(p^k)}(\beta_k) + S_{i+f/2}^{(p^k)}(\beta_k) \neq 1$ *for* $k > 1$.

(v)

$$|\{j \mid S_i^{(p^k)}(\beta_k) = 0 \text{ (or } 1), \ j = 1,2,\dots,p^k - 1)\}|$$
$$= |\{j \mid S_i^{(p)}(\beta_1^j) = 0 \text{ (or } 1), j = 1,2,\dots,p-1\}| = r_p \cdot \mathrm{ord}_p(2),$$

where r_p is an integer satisfying inequality $0 \le r_p \le \frac{p-1}{2\,\mathrm{ord}_p(2)}$.

Similarly, $S_c^{(q^m)}(X) + 1$ is the generating polynomial of sequence defined by (5) for $v = q^m$. Hence, the properties of $S_c^{(q^m)}(X)$ are the same as those of $S_b^{(p^n)}(X)$ (of course, we need to use q^m instead of p^n).

3.1 The Values of Subsidiary Polynomials

In this subsection we will show that the values of subsidiary polynomials define the values of $S(\alpha^j)$. Here and further we always suppose that $\gcd(p, q-1) = \gcd(p-1, q) = 1$.

Lemma 3. *Let $v = p^k q^l, k = 1,2,\dots,n;\ l = 1,2,\dots,m$ and $j \in \mathbb{Z}_N, j \neq 0$. Then*

$$\sum_{i \in \frac{N}{v} C_1^{(v)}} \alpha^{ij} = \begin{cases} T_b^{(p^k)}(\beta^{jp^{n-k}q^{m-l}}), & \text{if } j \equiv 0 \pmod{q^{l-1}} \text{ and } j \not\equiv 0 \pmod{q^l}, \\ 0, & \text{otherwise.} \end{cases}$$

Proof. According to the choice of α, β, γ we obtain that

$$\sum_{i \in p^{n-k}q^{m-l}C_1^{(v)}} \alpha^{ij} = \sum_{u \in C_1^{(v)}} \beta^{ujp^{n-k}q^{m-l}} \gamma^{ujp^{n-k}q^{m-l}}.$$

Since by Lemma 1 $C_1^{(v)} \bmod p^k = C_1^{(p^k)}$ and $C_1^{(v)} \bmod q^l = \mathbb{Z}_{q^l}^*$, it follows that

$$\sum_{i \in p^{n-k}q^{m-l}C_1^{(v)}} \alpha^{ij} = \sum_{u \in C_1^{(p^k)}} \beta^{ujp^{n-k}q^{m-l}} \cdot \sum_{u \in \mathbb{Z}_{q^l}^*} \gamma^{ujp^{n-k}q^{m-l}}.$$

Denote $\gamma^{p^{n-k}q^{m-l}}$ by $\tilde{\gamma}_l$. Then $\tilde{\gamma}_l$ is a primitive q^l-th root of unity since $\gcd(p-1, q) = 1$.
Let $A_l = \sum_{u \subset \mathbb{Z}_{q^l}^*} \tilde{\gamma}_l^{uj}$. It is clear

$$A_1 \pmod 2 = \begin{cases} 1, & \text{if } j \not\equiv 0 \pmod q, \\ 0, & \text{if } j \equiv 0 \pmod q. \end{cases}$$

Suppose $l > 1$; then

$$A_l = \sum_{u\in\mathbb{Z}_{q^l}} \tilde{\gamma}_l^{uj} - \sum_{u\in q\mathbb{Z}_{q^{l-1}}} \tilde{\gamma}_l^{uj} = \sum_{u\in\mathbb{Z}_{q^l}} \tilde{\gamma}_l^{uj} - \sum_{u\in\mathbb{Z}_{q^{l-1}}} \tilde{\gamma}_l^{ujq}.$$

We consider the following three cases.

(i) Let $j \not\equiv 0 \pmod{q^{l-1}}$. Obviously here $A_l = 0$.
(ii) Let $j \equiv 0 \pmod{q^{l-1}}$ and $j \not\equiv 0 \pmod{q^l}$. In this case $A_l \equiv 0 - q^{l-1} \equiv 1 \pmod 2$.
(iii) Suppose $j \equiv 0 \pmod{q^l}$; then $A_l = q^l - q^{l-1}$ and $A_l \bmod 2 = 0$.

This completes the proof of this lemma. □

Lemma 4. *Let $j = q^a j_0$ and $\gcd(j_0, q) = 1$, $0 \le a \le m$. Then*

$$\sum_{k=1}^{n}\sum_{l=1}^{m}\sum_{i\in p^{n-k}q^{m-l}C_1^{(p^kq^l)}} \alpha^{ij} = \begin{cases} S_b^{(p^n)}(\beta^{jq^{m-a-1}}), & \text{if } a < m,\\ 0, & \text{if } a = m.\end{cases}$$

Proof. If $a = m$ then $j \equiv 0 \pmod{q^l}$ for $l = 1, 2, \dots, m$ and by Lemma 3 we observe that $\sum_{l=1}^{m}\sum_{i\in p^{n-k}q^{m-l}C_1^{(p^kq^l)}} \alpha^{ij} = 0$.

Let $a < m$. In this case $j \equiv 0 \pmod{q^a}$ and $j \not\equiv 0 \pmod{q^{a+1}}$. Then again by Lemma 3 we have

$$\sum_{l=1}^{m}\sum_{i\in p^{n-k}q^{m-l}C_1^{(p^kq^l)}} \alpha^{ij} = \sum_{i\in p^{n-k}q^{m-a-1}C_1^{(p^kq^{a+1})}} \alpha^{ij} = T_b^{(p^k)}(\beta^{jp^{n-k}q^{m-a-1}}).$$

Thus, by definitions of $T_b^{(p^k)}(X)$ and $S_b^{(p^n)}(X)$ we see that

$$\sum_{k=1}^{n}\sum_{l=1}^{m}\sum_{i\in p^{n-k}q^{m-l}C_1^{(p^kq^l)}} \alpha^{ij} = \sum_{k=1}^{n} T_b^{(p^k)}(\beta^{jp^{n-k}q^{m-a-1}}) = S_b^{(p^n)}(\beta^{jq^{m-a-1}}).$$

□

Lemma 5. *Let $j = q^a j_0$ and $\gcd(j_0, q) = 1$, $0 \le a \le m$. Then*

(i) $S(\alpha^j) = S_b^{(p^n)}(\beta^{jq^{m-a-1}}) + S_b^{(p^n)}(\beta^{jq^m}) + S_c^{(q^m)}(\gamma^{jp^n}) + 1$ *for* $a < m$,
(ii) $S(\alpha^j) = S_b^{(p^n)}(\beta^{jq^m}) + (q^m + 1)/2$ *for* $a = m$.

Proof. By (4) and (5) we see that

$$S(\alpha^j) = \sum_{k=1}^{n}\sum_{l=1}^{m}\sum_{i\in p^{n-k}q^{m-l}C_1^{(p^kq^l)}} \alpha^{ij} + \sum_{k=1}^{n}\sum_{i\in p^{n-k}q^{m}C_1^{(p^k)}} \alpha^{ij} + \sum_{l=1}^{m}\sum_{i\in p^{n}q^{m-l}C_1^{(q^l)}} \alpha^{ij} + 1.$$

The first sum in the last relation is studied in Lemma 4. Using the definition of subsidiary polynomials and equality $\alpha = \beta\gamma$ we get that

$$\sum_{k=1}^{n} \sum_{i\in p^{n-k}q^m C_1^{(p^k)}} \alpha^{ij} = \sum_{k=1}^{n} \sum_{j\in C_1^{(p^k)}} \beta^{jp^{n-k}q^m} \gamma^{jp^{n-k}q^m} = \sum_{k=1}^{n} \sum_{i\in C_k^{(p^k)}} \beta^{ip^{n-k}q^m} = S_b^{(p^n)}(\beta^{jq^m}).$$

and

$$\sum_{l=1}^{m} \sum_{i\in p^n q^{m-l} C_1^{(q^l)}} \alpha^{ij} = \sum_{l=1}^{m} \sum_{j\in C_1^{(q^l)}} \beta^{jp^n q^{m-l}} \gamma^{jp^n q^{m-l}} = \sum_{l=1}^{m} \sum_{i\in C_1^{(q^l)}} \gamma^{ip^n q^{m-l}} = S_c^{(q^m)}(\gamma^{jp^n}).$$

Then the statement of this lemma follows from Lemma 4. □

3.2 The Values of Generating Polynomial

Here and further we will always suppose that $2^{p-1} \not\equiv 1 \pmod{p^2}$, $2^{q-1} \not\equiv 1 \pmod{q^2}$ and $\gcd(p, q-1) = \gcd(p-1, q) = 1$.

As usual, we denote by $\mathbb{F}_2(\beta_k)$ a simple extension of $\mathbb{F}_2$ obtained by adjoining an algebraic element β_k and by $[\mathbb{F}_2(\beta_k) : \mathbb{F}_2]$ the dimension of the vector space $\mathbb{F}_2(\beta_k)$ over $\mathbb{F}_2$ [2]. Here $\beta_k = \beta^{p^{n-k}}, k = 1, \ldots, n$ and $\gamma_l = \gamma^{q^{m-l}}, l = 1, \ldots, m$, as before. It is well known that if $r_1 = [\mathbb{F}_2(\beta_1) : \mathbb{F}_2]$ then r_1 divides $p-1$ and if $t_1 = [\mathbb{F}_2(\gamma_1) : \mathbb{F}_2]$ then t_1 divides $q-1$ [2]. Let $\mathbb{K} = \mathbb{F}_2(\beta_1) \cap \mathbb{F}_2(\gamma_1)$. Then $\mathbb{K}$ is a finite field and $[\mathbb{K} : \mathbb{F}_2] = \gcd(r_1, t_1)$.

Lemma 6. *With notations as above, we have $\mathbb{F}_2(\beta_k) \cap \mathbb{F}_2(\gamma_l) = \mathbb{K}$ for $k = 1, \ldots, n; l = 1, \ldots, m$.*

Proof. Let $\mathbb{F} = \mathbb{F}_2(\beta_k) \cap \mathbb{F}_2(\gamma_l)$. Then $[\mathbb{F} : \mathbb{F}_2]$ divides $[\mathbb{F}_2(\beta_k) : \mathbb{F}_2]$ and $[\mathbb{F}_2(\gamma_l) : \mathbb{F}_2]$. According to [16] we know that $[\mathbb{F}_2(\beta_k) : \mathbb{F}_2] = p^{k-1} r_1$ and $[\mathbb{F}_2(\gamma_l) : \mathbb{F}_2] = q^{l-1} t_1$ for p, q such that $2^{p-1} \not\equiv 1 \pmod{p^2}$, $2^{q-1} \not\equiv 1 \pmod{q^2}$. Hence $[\mathbb{F} : \mathbb{F}_2]$ divides $\gcd(p^{k-1} r_1, q^{m-1} t_1)$. By the condition $\gcd(p, q-1) = \gcd(p-1, q) = 1$, then $[\mathbb{F} : \mathbb{F}_2]$ divides $\gcd(r_1, t_1)$. Thus, we get $[\mathbb{F} : \mathbb{F}_2]$ divides $[\mathbb{K} : \mathbb{F}_2]$. Since $\mathbb{K} \subset \mathbb{F}$, this completes the proof of this lemma.

□

Lemma 7. *Let notations be as above and $S_c^{(q^m)}(\gamma^j) \in \mathbb{K}$ for $m > 1$. Then $j \equiv 0 \pmod{q^{m-1}}$.*

Proof. By Lemma 2 (i) it is clear that without loss of generality it is enough to consider the case $c = 0$. Let u be an integer such that $2 \equiv g^u \pmod{q^m}$. Denote by r degree $[\mathbb{K} : \mathbb{F}_2]$. Since $S_c^{(q^m)}(\gamma^j) \in \mathbb{K}$, it follows that $S_0^{(q^m)}(\gamma^j) = S_0^{(q^m)}(\gamma^j)^{2^r}$. Then again by Lemma 2 (i) we get

$$S_0^{(q^m)}(\gamma^j) = S_0^{(q^m)}(\gamma^j)^{2^{ir}} = S_0^{(q^m)}(\gamma^{j2^{ir}}) = S_{iur}^{(q^m)}(\gamma^j) \text{ for } i = 0, 1, \ldots \tag{6}$$

Let $w = \gcd(q^{m-1}h, ur)$, where $h = (q-1)/e$ as before. There exist integers x, y such that $xq^{m-1}h + yur = w$. Then we see from (6) and Lemma 2 (i) that $S_0^{(q^m)}(\gamma^j) = S_{iw}^{(q^m)}(\gamma^j)$ for $i = 0, 1, \ldots$. By [16] $\gcd(u, q) = 1$ for $2^{q-1} \not\equiv 1 \pmod{q^2}$ and $\gcd(q, r) = 1$ here. Hence $w = \gcd(h, ur)$ and w divides h. So, we observe that $S_0^{(q^m)}(\gamma^j) = S_{ih}^{(q^m)}(\gamma^j)$ for $i = 0, 1, \ldots$.

Further, $S_t^{(q^m)}(\gamma^j) + S_{t+q^{m-1}h/2}^{(q^m)}(\gamma^j) = 1$ for $t = 0, 1, \ldots$ by Lemma 2 (ii). Thus,

$$S_{q^{m-1}h/2}^{(q^m)}(\gamma^j) = S_{q^{m-1}h/2+ih}^{(q^m)}(\gamma^j)$$

for $i = 0, 1, \ldots$. Since

$$S_{q^{m-1}h/2+(q^{m-1}+1)h/2}^{(q^m)}(\gamma^j) = S_{q^{m-1}h+h/2}^{(q^m)}(\gamma^j) = S_{h/2}^{(q^m)}(\gamma^j),$$

it follows that $S_0^{(q^m)}(\gamma^j) + S_{h/2}^{(q^m)}(\gamma^j) = 1$. According to Lemma 2 (iii), in this case $j \equiv 0 \pmod{q^{m-1}}$.

□

Let k_0 be the same as before, i.e., $k_0 = 0$ or $k_0 > 0$ is the largest integer such that $q^m \in D^{(p^{k_0})}$ or $q^m \in \eta^{p^{k_0-1}f/2} D^{(p^{k_0})}$.

Lemma 8. *Let $j \in p^{n-k}q^{m-1}\mathbb{Z}^*_{p^k q^{m-1}}, 1 \le k \le n$ and $S_b^{(p^n)}(\beta^j) + S_b^{(p^n)}(\beta^{jq^m}) \in \mathbb{K}$ for $n > 1$. Then $j \equiv 0 \pmod{p^{n-1}}$ or $k \le k_0$.*

Proof. Without loss of generality it is enough to consider the case $b = 0$. Let $j = p^{n-k}q^{m-1}t$, where $\gcd(t, pq) = 1$. If $k = 1$ then $j \equiv 0 \pmod{p^{n-1}}$. So, this lemma is right for $k = 1$.

Let $k > 1$ and denote $\beta^{p^{n-k}q^{m-1}t}$ by $\tilde{\beta}_k$. Then $\tilde{\beta}_k$ is a primitive p^k-th root of unity and $S_0^{(p^k)}(\tilde{\beta}_k) + S_0^{(p^k)}(\tilde{\beta}_k^{q^m}) \in \mathbb{K}$ by Lemma 2 (i).

Suppose $k > k_0$; then $q^m \in \eta^z D^{(p^k)}$ for $z \ne 0$ and $z \ne p^{k-1}f/2$. By Lemma 2 (i) we get that $S_0^{(p^k)}(\tilde{\beta}_k) + S_z^{(p^k)}(\tilde{\beta}_k) \in \mathbb{K}$.

We can show in the same way as in Lemma 7 that

$$S_0^{(p^k)}(\tilde{\beta}_k) + S_z^{(p^k)}(\tilde{\beta}_k) = S_{f/2}^{(p^k)}(\tilde{\beta}_k) + S_{z+f/2}^{(p^k)}(\tilde{\beta}_k).$$

Using the definitions of $S_i^{(p^k)}(X)$ and $T_i^{(p^k)}(X)$ we obtain that

$$T_0^{(p^k)}(\tilde{\beta}_k) + T_{f/2}^{(p^k)}(\tilde{\beta}_k) + T_z^{(p^k)}(\tilde{\beta}_k) + T_{z+f/2}^{(p^k)}(\tilde{\beta}_k) \in \mathbb{F}_2(\beta_{k-1}).$$

Let $\mathscr{D} = D^{(p^k)} \cup \cdots \cup \eta^{f/2-1} D^{(p^k)} \cup \eta^{p^{k-1}f/2} D^{(p^k)} \cup \cdots \cup \eta^{p^{k-1}f/2+f/2-1} D^{(p^k)}$ and $\mathscr{C} = \eta^z \mathscr{D}$. Then

$$T_0^{(p^k)}(\tilde{\beta}_k) + T_{f/2}^{(p^k)}(\tilde{\beta}_k) = \sum_{i \in \mathscr{D}} \tilde{\beta}_k^i$$

and

$$T_z^{(p^k)}(\tilde{\beta}_k) + T_{z+f/2}^{(p^k)}(\tilde{\beta}_k) = \sum_{i \in \mathscr{C}} \tilde{\beta}_k^i.$$

It is clear that $|\mathscr{D}| = |\mathscr{C}| = p-1$ and $\mathscr{D} \pmod p = \mathscr{C} \pmod p = \mathbb{Z}_p^*$.

Denote by $x_i \in \mathscr{D}$ and $y_i \in \mathscr{C}$, respectively, such that $x_i \mod p = y_i \mod p = i$, $i = 1, \ldots, p-1$. Then

$$\sum_{i \in \mathscr{D}} \tilde{\beta}_k^i = \sum_{i=1}^{p-1} \tilde{\beta}_{k-1}^{(x_i - i)/p} \cdot \tilde{\beta}_k^i \text{ and } \sum_{i \in \mathscr{C}} \tilde{\beta}_k^i = \sum_{i=1}^{p-1} \tilde{\beta}_{k-1}^{(y_i - i)/p} \cdot \tilde{\beta}_k^i.$$

Suppose that for any i we have $\tilde{\beta}_{k-1}^{(x_i-i)/p} = \tilde{\beta}_{k-1}^{(y_i-i)/p}$. Then $x_i \equiv y_i \pmod{p^k}$ for $i = 1, 2, \ldots, p-1$. Hence $\mathscr{D} = \mathscr{C}$. Then $z = 0$ or $z = p^{k-1} f/2$. This is impossible because $k > k_0$. Thus, we have that the polynomial $f(X) = \sum_{i=0}^{p-1} (\tilde{\beta}_{k-1}^{(x_i-i)/p} + \tilde{\beta}_{k-1}^{(y_i-i)/p}) X^i$ has at least one nonzero coefficient and $f(\tilde{\beta}_k) \in \mathbb{F}_2(\beta_{k-1})$. This is impossible since $\deg f(X) < p$ and $[\mathbb{F}_2(\beta_k) : \mathbb{F}_2(\beta_{k-1})] = p$ for $k > 1$. So, $k \le k_0$. □

4 The Proof of Main Theorem

In this section we finish the proof of Theorem 1 in the following two Lemmas.

Lemma 9. *Let notation be as above and $k_0 > 0$. Let $2^{p-1} \not\equiv 1 \pmod{p^2}$, $2^{q-1} \not\equiv 1 \pmod{q^2}$, $\gcd(p, q-1) = \gcd(p-1, q) = 1$ and let s^∞ be defined by* (5). *Then*

$$\mathrm{LC}(s^\infty) = N - r_p \cdot \mathrm{ord}_p(2) - p^{k_0} r_q \cdot \mathrm{ord}_q(2) - \delta,$$

where

$$\delta = \begin{cases} 1, & \text{if } (p^n q^m + 1)/2 \text{ is even}, \\ 0, & \text{if } (p^n q^m + 1)/2 \text{ is odd} \end{cases}$$

and $0 \le r_p \le \frac{p-1}{2\,\mathrm{ord}_p(2)}$, $0 \le r_q \le \frac{q-1}{2\,\mathrm{ord}_q(2)}$.

Proof. As noted before we have

$$\mathrm{LC}(s^\infty) = N - \left| \{ j \mid S(\alpha^j) = 0,\ j = 0, 1, \ldots, N-1 \} \right|.$$

First of all we note that by definition $S(1) = (p^n q^m + 1)/2$. Further, according to Lemma 5 we have

$$S(\alpha^j) = S_b^{(p^n)}(\beta^{jq^{m-a-1}}) + S_b^{(p^n)}(\beta^{jq^m}) + S_c^{(q^m)}(\gamma^{jp^n}) + 1 \tag{7}$$

for $j = q^a j_0$, $a < m$, $\gcd(j_0, q) = 1$ and

$$S(\alpha^j) = S_b^{(p^n)}(\beta^{jq^m}) + (q^m + 1)/2$$

for $j = q^m j_0$.

Let $S(\alpha^j)=0,\ 1\le j\le N-1$. We consider a few cases.

(i) Suppose $j\equiv 0 \pmod{q^m}$; then $S(\alpha^j)=S_b^{(p^n)}(\beta^{jq^m})$ for even $(q^m+1)/2$ and $S(\alpha^j)=S_b^{(p^n)}(\beta^{jq^m})+1$ for odd $(q^m+1)/2$. By Lemma 2 (v) we get

$$|\{j|\ S(\alpha^j)=0,\ j=q^m,\dots,(p^n-1)q^m\}|=$$
$$|\{j|\ S_b^{(p)}(\beta_1^j)=0,\ (\text{ or } 1), j=1,2,\dots,p-1\}|=r_p\cdot \mathrm{ord}_p(2).$$

(ii) Let $j\not\equiv 0 \pmod{q^m}$. According to (7) we see that

$$S_b^{(p^n)}(\beta^{jq^{m-a-1}})+S_b^{(p^n)}(\beta^{jq^m})=-S_c^{(q)}(\gamma^{jp^n})-1.$$

Hence $S_c^{(q^m)}(\gamma^{jp^n})\in\mathbb{F}_2(\beta)$. Then by Lemma 6 we get

$$S_c^{(q^m)}(\gamma^{jp^n})\in\mathbb{K}=\mathbb{F}_2(\beta_1)\cap\mathbb{F}_2(\gamma_1).$$

In this case, by Lemma 7 we have $j\equiv 0 \pmod{q^{m-1}}$. Hence $j\in p^{n-k}q^{m-1}\mathbb{Z}^*_{p^kq}$ for $k:\ 1\le k\le n$ and the sum $S_b^{(p^n)}(\beta^j)+S_b^{(p^n)}(\beta^{jq^m})$ also belongs to $\mathbb{K}$. Further by Lemma 8 we get $k\le k_0$ or $j\equiv 0 \pmod{p^{n-1}}$. If $j\equiv 0 \pmod{p^{n-1}}$ then $k=1$ and since $k_0>0$, it follows that $k\le k_0$ in any case.

By choosing k_0 we see that $q^m\in D^{(p_0^k)}$ or $q^m\in \eta^{p^{k_0-1}f/2}D^{(p_0^k)}$. Hence for any j : $j\equiv 0 \pmod{p^{n-k_0}}$ we have

$$S_b^{(p^n)}(\beta^{jq^m})=\begin{cases}S_b^{(p^n)}(\beta^j), & \text{if } q^m\in D^{(p_0^k)},\\ S_{b+p^{k-1}f/2}^{(p^n)}(\beta^j), & \text{if } q^m\in\eta^{p^{k_0-1}f/2}D^{(p_0^k)}.\end{cases}$$

In any case, by Lemma 2 (ii) $S_b^{(p^n)}(\beta^j)+S_b^{(p^n)}(\beta^{jq^m})$ is equal to 0 or 1 for all $j\in p^{n-k_0}q^{m-1}\mathbb{Z}_{p^{k_0}q}$ and $j\not\equiv 0 \pmod{q^m}$.

Then, according to (7) we obtain $S_c^{(q^m)}(\gamma^{jp^n})=S_c^{(q)}(\gamma_1^{j_0p^n})\in\{0,1\}$ where $j=q^{m-1}j_0$, $\gcd(j_0,q)=1$. In this case, by Lemma 2 (v) we have

$$|\{j|\ S_b^{(p)}(\beta_1^{j_0})=0,\ (\text{ or } 1), j_0=1,2,\dots,p-1\}|=r_p\cdot \mathrm{ord}_p(2).$$

For fixed j_0 we have p^{k_0} numbers from $\mathbb{Z}_{p^{k_0}q}$ with the same residue modulo q. Thus, we get

$$|\{j\ \mid\ S(\alpha^j)\ =\ 0,\quad j\ =\ 1,2,\dots,N,\quad j\not\equiv 0 \pmod{q^m}\}|\ =\ p^{k_0}r_q\cdot \mathrm{ord}_q(2),$$

where $0\le r_q\le \frac{q-1}{2\,\mathrm{ord}_q(2)}$.

Finally, we get $|\{j\mid S(\alpha^j)=0,\ j=1,2,\dots,N,\}|=r_p\cdot\mathrm{ord}_p(2)+p^{k_0}r_q\cdot\mathrm{ord}_q(2)$.

□

We consider a few examples with different values r_p, r_q and $k_0=1$.

Example 1. (i) $p = 19, q = 7, e = 3$, in this case $7 \in D^{(19)}$ and $r_p = 0, r_q \cdot \text{ord}_7(2) = 3$. Hence $LC(s^\infty) = N - 19 \cdot 3$ for $n = 1,2; m = 1,2$.

(ii) $p = 7, q = 43, e = 3$, here $43 \in D^{(7)}$, but $r_p = 1, r_q = 0$. Hence $LC(s^\infty) = 301 - 1 \cdot 3 = 298$.

(iii) $p = 43, q = 7, e = 3$, here $f = 14$, $7 \in \eta^7 D^{(43)}$ and $r_p = 0, r_q \cdot \text{ord}_7(2) = 3$. Hence $LC(s^\infty) = 301 - 43 \cdot 3 = 172$. Similarly, for $n = 1,2; m = 1,2$.

(iv) $p = 41, q = 31, e = 5, f = 8$, $31 \in \eta^4 D^{(41)}$, $r_p = 0, r_q \cdot \text{ord}_3 1(2) = 15$. Finally, $LC(s^\infty) = 1271 - 41 \cdot 15 - 1 = 655$.

(v) $p = 7, q = 73, e = 3$, here $f = 2$, $73 \in \eta D^{(7)}$ and $r_p = 1, r_q \cdot \text{ord}_{73}(2) = 7 \cdot 18$. Hence $LC(s^\infty) = 511 - 7 \cdot 18 - 3 - 1 = 381$.

Lemma 10. *Let notation be as above and* $k_0 = 0$. *Let* $2^{p-1} \not\equiv 1 \pmod{p^2}$, $2^{q-1} \not\equiv 1 \pmod{q^2}$, $\gcd(p, q-1) = \gcd(p-1, q) = 1$ *and let* s^∞ *be defined by* (5). *Then*

$$\text{LC}(s^\infty) = N - r_N \cdot \text{ord}_{pq}(2) - r_p \cdot \text{ord}_p(2) - r_q \cdot \text{ord}_q(2) - \delta,$$

where $0 \le r_N \le \frac{(p-1)(q-1)}{2\,\text{ord}_{pq}(2)}$ *and*

$$\delta = \begin{cases} 1, & \text{if } (p^n q^m + 1)/2 \text{ is even}, \\ 0, & \text{if } (p^n q^m + 1)/2 \text{ is odd}. \end{cases}$$

Proof. Let $S(\alpha^j) = 0, j \neq 0$. As in Lemma 9 we obtain that $|\{j \mid S(\alpha^j) = 0,\ j = q^m, \dots, (p^n - 1)q^m\}| = r_p \cdot \text{ord}_p(2)$ and if $j \not\equiv 0 \pmod{q^m}$ then $S_c^{(q^m)}(\gamma^{jp^n}) \in \mathbb{K}$ and $S_b^{(p^n)}(\beta^j) + S_b^{(p^n)}(\beta^{jq^m}) \in \mathbb{K}$.

In the last case, according to Lemma 7 and 8 we get that $j \equiv 0 \pmod{q^{m-1}}$ and $j \equiv 0 \pmod{p^{n-1}}$. Further, if $j \equiv 0 \pmod{p^n}$ then by Lemma 5 we have $S(\alpha^j) = S_c^{(q)}(\gamma^{jp^n}) + 1$ and in this case we observe that $|\{j \mid S(\alpha^j) = 0,\ j = p^n, \dots, (q^m - 1)p^n\}| = r_q \cdot \text{ord}_q(2)$ by Lemma 2 (v).

Suppose $j = p^{n-1} q^{m-1} t$ and $\gcd(t, pq) = 1$. Then by Lemma 7 and Lemma 2 (i)

$$S(\alpha^j) = S_b^{(p)}(\beta_1^{tq^{m-1}}) + S_b^{(p)}(\beta_1^{tq^{2m-1}}) + S_c^{(q)}(\gamma_1^{tp^{2n-1}}) + 1.$$

It is clear that if $S(\alpha^j) = 0$ then $S(\alpha^j)^{2^u} = 0$ for $u = 0, 1, \dots, \text{ord}_{pq}(2)$. Hence, $|\{j : S(\alpha^j) = 0,\ j \in \mathbb{Z}_{pq}| = r_N \cdot \text{ord}_{pq}(2)$ for the some r_N.

Let $w = \zeta_p^{f/2} \zeta_q^{h/2}$. Then $w \equiv \eta^{f/2} \pmod{p}$ and $w \equiv \xi^{h/2} \pmod{q}$. So, by Lemma 2 (i) we obtain

$$S(\alpha^{wj}) = S_{b+f/2}^{(p)}(\beta_1^{tq^{m-1}}) + S_{b+f/2}^{(p)}(\beta_1^{tq^{2m-1}}) + S_{c+h/2}^{(q)}(\gamma_1^{tp^{2n-1}}) + 1.$$

Hence, by Lemma 2 (ii) we see that $S(\alpha^{wj}) = S(\alpha^j) + 1$. Thus, $0 \le r_N \le \frac{(p-1)(q-1)}{2\,\text{ord}_{pq}(2)}$. This completes the proof of this lemma. □

The statement of Theorem 1 follows from Lemmas 9 and 10.

The following examples show that in this case the value r depends on N, so we are using a denotation r_N.

Example 2. (i) Let $p = 73, q = 7, e = 3, b = 0, c = 0$. Here $7 \in \eta^9 D^{(73)}$, $r_p \cdot \mathrm{ord}_p(2) = 18, r_q \cdot \mathrm{ord}_q(2) = 3$ and $\mathrm{ord}_{511}(2) = 9$.

For $n = m = 1$ we get $LC(s^\infty) = 435 = 511 - 76$, in this case $r_N = 5$.

For $n = 1, m = 2$ we have $LC(s^\infty) = 3577 - 18 - 3 = 3556$, i.e., $r_N = 0$.

(ii) Let $p = 41, q = 11, e = 5, n = 2, m = 1, b = 0, c = 0$.

Here $\mathrm{ord}_{pq}(2) = 20$. For $n = 1,2; m = 1,3$ we get $LC(s^\infty) = N - 101$ ($r_N = 5$), and $LC(s^\infty) = N - 200$ ($r_N = 10$) if $n = 1,2; m = 2$.

5 Conclusions

Pseudorandom sequences are widely used in communication, radar navigation, cryptography and some other scenarios. By using the new generalized cyclotomy presented by Zeng et al., we constructed a new kind of generalized cyclotomic sequences with period $p^n q^m$ where p, q are odd distinct primes and n, m are natural numbers. Thus, we generalized the results obtained in [15–17].

Our results show that such sequences have high linear complexity and are suitable for applications. To illustrate the results, some examples are presented. For further study, the k-error linear complexity, autocorrelation, 2-adic complexity and some other cryptographic properties of these sequences may be interesting topics.

References

1. Golomb, S.W.: Shift Register Sequences. Holden-Day, San Francisco (1967)
2. Lidl, R., Niederreiter, H.: Finite Fields. Encyclopedia of Mathematics and Its Applications, vol. 20. Addison-Wesley (1983)
3. Ding, C., Helleseth, T.: New generalized cyclotomy and its applications. Finite Fields Their Appl. **4**(2), 140–166 (1998)
4. Whiteman, A.L.: A family of difference sets. Illinois J. Math. **6**, 107–121 (1962)
5. Ding, C., Helleseth, T.: Generalized cyclotomic codes of length $p_1^{e_1} \cdots p_t^{e_t}$. IEEE Trans. Inf. Theory **45**(2), 467–474 (1999)
6. Fan, C., Ge, G.: A unified approach to Whiteman's and Ding-Helleseth's generalized cyclotomy over residue class rings. IEEE Trans. Inf. Theory **60**(2), 1326–1336 (2014)
7. Du, X., Chen, Z.: A generalization of the Hall's sextic residue sequences. Inf. Sci. **222**, 784–794 (2013)
8. Yan, T., Li, S., Xiao, G.: On the linear complexity of generalized cyclotomic sequences with the period p^m. Appl. Math. Lett. **21**(2), 187–193 (2008)
9. Chen, X., Chen, Z., Liu, H.: A family of pseudorandom binary sequences derived from generalized cyclotomic classes modulo $p^{m+1}q^{n+1}$. Int. J. Netw. Secur. **22**(4), 610–620 (2020)
10. Hu, L., Yue, Q., Wang, M.H.: The linear complexity of Whiteman's generalized cyclotomic sequences of period $p^{m+1}q^{n+1}$. IEEE Trans. Inf. Theory **58**(8), 5533–5543 (2012)
11. Kim, Y.J., Song, H.Y.: Linear complexity of prime n-square sequences. In: 2008 IEEE International Symposium on Information Theory, pp. 2405–2408 (2008)
12. Ke, P., Zhang, J., Zhang, S.: On the linear complexity and the autocorrelation of generalized cyclotomic binary sequences of length $2p^m$. Des. Codes Cryptogr. **67**(3), 325–339 (2013)
13. Zeng, X., Cai, H., Tang, X., Yang, Y.: Optimal frequency hopping sequences of odd length. IEEE Trans. Inf. Theory **59**(5), 3237–3248 (2013)

14. Xu, S., Cao, X., Mi, J., Tang, C.: More cyclotomic constructions of optimal frequency-hopping sequences. Adv. Math. Commun. **13**(3), 373–391 (2019)
15. Xiao, Z., Zeng, X., Li, C., Helleseth, T.: New generalized cyclotomic binary sequences of period p^2. Des. Codes Cryptogr. **86**(7), 1483–1497 (2018)
16. Edemskiy, V., Li, C., Zeng, X., Helleseth, T.: The linear complexity of generalized cyclotomic binary sequences of period p^n. Des. Codes Cryptogr. **87**(5), 1183–1197 (2019)
17. Ye, Z., Ke, P., Wu, C.: A further study of the linear complexity of new binary cyclotomic sequence of length p^n. Appl. Algebra Eng. Commun. Comput. **30**(3), 217–231 (2019)
18. Ouyang, Y., Xie, X.: Linear complexity of generalized cyclotomic sequences of period $2p^m$. Des. Codes Cryptogr. **87**(5), 1–12 (2019)
19. Edemskiy, V., Wu, C.: On the linear complexity of binary sequences derived from generalized cyclotomic classes modulo $2^n p^m$. WSEAS Trans. Math. **18**, 197–202 (2019)
20. Ireland, K., Rosen, M.: A Classical Introduction to Modern Number Theory. Graduate Texts in Mathematics. Springer, New York (1990). https://doi.org/10.1007/978-1-4757-2103-4
21. Cusick, T., Ding, C., Renvall, A.: Stream Ciphers and Number Theory. Elsevier, North-Holland mathematical library (2004)

On the 2-Adic Complexity of Cyclotomic Binary Sequences with Period p^2 and $2p^2$

Fuqing Sun, Qin Yue(✉), and Xia Li

Department of Mathematics, Nanjing University of Aeronautics and Astronautics, Nanjing 211100, People's Republic of China
yueqin@nuaa.edu.cn

Abstract. Let p be a prime. In this paper, we obtain the 2-adic complexity of all almost balanced cyclotomic binary sequence of order two with period p^2; and also show the 2-adic complexity of several non-trivial balanced cyclotomic binary sequences of order two with period $2p^2$.

Keywords: Binary sequence · Autocorrelation · 2-adic complexity · Cyclotomic numbers

1 Introduction

Sequences are widely used in various fields, such as code-division multiple access (CDMA) communications and stream ciphers in [6,8,11,16]. In the past few decades, many efforts have been devoted to construct optimal sequences with respect to the theoretic bounds. As a result, numerous algebraic and combinatorial constructions of sequences with low correlation have been reported, see [2–5,8,14,15,21,22,27–30,32–36]. In cryptography, binary sequences, as candidates of keys in stream cipher system, are required to have high complexity and good balance. There are many works on the linear complexity of binary sequences. Both linear feedback shift registers (LFSRs) in [7] and feedback with carry shift registers (FCSRs) in [12,13] are important pseudo-random sequence generators. Linear complexity and 2-adic complexity of a sequence are defined as the length of the shortest LFSR or FCSR respectively, which is capable of generating a given sequence. Since the end of the last century, the 2-adic complexity has been viewed as one of the important security criteria of sequences. The sequences with 2-adic complexity n can be generated by a feedback (with carry) shift register of length n [13].

Let $\mathbf{s} = \{s_\lambda\}_{\lambda=0}^{\infty}$ be a binary sequence with period N. The autocorrelation function of the sequence $\mathbf{s}$ is defined by

$$C_{\mathbf{s}}(\tau) = \sum_{\lambda=0}^{N-1} (-1)^{s_{\tau+\lambda} - s_\lambda} \in \mathbb{Z} \ (0 \le \tau \le N-1).$$

2010 *Mathematics Subject Classification.* 94B05

The paper was supported by National Natural Science Foundation of China (No. 62172219).

S. Mesnager and Z. Zhou (Eds.): WAIFI 2022, LNCS 13638, pp. 334–347, 2023.
https://doi.org/10.1007/978-3-031-22944-2_22

For $\tau = 0, C_{\mathbf{s}}(0) = N$. For many applications in communication, the value of $\max\{|C_{\mathbf{s}}(\tau)| : 1 \leq \tau \leq N-1\}$ is required as small as possible. Therefore, the study of autocorrelation value of periodic sequence is also one of the hotspots of pseudo-random sequence research at this stage. Several classes of binary sequences with three-level autocorrelation and new families of binary sequences with optimal three-level autocorrelation have been found (see [3,4]).

Let $\mathbf{s} = (s_\lambda)_{\lambda=0}^{\infty}$ be a binary sequence of period N, where each $s_\lambda \in \{0,1\}$. Let $S = \{0 \leq \lambda \leq N-1 : s_\lambda = 1\}$ be the *support set* of the sequence $\mathbf{s}$, then $\mathbf{s}$ is called *balanced* if $|S| = \frac{N}{2}$; $\mathbf{s}$ is called *almost balanced* if $|S| = \frac{N \pm 1}{2}$. In fact, balanced or almost balanced binary sequences are widely used in communication systems, software testing, and stream ciphers.

Let

$$\mathbf{s}(x) = \sum_{\lambda=0}^{N-1} s_\lambda x^\lambda \in \mathbb{Z}[x], \tag{1.1}$$

$d = \gcd(\mathbf{s}(2), 2^N - 1)$. The 2-adic complexity of $\mathbf{s}$ can be defined as the real number:

$$\phi_2(\mathbf{s}) = \log_2(\frac{2^N - 1}{d}). \tag{1.2}$$

The 2-adic complexity of m-sequences was determined in [23]. Later, the 2-adic complexity of all known binary sequences with ideal two-level autocorrelation was determined in [26]. Hu introduced a simple method to compute the 2-adic complexity of any periodic binary sequence with ideal two-level autocorrelation [10]. Their 2-adic complexity reaches the maximum value $\log_2(2^N - 1)$. For some other sequences with good autocorrelation, the 2-adic complexity is determined or estimated by a nice lower bound [9,18–20,25].

In [10,26], Xiong and Hu can compute the 2-adic complexity of binary sequences with optimal autocorrelation of Legendre sequences and Ding-Helleseth-Lam sequences. In [31], Zhang can compute the 2-adic complexity of binary sequences with optimal autocorrelation of DHM sequences. In this paper, we shall complement Xiao's results in [25] and generalize Zhang's results in [31]. We use our method to determine the maximum evaluation of the 2-adic complexity in all almost cyclotomic binary sequences of order two with period $N = p^2$ and then the 2-adic complexity of several non-trivial balanced cyclotomic binary sequences of order two with period $N = 2p^2$.

The paper is organized as follows. In Sect. 2, we introduce some basic concepts and related results. In Sect. 3, we obtain that all almost balanced cyclotomic binary sequences of order two with period p^2 have the maximum 2-adic complexity $\log_2(2^{p^2} - 1)$. In Sect. 4, we mainly determine the 2-adic complexity of several non-trivial balanced cyclotomic binary sequences of order two with period $N = 2p^2$. In Sect. 5, we make a conclusion.

2 Preliminaries

Definition 2.1. Let p be an odd prime and let g be a primitive element of the finite field $\mathbb{F}_p$, i.e., $\mathbb{F}_p^* = \langle g \rangle$. Let $p = 2f + 1$ and $D_0 = \langle g^2 \rangle$. Then D_0 is a cyclic subgroup of $\mathbb{F}_p^*$ and there are two cosets in $\mathbb{F}_p^*$:

$$D_i = g^i D_0 = \{g^{i+2k} : 0 \le k \le f-1\}, i = 0, 1,$$

which are called *cyclotomic classes of order two* in $\mathbb{F}_p$.

The generalized cyclotomic classes of order 2 with respect to p^2 are defined by

$$D_i^{(p^2)} = \{a + bp \mid a \in D_i, 0 \le b \le p-1\}, i = 0, 1.$$

Let

$$R = \{0\},\ pD_i = \{ph \mid h \in D_i\}\ for\ i = 0, 1.$$

$$C_0 = pD_0 \cup D_0^{(p^2)}, C_1 = pD_1 \cup D_1^{(p^2)}. \tag{2.1}$$

Then $C_0 \cup C_1 \cup R = \mathbb{Z}_{p^2}$, $C_0 \cap C_1 = \emptyset$.

Definition 2.2. The cyclotomic numbers of order two in $\mathbb{F}_p$ are defined as

$$(i,j)_2 = |(D_i + 1) \cap D_j|, 0 \le i, j \le 1.$$

The evaluations of $(i,j)_2$ have been computed (see [1,17]).The following conclusion is well-known.

Lemma 2.3 [17]. *If* $p \equiv 3 \pmod 4$, *then*

$$(1,0) = (0,0) = (1,1) = \frac{p-3}{4},\ (0,1) = \frac{p+1}{4}.$$

If $p \equiv 1 \pmod 4$, *then*

$$(1,0) = (0,1) = (1,1) = \frac{p-1}{4},\ (0,1) = \frac{p-5}{4}.$$

Let p be an odd prime, n a positive integer, and ξ a primitive p-th root of unity in a ring $\mathbb{Z}/n\mathbb{Z}$. Define

$$\eta_0^{(n)} = \sum_{k \in D_0} \xi^k, \eta_1^{(n)} = \sum_{k \in D_1} \xi^k. \tag{2.2}$$

Then $\eta_0^{(n)} + \eta_1^{(n)} = -1$.

Lemma 2.4. *If* $p \equiv 3 \pmod 4$, *then* $\eta_0^{(n)}\eta_1^{(n)} = \frac{3-p}{4}$; *if* $p \equiv 1 \pmod 4$, *then* $\eta_0^{(n)}\eta_1^{(n)} = \frac{1-p}{4}$.

Proof. If $p \equiv 3 \pmod 4$, then

$$\begin{aligned}\eta_0^{(n)}\eta_1^{(n)} &= \sum_{h \in D_0} \xi^h \sum_{k \in D_1} \xi^k = (1,1) \sum_{t \in D_0} \xi^t + (0,0) \sum_{t \in D_1} \xi^t + \frac{p-1}{2} \\ &= \frac{p-3}{4}\left(\sum_{t \in D_0} \xi^t + \sum_{t \in D_1} \xi^t\right) + \frac{p-1}{2} = \frac{p+1}{4}.\end{aligned}$$

If $p \equiv 1 \pmod 4$, then

$$\eta_0^{(n)}\eta_1^{(n)} = \sum_{h\in D_0} \xi^h \sum_{k\in D_1} \xi^k = (1,0)\sum_{t\in D_0} \xi^t + (0,1)\sum_{t\in D_1} \xi^t$$
$$= \frac{p-1}{4}(\sum_{t\in D_0} \xi^t + \sum_{t\in D_1} \xi^t) = \frac{1-p}{4}.$$

□

3 2-Adic Complexity of Almost Balanced Sequences with Period p^2

In this section, we always assume that p is an odd prime. We shall determine the 2-adic complexity of all almost balanced cyclotomic binary sequences of order two with period p^2.

In the following, we define almost balanced cyclotomic binary sequences of order two.

Definition 3.1. The two cyclotomic binary sequences $\mathbf{s}_i = \{s_\lambda^{(i)}\}_{\lambda=0}^{\infty}$, $i = 0, 1$, with period $N = p^2$ are defined as follows

$$s_\lambda^{(i)} = \begin{cases} 1, & \text{if } \lambda \pmod{p^2} \in C_i \cup R \\ 0, & \text{otherwise,} \end{cases} \tag{3.1}$$

where each C_i is defined as (2).

Theorem 3.2. *Let $\mathbf{s}_i$, $i = 0, 1$, be two almost balanced cyclotomic binary sequences of period p^2 in (3.1).*

(1) If $p \equiv 1 \pmod 4$, then the two sequences $\mathbf{s}_0$ and $\mathbf{s}_1$ have the 2-adic complexity:

$$\phi_2(\mathbf{s}_i) = \log_2(\frac{2^{p^2}-1}{q}), \phi_2(\mathbf{s}_{i+1}) = \log_2(2^{p^2}-1),$$

if and only if

$$q = p^2 + p + 1 \text{ is a prime and } q|(2^p - 1). \tag{3.2}$$

Moreover, the two sequences have the maximum 2-adic complexity, i.e., $\phi_2(\mathbf{s}_i) = \log_2(2^{p^2}-1)$, $i = 0, 1$, if and only if (3.2) does not hold.

(2) If $p \equiv 3 \pmod 4$, then the two sequences $\mathbf{s}_0$ and $\mathbf{s}_1$ have the 2-adic complexity:

$$\phi_2(\mathbf{s}_i) = \log_2(\frac{2^{p^2}-1}{q}), \phi_2(\mathbf{s}_{i+1}) = \log_2(2^{p^2}-1),$$

if and only if

$$q = p^2 - p + 1 \text{ is a prime and } q|(2^p - 1). \tag{3.3}$$

Moreover, the two sequences have the maximum 2-adic complexity, i.e., $\phi_2(\mathbf{s}_i) = \log_2(2^{p^2}-1)$, $i = 0, 1$, *if and only if (3.3) does not hold.*

Proof. Let $\mathbf{s}_i(2) = \sum_{\lambda=0}^{p^2-1} s_\lambda^{(i)} 2^\lambda$ and

$$d_i = \gcd(\mathbf{s}_i(2), 2^{p^2}-1), i = 0, 1,$$

then the 2-adic complexity of each $\mathbf{s}_i$ is $\phi_2(\mathbf{s}_i) = \log_2(\frac{2^{p^2}-1}{d_i})$.

In the following, we need to estimate the values d_i. By

$$2^{p^2}-1 = (2^p-1)(\sum_{b=0}^{p-1} 2^{pb}), \gcd(2^p-1, \sum_{b=0}^{p-1} 2^{pb}) = 1,$$

$d_i = \gcd(\mathbf{s}_i(2), 2^{p^2}-1) = \gcd(\mathbf{s}_i(2), 2^p-1) \cdot \gcd(\mathbf{s}_i(2), \sum_{b=0}^{p-1} 2^{pb})$, $i = 0, 1$.

By the definition of each sequence $\mathbf{s}_i$,

$$\mathbf{s}_i(2) = \sum_{\lambda=0}^{p^2-1} s_\lambda^{(i)} 2^\lambda \equiv 1 + \sum_{k \in pD_i} 2^k + \sum_{k \in D_i^{(p^2)}} 2^k$$

$$\equiv 1 + \sum_{k \in D_i} 2^{pk} + \sum_{k \in D_i} 2^k \sum_{b=0}^{p-1} 2^{pb} \pmod{2^{p^2}-1}.$$

Then

$$\mathbf{s}_i(2) \equiv \frac{p+1}{2} + p \sum_{k \in D_i} 2^k \equiv \frac{p+1}{2} + p\eta_i^{(2^p-1)} \pmod{2^p-1}$$

and $\gcd(\mathbf{s}_i(2), 2^p-1) = \gcd(\frac{p+1}{2} + p\eta_i^{(2^p-1)}, 2^p-1)$;

$$\mathbf{s}_i(2) \equiv 1 + \sum_{k \in D_i} 2^{pk} \pmod{\sum_{b=0}^{p-1} 2^{pb}},$$

$\gcd(1 + \sum_{k \in D_i} 2^{pk}, 2^p-1) = 1$, and $\gcd(\mathbf{s}_i(2), \sum_{b=0}^{p-1} 2^{pb}) = \gcd(1 + \sum_{k \in D_i} 2^{pk}, \sum_{b=0}^{p-1} 2^{pb}) = \gcd(1 + \eta_i^{(2^{p^2}-1)}, 2^{p^2}-1)$. For convenience, let

$$\Delta = \mathbf{s}_0(2)\mathbf{s}_1(2).$$

(1) If $p \equiv 1 \pmod 4$, then by Lemma 2.4,

$$\Delta \equiv \frac{(p+1)^2}{4} - \frac{p+1}{2}p + p^2\frac{1-p}{4} \equiv \frac{1-p^3}{4} \pmod{2^p-1}.$$

In fact, the least prime divisor q of 2^p-1 has $q > p+2$ in [26].

Suppose that

$$D = \gcd(\Delta, 2^p - 1) = \gcd(p^3 - 1, 2^p - 1) = \gcd(p^2 + p + 1, 2^p - 1) > 1.$$

Then there is an odd prime divisor q of D and $\mathrm{ord}_q(2) = p$, so $p|(q-1)$. Let $q = pt+1, t \in \mathbb{N}$, then $(pt+1) \mid (p^2+p+1)$ and $p^2+p+1 = qs = pts + s$, $s \in \mathbb{N}$. Hence $s = 1$ and $t = p+1$, so $q = p^2+p+1$ is a prime. Conversely, if $q = p^2+p+1$ is a prime and $q|(2^p-1)$, then $\mathbf{s}_0(2)+\mathbf{s}_1(2) \equiv 1 \pmod{2^p-1}$ and $\gcd(\Delta, 2^p-1) = q$. Hence $\gcd(\mathbf{s}_i(2), 2^p-1) = q$ and $\gcd(\mathbf{s}_{i+1}(2), 2^p-1) = 1$.

On the other hand, by Lemma 2.4,

$$(1+\eta_0^{(2^{p^2}-1)})(1+\eta_1^{(2^{p^2}-1)}) \equiv \frac{1-p}{4} \pmod{2^{p^2}-1}.$$

Suppose that q is a common prime divisor of $p-1$ and $2^{p^2}-1$. Then by Fermat Theorem, $\gcd(q-1, p^2) > 1$, which is contradictory. Hence $\gcd(\Delta, \sum_{b=0}^{p-1} 2^{pb}) = 1$.

In summary, if $q = p^2+p+1$ is a prime and $q|(2^p-1)$, then there is a sequence $\mathbf{s}_i$ such that

$$\phi_2(\mathbf{s}_i) = \log_2(\frac{2^{p^2}-1}{q}), \phi_2(\mathbf{s}_{i+1}) = \log_2(2^{p^2}-1);$$

otherwise, the two sequences have the maximum 2-adic complexity, i.e., $\phi_2(\mathbf{s}_i) = \log_2(2^{p^2}-1)$, $i = 0, 1$.

(2) If $p \equiv 3 \pmod 4$, then by Lemma 2.4,

$$\Delta \equiv \frac{(p+1)^2}{4} - \frac{p+1}{2}p + p^2\frac{p+1}{4} \equiv \frac{1+p^3}{4} \pmod{2^p-1},$$

Suppose that

$$D = \gcd(\Delta, 2^p - 1) = \gcd(p^3 + 1, 2^p - 1) = \gcd(p^2 - p + 1, 2^p - 1) > 1.$$

Then there is an odd prime divisor q of D and $\mathrm{ord}_q(2) = p$, so $p|(q-1)$. Let $q = pt+1, t \in \mathbb{N}$, then $(pt+1) \mid (p^2-p+1)$ and $p^2-p+1 = qs = pts + s$, $s \in \mathbb{N}$. Hence $s = 1$ and $t = p-1$, so $q = p^2-p+1$ is a prime. Conversely, if $q = p^2-p+1$ is a prime and $q|(2^p-1)$, then $\mathbf{s}_0(2)+\mathbf{s}_1(2) \equiv 1 \pmod{2^p-1}$ and $\gcd(\Delta, 2^p-1) = q$. Hence $\gcd(\mathbf{s}_i(2), 2^p-1) = q$ and $\gcd(\mathbf{s}_{i+1}(2), 2^p-1) = 1$.

Similarly, we get the result. □

In fact, those results in [25] are contained in Theorem 3.2.

Using the magma system, when $p \equiv 1 \pmod 4$ is a prime and $p < 2^{23}$, we only find two primes $p = 5$ and 7253 such that $q = p^2+p+1$ is a prime and $q|(2^p-1)$ in Theorem 3.2.

Example 3.3. *Let* $\mathbf{s}_i$ *be two cyclotomic binary sequences of period* $N = 2p^2$ *defined as* (3.1) *and* $p = 5$. *Then* 2 *is a primitive element of the finite field* $\mathbb{F}_5$, *i.e.,* $\mathbb{F}_5^* = \langle 2 \rangle$. *Hence* $D_0 = \{1, 4\}$, $D_1 = \{2, 3\}$, $pD_0 = \{5, 20\}$, $pD_1 = \{10, 15\}$, $D_0^{(p^2)} = \{1, 4, 6, 9.11, 14, 16, 19, 21, 24\}$, $D_1^{(p^2)} = \{2, 3, 7, 8, 12, 15, 17, 18, 22, 23\}$.

By Theorem 3.2,

$$\mathbf{s}_0(2) \equiv 93 \pmod{2^5-1},\ \mathbf{s}_0(2) \equiv 1 + 2^5 + 2^{20} \pmod{\frac{2^{25}-1}{2^5-1}};$$

$$\mathbf{s}_1(2) \equiv 63 \pmod{2^p-1},\ \mathbf{s}_1(2) \equiv 1 + 2^{10} + 2^{15} \pmod{\frac{2^{25}-1}{2^5-1}}.$$

and $q = 5^2 + 5 + 1 = 31$ *is a prime. By* $31 \mid 93$, *then* $31 \mid \mathbf{s}_0$. *Similarly, we also get* $31 \nmid \mathbf{s}_1$, $\frac{2^{25}-1}{2^5-1} \nmid \mathbf{s}_0$, $\frac{2^{25}-1}{2^5-1} \nmid \mathbf{s}_1$. *Therefore we determine*

$$\phi_2(\mathbf{s}_0) = \log_2(\frac{2^{25}-1}{2^5-1}), \phi_2(\mathbf{s}_1) = \log_2(2^{25}-1).$$

4 2-Adic Complexity of Balanced Sequences with Period $N = 2p^2$

In this section, we always assume that p is an odd prime and $N = 2p^2$. We shall determine the 2-adic complexity of several non-trivial balanced cyclotomic binary sequences of order two with period N.

There is an isomorphism of two rings:

$$\varphi : \mathbb{Z}_2 \times \mathbb{Z}_{p^2} \cong \mathbb{Z}_{2p^2} = \mathbb{Z}_N, (a, b) \mapsto (p^2+1)b + p^2 a.$$

In the following, we consider a class of non-trivial balanced cyclotomic binary sequences of order two with period N. Let

$$S_i = \varphi((\{0\} \times (pD_i \cup D_i^{(p^2)})) \cup (\{1\} \times (pD_i \cup D_{i+1}^{(p^2)}))) \cup \{0\}, i = 0, 1, \quad (4.1)$$

be subsets of $\mathbb{Z}_N$, then $|S_i| = p^2$.

Definition 4.1. There are two non-trivial balanced cyclotomic binary sequences $\mathbf{s}_i = \{s_\lambda^{(i)}\}_{\lambda=0}^{\infty}$, $i = 0, 1$, as follows:

$$s_\lambda^{(i)} = \begin{cases} 1, & \text{if } \lambda \pmod N \in S_i, \\ 0, & \text{otherwise,} \end{cases} \quad (4.2)$$

where each S_i is the support set of each sequence $\mathbf{s}_i$ as (4.1).

Theorem 4.2. *Let* $\mathbf{s}_i$, $i = 0, 1$, *be two non-trivial balanced cyclotomic binary sequences of period* N *in* (4.2). *Then*

$$\phi_2(\mathbf{s}_i) = \log_2((2^{p^2}+1)\sum_{b=0}^{p-1} 2^{pb}),\ i = 0, 1.$$

Proof. Let $\mathbf{s}_i(2) = \sum_{\lambda=0}^{N-1} s_\lambda^{(i)} 2^\lambda$ and

$$d_i = \gcd(\mathbf{s}_i(2), 2^N - 1), i = 0, 1,$$

then the 2-adic complexity of each $\mathbf{s}_i$ is $\phi_2(\mathbf{s}_i) = \log_2(\frac{2^N-1}{d_i})$.

In the following, we need to estimate the values d_i. By

$$2^N - 1 = (2^{p^2} - 1)(2^{p^2} + 1) = (2^p - 1)(\sum_{b=0}^{p-1} 2^{pb})(2^p + 1)(\sum_{b=0}^{p-1}(-2)^{pb}),$$

$$\gcd(2^{p^2} - 1, 2^{p^2} + 1) = \gcd(2^p - 1, \sum_{b=0}^{p-1} 2^{pb}) = \gcd(2^p + 1, \sum_{b=0}^{p-1}(-2)^{pb}) = 1,$$

$d_i = \gcd(\mathbf{s}_i(2), 2^N - 1) = \gcd(\mathbf{s}_i(2), 2^p - 1) \cdot \gcd(\mathbf{s}_i(2), \sum_{b=0}^{p-1} 2^{pb}) \cdot \gcd(\mathbf{s}_i(2), 2^p + 1) \cdot \gcd(\mathbf{s}_i(2), \sum_{b=0}^{p-1}(-2)^{pb}),\ i = 0, 1.$

By the definition of each sequence $\mathbf{s}_i$,

$$\mathbf{s}_i(2) = \sum_{\lambda=0}^{N-1} s_\lambda^{(i)} 2^\lambda \equiv 1 + \sum_{k \in pD_i \cup D_i^{(p^2)}} 2^{(p^2+1)k} + \sum_{k \in pD_i \cup D_{i+1}^{(p^2)}} 2^{p^2+(p^2+1)k} \pmod{2^N - 1}.$$

On the one hand,

$$\mathbf{s}_i(2) \equiv 1 + \sum_{k \in pD_i \cup D_i^{(p^2)}} 2^k + \sum_{k \in pD_i \cup D_{i+1}^{(p^2)}} 2^k$$

$$\equiv 1 + 2 \sum_{k \in D_i} 2^{pk} + \sum_{k \in D_i} 2^k \sum_{b=0}^{p-1} 2^{pb} + \sum_{k \in D_{i+1}} 2^k \sum_{b=0}^{p-1} 2^{pb} \pmod{2^{p^2} - 1}.$$

Then

$$\mathbf{s}_i(2) \equiv 0 \pmod{2^p - 1}, \gcd(\mathbf{s}_i(2), 2^p - 1) = 2^p - 1;$$

$$\mathbf{s}_i(2) \equiv 1 + 2 \sum_{k \in D_0} 2^{pk} \pmod{\sum_{b=0}^{p-1} 2^{pb}},$$

$\gcd(1 + 2\sum_{k \in D_0} 2^{pk}, 2^p - 1) = 1$, and

$$\gcd(\mathbf{s}_i(2), \sum_{b=0}^{p-1} 2^{pb}) = \gcd(1 + 2\sum_{k \in D_i} 2^{pk}, \sum_{b=0}^{p-1} 2^{pb}) = \gcd(1 + 2\eta_i^{(2^{p^2}-1)}, 2^{p^2} - 1).$$

For convenience, let

$$\Delta = \mathbf{s}_0(2)\mathbf{s}_1(2).$$

If $p \equiv 1 \pmod 4$, then by Lemma 2.4,

$$\Delta \equiv (1 + 2\eta_0^{(2^{p^2}-1)})(1 + 2\eta_1^{(2^{p^2}-1)}) \equiv -p \pmod{2^{p^2} - 1}.$$

By the proof of Theorem 3.2,

$$\gcd(\mathbf{s}_i(2), 2^{p^2} - 1) = \gcd(-p, 2^{p^2} - 1) = 1.$$

Similarly, if $p \equiv 3 \pmod 4$, then $\gcd(\mathbf{s}_i(2), 2^{p^2} - 1) = 1$.

On the other hand, without loss of generality, let $2 \in D_0$, i.e., $p \equiv \pm 1 \pmod 8$,

$$\begin{aligned}
\mathbf{s}_i(2) &\equiv 1 + \sum_{k \in D_i^{(p^2)}} 2^{(p^2+1)k} - \sum_{k \in D_{i+1}^{(p^2)}} 2^{(p^2+1)k} \\
&\equiv 1 + \sum_{2k \in D_i} 2^{(p^2+1)2k} \sum_{b=0}^{p-1} (-2)^{pb} - \sum_{2k \in D_{i+1}} 2^{(p^2+1)2k} \sum_{b=0}^{p-1} (-2)^{pb} \\
&\equiv 1 + \sum_{k \in D_i} 2^{(p^2+1)2k} \sum_{b=0}^{p-1} (-2)^{pb} - \sum_{k \in D_{i+1}} 2^{(p^2+1)2k} \sum_{b=0}^{p-1} (-2)^{pb} \\
&\equiv 1 + \sum_{k \in D_i} 2^{2k} \sum_{b=0}^{p-1} (-2)^{pb} - \sum_{k \in D_{i+1}} 2^{2k} \sum_{b=0}^{p-1} (-2)^{pb} \pmod{2^{p^2} + 1}.
\end{aligned}$$

Then

$$\mathbf{s}_i(2) \equiv 1 + \sum_{k \in D_i} 2^{2k} - \sum_{k \in D_{i+1}} 2^{2k} \pmod{2^p + 1},$$

and $\gcd(\mathbf{s}_i(2), 2^p + 1) = \gcd(1 + \sum_{k \in D_i} 2^{2k} - \sum_{k \in D_{i+1}} 2^{2k}, 2^p + 1)$;

$$\mathbf{s}_i(2) \equiv 1 \pmod{\sum_{b=0}^{p-1} (-2)^{pb}}, \textit{ and } \gcd(\mathbf{s}_i(2), \sum_{b=0}^{p-1} (-2)^{pb}) = 1.$$

For convenience, let

$$\widetilde{\Delta} = (1 + \sum_{k \in D_0} 2^{2k} - \sum_{k \in D_1} 2^{2k})(1 + \sum_{k \in D_1} 2^{2k} - \sum_{k \in D_0} 2^{2k}).$$

If $p \equiv 1 \pmod 4$, then by Lemma 2.4,

$$\begin{aligned}
\widetilde{\Delta} &= (1 + \sum_{k \in D_0} 2^{2k} - \sum_{k \in D_1} 2^{2k})(1 + \sum_{k \in D_1} 2^{2k} - \sum_{k \in D_0} 2^{2k}) \\
&\equiv 1 - (\eta_0^{(2^{2p}-1)} + \eta_1^{(2^{2p}-1)})^2 + 4\eta_0^{(2^{2p}-1)} \eta_1^{(2^{2p}-1)} \\
&\equiv 1 - p \pmod{2^{2p} - 1}.
\end{aligned}$$

By the proof of Theorem 3.2, $\gcd(\widetilde{\Delta}, 2^{2p} - 1) = \gcd(1 - p, 2^{2p} - 1) = \gcd(\widetilde{\Delta}, 2^p + 1) = 1$. Hence $\gcd(\mathbf{s}_i(2), 2^p + 1) = 1$.

By $\gcd(\mathbf{s}_i(2), \sum_{b=0}^{p-1}(-2)^{pb}) = 1$, $\gcd(\mathbf{s}_i(2), 2^{p^2}+1) = 1$.
Similarly, if $p \equiv 3 \pmod 4$, then $\gcd(\mathbf{s}_i(2), 2^{p^2}+1) = 1$.
In summary, $d_i = \gcd(\mathbf{s}_i(2), 2^N - 1) = 2^p - 1$, $i = 0, 1$, then

$$\phi_2(\mathbf{s}_i) = \log_2((2^{p^2}+1)\sum_{b=0}^{p-1} 2^{pb}).$$

□

In the following, we consider another class of non-trivial balanced cyclotomic binary sequences of order two with period N. Let

$$\widehat{S}_i = \varphi((\{0\} \times (pD_{i+1} \cup D_i^{(p^2)})) \cup (\{1\} \times (pD_i \cup D_i^{(p^2)}))) \cup \{0\}, i = 0, 1, \quad (4.3)$$

be subsets of $\mathbb{Z}_N$, then $|\widehat{S}_i| = p^2$.

Definition 4.3. There are two non-trivial balanced cyclotomic binary sequences $\widehat{\mathbf{s}}_i = \{s_\lambda^{(i)}\}_{\lambda=0}^{\infty}$, $i = 0, 1$, as follows:

$$\widehat{s}_\lambda^{(i)} = \begin{cases} 1, & \text{if } \lambda \pmod N \in \widehat{S}_i, \\ 0, & \text{otherwise}, \end{cases} \quad (4.4)$$

where each $\widehat{S}_i$ is the support set of each sequence $\widehat{\mathbf{s}}_i$ as (4.3).

Theorem 4.4. *Let $\widehat{\mathbf{s}}_i$, $i = 0, 1$, be two non-trivial balanced cyclotomic binary sequences of period N in (4.4). Then*

$$\phi_2(\widehat{\mathbf{s}}_i) = \log_2((2^{p^2}+1)(2^p-1)),\ i = 0, 1.$$

Proof. Let $\widehat{\mathbf{s}}_i(2) = \sum_{\lambda=0}^{N-1} \widehat{s}_\lambda^{(i)} 2^\lambda$ and

$$d_i = \gcd(\mathbf{s}_i(2), 2^N - 1), i = 0, 1,$$

then the 2-adic complexity of each $\widehat{\mathbf{s}}_i$ is $\phi_2(\widehat{\mathbf{s}}_i) = \log_2(\frac{2^N-1}{d_i})$.
In the following, we need to estimate the values d_i. By

$$2^N - 1 = (2^{p^2}-1)(2^{p^2}+1) = (2^p-1)(\sum_{b=0}^{p-1} 2^{pb})(2^p+1)(\sum_{b=0}^{p-1}(-2)^{pb}),$$

$$\gcd(2^{p^2}-1, 2^{p^2}+1) = \gcd(2^p-1, \sum_{b=0}^{p-1} 2^{pb}) = \gcd(2^p+1, \sum_{b=0}^{p-1}(-2)^{pb}) = 1,$$

$d_i = \gcd(\widehat{\mathbf{s}}_i(2), 2^N-1) = \gcd(\widehat{\mathbf{s}}_i(2), 2^p-1) \cdot \gcd(\widehat{\mathbf{s}}_i(2), \sum_{b=0}^{p-1} 2^{pb}) \cdot \gcd(\widehat{\mathbf{s}}_i(2), 2^{p^2} + 1)$, $i = 0, 1$.

By the definition of each sequence $\widehat{\mathbf{s}}_i$,

$$\widehat{\mathbf{s}}_i(2)=\sum_{\lambda=0}^{N-1}\widehat{s}_\lambda^{(i)}2^\lambda\equiv 1+\sum_{k\in pD_{i+1}\cup D_i^{(p^2)}}2^{(p^2+1)k}+\sum_{k\in pD_i\cup D_i^{(p^2)}}2^{p^2+(p^2+1)k}\pmod{2^N-1}.$$

On the one hand,

$$\begin{aligned}\widehat{\mathbf{s}}_i(2)&\equiv 1+\sum_{k\in pD_{i+1}\cup D_i^{(p^2)}}2^k+\sum_{k\in pD_i\cup D_i^{(p^2)}}2^k\\&\equiv 1+\sum_{k\in D_i}2^{pk}+\sum_{k\in D_{i+1}}2^{pk}+2\sum_{k\in D_i}2^k\sum_{b=0}^{p-1}2^{pb}\pmod{2^{p^2}-1}.\end{aligned}$$

Then

$$\widehat{\mathbf{s}}_i(2)\equiv p+2p\sum_{k\in D_i}2^k\equiv p+2p\eta_i^{(2^p-1)}\pmod{2^p-1},$$

and $\gcd(\widehat{\mathbf{s}}_i(2),2^p-1)=\gcd(p+2p\eta_i^{2^p-1},2^p-1)$;
$\widehat{\mathbf{s}}_i(2)\equiv 0\pmod{\sum_{b=0}^{p-1}2^{pb}}$, and $\gcd(\widehat{\mathbf{s}}_i(2),\sum_{b=0}^{p-1}2^{pb})=\sum_{b=0}^{p-1}2^{pb}$.
For convenience, let

$$\Delta=\widehat{\mathbf{s}}_0(2)\widehat{\mathbf{s}}_1(2).$$

If $p\equiv 1\pmod 4$, then by Lemma 2.4,

$$\Delta\equiv(p+2p\eta_0^{(2^p-1)})(p+2p\eta_1^{(2^p-1)})\equiv -p^3\pmod{2^p-1}.$$

By the proof of Theorem 3.2,

$$\gcd(\Delta,2^p-1)=\gcd(-p^3,2^p-1)=\gcd(\widehat{\mathbf{s}}_i(2),2^p-1)=1,\ i=0,1.$$

Similarly, if $p\equiv 3\pmod 4$, then $\gcd(\widehat{\mathbf{s}}_i(2),2^p-1)=1$.
On the other hand, without loss of generality, let $2\in D_0$, then

$$\begin{aligned}\widehat{\mathbf{s}}_i(2)&\equiv 1+\sum_{k\in pD_{i+1}}2^{(p^2+1)k}-\sum_{k\in pD_i}2^{(p^2+1)k}\\&\equiv 1+\sum_{2k\in D_{i+1}}2^{(p^2+1)2pk}-\sum_{2k\in D_i}2^{(p^2+1)2pk}\\&\equiv 1+\sum_{k\in D_{i+1}}2^{(p^2+1)2pk}-\sum_{k\in D_i}2^{(p^2+1)2pk}\\&\equiv 1+\sum_{k\in D_{i+1}}2^{2pk}-\sum_{k\in D_i}2^{2pk}\pmod{2^{p^2}+1}.\end{aligned}$$

Then $\gcd(\widehat{\mathbf{s}}_i(2),1+\sum_{k\in D_{i+1}}2^{2pk}-\sum_{k\in D_i}2^{2pk})=\gcd(1+\sum_{k\in D_{i+1}}2^{2pk}-\sum_{k\in D_i}2^{2pk},2^{p^2}+1)$. In the following, we shall compute the value

$$\begin{aligned}\overline{\Delta}&=(1+\sum_{k\in D_0}2^{2pk}-\sum_{k\in D_1}2^{2pk})(1+\sum_{k\in D_1}2^{2pk}-\sum_{k\in D_0}2^{2pk})\\&\equiv 1-(\eta_0^{(2^{2p^2}-1)}+\eta_1^{(2^{2p^2}-1)})^2-4\eta_0^{(2^{2p^2}-1)}\eta_1^{(2^{2p^2}-1)}\pmod{2^{2p^2}-1}.\end{aligned}$$

If $p \equiv 1 \pmod 4$, then by Lemma 2.4,

$$\overline{\Delta} \equiv 1 - p \pmod{2^{2p^2} - 1} .$$

By the proof of Theorem 3.2, $\gcd(\overline{\Delta}, 2^{2p^2} - 1) = \gcd(1-p, 2^{2p^2} - 1) = \gcd(2^{p^2} - 1, 2^{p^2} + 1) = 1$. Hence $\gcd(\widehat{\mathbf{s}}_i(2), 2^{p^2} + 1) = 1$, $i = 0, 1$.

Similarly, if $p \equiv 3 \pmod 4$, then $\gcd(\widehat{\mathbf{s}}_i(2), 2^{p^2}+1) = 1$ $i = 0, 1$.

In summary, $d_i = \gcd(\mathbf{s}_i(2), 2^N - 1) = \sum_{b=0}^{p-1} 2^{pb}$, $i = 0, 1$, then

$$\phi_2(\widehat{\mathbf{s}}_i) = \log_2 \left((2^{p^2} + 1)(2^p - 1)\right).$$

5 Conclusion

In this paper, let p be an odd prime. We obtain the 2-adic complexity of all almost balanced cyclotomic binary sequence of order two with period p^2; determine the 2-adic complexity of several non-trivial balanced cyclotomic binary sequences of order two with period $2p^2$. In fact, we complement Xiao's results in [25] and generalize Zhang's results in [31].

References

1. Berndt, B.C., Evans, R.J., Williams, K.S.: Gauss and Jacobi Sums. Wiley, Hoboken (1998)
2. Cai, H., Zhou, Z., Yang, Y., Tang, X.: A new construction of frequency-hopping sequences with optimal partial hamming correlation. IEEE Trans. Inf. Theory **60**(9), 5782–5790 (2014)
3. Ding, C., Helleseth, T., Lam, K.Y.: Several classes of binary sequences with three-level autocorrelation. IEEE Trans. Inf. Theory **45**(7), 2606–2612 (1999)
4. Ding, C., Helleseth, T., Martinsen, H.: New families of binary sequences with optimal three-level autocorrelation. IEEE Trans. Inf. Theory **47**(1), 428–433 (2001)
5. Ding, C., Yang, Y., Tang, X.: Optimal sets of frequency hopping sequences from linear cyclic codes. IEEE Trans. Inf. Theory **56**(7), 3605–3612 (2010)
6. Cusick, T.W., Ding, C., Renvall, A.: Stream Ciphers and Number Theory. North-Holland/Elsevier, Amsterdam, The Netherlands (1998)
7. Golomb, S.W.: Shift Register Sequences. Holden-Day, San Francisco (1967)
8. Helleseth, T., Kumar, P.V.: Sequences with low correlation. In: Pless, Huffman (eds.) Handbook of Coding Theory, vol. II, pp. 1765–1854. Elsevier, Amsterdam, The Netherlands (1998)
9. Hofer, R., Winterhof, A.: On the 2-adic complexity of the two-prime generator. IEEE Trans. Inf. Theory **64**(8), 5957–5960 (2018)
10. Hu, H.: Comments on a new method to compute the 2-adic complexity of binary sequences. IEEE Trans. Inf. Theory **60**(9), 5803–5804 (2014)
11. Hu, L., Yue, Q.: Gauss periods and codebooks from generalized cyclotomic sets of order four. Des. Codes Cryptogr. **69**, 233–246 (2013)
12. Klapper, A., Goresky, M.: Cryptanalysis based on 2-adic rational approximation. Adv. Cryptol. **963**, 262–273 (1995)

13. Klapper, A., Goresky, M.: Feedback shift registers, 2-adic span, and combiners with memory. J. Cryptol. **10**(2), 111–147 (1997). https://doi.org/10.1007/s001459900024
14. Lempel, A., Cohn, M., Eastman, W.L.: A class of binary sequences with optimal autocorrelation properties. IEEE Trans. Inform. Theory **23**(1), 38–42 (1977)
15. Shi, X., Yan, T., Huang, X., Yue, Q.: An extension method to construct M-ary sequences of period 4N with low autocorrelation. IEICE Trans. Fundam. Electron. Commun. Comput. Sci. **104-A**(1), 332–335 (2021)
16. Simon, M.K., Omura, J.K., Scholtz, R.A., Levitt, B.K.: Spread Spectrum Communications, vol. 1. Computer Science Press, Rockville, MD (1985)
17. Storer, T.: Cyclotomy and Difference Sets. Markham Pub. Co. (1967)
18. Sun, Y., Wang, Q., Yan, T.: The exact autocorrelation distribution and 2-adic complexity of a class of binary sequences with almost optimal autocorrelation. Cryptogr. Commun. **10**(3), 467–477 (2018)
19. Sun, Y., Wang, Q., Yan, T.: A lower bound on the 2-adic complexity of the modified Jacobi sequence. Cryptogr. Commun. **11**(2), 337–349 (2019)
20. Sun, Y., Yan, T., Chen, Z.: The 2-adic complexity of a class of binary sequences with optimal autocorrelation magnitude. Cryptogr. Commun. **12**(4), 675–683 (2020)
21. Tang, X., Ding, C.: New classes of balanced quaternary and almost balanced binary sequences with optimal autocorrelation value. IEEE Trans. Inf. Theory **56**(12), 6398–6405 (2010)
22. Tang, X., Gong, G.: New constructions of binary sequences with optimal autocorrelation value/magnitude. IEEE Trans. Inf. Theory **56**(3), 1278–1286 (2010)
23. Tian, T., Qi, W.F.: 2-adic complexity of binary m-sequences. IEEE Trans. Inf. Theory **56**(1), 450–454 (2010)
24. Whiteman, A.L.: The cyclotomic numbers of order twelve. Acta Arith. **6**, 53–76 (1960)
25. Xiao, Z., Zeng, X., Sun, Z.: 2-adic complexity of two classes of generalized cyclotomic binary sequences. Int. J. Found. Comput. Sci **27**(7), 879–893 (2016)
26. Xiong, H., Qu, L., Li, C.: A new method to compute the 2-adic complexity of binary sequences. IEEE Trans. Inf. Theory **60**(4), 2399–2406 (2014)
27. Yang, Y., Huo, F., Gong, G.: Large zero odd periodic autocorrelation zone of Golay sequences and QAM Golay sequences, pp. 1024–1028. In: ISIT (2012)
28. Yang, Y., Tang, X., Zhou, Z.: The autocorrelation magnitude of balanced binary sequence pairs of prime period $N \equiv 1 \pmod 4$ with optimal cross-correlation. IEEE Commun. Lett. **19**(4), 585–588 (2015)
29. Yang, Y., Tang, X., Peng, D., Parampalli, U.: New bound on frequency hopping sequence sets and its optimal constructions. IEEE Trans. Inf. Theory **57**(11), 7605–7613 (2011)
30. Zeng, X., Cai, H., Tang, X., Yang, Y.: Optimal frequency hopping sequences of odd length. IEEE Trans. Inf. Theory **59**(5), 3237–3248 (2013)
31. Zhang, L., Zhang, J., Yang, M., Feng, K.: On the 2-adic complexity of the Ding-Helleseth-Martinsen binary sequences. IEEE Trans. Inf. Theory **66**(7), 4613–4620 (2020)
32. Zhou, Z., Helleseth, T., Parampalli, U.: A family of polyphase sequences with asymptotically optimal correlation. IEEE Trans. Inf. Theory **64**(4), 2896–2900 (2018)
33. Zhou, Z., Tang, X., Gong, G.: A new class of sequences with zero or low correlation zone based on interleaving technique. IEEE Trans. Inf. Theory **54**(9), 4267–4273 (2008)

34. Zhou, Z., Tang, X., Niu, X., Parampalli, U.: New classes of frequency-hopping sequences with optimal partial correlation. IEEE Trans. Inf. Theory **58**(1), 453–458 (2012)
35. Zhou, Z., Tang, X., Peng, D., Parampalli, U.: New constructions for optimal sets of frequency-hopping sequences. IEEE Trans. Inf. Theory **57**(6), 3831–3840 (2011)
36. Zhou, Z., Zhang, D., Helleseth, T., Wen, J.: A construction of multiple optimal ZCZ sequence sets with good cross correlation. IEEE Trans. Inf. Theory **64**(2), 1340–1346 (2018)

Author Index